Air Pollution
and
Plant Life

ENVIRONMENTAL MONOGRAPHS AND SYMPOSIA

A series in the Environmental Sciences

Convener and General Editor
NICHOLAS POLUNIN, CBE
Geneva, Switzerland

MODERNIZATION OF AGRICULTURE IN DEVELOPING
COUNTRIES: Resources, Potentials, and Problems

I. ARNON, *Hebrew University, Jerusalem, Agricultural Research Service,
Bet Dagan, and Settlement Study Centre, Rehovot, Israel*

STRESS EFFECTS ON NATURAL ECOSYSTEMS
Edited by
G. W. BARRETT, *Institute of Environmental Sciences and Department
of Zoology, Miami University, Oxford, Ohio, USA*
and
R. ROSENBERG, *Fishery Board of Sweden, Institute of Marine Research,
Lysekil, Sweden*

AIR POLLUTION AND PLANT LIFE
Edited by
MICHAEL TRESHOW, *Department of Biology, University of Utah,
Salt Lake City, Utah, USA*

Air Pollution and Plant Life

Edited by
Michael Treshow
Department of Biology,
University of Utah,
Salt Lake City, Utah, USA

A Wiley–Interscience Publication

JOHN WILEY & SONS
Chichester · New York · Brisbane · Toronto · Singapore

Library of Congress Cataloging in Publication Data:
Main entry under title:

Air pollution and plant life.
  (Environmental monographs and symposia)
  'A Wiley–Interscience publication.'
  Includes index.
  1. Plants, Effect of air pollution on.   2. Air—
Pollution—Environmental aspects.   I. Treshow, Michael.
II. Series.
QK751.A36   1983   581.5'222   83–5905

ISBN 0 471 90103 2

British Library Cataloguing in Publication Data:

Air pollution and plant life.—(Environmental
  monographs and symposia)
  1. Plants, Effect of air pollution on.
  I. Treshow, Michael   II. Series
  581.5'222   QK751

ISBN 0 471 90103 2

Typeset by Preface Ltd, Salisbury, Wilts and printed
by Page Bros. (Norwich) Ltd, Norwich

Contents

List of Contributors

ANDERSON, FRANKLIN K.   *Ford, Bacon and Davis, Inc., Salt Lake City, Utah 84108, USA*

BENNETT, JESSE H.   *Plant Stress Laboratory, Agricultural Research Northeastern Region, Beltsville Agriculture Research Center, Beltsville, Maryland 20705, USA.*

BIAŁOBOK, STEFAN   *Polish Academy of Sciences, Institute of Dendrology, 63–120 Kornik, Poland.*

DOVLAND, HARALD   *Norwegian Institute for Air Research, PO Box 130, N-2001 Kjellev-Lillestrøm, Norway.*

GILLETTE, DONALD G.   *United States Environmental Protection Agency, Research Triangle Park, North Carolina 27711, USA.*

HALBWACHS, GOTTFRIED   *Botanisches Institut der Universität für Bodenkultur, Gregor Mendel-strasse 33, A-1180 Wien, Austria.*

HEGGESTAD, HOWARD E.   *Plant Stress Laboratory, Agricultural Research Northeastern Region, Beltsville Agriculture Research Center, Beltsville, Maryland 20705, USA.*

HUTTUNEN, SATU   *Department of Botany, University of Oulu, PO Box 191, SF-90101 Oulu 10, Finland.*

KARENLAMPI, LAURI   *Ecological Laboratory, Department of Environmental Hygiene, University of Kuopio, PO Box 138, SF-70101 Kuopio 10, Finland.*

KHAN, A. A.   *Alberta Environmental Center, Vegreville, Alberta, Canada T0B 4L0.*

MALHOTRA, S. S.   *Canadian Forestry Service, Environment*
                          Canada, Northern Forest Research Center,
                          5320 122nd Street, Edmonton, Alberta,
                          Canada, T6H 3S5.

MATERNA, JAN   *Institute of Forestry, Zbraslav, Prague,*
                          Czechoslovakia.

ORMROD, DOUGLAS P.   *Department of Horticultural Science,*
                          University of Guelph, Guelph, Ontario,
                          Canada N1G 2W1.

OTTAR, BRYNJULF   *Norwegian Institute for Air Research, PO Box*
                          130, N-2001 Kjellev-Lillestrøm, Norway.

POSTHUMUS, ADAM C.   *Research Institute for Plant Protection,*
                          Binnenhaven 12, Wageningen, Netherlands.

ROBINSON, ELMER   *College of Engineering, Washington State*
                          University, SW 834 Crestview, Pullman,
                          Washington 99164, USA.

RUNECKLES, V. C.   *Department of Plant Science, University of*
                          British Columbia, 2075 Wesbrook Mall,
                          Vancouver, British Columbia, Canada V6T 2A2.

SEMB, ARNE   *Norwegian Institute for Air Research, PO Box*
                          130, N-2001 Kjellev-Lillestrøm, Norway.

SMITH, WILLIAM H.   *School of Forestry and Environmental Studies,*
                          Yale University, New Haven, Connecticut 06511,
                          USA.

SOIKKELI, SIRKKA   *Ecological Laboratory, Department of*
                          Environmental Hygiene, University of Kuopio,
                          PO Box 138, SF-70101 Kuopio 10, Finland.

TAYLOR, O. CLIFTON   *Department of Botany and Plant Sciences,*
                          University of California, Riverside, California
                          92502, USA.

TRESHOW, MICHAEL   *Department of Biology, University of Utah,*
                          Salt Lake City, Utah 84112, USA.

Series Preface

For civilization to survive in anything like its present form in the face of the world's ever-increasing population-pressures and demands on its limited resources, we human components will need to increase and continuingly widen our knowledge of the environment. Moreover, this knowledge will need to be closely followed by concomitant action to safeguard the biosphere and so maintain the framework and chief structures of our life-support system. This *increase* in knowledge and awareness must come through observation, research, and applicational testing, its *widening* through environmental education, and the necessary concerted *action* through duly organized application of the knowledge that has been thus acquired and disseminated.

The environmental movement has long been an undefined but widely effective vehicle for increasing appreciation of the vital nature and fundamental importance of Man's environment. It is hoped that the now established *World Campaign for The Biosphere* will focus attention on the fragility of this 'peripheral envelope of Earth together with its surrounding atmosphere in which living things exist naturally', on our utter dependence on its health as it constitutes our only life-support system, and on the necessity to foster it in every possible way.

To help to encourage such ideals and guide appropriate actions which in many cases are imperatives for Man and Nature, as well as to distil and widen knowledge in component fields of scientific and other environmental endeavour, we founded and are now fostering an open-ended series of *Environmental Monographs and Symposia*. This emanated from an invitation by the international publishers John Wiley & Sons, and consists of authoritative volumes of two main kinds: monographs in the full sense of being detailed treatments of particular subjects by from one to three leading specialists, and symposia by more than three specialist authors covering a particular subject between them under the guidance and editorship of a suitable specialist or up to three specialists (whether such a volume results in part or wholly from an actual 'live' symposium or consists entirely of 'contributed' papers conforming to an agreed plan).

There seems to be virtually no end to the possibilities for our series; we are constantly getting or being given new ideas, and now have very many to think about and, in chosen cases, to work on. At the same time, we hope to complement the existing SCOPE Reports, emanating from what in a sense is the world's environmental 'summit'. In addition to the present work and the already published *Modernization of Agriculture in Developing Countries* by Professor I. Arnon and *Stress Effects on Natural Ecosystems* edited by Professor Gary W. Barrett and Dr Rutger Rosenberg, future volumes of *Environmental Monographs and Symposia* will include *Ecosystem Theory and Application* edited by Professor George A. Knox, President of INTECOL, and myself, *The Stratospheric Ozone Shield* by Drs Byron W. Boville and Rumen D. Bojkov, and *Ecological Management of Impounded Rivers*, by Dr Geoffrey E. Petts.

Whether or not this series will in time come to cover, in however general a manner, the entire vast realm of environmental scientific and allied endeavour, must remain to be seen, though this was the gist of the publishers' original invitation and poses a challenge that we can scarcely forget. Meanwhile we believe we have decided on a constructive compromise with this propitious and promising series, in which we look forward to the effective participation of more and more of the world's leading environmentalists.

NICHOLAS POLUNIN
(Convener and General Editor of the Series)
Geneva, Switzerland.

Preface

The progress made in air pollution control technology, and the incorporation of many control measures and air pollution regulations, might lead one to believe that emissions and their adverse effects are lessening. But such is not the case. Production losses have continued to increase even during the past decade, and by 1980 reached 1.8 billion dollars in the United States alone. This was attributed largely to ozone, a pollutant scarcely recognized just two decades earlier. This loss does not include the impacts on forests or natural ecosystems—impacts for which a monetary value is almost impossible to estimate.

Pollutant sources are not restricted to a few industrial nations, nor are the emissions confined to some local region. Emissions from many sources, whether stationary or mobile, drift together and may be transported vast distances before ultimately settling to earth. Thus, the consequences may be far-reaching, in distance as well as in direct and indirect effects.

Each year knowledge about these effects increases substantially. But the knowledge is dispersed widely in the literature, and is often difficult to assemble. A single source of collected knowledge having long been lacking; it is the intention of this volume to provide such a common source.

In order to obtain the most thorough, current, and authoritative information possible on the effects of air pollutants on plants, we have gone to the research workers who have been most active in discovering and understanding these effects. These authors represent many, but by no means all, of the groups conducting air pollution research in many of the industrial countries of Europe and North America.

We have chosen to organize this knowledge in a manner that strives to develop concepts on the way pollutants act on plants. We have done this by treating the major air pollutants first at the chemical level of action, and then leading to the organismal and ecological consequences. Related subjects include the dispersion, long-range transport, and monitoring, of pollutants, as well as diagnosis and interaction with other abiotic stresses, diseases, and insects. Separate chapters also are devoted to pollutant combinations, trace-element pollution, the impact on agriculture and natural ecosystems, and pollutant uptake and control. We have attempted to be comprehensive in our treatments, and to provide sufficient detail to equip the reader with a depth as well as a breadth of knowledge.

This book should serve as a reference work as well as a text for students in the air pollution field and the many interacting and related disciplines. It should be equally valuable to those in the many government agencies, industry, and universities, and to others concerned with air pollutant effects who hope to understand them better than heretofore.

I am most indebted to the distinguished collaborators of this volume for their valuable contributions and the sharing of their expertise. Their enthusiasm, patience, and often prompt response to deadlines, is highly appreciated. I especially value the stenographic assistance of Mrs Maurine Vaughan in the preparation of this volume.

MICHAEL TRESHOW
Salt Lake City

Air Pollution and Plant Life
Edited by M. Treshow
© 1984 John Wiley & Sons Ltd.

CHAPTER 1

Introduction

MICHAEL TRESHOW

Professor of Biology, Department of Biology, University of Utah, Salt Lake City, Utah 84112, USA

Air is never pure, but neither does it have to be polluted. Air is a fascinating mix of gases and vapours containing also minute particles of many kinds. Dusts blown free from desert soils or wastelands, ash and gases spewed from volcanoes or rising from burning forests, even pleasantly odiferous organic vapours released from forest trees and other plants, together with pollens and spores, and the fresh aroma of the sea near the coasts—all contribute to the character of natural air.

Not every contaminant in the air is an air pollutant. Air pollutants are generally defined as aerial substances that have some adverse effects on plants, animals (including Man), or materials. The effects could be some subtle, uptoward physiological stress or an unpleasant odour, or they may be outright dangerous and present a measurable threat to the survival or well-being of an organism.

Every living terrestrial plant and animal is exposed to this mixture, and, over the course of evolution, each has evolved in harmony with it. In the past two centuries, though, the human species has altered the composition of the atmosphere both locally and globally. We continue to change it. We have burned coal and other fossil fuels by the millions of tons, and we have smelted metallic ores, so adding more and more wastes to the atmosphere. The automobile spawned another era of pollution, and as the numbers of vehicles soared into the millions, their exhaust fumes reacting with sunlight produced another major source of pollution. More recent years have seen a still wider array of pollutants.

The diverse array of wastes was often harmful to plants; a few classic pollutants stand out as most noteworthy. These are sulphur oxides from fossil fuels and smelters, often including fine particles of heavy-metal wastes, fluorides from aluminium reduction and phosphate production, and photo-

1

chemical pollutants, hydrocarbons, and carbon monoxide, from automobiles and other combustion sources.

Of these pollutants, the sulphur dioxide emissions from the burning of coal are among the most harmful to plants. As oil becomes increasingly expensive, and as the supply diminishes, more and more coal is being substituted to generate electrical energy. Some 19 million tons of sulphur dioxide are discharged from the US sources each year, and another 14 million tons from Canada. This trend is likely to persist for the next few decades, making pollution from sulphur and nitrogen oxides and acid precipitation increasingly critical. And yet little is really known about the long-range transport and global impact of these air pollutants.

Air pollution concerns us all. Each of us must breathe whatever the air harbours, and the crops on which we depend for food, and landscapes from which we gain so much pleasure, are continually subjected to a complex array of chemicals that can be harmful to their growth and production. The much-publicized sulphur oxides, photochemical pollutants, and fluorides, that provide the main hazards, have no monopoly. Chlorides, ammonia, nitrogen oxides, pesticides, dusts, ethylene, and combinations of all of these, can impair plant health. Each pollutant acts in its own distinctive way. Each becomes harmful at different concentrations, and each plant species responds differently to every pollutant. Furthermore, every response can be mitigated or intensified by every parameter of the physical environment. Thus the possible combinations of pollutant concentrations, and durations of exposure that can become harmful, are infinite.

One might suspect that a problem which had been so serious for so long would be amply treated in the literature. Yet air pollution science is such a rapidly developing field that only recently have we been coming to understand the various ways in which air pollutants affect plants, and the concentrations at which some effect may be expected to occur.

Each year sees the publication of proceedings of symposia treating important issues in air pollution biology. Each discusses some specialized and often technical treatment of the subject. Many are excellent, but none provides an overall and comprehensive treatment. The literature is directed mainly towards the specialist and so is justifiably technical. There is no compilation of the 'state of the art', intended for the student or general reader who might be concerned more with the broader aspects and general understanding of air pollution biology.

We hope that the present volume will fill this void. In our attempt to do as much, we shall first provide a brief background to the origins of air pollution problems and their significance to society and agriculture. We shall then discuss the dispersion and fate of atmospheric pollutants together with their transport and measurements, as these are integral parts of understanding air pollution.

As pollutants settle out from the atmosphere, much is deposited on vegetation. Pollutants then have the potential of entering the plant and its intercellular spaces, whereupon they are absorbed by the plant cells and may react with the cell components. It is then that the potential toxicity of the pollutants may be expressed.

In practically every population of plants there are a few individuals that can tolerate air pollution. Many cannot, while a few are so acutely sensitive that they may be harmed by pollutant concentrations that are scarcely above the background levels. More tolerant individuals may struggle for life for a few years before gradually succumbing to a stress to which they have evolved no defence. Some may survive, having been fortuitously embued with some degree of genetic tolerance. The vast majority of plant species possess varying degrees of defence such that, while not killed outright, or showing any obvious or visible sign of stress, neither are they in sound health.

Toxicity may be expressed in any number of ways. Often it involves the interference of any of a multitude of metabolic pathways. Each pollutant interacts in its own way with the biochemical and physiological processes of the plant. This interference by pollutants, or varied combinations of different pollutants, often leads to a loss of the normal green colour of the foliage, a killing of tissues, or some other clearly visible expression. The way in which the major pollutants act, and how the plant responds, comprise a large part of this volume. These basic principles, treating the mode of action of pollutants and the resulting physiological response and symptoms, are too often slighted in the air pollution literature.

Diagnosing air pollution injury is an especially exciting yet elusive dimension of air-pollution biology. Correct diagnosis is vital and yet rarely discussed. Injury caused by pollutants can easily be mistaken for injury caused by other stresses; or, just the opposite, injury symptoms from adverse temperature or moisture relations may resemble, and can be incorrectly attributed to, air pollutants. Air pollutants are 'environment pathogens' affecting the health of plants, and a knowledge of air pollution biology is necessary to distinguish the impact of pollutants from other environmental parameters. Sometimes related to this are the interactions of these other agents, including such biotic pathogens as fungi and insects, with air pollutants. These interactions, as well as the 'mimicking' symptoms that can so closely resemble air pollution injury, will be fully treated.

Lichens and bryophytes are unique in the plant world. Both lack the normal leaf and root systems of the flowering and cone-bearing plants, and therefore respond somewhat differently to air pollutants. Some of them are among the most sensitive of all organisms to the harmful effects of air pollutants. The fascinating story of their disappearance near large cities, first observed early in the nineteenth century, their absence many kilometres from large industries, and the reasons for this extreme sensitivity, will be developed.

Another area of concern involves the interactions of air pollutants in the rhizosphere—namely the soil and roots, and the microflora and fauna that inhabit both. Nitrogen-fixing bacteria provide an important example. Preliminary studies suggest that some pollutants may affect these bacteria which are so important to the normal development of agricultural crops (such as soybeans and alfalfa) and just as vital to the welfare of certain desert grasses in natural communities.

Finally, the ultimate impact of air pollutants on agriculture, forests, and natural ecosystems, will be explored. Methods for assessing crop losses and changes in the ecosystem and natural land-use will be discussed.

Most significantly, the ways in which air pollutants are being controlled, and how the effects of the remaining pollutants can be mitigated, are explained. The methods of control at the source are only touched upon, as this is largely an engineering matter, more attention being paid to chemical treatments and the use of tolerant and resistant plants.

There is nothing new about air pollution. The dense choking smoke from fires has been an offensive but necessary tradeoff for heat and cooking since the earliest times. The sulphurous mines and roasting ovens of Egypt and Sumaria provided even more offensive and widespread stenches. Pollution also included smoke from forest fires and volcanoes—all part of the natural environment—but while they were potentially offensive, they tended to be relatively local in significance.

The growth of cities in Europe, intensifying in the Middle Ages with their swelling populations dependent on wood for the hearth, generated air pollution of modest concern. But as the wood supplies diminished, new kinds of fuel were sought, and before long the potential energy from coal was discovered and harnessed. As coal gradually replaced wood, pollution became a critical environmental hazard. The acrid fumes were offensive; but in northern Europe, the alternative was a chilly or freezing home.

When the steam engine was discovered and became a practical force to lighten Man's work, helping to spawn the Industrial Revolution, coal was burned as never before. A centre of industry with a vast population, London was generally conceded to be the most polluted of cities. Notable accounts of air pollution appear as early as the reign of Edward I of England in the thirteenth century. The offensive fumes from burning coal attracted sufficient concern that when once it came into general use, regulations and restrictions to control its use were imposed—usually with little effect. The classic pamphlet written by John Evelyn, which was published at the command of Charles II in 1661, provided a remarkably thorough treatise on how coal smoke was affecting the populace of London, especially the children. Repeated occurrences of pollution episodes were described, which became increasingly serious to human health.

Evelyn also provided an early account of the impact of this smoke on plant health: 'This coale . . . flies abroad, . . . and in the Spring-time besoots all the

Leaves, so as there is nothing free from its universal contamination . . .'. Later: '. . . and kills our Bees and Flowers abroad, suffering nothing in our Gardens to bud, display themselves, or ripen.'

Evelyn went so far as to suggest various measures to control the smoke, as did many who followed him in the eighteenth and nineteenth centuries. But despite continued lethal episodes, many decades passed before any significant control measures reached fruition.

It was not until after December, 1952, when 4,000 persons died in London in a few days, that real gains in control emerged. Federal air pollution legislation was introduced in the United States by 1955, and in England only a year later. The next decade saw the passage of similar legislation over much of the industrial world.

Despite a gradual shift to the cleaner-burning gas and oil in subsequent years, sulphur dioxide still persists in concentrations that are harmful to forests, gardens, and crops. Now we learn that plants can be affected even in winter, when they are mostly dormant: Their growth in the following spring may be seriously impaired.

The increased use of petroleum products was by no means the ultimate solution to air pollution. Their combustion not only releases some sulphur, but the hydrocarbons and other products can in themselves be toxic. Far more significant, though, are the secondary pollutants that are formed when the energy of sunlight converts these chemicals to more toxic products. Thus we see the concentrations of ozone, nitrogen oxides, and peroxyacetyl nitrates that are produced, continually rising. These secondary pollutants are most often associated with the automobile, but any combustion of petroleum produces them.

When electricity was generated for the first time at the Pearl Street Station in New York in 1882, the energy came from coal, and ever since then the generation of electrical energy has largely relied on this fuel. As the polluting nature of coal, and problems of its transport and equipment maintenance became recognized, there was some departure from this. A change-over to the cleaner-burning and more efficient natural gas and oil prevailed in the 1970s, but the tremendous rise in the costs of oil within the decade soon encouraged a shift back to coal.

Economic consideration will probably make the use of coal for power generation increasingly important for at least the next two or three decades. After this, nuclear fusion and solar power may introduce another shift. For some years, though, even with modern scrubbers partially to curb the emissions, and stacks over 800 feet (244 m) high to disperse the remaining effluents, sulphur and nitrogen oxides will continue to be dispersed over the land. Air pollution is not likely to disappear in the near future. Sulphur and nitrogen gases will continue to be dispersed over distances of up to many hundreds of kilometres from their sources.

Ultimately they will come down, descending in a gaseous or aerosol form—or more likely they will have been oxidized to form sulphates or nitrates, combine with water, and be washed out as an acid mist or acidified deposition.

The way in which acidity may affect biological systems is only one of the many questions that are currently being asked. We know that fish populations in many lakes have been killed out, the acidity of the lakes being rendered too great to support such aquatic life. Such stresses on aquatic systems have been described in Norway and Sweden, where an estimated fallout of 800,000 tons of SO_2 annually is estimated, as well as in the northeastern United States and Canada. The possible impact on terrestrial systems is not yet well established, but many consider it to be at least as serious. Every plant species responds differently to acidity. Evolution has attuned each to its own range of preference and tolerance. Should the acidity go beyond this range, the plant cannot function normally. Not only is growth impaired, but the plant is rendered more sensitive to other environmental stresses—temperature extremes, disease, and insect pests.

Acid rain may also adversely alter the soil rhizosphere. The trend towards acidification has been increasing noticeably for at least three decades, particularly at northern latitudes where much of the world's remaining forest is found.

The great increase in carbon dixoide resulting from the burning of fossil fuels may also influence the acidity of precipitation and thereby act indirectly as an air pollutant. The removal of vast forests that once utilized much of the carbon dioxide to produce carbohydrates, may be ever more significant than such burning in altering the atmospheric carbon dioxide.

Knowledge of air pollution biology is of obvious concern to the field pathologist who must diagnose and assess any injury and potential losses. It is one more environmental factor influencing plant development, and must be understood correctly to interpret ecological interactions. Similarly, some knowledge of air pollution biology is essential to the forester, horticulturist, agronomist, and many other plant biologists.

It is equally critical to the decision makers in government, who must judge the costs of air pollution and the value of the benefits to be gained from its control. From these data they may establish the needs for the control of emissions, set the desired air quality standards, and see that they are attained. The government agencies, with the help of biologists and economists, must utilize all the data to establish such cost–benefit values. First, though, we must be able to recognize what air pollutants do to plants, and the concentrations that have varying degrees of impact. Only then can we determine the benefits to accrue from each degree of control. We hope that this book will provide some help to those who have to make these decisions.

Air Pollution and Plant Life
Edited by M. Treshow
© 1984 John Wiley & Sons Ltd.

CHAPTER 2

Concern about Atmospheric Pollution

DONALD G. GILLETTE

United States Environmental Protection Agency, Research Triangle Park, North Carolina 27711, USA

AIR POLLUTION—A SOCIAL PROBLEM

Air pollution is a social disease—a disease generated primarily from the activities of Man, adversely affecting his health and welfare. It is also a global problem, transcending natural and political boundaries and extending from the more densely populated and industrialized areas into the more remote and sparsely populated areas. It is a disease of industrial progress and affluence, a disease more prevalent in countries with higher *per capita* incomes, larger gross national products, and higher rates of energy consumption. It is a disease that is less tolerable among the rich than the poor, who are more concerned with the basic and immediate necessities of food, shelter, and clothing, and less with disamenities and possible longer-run and more subtle effects of dirty air.

During the past 30 years, the rapid economic growth which has occurred in the United States and many other countries, was manifested by the explosive increase in the volume and diversity of goods and services available to the general public. Accompanying this increase of goods and services were the by-products generated in the production and consumption processes. Having little or no economic value, most of the by-products are literally dumped into the environment, in order to minimize the cost of disposal to the firm that is responsible for generating them. In the manufacturing or consumption process, thousands of millions (i.e. the U.S. billion, 1×10^9; the European billion, 1×10^{12} will not be used in this volume) of tons of pollutants are produced annually in the United States as raw materials, and substances are transformed from one form to another.

The annual damage attributable to air pollutants, which would have occurred in the United States if there had not been government action to control pollution—based upon 1970 projected air-quality levels and 1978 prices—has been estimated to range from approximately $10 billion to more than $50 billion (Freeman, 1979). With government regulations, the improvement in air quality has been estimated to reduce the amount of potential damage by approximately 50%. Most of this reduction may be attributable to the reduction of observed health effects which was accomplished by meeting the primary air quality standards in the major urban areas. Benefits were also obtained in terms of reduced material damage, improved visibility, and some reduction in agricultural losses.

Much uncertainty remains as to the total amount of annual damage that is still occurring as the result of air pollution. That the compliance with primary air quality standards, based upon health effects threshold data and a relatively large margin of safety, completely eliminates any possible health effects that might be attributable to air pollution, is a dubious premise. Furthermore, the problems of measuring air quality, evaluating the spatial and temporal distribution of different combinations and types of pollutants, and defining the receptors exposed, are extremely difficult—particularly in remote areas and for mobile receptors.

Much as with health effects, measurement of the impact of air pollution on plant life and on ecosystems is also difficult—particularly at the lower levels of pollution. Air pollution damage to primarily commercial crop production in the United States has been estimated to range from $150 millions to more than $500 millions annually (Agricultural Research Service, 1965). If all of the hidden damages and the effects of air pollution on the natural environment and the biosphere with all its myriad forms of plant and animal life was measurable, the amount of damage to the natural habitat of Man could conceivably exceed several billion dollars annually. Although most of the allegations as to the seriousness of the air pollution problem with respect to the biosphere remain unsubstantiated, the possibility of serious consequences if air pollution is not abated remains a formidable threat. Indirectly, this threat is just as serious as are those which directly affect our health, for it can destroy the infrastructure from which our sustenance is derived.

Reported incidence and investigations of alleged air pollution damage to vegetation have been documented for more than a century. Most of the earlier-reported incidences occurred around specific industrial sites or other 'point sources'. Since the early 1940s, most of the damage to vegetation from air pollution has resulted from photochemical oxidants or smog, produced in the atmosphere and transported over wide geographical areas. More recently the problem of acid precipitation, initially noted in the northern European countries and more currently a recognized problem in the United States and adjacent Canada, has received increasing attention from scientists as a global problem with international implications. Both photochemical oxidants and acid precipitation are formed in the atmosphere from chemical reactions between primary pollutants emitted from fossil-fuel combustion or other sources and natural atmospheric elements. These reaction processes, depending upon meteorological and other conditions, may occur over a relatively short time-span and geographical area, or over a relatively longer period and larger geographical area. Consequently the resulting damage from these pollutants may be extremely widespread and subtle, and may occur at a considerable distance from the initial source of the pollution.

Prior to World War II, only a few scientists were involved in investigating air pollution effects on vegetation. In most cases, the reported incidences provided dramatic illustrations of the exploitation by private industry of so-called 'free goods'. Using the atmosphere as a free depository for its waste products, industry transformed systems and destroyed flora and fauna. But denudation and erosion of soils, destruction and extinction of plant life, and long-term ecological and economic impacts of emissions generated by certain industrial operations, received attention from few environmentalists.

It was not until after World War II, however, that rapid industrial growth in the more densely populated areas, and the resulting increased concentration of waste products that were being dumped into the atmosphere, brought forth greatly increased interest and concern about the possible environmental effects that such actions, if unabated, might produce.

PUBLIC OPINION AND RESPONSE

Before the mid-1950s, several local air pollution control agencies were established in the United States in areas where increased incidences of air-pollution damage from photochemical oxidants, sulphur oxides, or other pollutants, were being documented. In 1955, Public Law 84–194 was enacted by the US Congress to provide research support and technical assistance to State and local governments to develop an improved understanding and control of the problem (US Congress, 1963). Later on the Clean Air Act of 1967 was passed by Congress to enable the Federal Government to engage more directly in the regulation of air quality and air pollution research (US Congress, 1967). Prior to 1967 the Federal Government's role was passive and limited primarily to providing technical assistance and research support to State and local agencies, but since 1967 the role of the Federal Government has become an active one. This role necessitated a much stronger research programme than formerly existed, with the establishment of defensible air-quality standards as well as the means to see that these standards are met by Federal as well as State and local regulatory authorities.

During most of the period from the early 1950s through the early 1970s, public support for a stronger environmental programme was clearly evident in the United States. Enabling legislation to provide both Federal and State governments with the authority to implement legislation to clean up the environment was passed. Considerable publicity about the seriousness of the air pollution problem and the threat of its toxic wastes could be found in the newspapers. According to Gallup polls conducted in 1965 and 1970, the percentage of the population surveyed that selected reducing air and water pollution as one of the three most important national problems from a list of ten, increased from slightly less than 20% in 1965 to more than 50% in 1970 (Mitchell, 1979). This rise in public concern was also reflected in other polls.

Since the early 1970s, other major national problems such as inflation, the high cost of energy, the high level of unemployment, and high taxes, may have diminished the popularity of the environmental movement in the United States. Contrary to the expectations of many people, however, a survey conducted by Resources for the Future during the summer of 1978 indicated that public support for environmental protection remains strong—despite the strong publicity over Proposition 13, an anti-tax movement which began in the Far West (Ukup, 1979). Despite the energy crisis, high rates of inflation, taxes, and unemployment levels, the results of the poll indicated that people usually judged environmental problems to be serious, and expressed a willingness to increase government spending to mitigate such problems.

Each year the National Opinion Research Center solicits responses from a national sample to determine whether they think the Government is spending too much, too little, or about the right amount on a set of eleven social problems (Ukup, 1979). The set of problems listed in the surveys includes crime, drugs, health, education, environment, national defense, city, racial problems, welfare, space, and foreign aid. From 1972 to 1973, the annual percentage of the populations surveyed which indicated too little was being spent on environmental programmes, ranged from 65% in 1973 to approximately 52% in 1978. Of the eleven social problems mentioned, there were only three (crime, drugs, and health) that were ranked above the environmental problem in terms of too little being spent on the programme.

BENEFITS AND COSTS OF PLANT EFFECTS RESEARCH

More than $5 millions are spent annually in the United States to study the effects of air pollution on plant life (Gillette, 1979). Most of this research, conducted at university or research institutes, is financed by either the US Environment Protection Agency, the US Department of Agriculture, or the US Department of Energy. Less than one-fourth of the research funds are estimated to come from private industry. The amount spent on research on air pollution effects on vegetation each year in the United States is less than 3% of the estimated average annual damage to, or economic loss of, vegetation from air pollution which has been occurring in the United States in recent years.

Studies relating air pollution to vegetation injury or plant damage may include (1) field surveys or investigations, (2) controlled fumigation or laboratory experiments, and (3) longer-term ecological studies. Whereas most of the earlier studies were primarily field investigations, where plant injuries around industrial sites were documented and measured, the trend towards conducting larger-scale ecosystem studies has become increasingly evident. Laboratory or fumigation studies in which many of the environmental factors are controlled, remain important in verifying and understanding the underlying factors affecting plants. More emphasis, however, is now being placed on studies that look ·at the ecosystem as a whole, and on the more subtle effects of air pollution that are likely to occur at existing ambient pollutant levels. In contrast to earlier controlled laboratory experiments, many of the more recent studies are attempts to quantify the losses or plant injury that might occur in the field.

Prior to 1950, research on air pollution effects on vegetation was limited to the effects of sulphur oxides, fluorides, ethylene, chlorides, etc.—primary pollutants emitted from industrial sources that directly affect vegetation. After the widespread incidence of smog or photochemical oxidant, plant damage which occurred in the Los Angeles Basin in the mid-1940s and, nearly a decade later, in the northeastern States, photochemical oxidants, became the major pollutant of concern. In the mid-1970s, another type of pollutant—acid precipitation—received increased attention. It has been estimated that photochemical oxidant damage accounts for 80–90% of the air pollution damage to vegetation. This estimate, however, did not take into consideration the problem of acid rainfall, whose significance in terms of crop loss and damage to the natural environment is still largely unknown and a subject of wide speculation.

Despite the public support for environmental programmes as expressed in recent opinion polls, both the Federal and State governments were being urged in more recent years (1979–80) to reduce spending and to unshackle business from unnecessary regulations. More and better research on the effects of air pollution on vegetation may be essential to preserve or establish air quality standards and regulations to protect the environment. As a non-health or welfare effect, the economic benefits in terms of actual plant damage and possible reduction in such losses, as well as the costs incurred in obtaining such reductions, become important factors in this decision process.

The world of the plant scientist is much greater than the domain from which he draws his knowledge. Of the millions of different plant species existing on Earth, only a few have been especially studied when exposed to air pollutants. Unfortunately, many of the few which have been studied are of little commercial value and economic significance. Furthermore, Nature itself enters the selection process, eliminating those species that cannot survive in polluted atmospheres and replacing them with more resistant species. Assisting Nature is the geneticist who is engaged in propagating higher-yielding cultivars or plant species while perhaps not even knowing that air pollution resistance *per se* may be the major factor accounting for the higher yields. Unwittingly assisted by Man, Nature may gradually alter the world's plant population over time in response to the often difficult-to-measure and subtle changes in environmental conditions.

Much uncertainty exists as to the actual magnitude of the economic losses and plant damage that can be attributed to air pollutants in the United States. Perhaps more important than the current estimate of losses are the projected losses which will depend upon the ability of both Nature and Man to adjust to changing environmental conditions. What are the long-term consequences of current pollution levels? What is the long-term impact of acid precipitation on the future productivity of soil and the composition and health of its vegetative cover and inhabitants? Are the damages reversible? Are we as a society over-regulated? Are we paying too much to reduce risk and uncertainty?

To protect the environment adequately, and to provide the protection at the minimum cost to society, more and better data are needed on the effects of air pollution on vegetation. Existing air quality standards, and regulations enacted to protect the environment are still based on very limited quantities of effects data—whether they be health effects, materials, or vegetation. To arrive at air quality standards, and subsequently regulations to protect the environment, requires much wisdom and knowledge that few possess. Utilizing existing studies to derive inferences for other studies or population groups, whether they involve general geographical areas or plant species, is a common and necessary practice to derive estimates of economic loss or damage. Yet it is a questionable practice and formidable task. To reduce this uncertainty and improve the available estimates of the losses suffered by society, much more information on plant effects is needed. Perhaps, by providing that information, the present costs of protecting the environment can be made more commensurate with the benefits that society should receive.

Summary

Air pollution is a social disease that is brought on primarily by human activities, as a result of industrial development and affluence. It is less tolerable among the people of the wealthy nations than of the poor ones, who are concerned more with such immediate necessities as food and shelter. The adverse health effects and impacts on plant life and The Biosphere are enormous, but difficult to quantify.

Photochemical oxidants and sulphur oxides are the air pollutants that are most broadly damaging to plants. Fluorides, chlorine, ammonia, and others may be significant locally. Public concern during the 1950s and 1960s precipitated legislation that led to reducing these pollutants nationally and globally. But to protect the environment adequately, more and better knowledge is needed concerning the effects of air pollution on plants, and how to minimize their response.

REFERENCES

Agriculture Research Service (1965). Losses in Agriculture. *Agricultural Handbook No. 291,* United States Department of Agriculture, 120 pp.

Freeman, A. M. (1979). *The Benefits of Air and Water Pollution Control: A Review and Synthesis of Recent Estimates.* A report prepared for the Council on Environmental Quality, Washington, DC, USA: xv + 174 pp.

Gillette, D. G. (1979). Research costs and air pollution effects: plants. *Proceedings of the 72nd Annual Meeting of the Air Pollution Control Association.* Paper 79–46.5 (unpub.).

Mitchell, R. C. (1979). Silent Spring/Solid Majorities. *Public Opinion,* August/September Issue. 2,(4), 16–20.

Ukup, K. (1979). Environmental public opinion: trends and trade-offs. *Resources for the Future Discussion Paper D-36.*

US Congress (1963). *A Study of Pollution—Air.* Staff Report to the Senate Committee on Public Works. 62 pp.

US Congress (1967). *Clean Air Act of 1967.* Public Law 90–148. 81 Stat. 485.

Air Pollution and Plant Life
Edited by M. Treshow
© 1984 John Wiley & Sons Ltd.

CHAPTER 3

Dispersion and Fate of Atmospheric Pollutants

Elmer Robinson

Laboratory for Atmospheric Research, College of Engineering, Washington State University, SW 834, Pullman, Washington 99164, USA

Introduction

By definition, an air pollutant must be transported by the atmosphere from its source to the receptor. The concentration at which the pollutant reaches the receptor is to a large measure dependent on the meteorological characteristics of the transporting layer of the atmosphere. In this chapter we shall examine the main factors that influence the dispersion of pollutants as they are transported by the atmosphere. This transport process will be divided into a number of successive phases—ranging from the initial emission from the stack or other source, to the final atmospheric scavenging processes of chemical reaction, precipitation rainout or washout, or dry deposition.

This discussion is designed to give the reader a general overview of the dispersion, transport, and scavenging, processes. For detailed presentations on the numerous dispersion models and methods of their calculation, the reader is referred to one of the excellent books or monograph chapters dealing with this complex topic, e.g. Slade (1968), or ASME (1973), or Pasquill (1974).

The dispersion of air pollutants into the atmosphere downwind from a source is the initial meteorological phase of an air pollution cycle. Within the general context of 'transport and dispersion', there are several phases of development which together make up the atmospheric cycle. These are: (1) source emission characteristics, which include the initial dilution at the source and the plume's rise to an equilibrium transport layer; (2) dilution by eddy diffusion in proportion to the atmospheric turbulence-field during travel with the average wind-field; (3) long-range transport out of the local area accompanied by air-mass contamination and also dilution due to eddy diffusion and wind-shear effects between pollutant-carrying layers; (4) atmospheric reactions of the pollutants in the plume to form secondary pollutant compounds and to alter pollutant concentrations; (5) precipitation scavenging processes that alter precipitation formation and cloud droplet chemistry, and provide a mechanism to carry pollutants to the Earth's surface; and (6) dry deposition scavenging processes that remove pollutants from the atmosphere to the Earth's surface.

These several aspects of the air pollution atmospheric cycle will be discussed briefly in the following sections of this chapter. This discussion will emphasize the point of view of the interested scientist or engineer who is not a meteorologist but who needs to understand the basic features of the atmospheric transport and dispersion problem.

SOURCE EMISSION AND INITIAL PLUME RISE

The physical characteristics of the source of an air pollutant can play an important role in the subsequent dispersion of the material. The main problem at the source point is whether the emission escapes cleanly from the vicinity of the source or whether it is trapped in the aerodynamic circulations around the stack and other source structures. There are two problem areas that are related to the source. First, a low stack-emission velocity may permit downwash behind the stack. Second, stack downwash or a low stack-height can cause effluents to be trapped in aerodynamic eddies in the lee of the building, and so produce relatively high pollutant concentrations at the Earth's surface.

The problem of stack downwash is illustrated by Fig. 1, from Slade (1968). The cause of this downwash is eddy formation behind the stack due to the wind moving around the stack. This flow-field develops eddies that can move pollutants vertically downwards towards the ground in the lee of the stack. With this downward motion the effect of stack height may be severely limited, and the plume can come close to the ground relatively near to the stack. The result of this downwash will be relatively high concentrations occurring at ground-level, where most receptors are, and a high potential for damage to occur from these excessive concentrations of stack emissions. Experience both in the field and with wind-tunnel experiments has shown that the stack downwash may be avoided when the stack emission velocity exceeds the ambient wind-speed by a factor of about 1.5. Thus stack downwash may be avoided by proper design of the stack and setting the stack emission speed in relation to the expected wind conditions at the plant-site.

Downwash of pollutant emissions may also occur in the downwind area from buildings and other plant structures if the pollutant emission does not escape the aerodynamic eddies that form in the downwind area or lee of a building. This aerodynamic downwash is illustrated in Fig. 2 (Slade, 1968).

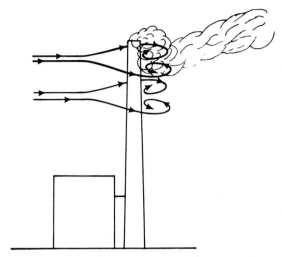

Fig. 1   Stack downwash (from Slade, 1968).

Note that the eddy which forms behind the building has a reverse flow back towards the building, so that pollutants which are entrained in this eddy may be carried towards the building and adversely affect ventilation systems or other activities close to the building. The presence of pollutants at the ground surface will also increase the probability of relatively high concentrations occuring in the vicinity of the plant, and thus increase the probability of damage occurring from pollutant emissions. Again, with this downwash situation the proper design and engineering of the stack emission-point can avoid this eddy entrainment problem. Experience and wind-tunnel tests have shown that a stack approximately 2.5 times the height of the building or other interfering structure is unlikely to introduce the pollutant emissions into the eddies forming behind the building.

For agricultural damage problems this eddy entrainment question is sometimes quite significant, because aluminium plants which produce fluoride emissions in their pollutant gases are typically characterized by low stacks or

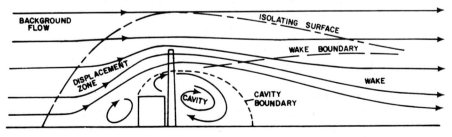

Fig. 2   Downwash due to lee eddies (after Slade, 1968).

emissions that escape at building roof-level through ductwork vents or roof monitors, and with little thermal buoyancy. These types of emission points are typical of the designs that would be expected to be entrained in downwind eddies, and therefore provide maximum exposure of nearby surface receptors to the emissions.

DISPERSION IN THE LOCAL AREA

When once the pollutant plume has left the vicinity of the source, it is carried and dispersed by the atmosphere. This dispersion process is dependent upon the turbulence and wind-field that exists in the atmosphere at the time of the emission. The travel of a plume in the atmosphere is typically described as a Gaussian plume, because the average cross-section of the plume resembles a Gaussian or bell-shaped distribution. Fig. 3 illustrates a Gaussian plume in an ideal sense (Slade, 1968). Note that the cross-section of the plume both vertically and horizontally has the typical Gaussian or bell-shaped distribution. For calculation purposes it is generally assumed that the x axis follows the direction of the average plume, while the y axis has a cross-wind direction and the z axis is vertical. These computation axes are also illustrated in Fig. 3. The turbulent dispersion of pollutants from a continuous point-source is

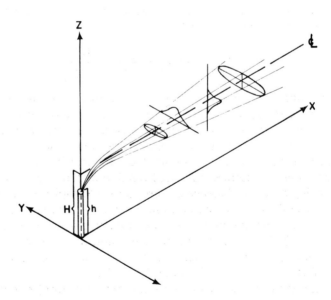

Fig. 3.   Ideal Gaussian plume, in x, y, z coordinates, where h is stack height and H is effective stack height. The plume spreads through an expanding oval cross-section along the centre line (℄). Concentrations have 'bell shaped' distributions about the centre line. (After Slade, 1968.)

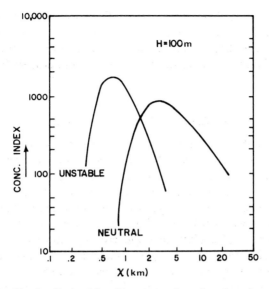

Fig. 4   Ground-level concentration of stack emission along the plume's axis.

frequently modelled by the Gaussian equation shown as Equation (1) (for explanation of symbols etc. *see* text):

$$C = \frac{Q}{\pi \sigma_y \sigma_z U} \ \exp\left[-\frac{y^2}{2\sigma_z^2} - \frac{H^2}{2\sigma_y^2} \right] \tag{1}$$

Fig. 4 illustrates the pattern of pollutant concentrations along the plume's axis. Close to the stack, concentrations are essentially zero because the pollutant has not had time to reach the ground, whereas at greater distances the concentrations increase rapidly to a maximum value and then decrease gradually at increasing distances from the stack. The maximum concentration is expected to occur at distances ranging from 5 to 20 times the height of the stack, depending on the turbulence pattern in the atmosphere. By following through the several terms that are given in Equation (1), we can examine the impact of the turbulent atmosphere on the dispersing pollutant plume.

In equation (1), C stands for the concentration of the effluent. This is calculated typically in mass per unit volume and is expressed frequently as parts per million or parts per billion (i.e. 1×10^9).

On the right-hand side of the equation the denominator Q stands for the emission factor from the source. This emission factor is in mass per unit time and illustrates the fact that the concentration emitted from the source is less important than is the total emission mass from the source. In the denominator of the Gaussian expression, π is the familiar mathematical constant.

The two terms identified as σ_y and σ_z are the lateral and vertical standard deviations of the diffusing plume. These are the statistical representations of the bell-shaped curves that are illustrated in the downwind plume in Fig. 3. In the atmosphere, σ_y and σ_z are dependent upon the turbulence field of the dispersing atmosphere and its wind-field. If turbulence is strong, the plume expands rapidly and σ_y and σ_z have relatively large values. As can be seen in Fig. 3, σ_y and σ_z also increase with increasing distance from the effluent source.

The term U in the denominator of Equation (1) stands for the average wind-speed in the plume-carrying layer and occurs as a direct factor in the denominator. This illustrates the fact that concentrations are inversely proportional to the wind-speed, and thus doubling the wind-speed will be expected to reduce concentrations occurring downwind from a source by a factor of 2. Wind-speed also is a factor in determining the values of the two standard deviation terms, σ_y and σ_z. This is due to the fact that wind-speed is related to the turbulence structure of the atmosphere and to the amount of mixing that turbulence can contribute to the plume.

In Equation (1), 'exp' stands for the mathematical term e or the base of the natural logarithms raised to the power indicated by the terms in the brackets. Within the brackets on the right-hand side of Equation (1), the first ratio, $y^2/2\sigma_y^2$, relates to the cross-wind dispersion of the concentration pattern and provides for the reduction of concentrations as a receptor moves away from the axis of the plume. The larger the dispersion in the y direction is, the larger will be the value of σ_y, and thus the more influence will this term have in reducing the concentration at the expected receptor. The second term within the bracket, $H^2/2\sigma_z^2$, related to the effective height of the plume and the turbulent transport of effluents from that height to the ground surface. This turbulence transport is related to the turbulence that is defined by the standard deviation on the vertical axis of the plume's dispersion, the σ_z. Note that the power of the exponential is negative, and thus the impact of the exponential term decreases as the distance from the stack increases.

The effective height of the plume, H, is determined not only by the physical height of the stack but also by the buoyancy or momentum rise of the plume itself. Fig. 3 also illustrates the height of the stack and the effective height of the plume, with h indicating the stack height and H indicating the total plume height which, as shown in Fig. 3, consists of the physical height of the stack *plus* an additional plume-rise.

Plume-rise is an important factor in determining the downwind concentration pattern calculated from a Gaussian model such as that given by Equation (1). Thus it is useful to consider the most important units or factors that go into determining plume-rise. Plume-rise occurs within the emission plume relatively close to the source, where the plume buoyancy has not yet been fully diluted and compensated for by the atmosphere, and thus the path of the plume is determined by the momentum and heat characteristics of the emission gases. Plume-rise or ΔH is dependent upon at least six factors (Briggs, 1969). The first of these is the total heat-emission from the source. The heat content of the gases forms the emission-providing buoyancy for the gases, with increased buoyancy leading to a higher plume-rise and thus a larger value for the effective plume-height in the Gaussian equation.

The plume-rise also depends upon the velocity of the emitting gases, emissions having a high velocity rise higher in the atmosphere and result in an increased plume-rise. The area of the source also plays a role in combination with the other factors, because a large source with a rapid exit-speed and a high heat-content will also be more massive, will persist in the atmosphere for a longer time, permitting buoyancy forces to persist, and provide a large plume-rise. Wind-speed is a factor in the plume-rise equation because wind-speed introduces turbulent mixing and entrainment into the plume. Strong winds will thus reduce the height of the plume-rise and reduce the effect of buoyancy. Stability of the atmosphere in terms of its tendency to restrict vertical motion may also limit the buoyant rise of the plume. Stability is frequently measured by the vertical temperature lapse-rate $\Delta T/\Delta Z$. The plume-rise is also a function of the downwind distance, x, because plume-rise occurs over a period of time after emission, and during this time the normal wind-field carries the plume some distance downwind in the x direction from the source itself.

In the Gaussian expression of Equation (1), probably the most important set of factors is the turbulence field introduced in the equation by the values of σ_y and σ_z. Turbulence in the atmosphere is a very complex subject, and a detailed treatment of all of its ramifications is beyond this particular discussion; however, there are a few factors that are worthy of being pointed out in order to guide the observer who needs to understand in general how pollutants are dispersed.

Turbulence can be separated into mechanical turbulence and convective turbulence. Mechanical turbulence is the mixing that is introduced into the atmosphere by the flow of the wind across the rough surface of the Earth. Roughness elements such as trees, forests, hills, valleys, and city buildings, introduce eddies, such as were illustrated behind the building in Fig. 2, into the flowing wind-field. This turbulence is dependent upon the state of the wind and the roughness of the surface—the rougher the surface is, the more turbulence will be introduced.

The second type of turbulence is convective turbulence, which has its source in the heating of the ground surface by the sun followed by the development of rising currents over areas of the Earth where heating is relatively strong—such as ploughed fields or large paved areas. These warm areas cause the air to be heated and become more buoyant, and to rise vertically in eddy currents and large vertical circulations. These 'convective cells' or 'convective turbulent eddies' serve to disperse pollutants and to provide an additional turbulence process to disperse pollutants. Convective turbulence is dependent on the degree of heating that occurs at the ground surface due to the sun. It is in turn dependent not only on the intensity of solar radiation and the nature of the underlying surface, but also on the presence of clouds and the height of the sun above the horizon.

The intensity with which convective or mechanical turbulence develops in the atmosphere is also dependent upon the atmospheric property called stability. This is the thermodynamic property of the atmosphere that may either inhibit or accelerate vertical motions which are generated initially by turbulence. A very stable atmosphere, characterized by an actual increase of temperature with altitude and called an inversion, will effectively prevent the development of vertical motion. Because of this restricting property, any inversion will prevent the rapid vertical dispersion or dilution of pollutant effluent. As the vertical temperature lapse-rate, $\Delta T/\Delta Z$, decreases and becomes negative to indicate a decrease of temperature with height, vertical motion becomes more likely to occur and thus, as the lapse-rate decreases, turbulence becomes favoured and dilution becomes more rapid.

When conditions are such that a temperature decrease of approximately 1 °C per 100 metres' rise occurs, which is the definition of an adiabatic process lapse-rate, turbulence is no longer inhibited and dilution proceeds in such an atmospheric layer at a relatively rapid rate. Under conditions of strong surface heating, the change of temperature with height becomes greater than the 1°C per 100 metres of the adiabatic lapse-rate, and turbulence is strongly favoured in this super-adiabatic or very unstable situation. Thus in a very unstable condition, dilution proceeds at a very rapid rate, and average concentrations decrease more rapidly with time under an unstable situation than in any other thermodynamic-type turbulence.

Figs 5, 6, 7, 8, and 9, illustrate typical plume patterns that occur downwind from a stack under various atmospheric stability conditions (ASME, 1973). Fig. 5 illustrates the plume type that is characteristic of a very unstable atmosphere. This plume is broken up rapidly into large eddies and has the characteristic or descriptive term 'looping' applied to it. The looping designation is descriptive, because the elements of a visible plume seem to move rapidly in the vertical direction touching the ground at one place or another and then rapidly being carried vertically to the top of the plume-layer as the effluent loops across the countryside. Part (a) of Fig. 5 illustrates this looping broken plume. Part (b) of Fig. 5 shows that the plume meanders and moves horizontally in the cross-wind direction as the well-developed eddies characteristic of an unstable atmosphere move the plume in the downwind direction. Part (c) of Fig. 5 illustrates the ground concentration of effluents that may occur during a looping plume. Note that very high but very short-period concentrations may occur as a relatively high-concentration portion of an

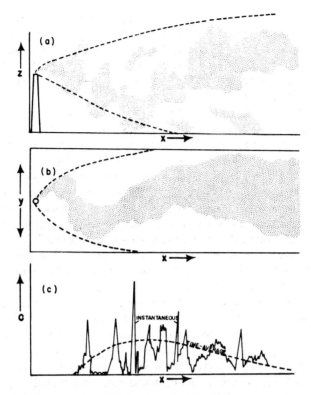

Fig. 5 Plume and dispersion characteristics during very unstable conditions—a 'looping' plume (adapted from ASME, 1973). For explanation, *see* text.

eddy reaches the ground at a particular location. The time-average line in Fig. 5(c) illustrates that the peak concentration occurs relatively close to the stack, compared with the distance from the stack in other turbulent situations.

Fig. 6 illustrates the characteristics of a plume dispersing from a stack under adiabatic or neutral conditions. This plume is dispersed by turbulent eddies that are significantly smaller in dimensions than those which disperse a looping plume under unstable conditions, and the plume assumes a more nearly regular outline as it disperses in the downwind direction; thus it is descriptively identified as a 'coning' plume. Fig. 6(a) illustrates the fact that the coning plume moves into a more regular pattern than does the looping plume, and Fig. 6(b) shows that the lateral dispersion characteristics are not nearly as marked as those which occurred in Fig. 5(b) for the unstable looping plume. The ground concentration pattern illustrated in Fig. 6(c) shows a much more uniform concentration occurring at a significantly greater distance downwind

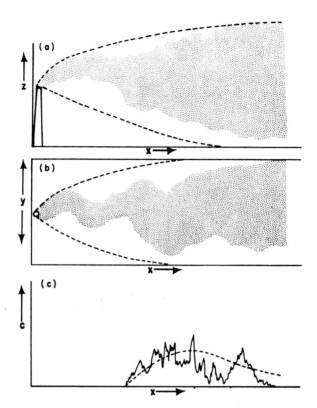

Fig. 6 Plume and dispersion characteristics during near-neutral conditions—a 'coning' plume (adapted from ASME, 1973). For explanation, *see* text.

from the source. The increased distance downwind for concentrations with a coning plume is due to the fact that the vertical turbulence eddies are not as strong, and thus it takes longer for the plume to be carried down to the ground from the height of the effective plume-axis.

Fig. 7 illustrates the characteristics of a plume emitted under strong stable situations such as engender inversions. Because the plume is not dispersed significantly in the vertical direction while, however, it may expand slowly in the horizontal direction, the plume is identified as 'fanning'. Fig. 7(a) shows that the plume is not expected to move towards the ground in this inversion or very stable situation. This is reasonable, because a stable situation does not contain vertical eddy motions that can entrain segments of the plume and carry these segments either downwards towards the ground or upwards. Fig. 7(b) shows the two horizontal characteristics of a fanning plume. First, a single constant wind direction, and thus a narrow ribbon in the upper part of

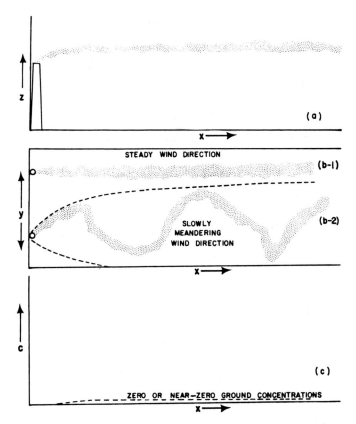

Fig. 7 Plume and dispersion characteristics during surface inversion conditions—a 'fanning' plume (adapted from ASME, 1973).

Fig. 7(b), or a slowly meandering plume spread out in the cross-wind direction as the wind slowly changes its direction. Fig. 7(c) shows that ground-level concentrations are essentially zero under fanning conditions; again this occurs because the vertical eddy turbulence is absent.

Fig. 8 shows one of the variations of plume dispersion characteristics that occurs when a turbulent layer containing the plume is capped or topped by a stable inversion layer. Under this situation the plume does not rise into the inversion layer but does disperse vertically below the inversion and reaches the ground in a manner similar to either a coning or a looping plume. Fig. 8(c) illustrates a higher concentration at the ground surface than occurred for either a looping or a coning plume, because the overlying or trapping inversion surface reduces the vertical spread of the plume, and thus higher concentrations occur at the ground surface.

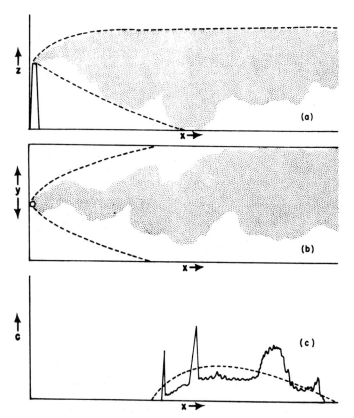

Fig. 8   Plume and dispersion characteristics during conditions
having an inversion above the plume axis—a 'limited mixing' or
'trapping' situation (adapted from ASME, 1973).

The diagrams of Fig. 8 are also illustrative of what may happen to a fanning
plume (*see* Fig. 7) during mid- or late-morning hours when the surface mixing
layer first becomes deep enough to intercept the fanning plume and mix it
down to the ground. Such a condition is called 'fumigation' and can be
responsible for occurrences of relatively high ground-level concentrations,
although fumigation conditions are expected to last for only a short period of
time (e.g. 30 minutes or so). Fumigation conditions lead to higher concentrations than the trapping situation, because the intercepted fanning plume leading to the fumigation will not have undergone as much lateral dispersion.

Fig. 9 shows the plume characteristics that occur when the inversion surface
occurs below the plume axis rather than above the plume axis (as was
described in the preceding figure). When the inversion surface occurs below
the dispersing plume, and as vertical turbulence is absent within the inversion

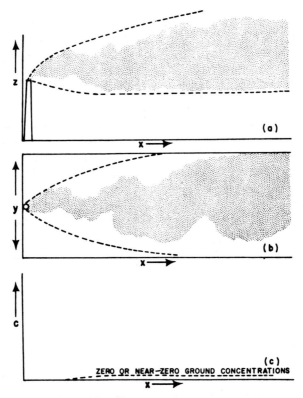

Fig. 9   Plume and dispersion characteristics during conditions having a surface inversion below the stack-top—a 'lofting plume (adapted from ASME, 1973).

surface, the plume is not transported to the surface from the effective axis of the plume. Thus dispersion may continue above the plume axis and in the layer above the top of the inversion, and there is diffusion downwind. However, as there is no vertical motion downwind towards the ground, surface concentrations at the ground are essentially zero, as illustrated in Fig. 9(c).

On the basis of typical plume-dispersion characteristics under different atmospheric stability conditions, it is possible to predict typical plume-dispersion features from a knowledge of the stability pattern that exists within a given area. Typically, night-time conditions are characterized by a fanning plume, and this plume will persist until one or two hours after sunrise—at which time surface heating will usually begin to produce convective turbulence. The onset of convective turbulence will produce a shallow convective layer, such as is illustrated in Fig. 8. There will follow a short period of relatively high concentrations beneath the plume as convection entrains the fanning plume. Concentrations then decrease as turbulence expands up into the stable layer above the plume.

As the convective layer becomes well developed in the late morning and early afternoon hours, a looping plume will be the characteristic pattern, and concentrations will be characterized by short periods of widely scattered but relatively high levels as turbulent elements move parcels of the pollutant across the ground surface. Late in the afternoon, convective turbulence may be expected to decrease; but as mechanical turbulence can persist, a coning plume may develop. As such a coning plume dissipates and an inversion develops at the Earth's surface, the lofting plume pattern takes over. Then, as the night-time stable air develops through an even deeper layer, a fanning plume develops and progresses into the diurnal pattern for the following day.

The preceding discussion of plume dispersion is only a very general account of what happens in the atmosphere as a pollutant plume is carried downwind from a source. Calculation procedures and plume dispersion models are abundant in the literature, and the interested reader is referred to one of the many books and monographs that describe these procedures. Examples include ASME, (1973); Briggs (1969); Pasquill (1974); Slade (1968). Several chapters in Stern, Volume 1 (1976) cover dispersion modelling in considerable detail also.

REGIONAL DISPERSION AND LONG-RANGE TRANSPORT

The preceding discussion described the dispersion of a plume in that area relatively close to the source where turbulent mixing of the atmosphere is producing relatively rapid changes in the dimensions and characteristics of the plume as it moves along with the atmosphere. At some distance downwind, the plume will effectively fill a layer or dispersing volume of the atmosphere. It will then disperse over further distances, but with apparently much less dispersion or dilution. This is the region which, for this discussion, we have characterized as regional or long-distance transport. This is the portion of the plume travel where the pollutants become quasi-stable constituents of the atmospheric air-mass and cease showing strong decreasing concentration gradients with increasing distance from the particular source.

Dispersion of pollutants within this regional or long-range transport area are dependent upon the meteorological characteristics that describe turbulence within the air-mass itself. These factors are the thermodynamic properties or the lapse-rate stability parameter, $\Delta T / \Delta Z$, the wind field of the dispersing air-mass characterized by the main wind pattern across the regional area, and the change of wind with height or the wind-shear.

Large, slow-moving high-pressure systems are frequently characterized by combinations of stability, wind, and wind-shear, such that the persistence of pollutants within the air-mass is favoured, and the air-mass meteorology can contribute to the accumulation of relatively high concentrations of pollutants.

These may persist within the air-mass for a considerable period of time, often as long as a week. For example, within a high-pressure system there is frequently an inversion surface, located at some distance above the ground surface, which can favour the formation of a trapping type of plume pattern and thus prevent significant vertical dispersion of air pollutant emissions. Along with the trapping inversion, weak winds are characteristic of many high-pressure systems. These will restrict the horizontal dispersion and dilution of the effluent plume, thus leading to further pollutant accumulation because of the reduction of horizontal dispersion.

High-pressure systems are also characteristically nearly cloud-free. This leads to strong sunlight and to favourable conditions for the development of photochemical pollutants with their accompanying adverse effects. High-pressure systems and clear skies are also not conducive to the formation of rainfall, and thus scavenging processes associated with rain are not prevalent. For these several reasons, a maximum concentration of pollutants may persist in the atmosphere for considerable periods of time when a slow-moving, large high-pressure system is present.

Long-range transport of air pollutants in the atmosphere typically refers to air-mass contamination by a relatively large number of sources—to form a pollutant cloud that can be carried by the moving air-mass for long distances. Long distances in this context may be a thousand kilometres or more, and thus the impacted area can be in another state or another country or even another continent. There are many political problems that may occur in a control programme because of such distances, and because of the problems that may occur in trying to link the source area with receptor damage.

Wind-shear, or the change of wind direction with height, becomes important in the long-range transport of pollutants, because the shear provides for a diluting mechanism within the pollutant air-mass. As the upper layers of the pollutant air-mass travel in one direction and the lower surface layers travel in a different direction and at a different speed, the wind-shear will spread out the plume gradually as the air-mass travels along. Thus, active wind-shear conditions can continue the dilution process even though other properties of the air-mass may be favourable for the persistence of relatively high concentrations.

One of the most commonly discussed consequences of long-range transport is the formation of acidic constituents from effluents of gaseous sulphur dioxide and nitrogen oxides. The formation of acidic components in the atmosphere during long-range transport has led to a lowering of pH in the rainfall or other precipitation and to the development of the pollution system that has come to be known as 'acid rain' or acidic precipitation. The subject of such precipitation as a by-product of long-range transport has many important aspects and is covered later in this volume (particularly in the following chapter).

It is within the context of long-range transport or the persistence of the pollutant material in the atmosphere for periods of a week or more, that removal or scavenging processes become important factors which impact on the downwind concentration-patterns of air pollutants. The following sections will therefore discuss the several mechanisms of atmospheric scavenging processes as they effect effluents within the plume, and in the time-period characterized by long-range transport.

CHEMICAL SCAVENGING PROCESSES

Many of our more serious air pollutants that are emitted from urban and industrial sources fall in the general chemical class of reactive gases. These include such common air-pollutants as sulphur dioxide, the nitrogen oxides, fluorides, and hydrocarbons. Other common pollutants that are not normally considered to be especially reactive are carbon monoxide, carbon dioxide, and fine particles. Because of the effects of atmospheric reactions on atmospheric concentrations, atmospheric dispersion-patterns of reactive gases can be strongly affected by chemical reactions or scavenging processes that can take place in the atmosphere. These chemical scavenging reactions serve to reduce the concentrations of the gaseous constituents and to increase the concentrations of reaction products.

Probably the most important chemical reaction processes in the atmosphere involve the oxidation of sulphur dioxide to form sulphate particles and sulphuric acid droplets. The oxidation of SO_2 may take place by at least two common mechanisms. The first of these, and probably the more important of the two SO_2-reaction mechanisms, is a photochemical reaction process. This process may take place to a significant degree in the summer and early autumn months where, under the influence of sunlight and other pollutants in the atmosphere, SO_2 is oxidized to form particulate materials, namely sulphate compounds and sulphuric acid. Various chemical scavenging processes are discussed by Butcher & Carlson (1972), and by Haagen-Smit & Wayne (1976). As these two types of reaction products are particulate materials, they will contribute to visibility problems and to summer and autumn haze situations. Recent investigations of both visibility and turbidity in the United States have shown that summer-time conditions are becoming increasingly characterized by particulate haze situations—especially in the southeastern portions of the United States, where both moisture and SO_2 emissions are abundant and can contribute to the formation of visible air-mass haze conditions.

Another reaction process that is active in scavenging sulphur dioxide from the atmosphere begins with liquid droplets, either as fog, cloud, or rain droplets, and the subsequent diffusion of atmospheric SO_2 into the liquid droplets. When SO_2 goes into solution in a liquid droplet, the oxidation or reaction of the SO_2 can be accelerated if ammonia or an oxidation catalyst from the atmosphere has also been dissolved in the liquid droplet. When both ammonia and SO_2 are present in the same liquid droplet, ammonium sulphate can be expected to form; the fact that this mechanism does occur frequently is shown by the observation that analyses of the compounds which are present within the suspended particles in the atmosphere frequently identify ammonium sulphate as a prominet constituent of the atmospheric aerosol particles.

Atmospheric photochemical reactions involving nitrogen oxides and sunlight *plus* hydrocarbons lead to the well-known Los Angeles photochemical smog reaction in which ozone and other oxidant gasses are produced along with particulate materials and various compounds that lead to severe eye irritation and crop damage. These photochemical scavenging products are secondary pollutants in the atmosphere, and are due to the combinations of pollutants and sunlight in the air-mass at the same time rather than to specific sources of ozone or eye irritants.

From the standpoint of crop damage and other vegetation effects, atmospheric reaction processes, acting on the primary pollutants to produce damaging secondary pollutants, have been important aspects of air pollution and agricultural impacts for many years. Ozone damage in Los Angeles and other areas of California, as well as along the east coast of the United States, is due not to emissions of ozone from individual sources but to the formation of ozone as a by-product of photochemical reactions during pollutant scavenging processes. These smog-forming processes develop from a combination of emissions—most specifically hydrocarbon, nitrogen oxide, and sulphur oxide, emissions—when sufficient sunlight is present to drive the photochemical smog reactions. Thus, from an agricultural standpoint, an investigator must be aware not only of the primary emissions that may affect his area, but also of the atmospheric scavenging reactions that may take place secondarily and cause additional pollutants to form in the atmosphere at significant, crop-damaging concentrations.

PRECIPITATION SCAVENGING

The occurrence of precipitation, either rain or snow, provides another mechanism by which pollutants may be removed from the atmosphere and deposited back on the ground surface. This occurs because pollutants either in gaseous or particulate form will be entrained in liquid or solid precipitation particles and thus carried with them to the ground. Junge (1963) describes precipitation scavenging mechanisms.

Entrainment may occur within the cloud as the raindrops or precipitation particles are formed. This phase of the precipitation scavenging process is called 'rainout'. In the rainout process, diffusion mechanisms transfer pollutants from the atmosphere to the cloud droplets in proportion to the concentration of the pollutants within the cloud-mass. As clouds are characteristically at relatively high altitudes above the ground surface, the rainout process, to be effective, must occur under conditions where the pollutants are well mixed within the total vertical extent of the air mass, including the layers containing the cloud mass. Thus, rainout processes will be most effective during long-range transport situations. Under these conditions, the pollutants will be distributed through deep layers in the atmosphere.

After pollutants have been absorbed into cloud droplets, they may undergo reaction processes—such as the SO_2–ammonia reaction to form ammonium sulphate, or other analogous chemical reactions. Cloud droplets typically form the initial phase of the larger precipitation particles, and thus cloud droplets with absorbed pollutants are formed into larger, precipitating particles. In this manner, pollutants are commonly removed from the cloud-mass and the atmosphere when precipitation falls to the ground.

Cloud droplets that do not become part of precipitating particles characteristically will evaporate, leaving behind the pollutants in the atmosphere—usually in the form of particulate materials which can then be entrained into other condensing cloud particles until eventually, as the cycle is repeated, the pollutant materials will be absorbed into precipitating particles so that the material is eventually removed from the atmosphere. Junge (1963) has estimated that cloud droplets may go through an average of ten or more condensation—evaporation cycles before they eventually become part of precipitating particles; thus there is considerable time for cloud droplets to absorb pollutant materials before precipitation removes the pollutants.

The second type of precipitation scavenging is called 'washout', and occurs in the air below a precipitating cloud when rain or snow falls through a pollutant-contaminated layer. In this process, as the precipitation particles fall through the pollutant cloud, pollutants impinge on the surface of the precipitation particles, where they may be absorbed and entrained within the particle. This also provides a mechanism by which the pollutants can be carried to the ground surface. However, this process typically is relatively inefficient on a time-basis, because of the relatively low density of precipitation particles within any given volume of air and the short contact-time over which the precipitation particles are exposed to the pollutant cloud. But where the pollutant layer has a relatively high concentration, and the precipitation rate is relatively heavy, the removal of pollutants may be important.

The incorporation of pollutants into precipitation leads to the question of precipitation chemistry. Mention has already been made of the problem of acid precipitation, which is caused by incorporation of excessive acidic materials, either sulphur oxides or nitrogen oxides, into rain or snow. Where such precipitation is deposited on the ground surface, excessively low pH may occur in the surface water. Areas where this has occurred to the apparent detriment of the biosphere include the northeastern states of the United

States, the southeastern provinces of Canada, and the Scandinavian countries in northern Europe. These several areas are downwind, in terms of long-range transport, from areas where there are large concentrations of sulphur oxide and nitrogen oxide emissions—primarily from heavy fuel combustion and energy-generating sources.

As precipitation will be expected to include representative concentrations of most of the pollutant materials that enter the atmosphere, patterns of precipitation chemistry can be used to identify areas of excessive emissions. Although in the case of pH changes, sufficient travel time must be factored into the transport model to allow pollutant oxidation to take place and produce the acidic products. Aerosols containing heavy metals, including lead particles from automobile fuel additives and other similar pollutant constituents, have been widely detected in snow and glacial ice as well as in rainfall.

SURFACE DEPOSITION SCAVENGING

Pollutants are also scavenged by the ground and vegetation surfaces in the absence of precipitation. This process is called 'dry deposition' (Chamberlain, 1960). For relatively large particles, with diameters of 20 or more micrometres, deposition on the ground surface occurs mainly by gravitational settling, and is usually described by Stokes' law processes. Stokes' law relates settling velocity to a ratio of gravitational force and aerodynamic drag (Slade, 1968). These include the well-recognized process of gravitational sedimentation, which accounts for the zone of heavy dust deposition around many rock quarries and sand and gravel operations, for example. However, dry deposition on the ground surface is not restricted to large particles and gravitational settling, and there are important dry deposition processes that scavenge both very fine particles and gases. For these latter materials the dry deposition process is governed by a deposition coefficient called deposition velocity, which is the ratio of the rate of deposition divided by the concentration of the material in the boundary layer immediately above the ground surface. This coefficient has been called 'deposition velocity', because if the units of deposition rate and concentration are combined, the result is a coefficient with the units of velocity; thus Chamberlain, who developed this dry deposition concept, denoted the coefficient between deposition and concentration as the deposition velocity. The mathematical expression is:

$$D = V_d C$$

where: D is deposition rate, C is pollutant concentration near the ground, usually about 1 metre, and V_d is the coefficient of deposition velocity.

Considerable study over the past years has shown that deposition velocity is dependent upon the chemical characteristics of the material that is being considered; higher deposition velocities accompany the more reactive materials and are identified with surfaces that will promote reactivity. Both gases and particles are subject to this process of dry deposition. Dry deposition occurs because, at the ground or boundary layer surface, there is a very thin layer of air 10 to 100 micrometres thick, which is essentially bound to the soil or vegetation surface. As this layer is, practically speaking, fixed to the surface, it allows gaseous or fine particles that are entrained in it to have the necessary time for transport by Brownian movement through the layer and potential reaction with the underlying surface.

Thus reactions take place and remove materials from this thin boundary layer. As the boundary layer becomes deficient in the reacting pollutant relative to the overlying turbulent air, more pollutant is transferred into the fixed layer from the turbulent layer above, whereupon these newly arrived constituents also go through the removal mechanisms at the underlying boundary surface. Thus there is a flux of material from the turbulent atmosphere, into the fixed boundary layer, and then onto the fixed surface. This is the dry deposition scavenging process (Wesley & Hicks, 1977).

As indicated above, the lower atmosphere and the surface boundary layer can, for the purposes of explaining the dry deposition process, be divided into three zones. The bulk of the atmosphere comprises the aerodynamic turbulence zone, where eddy turbulence mixes the pollutants and provides the mechanism that transports the material vertically down towards the ground surface. This turbulent mixing establishes a relatively uniform concentration in this lower atmospheric layer. Beneath the aerodynamically turbulent zone there is the fixed or laminar boundary layer, through which the pollutant materials are moved primarily by molecular or Brownian diffusion to the actual physical boundary of the underlying surface. At the boundary surface the pollutants are removed from the adjacent layer by reaction or adsorption onto the surface.

In the case of vegetation, the removal mechanisms from the boundary layer to the surface itself occurs not only on the surface but within the stomata of the plant material. Gases that are soluble, such as SO_2, will have significantly different deposition characteristics from those of the relatively insoluble gasses, such as CO, because they are drawn into the leaf structure through the stomata and then are efficiently removed by the plant's internal processes. Also, very moist surfaces, such as moist vegetation or water surfaces, can increase the dry deposition rate for soluble materials as compared with the deposition rate over dry desert or arid surfaces.

In terms of pollutant transport, the recognition of the importance of dry deposition processes began to be noticed about 1975, and studies carried out in both the United States and Europe since that time have shown the importance of the dry deposition process in the atmospheric cycles of pollutants. For example, for a typical coal-fired power-generating plant with a medium to tall stack, it has been shown that over a period of several days during the process of long-range transport, dry deposition processes will provide for the scavenging of approximately 50% of the emitted material. Chemical reaction processes within the atmosphere, namely gas to particle conversion and pre-cipitation scavenging, account for the other half of the scavenging of the pollutants. (Husar *et al.,* 1978).

At the present time one argument against the practice of construction of excessively tall stacks for power-generating and other major SO_2 emission-sources, is the contention that these tall stacks can reduce the likelihood of dry deposition and scavenging at the ground surface and thus will increase the scavenging of pollutants by precipitation. This, it is argued, can lead to increased problems due to precipitation acidity and changes in precipitation chemistry. Thus, within the 1970s, the increasingly widespread problems due to precipitation pH or 'acid rain' have been attributed by many investigators to the rapid introduction of excessively tall stacks for major pollutant sources. It is of course also recognized that the tall stacks were introduced to reduce the impact of pollutants in the area immediately surrounding the source, the region which was described above as the Gaussian plume dispersion region. The tall stacks have been quite effective in bringing about reductions in local, ground-level pollutant concentrations.

SUMMARY AND CONCLUSIONS

The preceding discussion has considered pollutant effluents in a step-wise fashion from the point of emission, through the near-by dispersion processes, and into situations of long-range transport where scavenging processes become important factors in determining the concentration pattern of the effluents. Within each dispersion zone there are conditons that are particu-larly important to the agriculturist or scientist who is concerned with vegeta-tion and other biological damage. Near the source itself, pollutants may be entrained in eddy currents around the source buildings that bring them to the ground quickly and close to a source. The typical design of an aluminium plant can favour such building-generated turbulence, and may be faced with zones of especially high concentrations of pollutants near the plant. These near-by contamination areas may result in important regions of vegetation damage.

Beyond the immediate environs of the pollutant source, the plume from a stack typically undergoes Gaussian dispersion, and the concentration pattern at the ground surface decreases rapidly with increasing distance from the source. This pattern of decrease is strongly dependent upon the local meteorology, which becomes a more and more important factor as the plume is dispersed by the atmosphere. Beyond the Gaussian plume zone the pollutant effluents become a relatively well-mixed part of the travelling air-mass into which they were dispersed, and concentrations become dependent upon both the meteorology of the air-mass and the scavenging processes that take place over a period of days during the air-mass travel. If the air-mass is characterized by a slow-moving, well developed high-pressure system, pollutant concentrations can persist at relatively high concentrations without significant dilution for a period of days. This meteorological situation leads to the occurrence of air pollution episodes.

Where the air-mass is characterized by clouds and storm systems, the pollutants may be incorporated into precipitation scavenging processes and make a significant contribution to changing the chemistry of the precipitation. Increased acidity in precipitation has, in recent years, brought about major problems due to low pH of rainfall or snowfall, and there are areas, such as the northeastern United States, southeastern Canada and southern Scandinavia, where the biosphere is especially sensitive to these changes.

These topic areas should all be investigated in considerable depth, but most investigators will probably be satisfied with a general account of the main points of these processes, such as has been provided in the preceding discussion. Meanwhile other chapters in this volume will consider these topics in more detail.

References

American Society of Mechanical Engineers (cited as ASME) (1973). *Recommended Guide for the Prediction of the Dispersion of Airborne Effluents*, 2nd ed. American Society of Mechanical Engineers. New York, NY, USA: xii + 85 pp., illustr.

Briggs, G. A. (1969). *Plume Rise.* US Atomic Energy Commission Critical Review Series, National Technical Information Service, TID-25075, Springfield, Virginia, USA: vi + 81 pp.

Butcher, S. S. & Carlson, R. J. (1972). *An Introduction to Air Chemistry.* Academic Press, New York, NY, USA: xiii + 241 pp., illustr.

Chamberlain, A. C. (1960). Aspects of the deposition of radioactive and other gases and particles. *Int. J. Air Pollut.,* **3**, pp. 63–88.

Haagen-Smit, A. J. & Wayne, L. G. (1979). Atmospheric reactions and scavenging processes. Chapt. 6 in *Air Pollution*, 3rd ed., Vol. I (Ed. A. C. Stern). Academic Press, New York, NY, USA: xviii + 715 pp., illustr.

Husar, R. B., Patterson, D. E., Husar, J. D., Gillani, N. V. & Wilson, W. E., Jr (1978). Sulfur budget of a power-plant plume. *Atmos. Environ.,* **12**, pp. 549–68.

Junge, C. E. (1963). *Air Chemistry and Radioactivity,* Academic Press, New York, NY, USA: xii + 382 pp., illustr.

Pasquill, F. (1974). *Atmospheric Diffusion,* 2nd ed. Ellis Harwood Ltd., Chichester, UK; Halsted Press, New York, NY, USA: xi + 429 pp., illustr.

Slade, D. (Ed.) (1968). *Meteorology and Atomic Energy.* US Atomic Energy Commission, National Technical Information Service, No. TID-24190, Springfield, Virginia, USA: x + 445 pp., illustr.

Stern, A. C. (Ed.) (1976). *Air Polution,* 3rd ed., Vol. 1. Academic Press, New York, xviii + 715 pp., illustr.
Wesley, M. L. & Hicks, B. B. (1977). Some factors that affect the deposition rates of sulfur dioxide and similar gases on vegetation. *J. Air Pollut. Control Assoc.,* **27**, pp. 1110–6.

Air Pollution and Plant Life
Edited by M. Treshow
© 1984 John Wiley & Sons Ltd.

CHAPTER 4

Long Range Transport of Air Pollutants and Acid Precipitation

BRYNJULF OTTAR, HARALD DOVLAND, & ARNE SEMB

Norwegian Institute for Air Research, PO Box 130, N-2001 Kjeller–Lillestrøm, Norway

INTRODUCTION

Until the middle of the present century, air pollution was largely considered a local problem, and the tall stacks were commonly used to reduce concentrations at ground level. For a long time this approach seemed to solve most of the air pollution problems. Towards the end of the 1960s, however, this situation was disturbed by the observation that serious acidification of the atmospheric precipitation was taking place in Europe (Odén, 1968).

The main acidifying agent was sulphuric acid, and the most likely source of it was the increasing emissions of sulphur dioxide from the use of fossil fuels. The sulphur pollution was accompanied by considerable amounts of soot and fly-ash which occasionally resulted in a dark staining of snow. The studies which followed confirmed these observations and showed that long-range transport of air pollutants is a phenomenon which concerns a number of different chemical substances.

The increasing concern for ecological effects from the long-range transport and deposition of air pollutants led to a number of national and international research programmes. On the initiative of the Nordic countries, the Organization for Economic Cooperation and Development (OECD) in July 1972 started a 'Cooperative technical programme to measure the long-range transport of air pollutants' (LRTAP). The objective of this programme was 'to determine the relative importance of local and distant sources of sulphur compounds in terms of their contribution to the air pollution over a region'. Eleven countries in northwestern Europe participated, and the programme was concluded in 1977 (OECD, 1977).

The OECD/LRTAP programme was based on the following elements: an emission inventory, daily measurements at ground-level stations, aircraft sampling, and atmospheric dispersion models to describe the connection between emissions and observed concentrations in air and precipitation. This programme made it possible for the first time to quantify the depositions within one country due to emissions in any other country. It showed that 'the air quality in any one European country is measurably affected by emissions from other European countries'. From an air-quality management point of view, this implies that 'if some countries find it desirable to reduce substantially the total deposition of sulphur within their borders, individual national control programmes can achieve only a limited improvement'. It was accordingly recommended that continued studies ought to include all European countries, and in 1977 the Cooperative Programme for Monitoring and Evaluation of Long-range Transmission of Air Pollutants in Europe' (EMEP) was organized under the United Nations Economic Commission for Europe (ECE) in cooperation with the United Nations Environmental Programme (UNEP) and the World Meteorological Organization (WMO). At present about 20 countries participate in the EMEP, and its design broadly follows that of the OECD/LRTAP programme.

In North America, a number of regional programmes sponsored by national government agencies and departments are in progress. While the European and Canadian studies have responded to ecological problems resulting from acid precipitation, the US emphasis was initially on health effects and reduction of visibility due to the increased concentration of sulphate aerosol. This has, however, changed markedly during recent years, and today a number of the US studies are responding to acid precipitation problems. Most of the North American programmes, as well as the OECD/LRTAP programme, are described in the proceedings of the Dubrovnik symposium (Husar *et al.*, 1978).

Acidification of rivers and lakes, and loss of fish populations, have been observed in Norway since the 1920s (Dahl, 1927), but it was not until 1959 that precipitation acidity was suggested by Dannevig (1959) as a probable source. A comprehensive interdisciplinary research programme to study the effect of air pollutants deposited in Norway from long-range transport, the SNSF project, was started in 1972 and is now being concluded (Overrein *et al.*, 1980). The studies have involved comprehensive monitoring of precipitation input, catchment studies, studies of cation leaching from soils, and studies of physiological effects on freshwater organisms.

The interactions between atmospheric input from precipitation and dry deposition, and run off from vegetation or soil and of surface waters, are complicated. It appears that the time-scale for effects to occur in a given watershed can be considerable, at least of the orders of years, although the 'response time' will be rather variable from one catchment to another.

Nevertheless, the correlation between acid precipitation (or the input of excess sulphate), and the water chemistry and loss of fish populations in the granitic bedrock areas of Southern Norway is overwhelming. At the same time, however, soil fertility and growth of conifers does not seem to be much affected even by excessive application of sulphuric acid, though subtle effects of long-term changes cannot be excluded.

Acidification of surface waters and depletion of fish stocks also occur in the USA and Canada, and extensive research programmes have been initiated to study the possible effects.

THE SOURCES OF AIR POLLUTION

The main sources of air pollution are the use of fossil fuels for heat and energy production, transportation vehicles, and various industrial production processes as well as the final use and disposal of many industrial products. Major components include sulphur dioxide and nitrogen oxides, as well as less specific pollutants such as particulates and hydrocarbons. Chemical reactions in the atmosphere lead to formation of secondary pollutants such as sulphuric acid, nitric acid, ozone, and photochemical oxidants.

Important minor constituents are trace elements such as zinc, lead, or cadmium, and polycyclic aromatic hydrocarbons. These are transported and deposited in much the same way as the sulphur and nitrogen compounds. Mercury and certain of the halogenated hydrocarbons can be re-emitted to the atmosphere after deposition, because of their high vapour-pressure and persistence.

The most important air pollution constituents in relation to the ecological impact of acid precipitation are the sulphur and nitrogen compounds. Some of the other primary and secondary air pollutants do, however, have an effect on the chemical and physical reactions of the sulphur and nitrogen components in the atmosphere.

Global atmospheric budgets have been widely used to compile the various sources and sinks involved in the cycling of sulphur and nitrogen compounds (Eriksson, 1963; Robinson & Robbins, 1970*a*, 1970*b*; Granat *et al.*, 1976). These studies generally conclude that on a global scale, emissions from fossil-fuel combustion and other man-related processes are of the same order of magnitude as the less well defined emissions from natural sources.

The Man-made emissions of sulphur and nitrogen compounds are largely confined to relatively small areas in the northern hemisphere, and in Europe and North America by far outweigh the natural emissions. These emissions will therefore have a dominating influence on the air and precipitation chemistry in these two regions.

Both sulphur dioxide and nitrogen oxides are produced by combustion of

fossil fuels, mainly coal and oil. There is a wealth of statistical information
available on the production, trade, and use, of these fuels—*see*, for instance,
Semb (1978, 1979).

Sulphur Dioxide Emissions

The emissions of sulphur dioxide can be calculated from consumption data
and the sulphur content of the different fuel-types, making allowances for the
retention of sulphur in slags and ashes, and the amount removed by flue-gas
cleaning. Sulphur in crude oil varies from 0.1–0.5% in North African and
North Sea crudes, to 1.5–3% in the Middle East crudes. During the refining
process, most of this sulphur ends up in the heavy distillates and residual oil
fractions.

The sulphur content of coals is also different in different geological
deposits, and may even show considerable variation within the same general
mining area. The average sulphur content in Western European coals is
1–1.5%, while the average sulphur content of coals consumed in the USA has
been estimated at 2.86% (ECE, 1976). Certain brown coals in the German
Democratic Republic and Czechoslovakia also have rather high sulphur
contents.

Emissions of sulphur dioxide from industrial processes come mainly from
roasting of sulphidic copper, nickel, lead, and zinc, ores, the manufacturing of
sulphuric acid, and the paper and pulp industry. On a global basis these
emissions may account for about 10% of the total Man-made emissions of
sulphur (Cullis & Hirschler 1979). In a few countries the share is larger. This
particularly applies to Canada (Katz, 1977), but also to Norway and Sweden.
Reasons for this high proportion are, partly, the utilization of large sulphidic
ore deposits. On the other hand, natural gas does not normally contain sulphur, though there are exceptions in the acid gas sources of western Alberta
in Canada and in the Lac district in France.

The relative contributions to the sulphur emission from different sectors
vary from country to country, according to the fuel consumption pattern and
sulphur content of the fuels. For most countries, however, electric power-
plants, refineries, and large industrial boilers, account for the main part of the
sulphur dioxide emissions related to fossil fuels. In the US, because of the
high sulphur content of the fuel, coal-fired power plants account for more
than half of the sulphur emissions. A summary of the emissions in OECD-
Europe and North America, for different sectors of consumption, is given in
Table I.

The world energy consumption of solid and liquid fuels in 1976 was about
2,700 and 3,730 million tonnes of coal equivalents, respectively. About 70%
of this is consumed within the economic and geographic regions of North
America, Eastern Europe and Western Europe.

Table I Emissions of sulphur dioxide (Tg [= 10^{12} grams] S/year) in different consumption sectors in 1975.

	OECD-Europe	North America
Power plants	4.2	8.7
Industrial/Commercial	4.0	2.5
Transport	0.3	0.4
Industrial processes	≈1	3.8
Other	1.6	—
Total	≈11	≈15

Geographical surveys of the anthropogenic sulphur dioxide emissions in Europe have been constructed in connection with the OECD/LRTAP Programme and the ECE/EMEP Programme, primarily for use in the mathematical modelling of the atmospheric transport of the pollutants. The EMEP emission survey (Dovland & Saltbones, 1979) covers all of Europe in a 150

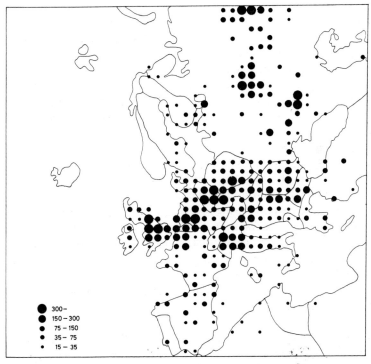

Fig. 1 Estimated sulphur dioxide in 150 km × 150 km grid squares for Europe in 1978. Unit = kilotonnes S/year.

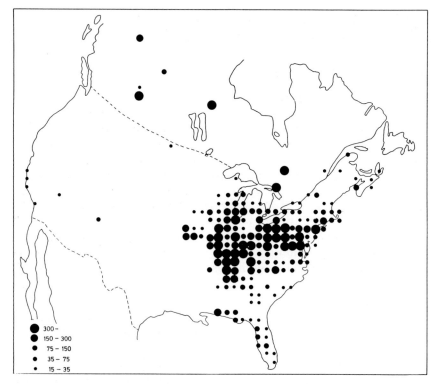

Fig. 2   Sulphur dioxide emissions in 127 km × 127 km grid squares in North
America, 1970–75. Unit = kilotonnes SO$_2$/year.

km × 150 km grid system (Fig. 1) and is primarily based on information
obtained from the participating countries, and on the spatial distribution of
the emissions. For areas where such data are not readily available, population
statistics and information on the location of industrial activities were used to
estimate the geographical distribution of the emissions. Similar surveys have
been worked out for North America, and Fig. 2 has been constructed from a
preliminary compilation by E. C. Voldner, Atmospheric Environment Service, Canada (personal communication).

   Attempts have also been made to quantify the seasonal variation of the SO$_2$
emissions in Europe. It was concluded that the variable component corresponds to about 30% of the annual emissions. As this part is mainly due to
space-heating, climatological considerations indicate that the peak will occur
in January. This seasonal variation is, however, only representative for northern Europe. In regions where the winter mean temperature is not significantly
below 18°C, the room-heating demand will be much less. Widespread use of

air conditioning in the summer season, as in the United States, may reverse the seasonal variation of the SO_2 emission (cf. Fig. 5).

Emission of Nitrogen Compounds

The anthropogenic nitrogen oxide emissions arise to a large extent from fossil fuel combustion—partly by oxidation of nitrogen compounds in the fuel, but mainly by dissociation and oxidation of elemental nitrogen from the combustion air. The emission depends on the combustion conditions, and is favoured by high temperatures and slightly oxidizing flame conditions. Emission factors (EPA, 1977) for different fuels and consumption sectors reflect this, and there is quite a wide range of emission factors for different types of sources within each sector—depending on combustion-chamber design and operating conditions.

If a sufficient breakdown of the consumption of fuels in types of fuel and consumption sectors is available, emission factors can be used to provide an estimate of the emission of nitrogen oxides from a country or region. An example of such an assessment is given in Table II, from which it can be seen that thermoelectric power plants and transport account for about 30 and 40% of the total emissions, respectively. Appropriate fuel consumption data and emission factors for these two sectors are therefore the most critical part of a nitrogen oxide emission survey.

Surveys of the spatial distribution of NO_x emissions within countries are already available in some cases (Semb, 1979), and are based on location and capacity of power-plants, fuel consumption by districts, and distribution of road traffic. Because of the intercorrelation between industrial activities, population density, and traffic, there is a high correlation between the emissions of sulphur dioxide and nitrogen oxides for large grid squares within a country. If major process emissions and regions burning fuel of particularly high or low sulphur content are accounted for, existing sulphur dioxide emission data can be used to approximate the spatial distribution of NO_x emissions within individual countries.

Fig. 3 shows how the estimated NO_x emissions are related to the total energy consumption for some European countries, deductions being made for hydro- and nuclear electricity production and the non-combustion use of solid fuels in the iron and steel industry. For most countries the average emission factor is within the range 5–10 kg NO_2/tonne coal equivalent (0.17–0.35 g N/MJ). This is a rather small range compared with the corresponding range for sulphur dioxide, which is about 3–20 kg S/tonne coal equivalent (0.1–0.7 g S/MJ).

Process emissions of nitrogen oxides are associated with the production of fertilizers, nitric acid, and explosives. On the regional scale, these sources are relatively insignificant.

Table II   Fuel consumption and estimated NO_x emission within OECD Europe in 1975 (Semb, 1979).

	Emission-factor kg NO_2/tonne fuel	Fuel consumption Tg = teragrams = 10^{12} g	NO_x-emission Tg NO_2
Hard coal			
Power-plants	9	133	1.2
Industry	6	22	0.1
Other	2	24	0.05
Brown coal			
Power-plants	4	137	0.5
Residual fuel oil			
Power-plants	12	69	0.8
Refineries	8	19	0.15
Industry	8	95	0.75
Other	6	27	0.16
Gas/diesel oil			
Industry	8	24	0.2
Other	4	121	0.5
Transport	36	46	1.7
Motor gas			
Transport	25	90	2.2
Natural gas			
Power-plants	1	336	0.3
Industry	0.3	642	0.2
Other	0.2	554	0.1
			9.0

While biological emissions of gaseous compounds to the atmosphere are generally more important in the nitrogen cycle than in the sulphur cycle, it appears that the principal components emitted by denitrification processes are N_2 and N_2O. Emission of NO is confined to acidic soils, which are generally nitrogen-deficient. Quantitative estimates can hardly be made at present, however.

A recent survey of the NO_x emissions from all sources has been made by Böttger *et al.* (1978), who have estimated latitudinal distributions. In addition to anthropogenic and biological emissions, generation of NO_x by thunderstorm activity and by stratospheric oxidation of NH_3 are also discussed. Again, quantitative figures are difficult to assign, but it appears that both sources have to be considered in global models.

Because of the implications for soil fertility, there is a wealth of information on the loss of ammonia from soils and from animal manure. The potential

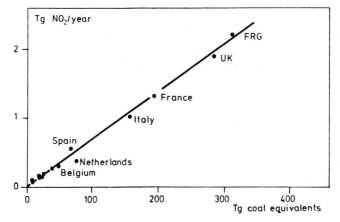

Fig. 3 Nitrogen oxide emissions and energy production from fossil fuels in some European countries (Semb, 1979). Tg = tera-grams = 10^{12} g.

emission to the atmosphere is assumed to be equivalent to 10–30% of the crop's nitrogen turnover or animal manure nitrogen content (Healy *et al*, 1970; Söderlund, 1977). However, these emissions occur close to the ground, and the uptake of ammonia by plant leaves is an efficient removal process. This may effectively reduce the amount of ammonia transported outside the source area, unless there is sulphur dioxide and enough humidity or oxidants for the formation of ammonium sulphate aerosol particles.

Emission of Other Compounds

A survey of reactive hydrocarbon emissions is at present being carried out by OECD in connection with the study of photochemical oxidants and their precursors. Emissions from motor vehicles, and storage and handling of gasoline and of solvents, are included. Although emission factors may be even more uncertain than for nitrogen oxides, this survey is an important step in the work to quantify the large-scale occurrence and transport of photo-chemical oxidants.

A number of human activities contribute to the emission of particulates to the atmosphere. Emission factors are generally given in terms of total weight, irrespective of composition. Additional specifications and size-distribution are therefore necessary for an emission survey relevant to the long-range atmospheric transport of particulate matter. In addition, soil dust particles are emitted both directly by wind erosion and in connection with cultivation, transport, etc. For the larger aerosol particles near the ground, sedimentation and deposition by impaction provide efficient removal-mechanisms. The fraction transported over long distances will depend on the size-distribution and the turbulent vertical rate of exchange, and is therefore not directly related to the emissions per unit area near the surface.

The situation is different for specific components of the submicrometre fraction, such as soot and certain trace-elements. Carbonaceous soot particles are formed by incomplete combustion of hydrocarbons, and it appears that small stationary combustion sources and motor vehicles account for a substantial part of the emissions. Traditionally, optical methods have been used to estimate the concentration of black particles or 'smoke' collected on filters. New and more specific methods for measuring carbonaceous particulates (Rosen & Novakov, 1978) will provide an improved basis for quantification of the emissions of this atmospheric pollutant.

Emission Trends

The historical development of the emission of sulphur and nitrogen compounds provides important information in relation to effects, particularly where historical records of precipitation chemistry are non-existent or of limited quality. Thus Fjeld (1976), using World Power Conference data and United Nations data on energy supplies, has estimated the European sulphur dioxide emissions for the period 1900–72 (Fig. 4). It appears that the emissions have increased in two steps since the industrial revolution: by the increased use of coal up to about 1930, and by the introduction of fuel oils since 1950. After 1973 the European emissions remained essentially constant. (Because of the simplified assumptions used in compiling this inventory, the emission data in Fig. 4 do not necessarily agree with other emission surveys.)

Because of the high sulphur content, coal combustion accounts for a high proportion of the sulphur dioxide emissions in the USA. Fig. 5 shows the trend in US coal consumption during 1940–80 (after Husar *et al.,* 1979). In 1974 the US winter coal consumption was well below, while the summer consumption was above, the 1943 peak. Since 1960 the average growth-rate of summer consumption was 5.8% per year, while the winter consumption increased at only 2.8% per year. In addition to this, crude oil consumption increased by 50% in the period 1959–73.

The emission of nitrogen oxides from fossil fuel combustion may be assessed on the basis of emission factors. However, during the last 30 years the nature of the emissions has changed from a prevalence of coal-fired steam engines and small heating appliances to thermal power-plants and much larger oil-fired boilers for industrial and space heating applications. It is therefore likely that the emissions of nitrogen oxides have increased relatively more than the fuel consumption, particularly in Europe.

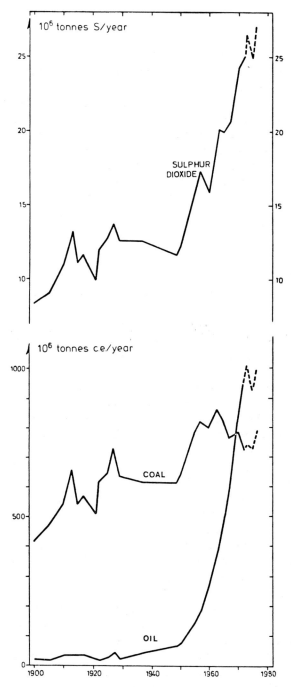

Fig. 4 Consumption of coal and oil in Europe, and estimates of the resulting sulphur dioxide emissions, since the year 1900. (c.e. = coal equivalent.)

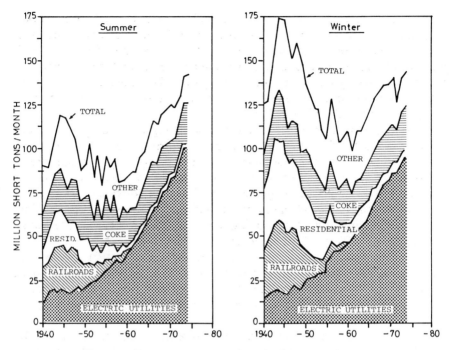

Fig. 5   US winter and summer coal consumption, 1940–74 (after Husar *et al.*, 1979).
(1 short ton = 0.907 [metric] tonne.)

MODELLING OF THE LONG-RANGE TRANSPORT

In order to quantify the long-range transport of air pollutants, it is necessary to formulate mathematical–physical models relating observed concentrations to emissions and meteorological conditions. Although such models are at an early stage of development, they form an indispensable part of the large projects which have been or are being undertaken, and they provide a useful tool for improving our understanding of the long-range transport. A review of the synoptic-scale transport modelling has been given by Eliassen (1980).

From a practical point of view, one may distinguish between models designed to calculate concentration and deposition with a time-resolution of one day or better, and models which give only average values over a longer period of time. In all models, the long-range transport of air pollution can only be described with limited spatial resolution; each point or grid element usu-ally represents an area of the order of 100 × 100 km. Within these grid elements, emissions, concentrations, meteorological data, and transformation and deposition rates can only be represented by total figures and averages.

Still, day-to-day calculations of concentration fields based on actual meteorological data require considerable numerical capacity in order to simulate the actual transport. The numerical advection may be carried out with either an Eulerian or a Lagrangian model. In the Eulerian model, the concentration of air pollutants is calculated as a function of time at fixed points by evaluating the exchange of air pollutants between the grid elements. In order to avoid systematic errors resulting from the numerical calculations, rather elaborate advection methods have to be used (Egan & Mahoney, 1972; Christensen & Prahm, 1976).

In the Lagrangian approach, the transport is represented by air parcels following the path of individual trajectories. This results in an irregular distribution of the air parcels, and new values at the regular grid-points; these are usually re-formed by interpolation after a time-period of 6–12 hours (Eliassen, 1978). With the Lagrangian model, trajectories can be used to estimate the contributions from sources in a given region to the concentration at any given point, without calculating the concentration field for the whole region (OECD, 1977; Eliassen, 1978).

In the Monte Carlo model of Husar & Patterson (1979), the source strength is represented by the number of identical quanta emitted per hour. While moving along the trajectories, these air parcels are exposed to random dispersion, and deposition and chemical reactions are introduced by statistical probability functions.

When the chemical reactions in the atmosphere can be satisfactorily described by first-order chemical reactions, as in the case of sulphur dioxide, the sources can be treated independently. Serious complications arise, however, when the reaction rates depend on concentration, humidity, and radiation intensity, as in the case at nitrogen oxides and the formation of nitrates, ozone, and other photochemical oxidants.

In addition to these 'event'-models, models have also been designed to calculate annual average concentrations—without entering into cumbersome day-to-day calculations—by using average wind-fields or trajectory statistics (Bolin & Persson, 1975; Fischer, 1978). For the central parts of Europe, similar results are obtained to those with the event-models. For more remote areas where high concentrations and deposition occur only under special weather conditions, the applicability of such calculations is limited.

The application of these models has been an important aid in the understanding of large-scale dispersion of air pollutants, particularly for sulphur dioxide. However, the complexity of the physical dispersion, as well as the transformation and deposition processes, pose severe limitations on the model calculations. These processes are discussed in more detail in the following subsection.

Wind Fields and Trajectories

It is convenient to consider the atmospheric dispersion of pollutants as a combination of advection and turbulent diffusion. The plume of pollutants is advected with the mean wind, and at the same time the turbulence that consists of eddies of different sizes, will disperse the pollutants in the vertical and lateral direction and thereby increase the size of the plume.

Both the vertical and lateral diffusions of a plume depend strongly on the meteorological conditions in the atmospheric boundary layer. This has been studied for a relatively long time, and prediction of ground-level concentrations around point sources is now almost a classical problem in air pollution meteorology (e.g. Pasquill, 1974). Turbulent diffusion of pollutants is of greatest importance during the first few tens of kilometres of transport downwind from the source. The vertical distribution will then generally be more-or-less uniform within the so-called mixing layer. After the pollutants are well mixed in the vertical, one may, as a first approximation, also neglect further dilution by the horizontal eddies, and assume that the polluted air-masses are advected by the mean wind, while subject to chemical transformations and removal processes.

This rather simple one-layer model is, however, not always a realistic one. For instance, under very stable conditions with strongly reduced turbulence, plumes may travel for long distances with very little dilution. The extent of long-range transport will therefore depend on the meteorological conditions, as shown by studies in both Europe and North America.

The height of the mixing layer within which the pollutants are assumed to be well-mixed is an important question. An inversion at a height of some hundred metres, capping a well mixed boundary layer, obviously provides an upper limit of the mixing layer. But in several cases no marked top exists, thus making it difficult to determine the height of the mixing layer from standard meteorological data. Moreover, the height of the mixing layer will normally have a diurnal variation, being low during the night and highest in the early afternoon (e.g. Smith & Hunt, 1978).

Diurnal variations of the mixing height with the build-up of a stable ground inversion during the night, prevents elevated plumes from reaching the ground. When the inversion breaks up during the day, convective mixing will take place. Situations of this type have been extensively investigated for city and power-plant plumes (Wilson & Gillani, 1979). A great variety of similar situations may arise, depending on weather conditions and the length of the day. At higher latitudes, large-scale inversions may have a profound effect on the long-range transport of the air pollutants.

Measurements of the vertical distribution of sulphur dioxide and sulphate particles in Europe show that the pollutants in general are contained within the lowest 2 km of the atmosphere, with a maximum concentration a few hundred metres above the ground (OECD, 1977; Gotaas, 1980). Both wind-speed and direction commonly change with height within this layer, which obviously complicates the choice of *one* advecting wind. The horizontal flux density of pollutants will vary with height according to the vertical distribution of the wind-vector and the concentration of pollutants. Ideally, one should therefore use a horizontal wind which describes the integrated transport.

Detailed information about the vertical profiles of wind and air pollution is, however, not readily available. In order to define the transport path or trajectory of an air parcel, it is necessary to make a somewhat arbitrary choice between the different wind data available from the meteorological network. In the OECD/LRTAP programme, trajectories computed from three different winds were compared: Observed 850-mb (1,200–1,500-m) winds, observed surface winds, and surface geostrophic winds calculated from surface pressure charts and backed about 5 degrees. It was concluded that the observed surface winds were not appropriate for studies of the long-range transport. The 850-mb and surface geostrophic trajectories would, over a period of a year or so, give approximately the same results. On some days, however, significant differences could occur, illustrating the difficulties in obtaining reliable trajectories on an event basis. These problems have been discussed by Pack *et al.* (1978) and Smith & Hunt (1978).

The determination of trajectories forms an essential part of the mathematical modelling of the long-range transport. In addition, trajectories are widely used for analysis of measured surface concentrations of pollutants. The observed data may be grouped into sectors, according to direction of origin as determined by the trajectories. As an example, Fig. 6 shows the mean concentrations in air and precipitation in 1978 for 8 transport sectors, as determined from 850-mb trajectories, for two Norwegian sites. For the more southerly site, at Birkenes, it is clearly demonstrated that the high mean concentrations are not randomly distributed, but directed towards the large emission-areas in Europe. The northern site, Kårvatn, does not receive precipitation with southerly winds because of intervening mountains, and this is clearly reflected in the sector analysis. The mean concentrations of sulphur dioxide are very low for all transport directions at Kårvatn, while the aerosol sulphates (which are secondary pollutants formed from sulphur dioxide) still show higher values for the southerly sectors.

Chemical Transformation in the Atmosphere

Air pollutants are emitted to the atmosphere as gases or aerosols. Chemical and physical processes in the atmosphere then transform sulphur dioxide to sulphuric acid and ammonium sulphate aerosol, while nitrogen oxide is oxidized to dioxide and further to nitric acid. A distinction is generally made between homogeneous gas phase oxidation of sulphur dioxide, and heterogeneous processes involving absorption and oxidation in the liquid phase of aerosols and cloud droplets. Both processes produce sulphate aerosols in the submicrometre size-range.

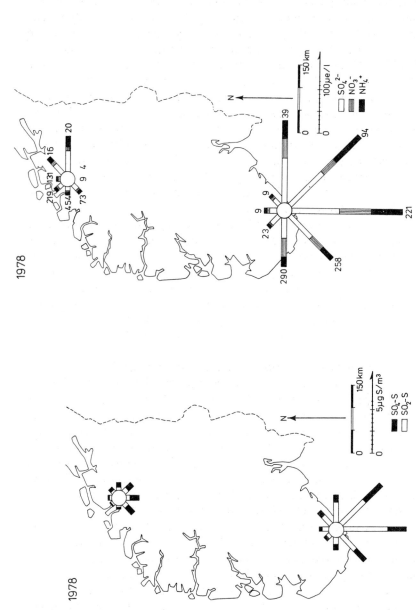

Fig. 6 Mean extractions in 1978 for 8 transport sectors at Birkenes and Kårvatn of (a) sulphur dioxide (μg SO$_2$-S/m^3) and aerosol sulphate (μg SO$_4$-S/m^3), (b) sulphate, ammonium, and nitrate (μe/l) in precipitation. The numbers give the amount of precipitation. The sulphate concentrations have been corrected for sea salt. The length of the lines are proportional to the mean concentrations in the indicated transport directions. (e = equivalents.) (Birkenes is the more southerly site; (a) to left; (b) on right.

Uncatalysed, liquid-phase oxidation of sulphur dioxide seems to be of little importance. The oxidation rate is increased by the presence of catalysts such as manganese and iron, and by strongly oxidizing agents such as ozone and hydrogen peroxide. The liquid-phase absorption rate of sulphur dioxide is reduced as the pH decreases due to the formation of sulphurous acid. The presence of ammonia will, however, reduce the acidity of the droplets and thus promote the liquid-phase oxidation. As a result of the equilibrium between gaseous ammonia and ammonium ions in aerosol particles, acid aerosol may pick up ammonia over agricultural areas, and this ammonia may be given off again when the air-mass passes over nitrogen-deficient forested areas. Thus, the liquid-phase oxidation of sulphur dioxide becomes dependent on seasons and properties of the areas passed by the polluted air-masses. The disagreement about the reaction rates indicates the complexity of the reactions (Beilke & Gravenhorst, 1978).

The homogeneous oxidation of sulphur dioxide is mainly due to reactions with OH and HO_2 radicals and other oxidants formed by photochemical reactions in the atmosphere. The reaction rates therefore depend on the sunlight intensity and the occurrence of other photochemical reactions in the atmosphere. Model predictions for a clean atmosphere (Altshuller, 1979) show that, at mid-latitudes, the reaction rate varies from insignificant during winter to about 1% per hour at midday in summer. In urban atmospheres, however, the reaction rate may be up to 5% per hour (Eggleton & Cox, 1978).

In studies of long-range transport of sulphur compounds, it is not possible at present to give a detailed quantitative description of each of the different transformation processes; this would require extensive information on the distribution of the other relevant reactants in the same grid-system. Instead, one has to use an overall transformation rate, as estimated by Eliassen & Saltbones (1975), by adjusting model calculations to observed concentrations. Their estimates ranged from 0.3 to 1.7% per hour, with a mean value of about 0.7% per hour. Aircraft measurements by Smith & Jeffrey (1975) off the east coast of England gave a considerably more rapid transformation, indicating that it may be faster close to the sources than at greater distances. Recent measurements indicate that, under winter conditions in the Arctic, the transformation rate is much below the values found at lower latitudes.

The homogeneous oxidation process leads to the formation of SO_3 molecules which may condense on existing nuclei or form stable clusters with water molecules. Stable clusters grow further by agglomeration and condensation to a mass median diameter of about 0.1–0.5 μm. Even the heterogeneous transformation process leads to particles of about this size, so that the bulk of the aerosol sulphates in long-range transported polluted air is in the so-called accumulation mode, with a mass median diameter in the range 0.3–1.0 μm—i.e. in the same size range as other combustion aerosols such as carbonaceous particle agglomerates, lead aerosols, etc. (Whitby, 1978). Aerosol particles which originate from mechanical processes, such as soil dust and sea-salt particles, will have mass median diameters in the range above 2 μms.

Because of the irreversible transformation of sulphur dioxide to non-volatile sulphate, the ratio between measurements of these compounds provide a measure of the age of the pollutants. A seasonal variation of this ratio is usually observed, with the lowest sulphur dioxide/sulphate ratio occurring during the summer. This may partly be explained by faster transformation during the summer, but such other effects as seasonal variation of the dry-deposition rates and the mixing height, will also contribute.

The chemistry of the nitrogen oxides is more complex than the sulphur ores, and the overall transformation rates are not well known. Nitrogen oxides play an important role in the formation of photochemical oxidants, and thus affect the photochemical oxidation rate of sulphur dioxide. The oxidation of nitrogen oxides produces nitrate aerosols and gaseous nitric acid, and there is an equilibrium between the content of nitrate in the aerosol phase and that of gaseous nitric acid. Thus, an acidification of the aerosol by sulphur dioxide will 'push' the nitrate into the gaseous phase.

As the photochemical reactions of the nitrogen oxides are not linear, the emissions from different grid elements cannot be treated independently in the models. The overall reaction-rate will depend on the actual concentration and its distribution within the grid element, radiation conditions, etc.

The Deposition Processes

Pollutants are removed from the atmosphere both during dry periods (sedimentation, absorption, and impaction) and with precipitation (rainout and washout). The relative importance of the dry- and wet-deposition depends mainly on the distance from the sources. In a heavily polluted area, the amounts removed by dry deposition processes are much larger than the wet-deposition, while in remote areas with much precipitation the wet-deposition will be the dominating process.

Turbulent motions will bring the gaseous or particulate pollutants into contact with the surface, where they may be removed by absorption, adsorption, or impaction. The rate of this dry deposition is usually assumed to be proportional to the concentration adjacent to the surface, the 'constant' of proportionality being called the deposition velocity. The deposition velocity is, however, highly dependent on a large number of factors, such as the type of pollutant (type of gas, particle size), the turbulence of the atmosphere close to the ground, and the characteristics of the underlying surface (type and status of the vegetation, ocean, snow, etc). Examples of efficient sinks for sulphur dioxide are generally water (unless very acid) and growing vegetation; a snow-covered surface, with an inversion immediately above, provides a far less efficient sink. For further discussion of dry deposition, reference is made to Garland (1978) and Fowler (1980).

Methods for direct and routine measurement of the dry-deposition are not available at present. Estimates of it therefore have to be based on measured air concentrations of the pollutants and experimentally determined deposition velocities. Owing to lack of detailed information, a constant overall deposition velocity is often used for each component in the model.

The concentration of pollutants in precipitation is the result of several complex processes both within and below clouds, and uncertainty still remains as to a detailed description of the scavenging efficiency of precipitation.

Sulphur dioxide may be absorbed both in the cloud droplets and in falling raindrops. The solubility of sulphur dioxide in water at pH < 5 is, however, rather small, and measurements of dissolved sulphur dioxide have shown that it generally contributes less than 10% of the total sulphur in precipitation.

The major wet removal mechanism for sulphur seems to be scavenging of sulphate aerosol particles. These are efficient cloud-condensation nuclei, and practically all the sulphate aerosol will be incorporated in the cloud droplets. In addition, the sulphate concentration of cloud droplets may increase, due to heterogeneous oxidation of sulphur dioxide.

Scott (1978) has presented a parameterization of the wet-removal process, for cold, warm, and convective, clouds assuming the in-cloud oxidation to be negligible. In cold clouds, which account for the main part of the scavenging in temperate regions, precipitation is triggered by relatively clean ice-crystals which are formed in the upper part of the cloud and rapidly grow to precipitation size, i.e. the Bergeron process. These hydrometeors will accrete sulphate-containing cloud droplets while falling through the cloud.

The content of sulphate in precipitation from this process is not readily predictable. Long records of air and precipitation chemistry data are available, but concentrations of sulphate aerosol in the air have generally been measured only at ground level, and the 24-hourly sampling periods for both air and precipitation samples do not give the necessary time-resolution. Obviously more detailed studies of precipitation scavenging, with measurements both at ground-level and by aircraft, are needed. However, as the amount of water drawn from each cubic metre of air involved in a rainstorm is typically about 1 ml, a concentration of 1 μmol/m^3 of air should produce a concentration of about 1 mmol/litre of precipitation—if the pollutants are reasonably well-mixed with height, and the scavenging is efficient. This crude relationship is fairly consistent with the observations; but the correlation for daily precipitation samples and 24-hourly air concentrations is not high, and is therefore of limited value for predictive purposes.

A typical set of data are shown in Fig. 7. The results from a large number of such data from Europe show that the content of sulphate in precipitation

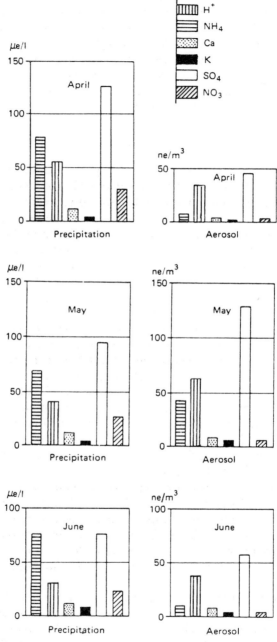

Fig. 7   April–June monthly mean concentrations of major ions at Birkenes, Norway, in 1974.

generally corresponds to the total amount of particulate sulphates *plus* a minor contribution from absorbed sulphur dioxide. Furthermore, the totality of nitrate in precipitation seems almost entirely due to uptake, in chloride droplets, of gaseous nitrogen oxides or nitric acid. The large variations in the amount of precipitation over relatively short distances, and the limited number of observations generally available, complicates the averaging and interpolation which are necessary to describe the wet deposition of air pollutants in connection with the long-range transport.

ACID PRECIPITATION

The ionic composition of precipitation is largely acquired by scavenging of the ammonium sulphate aerosol, sulphur dioxide, ammonia, and nitrogen oxides—including gaseous nitric acid. Hydrochloric acid from the burning of coal with high salt content, or from PVC or chlorinated solvents, may be an additional source of precipitation acidity on a local scale. In addition, contributions also come from sea-spray particles in marine environments, and soluble components of soil dust.

The mean concentrations of major ions in precipitation at the Hubbard Brook Experimental Watershed in New Hampshire, USA (Likens *et al.*, 1977), and at Birkenes in Norway, are shown in Fig. 8. The two data-sets are not quite comparable, the Hubbard Brook data having been derived from weekly bulk-precipitation samples, while the Birkenes site had daily precipitation collections. The time periods are also different. However, except for the content of sea-salt constituents at Birkenes, which is only 20 km from the coast, the similarity in ionic composition is rather close. This to a large extent is due to similar emission intensities of nitrogen oxide and sulphur dioxide upwind of the two sites. Ammonium concentrations were lower at Hubbard Brook, because emissions of ammonia were mainly associated with agricultural activities. In the USA, these are located west of the industrial regions in the northwestern and midwestern States.

Soil dust components, such as calcium and potassium, are of minor importance both at Birkenes and at Hubbard Brook. Soil dust may come from the immediate surroundings or be transported over long distances. Although 'red rain' from the Sahara has been observed in England (Anon., 1968), such long-distance transport of soil-dust apparently does not contribute much to the general chemical composition of precipitation in Scandinavia or New England.

Precipitation chemistry data frequently give unrealistic, high values for the concentrations of calcium and potassium, and too high pH, because, for convenience the sites are often chosen near to tilled fields and gravel roads.

A thorough study of the representativity of monthly bulk-precipitation samples was made by Granat (1974), who also showed that site selection was

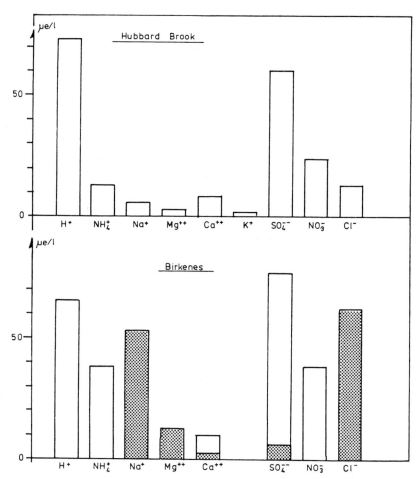

Fig. 8 Mean concentration of major ions in precipitation at Hubbard Brook, New Hampshire, USA, and at Birkenes, Norway. Precipitation chemistry data from Hubbard Brook are from Likens *et al.* (1977). Sea-water components at Birkenes are stippled; Birkenes is only 20 km from the coast and 190 m a.s.l., while Hubbard Brook is 116 km from the Atlantic coastline and 500–700 m a.s.l.

generally more critical for soil dust components such as calcium and potassium, than for sulphate or nitrate. However, if the sampling site was carefully selected in an area with permanent vegetation surrounded by forest, the soil dust component was virtually eliminated, even with monthly samples. Use of wet-only precipitation collectors, i.e. gauges with automatic lids which are only open during precipitation, is recommended by the World Meteorological Organization (WMO, 1978) for precipitation sampling in the

BAPMoN network. Daily precipitation sampling and rinsing of the precipitation collector also helps to reduce the soil dust contamination problem.

Even with these measures, precipitation in regions with significant emissions of soil dust from agricultural activities and from wind erosion, will contain more soil dust than will precipitation in humid and forested regions.

A comparatively dense network of precipitation stations, based on daily samples, was maintained during the OECD/LRTAP programme (*see* above). Although the analysis programme was limited to sulphate and strong acidity (or low pH), the resulting concentration patterns of Fig. 9 reveal regions with high pH relative to the concentrations of excess sulphate. This occurs in Denmark and in France, and there is reason to believe that the higher pH is due to neutralization by ammonia and soil dust from agricultural activities.

This interaction between sulphur and nitrogen oxides from fossil fuels, and soil dust and ammonia, is also seen in data from North America (Fig. 10). On the basis of old precipitation data, Lau & Charlson (1977) also identified an area with high partial pressure of ammonia in northwest USA, and extremely low equilibrium concentrations in the eastern States.

Within the regions of 'acid precipitation', e.g. in southern Norway, there is a high degree of correlation between 'acidity' and the concentrations of excess sulphate, *see* Fig. 11.

Obviously, the occurrence of acid precipitation depends strongly on wind direction in relation to emission sources. This particularly applies to remote locations, where air mass histories also seem to be important (cf. Fig. 6). Very high acidities are sometimes observed when air masses have remained over the sea for several days before precipitation is released. Under such circumstances, pH values down to 2.5 and 2.7 have been observed in precipitation in Scotland, on the western coast of Norway, and on Iceland.

The seasonal variations in precipitation chemistry are somewhat different from the assumed variation of the emissions of sulphur and nitrogen oxides. Fig. 12 shows the monthly mean concentrations of major constituents in precipitation for Birkenes in southern Norway. The most striking feature is the spring maximum for sulphate, which is similar to that found by Granat (1978) for a large number of stations in Northern Europe. The sulphur dioxide emissions in Europe are assumed to peak in January. The reasons for this phase-lag in precipitation sulphate concentrations have not been fully explained. Probably several factors contribute, including seasonal variations in the transport pattern. Homogeneous oxidation of sulphur dioxide will also become increasingly important during the spring, and the amount of precipitation is generally low.

In the northeastern USA and Canada, both deposition and sulphate concentrations are at a maximum in summer, reflecting the different climates and fuel consumption patterns on the European and the North American continents. Precipitation occurs in events, and a special definition of 'episodicity'

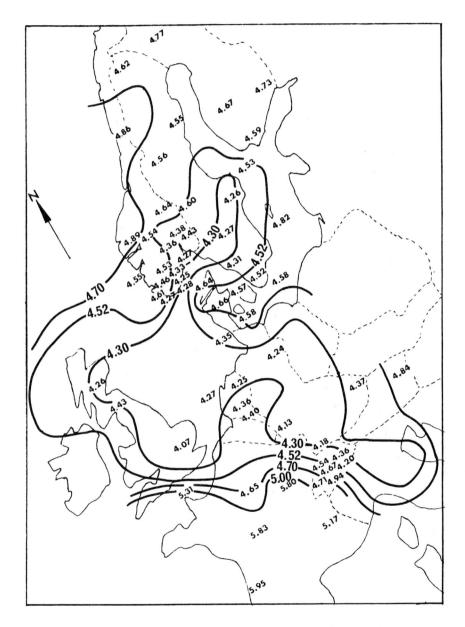

Fig. 9   Mean weighted concentrations of excess sulphate (left) and pH (right) from the LRTAP network, 1974. Isopleths are drawn from an objective analysis computer programme.

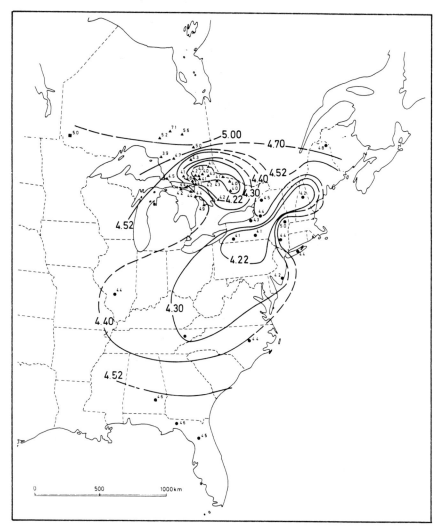

Fig. 10   Weighted mean pH in precipitation in North America 1972–75. This figure has been made by R. F. Wright on the basis of data from Kramer (1975) and Likens (1976).

has been used to quantify the variability in event deposition (OECD, 1977; Smith & Hunt, 1978). Highly episodic sulphate deposition typically occurs at sites which occasionally experience precipitation with strongly polluted air, while the sulphate deposition is less episodic at sites that are either always polluted or always very clean.

The relationship between emissions, transport, and deposition over a large

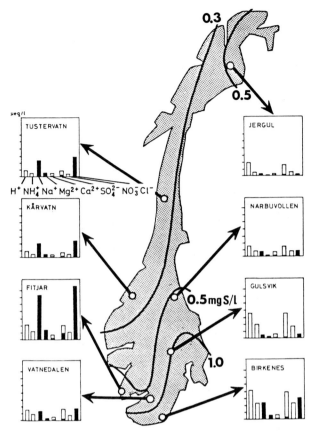

Fig. 11 Mean concentrations of major ions in precipitation in Norway during 1978–79, given in histograms and with isolines for sulphate. Black bars represent the sea-salt component.

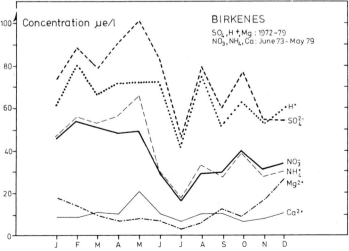

Fig. 12 Seasonal variations for major ions in precipitation at Birkenes, Norway.

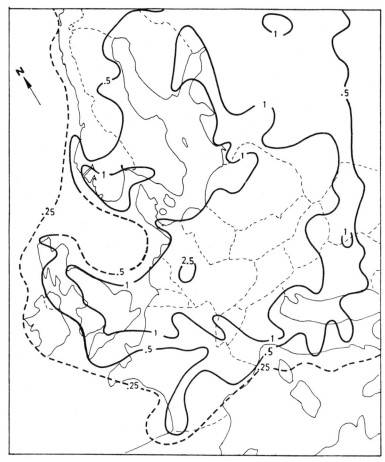

Fig. 13   Estimated wet sulphur deposition pattern for 1974. Unit: g S/m^2.

region has mainly been studied for sulphur in the atmosphere. In the OECD/LRTAP study, annual wet deposition was estimated from the calculated concentration of aerosol sulphate and the precipitation amounts, using an empirical relationship between calculated aerosol sulphate and precipitation sulphate concentrations. The resulting wet-deposition field is shown in Fig. 13. Compared with the emission field, this shows clearly the effect of geographically enhanced precipitation in southern Norway and in the Alps.

Dry deposition was estimated from the calculated concentrations of SO$_2$ and sulphate aerosol using average dry-deposition velocities of 0.8 and 0.2 cm/s respecively. The calculated total deposition is shown in Fig. 14. This represents 80% of the anthropogenic emissions of SO$_2$ in the area, 50% being estimated as dry deposition and the remaining 30% being deposited by

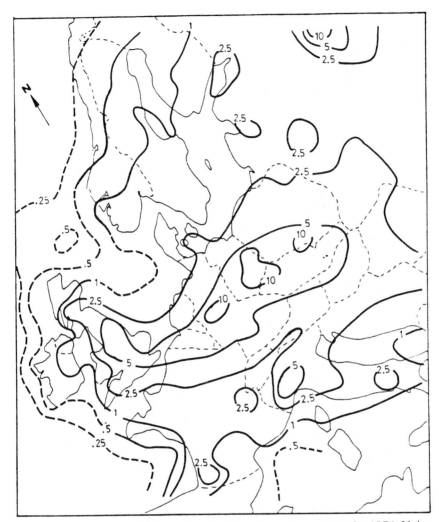

Fig. 14   Estimated total (dry and wet) sulphur deposition pattern for 1974. Unit:
g S/m².

precipitation. The remaining 20% is net transport out of the area. This overall
mass balance is not very sensitive to changes in the physical parameters. On a
smaller scale, however, these represent rather crude approximations and particularly the dry-deposition rate is strongly affected by changes in surface
cover and season.

A summary of data on the emission and observed deposition of nitrogen
oxides has recently been completed by Bónis *et al.* (1980). The regional
variation in content of nitrate and ammonium in precipitation can be

reasonably well explained on the basis of emissions and chemistry transport processes. Although emission inventories are relatively uncertain, model calculations may substantially improve our understanding of the large-scale dispersion of these compounds, particularly if measurements of representative concentrations are becoming available.

SUMMARY

Air pollution arises primarily from the use of fossil fuels for heat and energy production, including transportation. Sulphur oxides, the most prevalent pollutants, arise mainly from roasting sulphidic copper, nickel, lead, and zinc, ores, manufacturing of sulphuric acid, the paper and pulp industries, and burning coal for energy. Nitrogen oxides are associated with any combustion process and are intimately involved with photochemical reactions. Long-range transport of pollutants has been studied, using various mathematical–physical models. These generally are based on deposition with a time-resolution of one day or better, or on average values over a longer period of time. Models have provided an important aid in understanding large-scale dispersion of pollutants.

Dispersal is complicated not only by meteorological and depositional factors but also by atmospheric chemical changes, reactions, and interactions. Oxidation of SO_2 is due mainly to reactions with OH and HO_2 radicals formed by photochemical reactions. The most efficient sinks are generally water and vegetation. Models and actual measurements have shown that, over Europe, some 20% of the pollutants are transported out of their regions of origin, thus demonstrating that individual national control programmes can achieve only limited improvement in air quality.

REFERENCES

Altshuller, A. P. (1979). Model predictions of the rates of homogeneous oxidation of sulphur dioxide to sulphate in the troposphere. *Atmos. Environ.*, **13**, pp. 1653–61.

Anon. (1968). Red rain. *Nature* (London), **219**, p. 112.

Beilke, S. & Gravenhorst, G. (1978). Heterogeneous SO_2-oxidation in the droplet phase. *Atmos. Environ.*, **12**, pp. 231–40.

Bolin, B. & Persson, C. (1975). Regional dispersion and deposition of atmospheric pollutants with particular application to sulphur pollution over western Europe. *Tellus*, **27**, pp. 281–310.

Bónis, K., Meszaros, E. & Putsay, M. (1980). On the atmospheric budget of nitrogen compounds over Europe. *Idöjaras* (Budapest), **84**, pp. 57–68.

Böttger, A., Ehhalt, E. & Gravenhorst, G. (1978). *Atmosphärische Kreisläufe von Stick-oxiden und Ammonia*. Kernforschungsanlage, Jülich, West Germany: 157 pp.

Christensen, O. & Prahm, L. P. (1976). A pseudospectral model for dispersion of atmospheric pollutants. *J. Appl. Meteor.*, **15**, pp. 1284–94.

Cullis, C. F. & Hirschler, M. M. (1979). Emissions of sulphur into the atmosphere. *Atmos. Environ.*, **14**, pp. 83–8.

Dahl, K. (1927). The effects of acid water on trout fry. *Salmon and Trout Magazine*, **46**, pp. 35–43.

Dannevig, G. (1959). [Nedbørens innflytelse på vassdragenes surhet og på fiskebestanden.—in Norwegian.] [Influence of precipitation on the acidity of watercourses and on fish stocks.] *Jeger og Fisker*, **3**, pp. 116–8.

Dovland, H. & Saltbones, J. (1979). *Emission of Sulphur Dioxide in Europe in 1978*. (EMEP/CCC-Report 2/79.) Norwegian Institute for Air Research, Lillestrøm, Norway: 33pp.

Economic Commission for Europe (cited as ECE) (1976). *The Second Seminar on Desulphurization of Fuels and Combustion Gases*. Washington, DC, 11–17 November 1975. (ENV/SEM 4/3.) Economic Commission for Europe, Geneva, Switzerland: 4 vols.

Egan, B. & Mahoney, J. R. (1972). Numerical modelling of advection and diffusion of urban area source pollutants. *J. Appl. Meteor.*, **11**, pp. 312–22.

Eggleton, A. E. J. & Cox, R. A. (1978). Homogeneous oxidation of sulphur compounds in the atmosphere. *Atmos. Environ.*, **12**, pp. 227–30.

Eliassen, A. (1978). The OECD study of long-range transport of air pollutants: Long-range transport modelling. *Atmos. Environ.*, **12**, pp. 479–87.

Eliassen, A. (1980). A review of long-range transport modelling. *J. Appl. Met.*, **19**, pp. 231–40.

Eliassen, A. & Saltbones, J. (1975). Decay and transformation rates for SO_2 as estimated from emission data trajectories and measured air concentrations. *Atmos. Environ.*, **9**, pp. 425–29.

EPA (1977). *Compilation of Emission Factors* (3rd edn). US Environmental Protection Agency, Research Triangle Park, North Carolina, USA: AP-42, 2 vols.

Eriksson, E. (1963). The yearly circulation of sulfur in nature. *J. Geophys. Res.*, **76**, pp. 4001–19.

Fischer, B. E. A. (1978). The calculation of long-term sulphur deposition in Europe. *Atmos. Environ.*, **12**, pp. 489–501.

Fjeld, B. (1976). *Forbruk av fossilt brensel i Europa og utslipp av SO_2 i perioden 1900–1972*. (TN 1/76.) Norwegian Institute for Air Research, Lillestrøm, Norway: 6pp. (mimeogs).

Fowler, D. (1980). Removal of sulphur and nitrogen compounds from the atmosphere in rain and by dry deposition. Pp. 22–32 in *Ecological Impact of Acid Precipitation*. (Ed. D. Drabløs and E. Tollan), pp. 21–22, SNSF Project.

Garland, J. A. (1978). Dry and wet removal of sulphur from the atmosphere. *Atmos. Environ.*, **12**, pp. 349–61.

Gotaas, Y. (1980). OECD programme on long-range transport of air pollutants—measurements from aircraft. *Ann. N.Y. Acad. Sci.*, **338**, pp. 453–62.

Granat, L. (1974). On the variability of rainwater composition and errors in estimates of areal wet-deposition. Pp. 531–51 in *Precipitation Scavenging: Proceedings of a Symposium* (Ed. R. G. Semonin & R. W. Beadle), Champaign, Illinois, 14–18 October 1974. U.S. Department of Commerce, Springfield, Virginia, USA: 842 pp.

Granat, L. (1978). Sulfate in precipitation as observed by the European atmospheric chemistry network. *Atmos. Environ.*, **12**, pp. 413–24.

Granat, L., Rodhe, H. & Hallberg, R. O. (1976). The global sulfur cycle. Pp. 89–134 in *Nitrogen, Phosphorous, and Sulfur Global Cycles* (Ed. B. H. Svenson & R. Söderlund). (SCOPE 7.) Stockholm, Sweden: 192 pp.

Healy, T. V., McKay, H. A. C. & Pilbeam, A. (1970). Ammonium and Ammonium sulphate in the troposphere over the United Kingdom. *J. Geophys. Res.*, **75**, 2317–21.

Husar, R. B., Lodge, J. P. & Moore, D. J. (Eds) (1978). Sulfur in the atmosphere. *Atmos. Environ.*, **12**, 1–816.

Husar, R. B. & Patterson, D. E. (1979). Synoptic-scale distribution of Man-made aerosols. *WMO Symposium on Long-range Transport of Pollutants* (Sofia, Bulgaria, 1–5 October 1979). WMO No. 538—supplement. World Meteorological Organization, Geneva, Switzerland: XI, Ia–m + 415 pp.

Husar, R. B., Patterson, D. E., Halloway, J. H., Wilson, W. E. & Ellestad, T. G. (1979). Trends of eastern US haziness since 1948. Pp. 249–56 in *4th Symposium Atmos. Turb. Diffusion and Air Pollution, Amer. Meteorol. Soc.* 15–18 Jan. 1979, Reno, Nevada, USA: 676 pp.

Katz, M. (1977). The Canadian sulphur problem. Pp. 21–67 in *Sulphur and its Inorganic Derivatives in the Canadian Environment.* (NRCC No. 15015.) National Research Council, Ottawa, Canada: 426 pp.

Kramer, J. R. (1975). *Fate of Atmospheric Sulfur Dioxide and Related Substances as Indicated by the Chemistry of Precipitation.* Thesis. Department of Geology, McMaster University, Hamilton, Ontario, Canada: 92 pp., appendix.

Lau, Ngar-Cheung & Charlson, R. J. (1977). On the discrepancy between background atmospheric ammonia and measurements and the excistence of acid sulphates as a dominant atmospheric aerosol. *Atmos. Environ.,* **11**, pp. 475–8.

Likens, G. E. (1976). Acid precipitation. *Chem. Eng. News,* **54**, pp. 29–44.

Likens, G. E., Borman, F. H., Pierce, R. S., Eaton, L. S. & Johnson, N. M. (1977). *The Biogeochemistry of a Forested Ecosystem.* Springer-Verlag, New York, NY, USA: xii + 146 pp. illustr.

Odén, S. (1968). *Nederbördens och Luftens Försurning, dess Orsaker, Förlopp och Verkan i Olida Miljöer.* Statens Naturvetenskapliga Forskningråd, Ekologikommitten, Bull. No. 1, Stockholm, Sweden: 86 pp.

OECD (1977). *The OECD Programme on Long-range Transport of Air Pollutants: Measurements and Findings* (2nd edn). Organization for Economic Cooperation and Development, Paris, France.

Overrein, L. N., Seip, H. M. & Tollan, A. (1980). *Acid Precipitation—Effects on Forest and Fish.* Final report of the SNSF-project 1972–1980, Olso-Ås, Norway: 175 pp.

Pack, D. H., Ferber, G. J., Heffter, J. L., Telegadas, K., Angell, J. K., Hoecker, W. H. & Machta, L. (1978). Meteorology of long-range transport. *Atmos. Eviron.,* **12**, 425–44.

Pasquill, F. (1974). *Atmospheric Diffusion* (2nd edn). Ellis Horwood Ltd., Chichester, UK; Halsted Press, New York, NY, USA: xi + 429 pp., illustr.

Robinson, E. & Robbins, R. C. (1970a). Gaseous sulfur pollutants from urban and natural sources. *J. Air Pollut. Control. Ass.,* **20**, pp. 233–5.

Robinson, E. & Robbins, R. C. (1970b). Gaseous nitrogen compound pollutants from urban and natural sources. *J. Air Pollut. Control. Ass.,* **20**, 303–6.

Rosen, H. & Novakov, T. (1978). Identification of primary particulate carbon and sulphate species by Raman spectroscopy. *Atmos. Environ.,* **12**, 923–7.

Scott, B. C. (1978). Parameterization of sulphate removal by precipitation. *J. Appl. Meteorol.,* **17**, pp. 1375–89.

Semb, A. (1978). Sulphur emissions in Europe. *Atmos. Environ.,* **12**, pp. 455–60.

Semb, A. (1979). Emission of gaseous and particulate matter in relation to long-range transport of air-pollutants. Pp. Ia–m in *WMO Symposium on the Long-range Transport of Pollutants.* (Sofia, Bulgaria, 1–5 October 1979.) WMO No. 538. World Meteorological Organization, Geneva, Switzerland: XI, Ia–m, + 415 pp.

Smith, F. B. & Hunt, R. D. (1978). Meteorological aspects of the transport of pollution over long distances. *Atmos. Environ.,* **12**, 461–77.

Smith, F. B. & Jeffrey, G. H. (1975). Airborne transport of sulphur dioxide from the U.K. *Atmos. Environ.,* **9**, pp. 643–60.

Söderlund, R. (1977). NO_x pollutants and ammonia emissions: A mass balance for the atmosphere over NW Europe. *Ambio,* **6**, pp. 118–22.

Whitby, K. T. (1978). The physical characteristics of sulphur aerosols. *Atmos. Environ.,* **12**, pp. 139–9.

Wilson, W. E. & Gillani, N. V. (1979). Transformation during transport: A state of the art survey of the conversion of SO_2 to sulphate. Pp. 157–64 in *WMO Symposium on the Long-range Transport of Pollutants.* (Sofia, Bulgaria, 1–5 October 1979.) WMO No. 538. World Meteorological Organization, Geneva, Switzerland: XI, Ia–m + 415 pp.

WMO (1978). *International Operations Handbook for Measurement of Background Atmospheric Pollution.* (WMO—No. 491.) World Meteorological Organization, Geneva, Switzerland: 110 pp.

Air Pollution and Plant Life
Edited by M. Treshow
© 1984 John Wiley & Sons Ltd.

CHAPTER 5

Monitoring Levels and Effects of Air Pollutants

ADAM C. POSTHUMUS

Research Institute for Plant Protection, Binnenhaven 12, Wageningen, Netherlands

INTRODUCTION

Air pollution in relation to vegetation, including agricultural, horticultural, and forestry crops, may be characterized in different ways. First of all, the concentration levels of different pollutants in the air may be measured and their possible effects may be estimated if the exposure–effect relationships under the prevailing conditions are known. But a second, more direct, approach is the actual measurement of the effects of the ambient air pollutants on the plants in question.

For the first approach several physical and chemical methods have been developed already, and very sophisticated techniques with semi- or fully-automatic apparatus are now available. But data on exposure–effect relationships for many plant species and varieties, with differing degrees of sensitivity to the different air pollutants, are mostly lacking or are insufficient, and the influences of the external and internal environmental conditions of the plants are generally not well known. For these reasons it is easier under the present circumstances to study and measure the air pollution effects themselves on the naturally growing plants, or on cultivated crop plants, and it is even better to expose standardized indicator and/or accumulator plants that have been grown under well-defined conditions.

But the physical/chemical measurement of air pollution levels is needed for careful control of air pollution in industrial regions or countries, and also in rural or natural areas because of possible long range transport of pollutants. The information gathered may be essential for the formulation of policies and regulations by governments to protect Man, other animals, plants, and materials, to abate air pollution, and for effective enforcement and general 'control' of these regulations.

On-line measuring systems, producing quickly available data, can provide a valuable warning of dangerously high pollutant levels. In general the purposes of monitoring air pollution levels are:

(a) to study the presence and distribution of pollutants in time and space;
(b) to monitor the air pollution abatement measures;
(c) to perform research on air pollution trends;
(d) to study the meso-scale and long range transport of pollutants; and
(e) to study the contribution of local sources of pollutants.

Monitoring in this context involving continuous or anyway repeated measurements of the effects of air pollution on vegetation, may be very important—for example to warn of possible damage to natural or cultivated plants, and to raise the alarm for possible risks to Man, other animals, and materials, from specific air pollutants to which certain plants etc. may be particularly sensitive. In addition, continuous effect monitoring with the same indicator plants over areas of reasonable extent, may help the study of the presence and distribution of air pollutants in time and space, the operation of air pollution abatement measures, and the analysis of air pollution trends. It should be clear, however, that the emphasis of such studies is on the occurrence and intensity of effects, and that they do not replace measurements of the concentrations of air pollutants.

In this chapter, examples of monitoring of air pollution levels and their effects on plants will be discussed, and it will be stressed that both concentrations and effects of air pollutants should be monitored simultaneously, as they complement each other. Measurements of concentrations are reasonably accurate and objective, and may be helpful in predicting and explaining the occurrence of effects on plants; but they cannot replace measurements of effects, which integrate the influences of intrinsic plant properties, exposure times, concentrations of pollutants, and the external and internal conditions of the plants.

As an example of a national programme, the national monitoring network for air pollution in the Netherlands will be reviewed. Existing international cooperation will be mentioned, and the need for truly international air pollution monitoring networks will be stressed.

MONITORING OF LEVELS OF AIR POLLUTANTS BY CONTINUOUS MEASUREMENT OF CONCENTRATIONS

As mentioned in the introduction, measurements of pollutant levels have to be performed to acquire data for control measures and to sustain possible explanations of cases of damaging effects. This may be done along two different lines of approach: static measurements and dynamic measurements. The first may be performed by physical/chemical equipment at stationary measuring points in a network for air pollution-level monitoring. The apparatus may be fully automatic and integrated in a monitoring system, to produce hourly and daily average values of pollutant concentrations at ground-level. The second type of measurements is carried out by equipment carried in a mobile laboratory, either in a van or an aircraft. In this type of approach, in addition to the measurements of ground-level concentrations, it is possible to determine the total amount of pollutants above the moving measuring point (the so-called 'pollution burden') by remote sensing techniques. During travel along certain lines, flux measurements may be performed for the respective transects. In so doing, transport of pollution on the mesoscale may be determined.

Continuous measurement of concentrations of air pollutants in a monitoring network may be performed for all the compounds for which automatic monitors have been developed. This is the case, at the moment, for at least SO_2, O_3, NO_x, CO, HF, and some hydrocarbons, and may be extended in the future to others—for example, NH_3 and H_2S. The equipment used should be well adapted to measure the real concentrations at ground-level automatically, with no personnel present during some months. Also, calibration cycles should be included in the measuring programme, and the output of the monitors should be in an adequate form for easy and fast processing of the data. An important factor in respect to the optimal functioning of automatic monitors may be the climate in the housing of the monitors. For example, the temperature must not fluctuate too much. Monitors that do not involve any wet chemical reactions in measurement are preferable (e.g. chemiluminescent O_3- and NO_x-monitors) to others.

For dynamic measurements in cars or aircraft, the latest type of equipment (without any wet chemical reaction) is a prerequisite. In addition, for remote sensing special devices are needed. Correlation spectrometry is a very suitable method which has good possibilities for this purpose and others.

When designing a monitoring network, one should be aware of the statistics of the number and location of measuring points, and the locations should be selected very carefully so as not to monitor the special emissions caused by very local sources. The measuring points may be distributed on a baseline-grid, and in special cases extra points may be needed in heavily industrialized or densely populated areas.

It is particularly worth-while to have measurements of meteorological parameters from the same locations. Not only wind velocity and wind direction, but also temperature, solar irradiation, humidity of the air, and rainfall, are important factors in interpreting the air pollution concentration and effect measurements.

MONITORING OF BIOLOGICAL EFFECTS OF AIR POLLUTANTS BY USE OF
PLANTS AS INDICATORS AND/OR ACCUMULATORS

General Backgrounds of the Methods

Because of the relatively high sensitivity of plants in general to several
air pollutants, and the sometimes rather specific symptoms accompanying the
effects of different air pollutants on plants, certain plant species may be used
as indicators for the detection, recognition, and monitoring, of air pollution
effects. When the plants also accumulate the polluting compounds without
changing their chemical nature by metabolism, and the pollutants are easily
analysed in samples of plant material, they may also be used as accumulators.
If the accumulation of air pollutants by plants is also considered as an effect of
the atmospheric pollution, plants are very suitable for detecting, recognizing,
and monitoring, air pollution effects.

For this monitoring of effects of air pollution on plants it is very important
that the following conditions are fulfilled:

(a) the effects should be clear-cut responses of the plants to the air pollut-
    ants;
(b) the effects should be well reproducible, by using plants of genetically
    uniform populations (seeds or cuttings), to guarantee the uniformity of
    the results;
(c) the effects should be characterized by specific symptoms for special air
    pollutants to be recognized;
(d) the plants should be very sensitive to the air pollutants involved to show
    the effects at very low concentrations of the pollutants; and
(e) the plants should be easy to grow and manage in standardized conditions,
    without diseases or pests.

Further, it will be very useful if the effects on indicator and/or accumulator
plants are representative of effects on other plants in the practice of agriculture or nature management.

For general information about the use of plants for biomonitoring air pollutants, reference may be made to recent publications of Manning & Feder
(1980) and Posthumus (1980).

Several types of air pollution effects are known, and these may be divided
into effects of acute exposures to high concentrations over short periods of
time, and effects of chronic exposures to low concentrations over long time
periods. Examples of effects of acute exposures are: clearly visible chlorosis
and necrosis of leaf tissue; leaf, flower or fruit, abscission; and epinastic
curvatures of leaves and petioles (leaf stalks). Effects of chronic exposures to
low concentrations of pollutants may appear as retardation or disturbance of
normal growth and development (resulting in reduction of growth, yield, or

quality of agricultural and horticultural crop-plants), slow discoloration (chlorosis), and/or leaf tip necrosis; in the long run, total dieback of plant organs may be caused. In some cases the symptoms of effects of acute and chronic exposures may be rather specific for a special air pollutant or a combination of different pollutants.

Many different plant species may be useful as indicators and/or accumulators of air pollutants by showing special symptoms or effects. For example, species of lichens (mostly epiphytic), mosses, ferns, and higher vascular plants (mostly phanerogams), have been used for this purpose. Biological effect monitoring may be performed by using the natural vegetation and the crop plants present in the area studied; but differences in soil, water, and other (e.g. climatic), conditions may influence the effects and diminish the comparability of results between sites. For this reason it is better to use selected indicator and/or accumulator plants, cultivated in as far as possible standardized conditions of soil and watering.

For this purpose higher plants have been mainly used until now—for example, the tobacco cultivar Bel W_3 in the Netherlands (Floor & Posthumus, 1977) and in the United Kingdom (Ashmore *et al.*, 1978), although transplanted lichens were used in the Ruhr area in the Federal Republic of Germany (Schönbeck, 1969; Schönbeck *et al.*, 1970). Several species and cultivars of natural and culivated plants, which have been shown to be sensitive to one or more air pollutants, are in current use in effective monitoring networks. Table I gives examples of indicator and accumulator plants of differing degrees of sensitivity based on the experience of the national monitoring network for air pollution effects on plants in the Netherlands (Posthumus, 1976). On the basis of results of Curtis *et al.*, (1977), and of Francis & Curtis (1979), the White-flowering Dogwood (*Cornus florida* L.) may be used as a potential indicator and accumulator of Cl^-. As in the case of F^- injury to gladioli, Cl^- at toxic concentrations causes characteristic marginal or tip-burn symptoms in the Dogwood, often with a sharp dark red pigmented line separating the living green tissue from the necrotic zone.

Indicator plants in more or less standardized conditions in the open air may be used to detect, recognize, and measure, the effects of acute exposures to air pollutants and also to study the presence, intensity, and distribution, of these effects in space and time. Effects of chronic exposures may be determined by comparing growth, development, production, and quality of indicator plants in pairs of small greenhouses, one of each pair being ventilated with charcoal-filtered air and the other with unfiltered ambient air. Growth retardations and yield reductions (fresh or dry matter production) have been found in the unfiltered as compared with the charcoal-filtered air. In this way possible crop losses in polluted areas may be indicated, although the effects of air pollution on crop yield in practice are often quite different because of the different internal and external environments of the plants. The use of open-

Table I   Review of indicator and accumulator plants used for effect measurements of different air pollutants in the Netherlands.

Plant species and variety	Air pollutant	Symptoms/effects
Gladiolus (*Gladiolus gandavensis* L.) cultivars Snow Princess and Flowersong Tulip (*Tulipa gesneriana* L.) cultivars Blue Parrot and Preludium	Hydrogen fluoride (HF)	Leaf tip and marginal necrosis (Snow Princess and Blue Parrot) and F-concentrations in dry matter (Flowersong and Preludium
Tobacco (*Nicotiania tabacum* L.) cultivar Bel W$_3$ Spinach (*Spinacia oleracea* L.) cultivars Subito and Dynamo	Ozone (O$_3$)	Leaf upper surface speckle necrosis
Small Nettle (*Urtica urens* L.) Annual Meadow-grass (*Poa annua* L.)	Peroxyacetyl nitrate (PAN)	Leaf under surface band-forming necrosis
Lucerne (*Medicago sativa* L.) cultivar Du Puits Buckwheat (*Fagopyrum esculentum* Mönch)	Sulphur dioxide (SO$_2$)	Interveinal chlorosis and necrosis
Petunia (*Petunia nyctaginiflora* Juss.) cultivar White Joy	Ethylene (C$_2$H$_4$)	Flower bud abortion, small flowers
Italian Rye-grass (*Lolium multiflorum* Lam.) cultivar Optima	Fluoride and metal ions (F,Cd,Mn,Pb,Zn)	Ion concentrations in dry matter

top chambers (Heagle *et al.*, 1973; Mandl *et al.*, 1973) instead of closed greenhouses is an improvement, but still different from the open air situation. Plants grown in the open air are much more resistant to air pollution than plants cultivated in greenhouses. As a consequence, greenhouse crops are generally more vulnerable to pollutants than are crops in the field.

Accumulator plants may also be cultivated in a standardized way at several locations to study the presence and distribution of air pollution. In particular, the non-convertible, persistent air pollutants such as fluoride and heavy metals may be measured relatively easily by physical/chemical analysis. This method is applied in the Federal Republic of Germany, using standardized grass-cultures (Scholl, 1974), and has produced considerable information (Prinz & Scholl, 1975). It must be stressed, however, that the accumulation of the pollutants in the grass is strongly dependent on the growth of the plants and on the climatic conditions. So, in order to make comparisons between samples at different locations or periods of time, it is necessary to know the internal and external conditions of the plants. Other plant species, for example among the mosses, have also been used as accumulators, but with the same problems. Only tree-bark may be used for accumulation of pollutants without the problem of interference by rapid growth of the material (Lötschert & Köhm, 1978).

Definitions of Indicator and Accumulator Plants

For a better understanding of biological effect monitoring research with plants, the following definitions and descriptions will prove useful:

Indicator plants: These are plants that may show clear symptoms of effects indicating the possible presence of some pollutant(s). These symptoms may be rather specific and lead to the qualitative determination of a pollutant, but mostly they do not provide a definite identification and the presence of a specific pollutant must be proved by other methods. The indicator plants function to detect and recognize the effects of the pollutants, but these effects may also be measured quantitatively in order to monitor the intensity of any effects. The permanent measurements of effect intensities may be used for surveys of air quality in relation to plants. The influence of the internal and external conditions of the plants are already included in the effects.

Accumulator plants: These are plants that readily accumulate specific air polluting compounds. After some time these compounds may be analysed in the plant material by physical/chemical methods to identify the pollutants and to obtain a quantitative measurement of the pollution burden (total amount of pollutants accumulated over some period). As the uptake of pollutants by the plants (determined in washed samples of plant material) is regarded as an effect of the air pollution, this may provide a method for air pollution-effect monitoring by accumulator plants. The total pollution burden may also be monitored by measuring the pollutant concentrations in unwashed samples of plant material.

Sometimes the same plant species may act as both indicator and accumulator for a special pollutant, for example tulips and gladioli for hydrogen fluoride. These species are also examples of plants reacting to the pollutant (HF) with effects of both acute and chronic exposures, depending on the concentration and the exposure period. In the long run, enough fluoride may accumulate to cause symptoms of acute exposure (leaf tip and marginal necrosis), and, in addition, reductions in bulk or tuber yield may result (Spierings & Wolting, 1971).

Standardization of Indicator and Accumulator Plants

Standardization of indicator and accumulator plants for airborne pollution is a very important prerequisite to eliminate unnecessary variation in the effects studied. As the effects of air pollutants on plants are, apart from the influences of the nature and concentration of the pollutant and of the exposure-time, dependent on plant species and cultivar, developmental and physiological stage of the plants, and physical environmental conditions, it is quite understandable that the selection of the plant material and the growth conditions are very important. Seeds or other plant material used must be as genetically homogeneous as possible (for example, clones). Growth conditions ought to be optimal and the same at all monitoring sites to be compared. In principle, this is possible only when totally artificial growth cabinets are used for the cultivation and exposure of the plants, so that the only differences will result from the quality of the air passing through the cabinets. In practice, this procedure has rarely been used in air pollution-effect monitoring because of its high costs.

The use of plants grown in soil at different locations (with different soils and different preceding crops) has been practised frequently, but is not to be encouraged. Standardization as far as possible of the soil and water conditions of the plants is advisable, and costs relatively little. Differences in climatic factors cannot be excluded in this way, but these parameters can be measured on the spot, to be included in the comparison of the effects. Possible influences or effects of both biotic and abiotic pathogens should be excluded as far as possible by proper treatments, or should be known exactly.

Application of Monitoring Air Pollution Effects on Plants

Monitoring of biological effects of air pollutants by the use of plants as indicators and/or accumulators has been applied on local, regional, and national scales. On a local scale, indicator and accumulator plants have been used to survey the effects of air pollution from a single source or cluster of sources on special horticultural, agricultural, or forestry crops, or on natural vegetation (Raay, 1969). The same indicators and accumulators may be used to sustain possible claims for compensation of economic losses. On a larger, regional or even national scale the results of biological effect monitoring networks have been used to get some idea of the distribution of these effects in space and time (Floor & Posthumus, 1977). This may lead to conclusions about differences in the air pollution-effect burdens between different regions of a country.

In the long run it is also possible to study trends in incidences of air pollution effects, for example the occurrence of maximal values of ozone effect intensities in summer. On an international scale it would be possible to compare the air pollution-effect burden between adjacent regions of different countries and eventually to trace sources of pollutants, if necessary, across national frontiers. It could be a worthwhile source of information to have an extensive biological effects monitoring network spread over Western European countries to show the distribution of specific effects of air pollution on plants in space and time.

In every case, the selection of plants and conditions should be adapted to the specific aim of the study. When the pollutant to be studied is known, selection of specific sensitive indicator plants is the first step towards adequate results. In cases of unknown pollutants, a series of plant species, sensitive to different compounds, should be used simultaneously. For comparison of the effects of air pollutants on plants at all locations of a regional monitoring network, it would be preferable to use identical indicator and accumulator plants in standardized conditions. It is always advisable to expose tolerant varieties of the indicator plant species next to the sensitive ones—to distinguish possible adverse effects of other biotic or abiotic factors, such as pathogens (viruses, bacteria, and fungi), insect pests, frost, and possible nutritional deficiencies.

A major problem of using biological monitors is the possibility of interactions when more than one pollutant is present (which is often the case). Two or more different air polluting compounds may act on the plants additively. synergistically, or antagonistically. This means that the intensity of the effect of a combination of air pollutants may be respectively equal to, higher than, or lower than, the sum of the effect-intensities of the pollutants when these are applied at the same concentrations and conditions, but separately. Moreover, combination effects may be different for mixtures of the same pollutants at different exposure times, concentration levels, and ratios of the components. Combined effects of mixtures of pollutants are therefore very difficult to interpret when no information about the nature and concentration of the pollutants is available. The ideal indicator plant, for this reason, should be a plant that is sensitive to only one air polluting compound. As this is unlikely to be possible for all the different pollutants, it is worth while to seek indicator plants whose sensitivity is as specific as possible to one component of the total complex of air pollution. It is consequently advisable to expose several indicator plants with different sensitivities at the same place and time in order to discern the effects of different airborne pollutants.

EXAMPLE OF A NATIONAL MONITORING NETWORK FOR AIR POLLUTION

A nationwide automated monitoring network for air pollution was started in the Netherlands in 1973 under the auspices of the Directorate-General for Environmental Hygiene of the Ministry of Public Health and Environmental Hygiene. At first only SO_2 concentrations were measured by the National Institute of Public Health, and effects of several air pollutants on indicator plants were determined by the Research Institute for Plant Protection, in one part of the country. By 1976 the biological-effects-monitoring network was completed at 40 experimental fields, regularly spread all over the country with a concentration in the area west of Rotterdam (Fig. 1). This effects-monitoring network is operated in close cooperation with the nation-wide automated monitoring network for air pollution concentrations, directed by the National Institute of Public Health at Bilthoven. Results prior to 1979 have been published in Dutch in reports of the Bilthoven Institute (Anon., 1976, 1978*a*, 1978*b*, 1978*c*, 1979*a*, 1979*b*). In 1978 the construction of the physical-chemical monitoring network was finalized by connecting the last multi-component measuring station. In the meantime an additional dynamic (mobile) monitoring system was developed, and now a moving laboratory adds essential new information to the results of the fixed network.

At the moment, SO_2 concentrations are measured automatically at 220 stations. The network comprises 100 regular baseline-grid stations with interstation distances of about 20 km, and 120 additional monitoring stations in urban and industrial areas. SO_2 concentrations are obtained as hourly mean values, which are computed in regional data reduction systems from the analog signals which are received *via* telephone lines from the monitoring stations, where SO_2 is measured by means of a Philips PW9700 monitor. The data of nine regional sub-systems are sent by telegraph lines to a central data acquisition system in the National Institute of Public Health at Bilthoven.

Wind-direction and wind-speed are measured at 40 stations. The hourly data from these measurements may be analysed directly for planning additional mobile measurements. Based on the concentration field of the early-morning hours, a measurement route for the mobile unit is planned in order to quantify the transport of SO_2 over mesoscale distances of 20 to 200 km. While moving, this laboratory measures both ground concentrations and overhead gas burden. The gas burden is measured by means of a correlation spectrometer (Barringer-Cospec IV). This instrument is installed in a van or in an aircraft together with an SO_2-concentration analyser (Teco model 43). Dynamic measurements are made during travel over distances of up to 300 km, and recorded by a scanning system. The measurement results are combined with accurate information on the current spatial position as computed from the data of navigational instruments by an on-board computer. The results are analysed in the laboratory after the return of the mobile unit, and related to the data of the monitoring network.

Fig. 1 Location of the experimental sites of the national monitoring network for effects of air pollution on plants in the Netherlands during 1978. The numbers refer to regional codes.

After analysis of the results of the year's period from October 1976 until October 1977 (Egmond & Tissing, 1978), the mesoscale SO_2 transport appeared to be an important factor for the ultimate maximum concentrations which have to be compared with public health standards. Also, in order to investigate the role of the major SO_2 source-areas both inside and outside the country, mobile measurements are made over routes at right-angles to the transport zones. The gas burden profile gives the total mass of SO_2 which is transported, and corresponds to the ground concentration field as obtained from the stationary network. The SO_2 flux may be computed from the gas-burden profile and wind speed. The SO_2 residence time is estimated from the ground concentration gradients.

In addition to the SO_2 measurements, O_3, NO, NO_2, and CO, are measured automatically at a varying number of stations (multi-component measuring stations). Ozone is measured at 30 stations, NO_x at 92 stations, and CO at 41 stations. The results of these measurements are processed in more or less the same manner as the SO_2 measurement results. However, in the first year of operation of the multi-component system, a lot of technical problems disturbed operations, and no reliable results were obtained.

Biological-effect monitoring is performed only during the vegetation period of the indicator and/or accumulator plants (from April until November) at 37 stations of the national monitoring network for air-pollution concentrations and at 3 other locations. The methods and materials are the same at all 40 experimental fields, and these fields are visited weekly during the monitoring period. But as not all are visited the same day, this restriction should be taken into account when comparing the results of the concentration measurement with the biological effect measurements resulting from one week's exposures.

The basic concept of effect measurements is the exposure of sensitive indicator and/or accumulator plants in a standardized way. Use is made in the Netherlands of a special plant cultivation set, derived from a similar system that was developed and used in the Federal Republic of Germany (Scholl, 1969; Haut *et al*., 1972). Plants are grown in standard soil in plastic containers with ceramic filter 'candles' in the soil, for automatic watering. Three of these containers are placed in a larger container, which serves as a source of water. In these standardized cultivation sets (Fig. 2), all species of indicator and accumulator plants may be cultivated by adapting the number of filter candles per container to the water requirement of the plants.

Effects of both acute and chronic exposures on airborne pollutants are monitored. Effects of acute exposures are measured weekly on sensitive indicator plants in cultivation sets in the open air. Effects of chronic exposures are studied and measured by comparing the growth and productivity of sensitive plants after long-term exposures in cultivation sets outdoors and, at some locations, also in pairs of small greenhouses, one of which is ventilated with charcoal-filtered air and one with unfiltered air. In addition, standardized cultures of Italian Ryegrass (*Lolium multiflorum*) in the open air have been used to accumulate several air-polluting compounds during 14-day periods for analyses of these compounds in the harvested plant material. Examples will be given below of these three types of effect-monitoring. A list of indicator and accumulator plants used in our national network for air-pollution effects is presented in Table I, and could be extended now by others—for example, Red Clover (*Trifolium pratense* L.) and Garden or Field Pea (*Pisum sativum*) for monitoring of effects of sulphur dioxide (Posthumus, 1978).

Effects of acute exposures to ozone on the sensitive tobacco variety Bel W_3

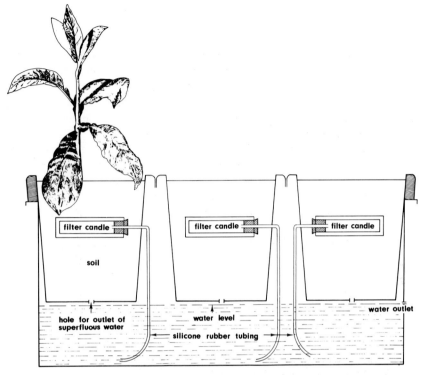

Fig. 2   Diagram of a standardized cultivation set for indicator and accumulator plants.

are determined by estimating, for all leaves of four plants per location, the percentage of leaf-area that is damaged by the specific speckle necrosis in the following classes: 0%, 0–5%, 10–25%, 25–50%, 50–75%, and 75–100%. From these weekly ratings the ultimate mean value for all leaves is calculated for every location. Thus a distribution of the O_3 effect-intensity in place and time during the vegetation period is obtained.

It appears from these results that the mean O_3 effect-intensity is higher in the western half of the Netherlands than in the eastern half (Fig. 3), and that there are weeks in every year, correlated with a sunny type of weather, when the effect of ozone is maximal (Floor & Posthumus, 1977). Fig. 4 shows the weekly variations of the mean leaf damage (in percentage leaf area) for tobacco Bel W_3 at the experimental field of the network in the northern and southern, and in the eastern and western, halves of the Netherlands for the monitoring periods from 1976 to 1978.

Effects of acute and chronic exposures to HF are studied on sensitive monocotyledonous ornamental plants such as tulips and gladioli. In the spring,

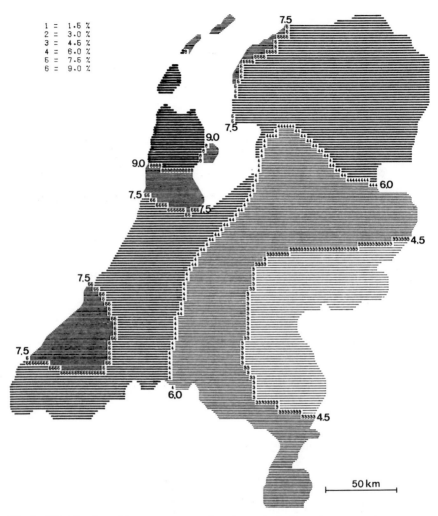

Fig. 3   Distribution of the mean ozone effect-intensity indicated by the large numbers on tobacco cultivar Bel W$_3$ (in percentage of leaf area damaged), measured weekly during the vegetation period from 6 June until 28 October 1977, in the Netherlands.

the tulip cultivar Blue Parrot, and in the summer the gladiolus cultivar Snow Princess, are grown in the cultivation sets outdoors. The necrosis of leaf-tips and -margins caused by hydrogen fluoride is rather specific, and may be assessed quantitatively by measuring the mean length of the necrotic tips, as a measure for the necrotic leaf area, after exposures for weeks or months. The fluoride content of leaf tips of tulips (5.0 cm long) and of gladioli (7.5 cm long) appeared to be very well correlated with the leaf necrosis. From the

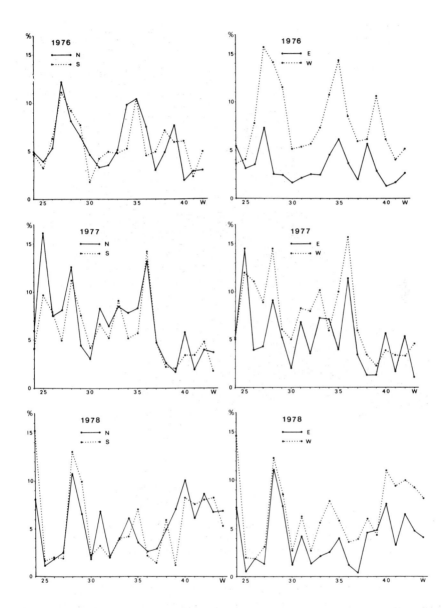

Fig. 4   Variation of the weekly mean ozone effect-intensities on tobacco cultivar Bel W_3 (in percentage of leaf area damaged) at the experimental fields of the national monitoring network for air-pollution effects in the northern (N) and southern (S), and in the eastern (E) and western (W), halves of the Netherlands during the vegetation periods of the years from 1976 to 1978 (the bottom-line indicates the week number from the beginning of each year).

fluoride effects monitoring network it was possible to show the distribution of fluoride pollution over the Netherlands, the southwestern part of the country being most polluted (Fig. 5).

Effects of chronic exposures to air pollution, indicated as growth- and yield reductions of horticultural crop plants, have been studied at several sites in the industrial area west of Rotterdam. Pairs of small greenhouses, ventilated with filtered and unfiltered air, have been used to grow tomato, lettuce, and other crops. Yield reductions of up to 20% in the unfiltered air, as compared with the filtered air, have been found. Differences in these reduc-

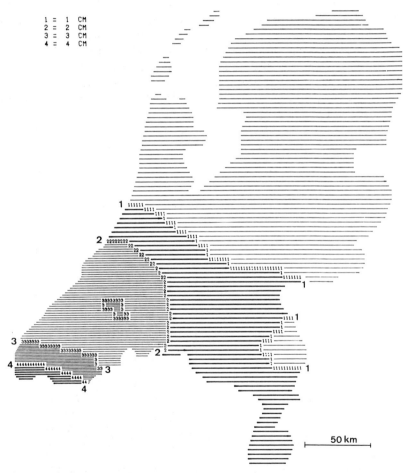

Fig. 5   Distribution of fluoride effect-intensity on the sensitive gladiolus cultivar Snow Princess (in length [cm] of necrotic leaf-tip), at the end of the vegetation period from 6 May until 26 August 1977, in the Netherlands.

tions for separate locations indicated the differences in air pollution effect burden. This type of result might be used for selection of areas where the air quality is unsuitable for horticulture.

Chemical/physical analysis of samples of standardized grass cultures for Pb, Cd, Zn, and F, have revealed possible differences in these air polluting compounds in different regions of the Netherlands. This may be important for the productivity and/or quality of some crops even when no effect is produced of acute exposure.

In the national monitoring network for air pollution in the Netherlands, it should be possible to compare results of the concentration measurements and biological-effect measurements. However, up to now this has not been done intensively, for several reasons. Sulphur dioxide concentrations have been measured at more than 200 sites very frequently, but the concentrations have been rather low during the vegetation period, and no visible effects of SO_2 have been measured on the indicator plants, except during a short air-pollution episode in 1978. Clear-cut effects of hydrogen fluoride have been found and measured quantitatively, but HF is not measured in the automated chemical/physical measuring network.

For ozone, very good evidence of effects on indicator plants has been gained, but O_3 concentration measurements in the network up to 1979 were rather few and unreliable. For 1979, a fairly good qualitative correlation was

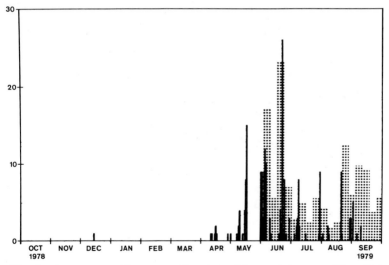

Fig. 6 Average weekly ozone effect-intensities on tobacco cultivar Bel W$_3$ (in percentage of leaf area damaged) and total daily frequencies of ozone concentration peak values (> 98-percentile values) in 1979 within the national monitoring network for air pollution in the Netherlands.

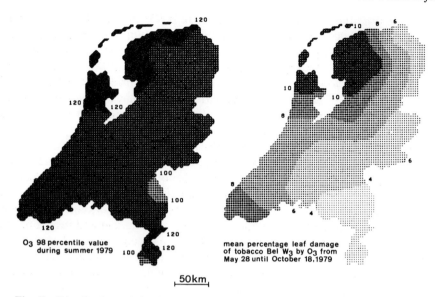

Fig. 7   Distribution of the 98-percentile daily ozone concentration values (in $\mu g/cm^3$) and the mean ozone effect-intensities on tobacco cultivar Bel W$_3$ (in percentage of leaf area damaged), measured weekly during the vegetation period from 28 May until 18 October 1979, in the Netherlands.

found between the O$_3$ effect-intensities on tobacco Bel W$_3$ and the peak O$_3$ concentration values (Fig. 6). Also, the spatial distribution of the mean values of effect-intensities and peak concentrations of O$_3$ during the summer of 1979 were well correlated (Fig. 7). In general, no more quantitative correlations between concentration and effect measurements have been established so far. This should surely be the aim of future research, but other factors such as climate will also have to be taken into account.

POSSIBLE INTERNATIONAL PROGRAMMES OF MONITORING NETWORKS FOR AIR POLLUTION

There has been some international cooperation in Europe towards monitoring air pollution effects on plants. Methods, materials, and results of this type of biological-effect monitoring have been exchanged between research workers, for example between the United Kingdom, the Federal Republic of Germany, Switzerland, Belgium, Denmark, Sweden, Finland, Poland, and the Netherlands. In addition to such forms of cooperation, the establishment of international (e.g. European) programmes for the monitoring of effects of airborne pollution on vegetation should be promoted. The need for this type of cooperation is quite clear when trans-frontier air pollution accounts for a great deal of the total pollution burden, as is the case in the Netherlands. Monitoring networks established under the auspices of an international organization would present great advantages for participating countries.

For the purpose of comparison of the effects of air pollution on plants at all points of an international air quality surveillance network, standardized methods including identical indicator and accumulator plants would be essential. To this end, standardized methods of cultivation, exposure, handling, and observation of selected plants must be developed, as has already been done for the European Economic Community (Posthumus, 1980).

Differences in climate between countries participating in international programmes for monitoring the effects of air pollution on plants may beset a problem in the need for selection of uniform collections of indicator and accumulator plants. It should be borne in mind, however, that the effects measured after some time of plant exposure are the integrated results of the influences of air pollutants and other (e.g. climatic) factors. As all these factors also act on natural vegetation, effects on indicator plants may help us to predict possible general risks. These risks may be compared for the plants in different regions or countries, and threatened zones may be designated.

DISCUSSION AND CONCLUSIONS

The first step towards controlling airborne pollutants is the monitoring of their concentrations and effects. Without data on concentrations and effects, there is no possibility of studying the occurrence and distribution of air pollutants and their effects in place and time, of setting standards for the protection of Man and his environment, and of checking the results of abatement measures. We need an alarm system in order to prevent the occurrence of excessive concentrations, resulting in hazardous effects on Man, other animals, plants, and materials.

For these purposes both physical/chemical concentration measurements and biological effect measurements with plants are needed. The concentrations measured by monitors are not sufficient to predict all possible effects of air pollutants, and indicator plants will never be able to give information about the identity and concentration of all polluting agents in the air. There is no question of replacing ambient air monitoring by physical/chemical methods with effect monitoring with plants: both should be used jointly. Ambient concentration measurements, effect-intensity measurements on plants (including chemical/physical leaf analysis), and measurements of meteorological parameters, may together produce the total picture of the pollution situation.

Concentration measurements may be performed in an automated, stationary monitoring network. But dynamic measurements have proved to be very important for the study of mesoscale transport of air pollutants. In particular, the measurement of the pollutant burden by remote sensing techniques is very useful to determine pollutant fluxes and to explain ground concentration levels by indicating the source areas.

Plants may be used as indicators and accumulators of air pollutants for detection, recognition, and monitoring purposes. An important problem remains in the representativeness of the effects on the indicator and accumulator plants for the vegetation as a whole, and for the separate species of natural and crop plants. The comparability of effects on indicator and accumulator plants with effects on other plants is not very well known, and much information about the exposure-effect relationships for many plant species in all possible conditions is still inadequate or even lacking. The occurrence of special effects of combinations of different air polluting compounds is also a great problem. Additive, synergistic, and antagonistic, combination effects have to be taken into account as far as possible. In the meantime the lack of specificity of indicator plants for a single air pollutant is a problem. Sometimes there exists no real specific sensitivity to only one pollutant, but it is possible to discriminate between different pollution situations by using sets of indicator plants having different sensitivities.

When plants are used for the monitoring of air pollution effects, a high degree of standardization of the plant material and of physical and chemical environmental conditions is a prerequisite. In international monitoring networks, geographical and climatological differences should be taken into account, but it seems to be better to use uniform starting material and to accept an integrated effect of air pollutants and environmental conditions, than to use different and incomparable plant material adapted to different local conditions. The risk of the effect of air pollutants on vegetation is largely determined by environmental factors, so these must be included in any effect-monitoring.

In the belief that there are many problems and uncertainties in the use of plants as biological indicators and accumulators of air pollution, it is worthwhile to summarize the advantages of this approach:

(a) it provides a direct method of studying the effects of the prevailing air pollution on living organisms;

(b) it provides a measure of the integrated effects of all environmental factors, including air pollutants and weather conditions;

(c) it is possible to study the relationships between concentrations and effects when both are measured at the same sites;

(d) it provides possibilities for determining spatial and temporal trends in the occurrence and intensity of effects of several air pollutants on natural and cultivated plants;

(e) sometimes it enables the analysis of polluting compounds by measuring accumulation within plants; and

(f) it acts as a sensitive early-warning system which may stimulate prophylactic measures to prevent or diminish disastrous effects of air pollution.

ACKNOWLEDGEMENTS

The author thanks Mr H. Floor, of the Research Institute for Plant Protection at Wageningen, and Mr N. D. van Egmond, of the National Institute of Public Health at Bilthoven, for kindly providing illustrative material and for their very helpful cooperation.

SUMMARY

After a short introduction about the philosophy of the use of plants as indicators and accumulators of air pollutants, next to physical-chemical monitoring of these pollutants, both types of monitoring of air pollution are described in more detail. Purposes, methods, and problems, in relation to the monitoring of air pollution and of its effects on plants, are discussed. It is stressed that both types of monitoring should be used at the same time complementarily.

Stress is also laid on the high level of standardization needed for adequate exposure of indicator and accumulator plants to enable the comparison of the effects at different places and times. Application of air pollution effects monitoring with plants at local, regional, national, and international, scale is mentioned.

As an example of a national monitoring network for air pollution (measuring concentrations of pollutants and effects on plants), the nation-wide monitoring network in the Netherlands is described. Some results of this monitoring network are given and the problems in relating the concentrations and effects measurements are discussed.

The general problems of specificity and representativeness of effects on indicator plants and of the possible combination effects of different air pollutants together are dealt with in the discussion. In the conclusions the advantages of the use of plants as biological indicators and accumulators of air pollutants are summarized.

REFERENCES

Anon. (1976). *National Meetnet voor Luchtverontreiniging: Overzicht van de Meetuitkomsten in de Periode 1 Oktober 1975–1 Oktober 1976.* Rijksinstituut voor de Volksgezondheid, Bilthoven, Netherlands: Rapport nr. 212 LMO, 45 pp.
Anon. (1978a). *Nationaal Meetnet voor Luchtverontreiniging: Verslag over de Periode 1 Oktober 1976–1 Oktober 1977.* Rijksinstituut voor de Volksgezondheid, Bilthoven, Netherlands: Rapport nr. 4/78 LMO, 77 pp.

Anon. (1978*b*). *Nationaal Meetnet voor Luchtverontreiniging: Verslag over de Periode 1 April 1977–1 April 1978.* Rijksinstituut voor de Volksgezondheid, Bilthoven, Netherlands: Rapport nr. 105/78, 88 pp.

Anon. (1978*c*). *Nationaal Meetnet voor Luchverontreiniging: Verslag over de Periode 1 Oktober 1977–1 Oktober 1978.* Rijksinstituut voor de Volksgezondheid, Bilthoven, Netherlands: Raport nr. 241/78, 21 pp.

Anon. (1979*a*). *Nationaal Meetnet voor Luchtverontreiniging: Verslag over de Periode 1 April 1978–1 April 1979.* Rijksinstituut voor de Volksgezondheid, Bilthoven, Netherlands: Rapport nr. 114/79, 116 pp.

Anon. (1979*b*). *Nationaal Meetnet voor Luchtverontreiniging: Verslag over de Periode 1 Oktober 1978–1 Oktober 1979.* Rijksinstituut voor de Volksgezondheid, Bilthoven, Netherlands: Rapport nr. 236/79, 88 pp.

Ashmore, M. R., Bell, J.N. B. & Reily, C. L. (1978). A survey of ozone levels in the British Isles using indicator plants, *Nature* (London), **276**, pp. 813–5.

Curtis, C. R., Lauver, T. L. & Francis, B. A. (1977). Foliar sodium and chloride in trees: seasonal variations, *Environ. Pollut.*, **14**, pp. 69–80.

Egmond, N. D. van & Tissing, O. (1978). SO_2 monitoring for testing public health criteria in the Netherlands, *VDI-Berichte,* **314,** pp. 73–7.

Floor, H. & Posthumus, A. C. (1977). Biologische Erfassung von Ozon-und PAN-Immissionen in den Niederlanden 1973, 1974 and 1975. *VDI-Berichte,* **270,** pp. 183–90.

Francis, B. A. & Curtis, C. R. (1979). Effect of simulated saline cooling-tower drift on tree foliage. *Phytopathology,* **69**, pp. 349–53.

Haut, H. van, Scholl, G. & Haut, G. van (1972). Ein doppelwandiges Vegetationsgefäss aus Kunststoff mit selbsttätiger Bewasserung. *Landwirtschaftl. Forschung,* **25**, pp. 42–7.

Heagle, A. S., Body, D. E. & Heck, W. W. (1973). An open-top field chamber to assess the impact of air pollution on plants. *J. of Environ. Quality,* **2**, pp. 365–8.

Lotschert, W. & Köhm, H. J. (1978). Characteristics of tree bark as an indicator in high-emission areas, II: Contents of heavy-metals. *Oecologia,* **37**, pp. 121–32.

Mandl, R. H., Weinstein, L. H., McCune, D. C. & Keveny, M. (1973). A cylindrical, open-top chamber for the exposure of plants to air pollutants in the field. *J. of Environ. Quality,* **2**, pp. 371–6.

Manning, W. J. & Feder, W. A. (1980). *Biomonitoring Air Pollutants with Plants.* Applied Science Publishers, London, England, UK: x + 140 pp., illustr.

Posthumus, A. C. (1976). The use of higher plants as indicators for air pollution in the Netherlands. Pp. 115–20 in *Proceedings of the Kuopio Meeting on Plant Damages Caused by Air Pollution* (Ed. Lauri Kärenlampi), Kuopio and Kuopio Naturalists' Society, Kuopio, Finland: IV + 160 pp., illustr.

Posthumus, A. C. (1978). New results from SO_2-fumigations of plants. *VDI-Berichte,* **314**, pp. 225–30.

Posthumus, A. C. (1980). *Elaboration of a Communitive Methodology for the Biological Surveillance of Air Quality by the Evaluation of the Effects on Plants.* Commission of the European Communities, Brussels, Belgium, Report EUR 6642 EN, 40 pp.

Prinz, B. & Scholl, G. (1975). Erhebungen über die Aufnahme und Wirkung gas- und partikelförmiger Luftverunreinigungen im Rahmen eines Wirkungskatasters. *Schriftenreihe der Landesanstalt für Immissions- und Bodennutzungsschutz des Landes Nordheim Westfalen,* Essen, **36**, pp. 62–86.

Raay, A. van (1969). The use of indicator plants to estimate air pollution by SO_2 and HF. Pp. 319–28 in *Proceedings of the First European Congress on the Influence of Air Pollution on Plants and Animals.* (Wageningen 1968.) Pudoc, Wageningen, Netherlands: iv + 415 pp., illustr.

Scholl, G. (1969). Ein Verfahren zur halbautomatischen Wasserversorgung von Pflanzenkulturen in Vegetationsfefässen. *Zeitschr. für Pflänzenernährung und Bodenkunde,* **124**, pp. 126–9.

Scholl, G. (1974). Ermittlung über die Belastung der Vegetation durch Schwermetalle in verschiedenen Immissionsgebieten. *Staub-Reinhaltung der Luft,* **34**, pp. 89–92.

Schönbeck, H. (1969). Eine Methode zur Erfassung der biologischen Wirkung von Luftverunreinigung durch transplantierte Flechten. *Staub-Reinhaltung der Luft,* **29**, pp. 14–8.

Schönbeck, H., Buck, M., Haut, H. van & Scholl, G. (1970). Biologische Messverfahren für Luftverunreinigungen. *VDI-Berichte,* **149**, pp. 225–36.

Spierings, F. H. & Wolting, H. G. (1971). Der Einfluss sehr niedriger HF-Konzentrationen auf die Länge der Blattspitzen-Schädigung und den Zwiebelertrag bei der Tulpenvarietät Paris. *VDI-Berichte,* **164**, pp. 19–21.

Air Pollution and Plant Life
Edited by M. Treshow
© 1984 John Wiley & Sons Ltd.

CHAPTER 6

Diagnosis of Air Pollution Effects and Mimicking Symptoms

MICHAEL TRESHOW

Department of Biology, University of Utah, Salt Lake City, Utah 84112, USA

ELEMENTS OF DIAGNOSIS

Diagnosis is the process of determining the nature and circumstances of an unhealthy condition, namely recognizing the cause of a disease or disorder. The condition may be associated with some organism, in which case it is said to be biotic. Or it may be associated with some physical stress, being then abiotic.

Diagnosis can be fairly simple and definitive when one is familiar with the symptoms of the condition, or when known pathogenic organisms are involved. When the condition is noted for the first time or appears in an unfamiliar area, it is often necessary for the observer to delve into the background of the disease. One must learn something about the environmental stresses, pathogens, and parasites, occurring in the area. And one must understand the basic procedures of diagnosis, what background information to seek, and how to interpret it.

Correct diagnosis involves many integrated elements. The symptoms—that is, what the conditions look like—are an especially vital element, and a knowledge of symptomatology is surely the best means of diagnosis (Jacobson & Hill, 1970). But the presence or absence of an organism, the plant parts and species affected, the distribution of affected plants, characteristics of the location and terrain, and the crop or ecosystem history, must all be considered. Together, they comprise the elements of diagnosis, and together they provide the total syndrome—the total disease pattern, including all the associated parameters (Treshow, 1970; Lacasse & Treshow, 1976).

97

Symptoms

A symptom is any perceptible change in the expected appearance, structure, or function, of an organism which suggests that it is unwell. Symptoms are the result of physical or biotic agents that cause stress. Unfortunately for the diagnostician, the plant may respond to a number of stresses in essentially the same way. But whatever the cause of stress may be, loss of the green chlorophyll pigments of the leaf, with concomitant adverse effects on the all-important process of photosynthesis, are often an early indication that something is wrong. The green colour gradually fades, rendering parts or all of the leaf chlorotic, or yellowed. Thus, chlorosis is a common, almost ubiquitous expression. Frequently, as the tissues' functioning continues to be impaired, the tissues die and become necrotic, turning various shades of brown. Such necrosis, expressed as tip-dieback, is especially characteristic of conifer needles and can be caused by almost any stress.

The appearance of chlorotic or necrotic flecks or spots on the leaf, often in the apparent absence of any other injury, is also fairly common. These lesions can be caused by certain air pollutants, but are more often associated with other pathogens.

Less often, leaves may be twisted, cupped, or afflicted with strange overgrowths. In such cases, insect or virus pathogens are the most likely culprits; but again, other pathogens may be responsible.

Symptoms are not always limited to leaves. Flowers, fruits, stems, roots, or any plant organ may be affected by a pathogen and respond with the appearance of characteristic symptoms. Cankers—local areas of dead tissue—provide the common symptoms on stems or roots. When a canker encircles a stem or root, the food, water, and nutrient supply, of herbaceous plants is disrupted beyond the affected area, and the affected tissues show secondary symptoms of starvation or drought.

In perennial plants, such as trees, trunk-girdling cankers prevent translocation of food materials to the roots, because the phloem has been killed. However, functional xylem remains, and so the portion of the tree distal to the canker may live for more years, as it will continue to be supplied with water and essential nutrients as long as the roots remain alive. The roots die when once their stored food reserves have become exhausted. However, this may take several years, especially if the species forms root-grafts readily, and the top will function normally until the roots die. Partial girdling weakens the distal tissues without necessarily killing them. Cankers are most often caused by fungi or bacteria.

Flowers and fruits are not often affected by air pollutants, although there are some notable exceptions that will be discussed later. As flower parts and fruits fundamentally are modified leaves, it is reasonable to expect them to respond to stress in similar ways, as indeed they do. But although necroses can sometimes occur in them, the flowers tend to be more tolerant of most pollutants. When symptoms appear on flowers or fruits, some biotic causal factor is most often responsible.

Presence of Organisms

If some sign of an organism, or especially the organism itself, can be found consistently associated with a disease or its symptoms, there is a strong likelihood that it is responsible. It is important to keep in mind that presence of an organism does not always mean that it caused the disease in question. Some organisms are secondary, invading only previously damaged tissues. Or, their presence may be coincidental and be of little consequence.

Conversely, the absence of some sign of an organism does not assure that the symptoms have a physical, abiotic origin. An insect may have come and gone—perhaps without leaving any skeletal remains or other signs. Fungi, while sometimes producing mycelia or reproductive structures on the leaves or stem, may just as likely develop only within the host tissues and be found only by isolation or microscopic examination. Bacteria are almost always difficult to find. Viruses and mycoplasmas are even smaller, and unless one uses an electron-microscope, they are recognized solely on the basis of the symptoms.

Plant Parts Affected

Diagnosis is further aided by noting which plant parts show injury, and the distribution of symptoms on the affected plant. Symptoms are most often apparent on the leaves, which are usually the most abundant and most obvious organ. Even though injury may occur on the stem, trunk, or roots, sooner or later it is the leaves that show the stress.

Several questions may be asked. Are affected leaves distributed uniformly over the plant, or limited to a single branch? If symptoms are evident on part of the plant only, how severe are they? Severe symptoms distributed in this way suggest that part of the root system may be damaged. Should the entire plant be severly affected, as when all the leaves are wilted, one again looks to the roots.

Digging around the main stem or trunk, look for lesions or cankers—some dead tissue. Often some fungus or bacterium will have caused the trouble, but this is hard to diagnose. It is much easier to determine when insect injury, or tooth-marks for example from Gophers or Ground-squirrels (commonly *Citellus tridecemlineatus*), are found.

The trunk itself may be damaged, again by fungi, bacteria, insects, or rodents, but occasionally Porcupines (*Erethizon dorsatum*), or even rubbing by livestock in some situations, can be harmful.

Where air pollutants are involved, the symptoms are almost always limited to the leaves. The exceptions occur with some peach varieties on which fluorides can cause a premature softening of the basal part of the suture area. Other fruits are far more tolerant and are rarely affected. Flowers are rarely injured by air pollutants—except for ethylene, which can cause abscission or wilting. Also with ethylene, petals may turn yellow and wither, buds may remain partly or wholly closed, and flowers may open slowly if at all.

Distribution of Plants

The distribution of affected plants in a field or geographic area can be very helpful to diagnosis. When affected plants are limited to a local area, some characteristic of that part of a field or orchard may be sought—perhaps some local feature of the terrain or soil, a low wet area, or a rocky, formerly raised area where the topsoil has been levelled off. A local invasion by some biotic pathogen, an insect or a fungus, is also a distinct possibility.

The effects of air pollutants are generally more widespread than those of other causes of disorders. Even when the source is relatively minor, the pollutant is not limited to a particular farm or field.

Species Affected

Knowing the species affected is especially useful. Certain diseases, and the agents that cause them, are often associated with particular species. A given virus may be restricted to certain species, while a given fungus occurs only on others. Differences in sensitivity among species are particularly apparent when abiotic stresses occur—not only adverse moisture or temperature, but, most notably of all, air pollutants.

Only a few different species are highly sensitive to any pollutant. These biological indicators are discussed in depth later in this chapter. If the symptoms that are being diagnosed appear on plants that are sensitive to an air pollutant, it may have been responsible. But then again, it may not have been. The appearance may have been coincidental. One then looks to other sensitive species and, of course, the symptoms.

If species known to be sensitive to an air pollutant are unaffected, and symptoms appear only on more tolerant plants, air pollutants can be discounted as a causal agent.

Crop or Ecosystem History

Background information is essential to any diagnosis. Even where a diagnosis may at first seem simple, some unsuspected cultural practice or climatic factor may be more basic and underlie the initially suspected stress. Winter freezing may have caused tree damage and decline that becomes apparent only years or even decades later. Pesticides may have a continuing impact long after the applications are all but forgotten.

Irrigation practices, pruning, and all other cultural practices, have an important bearing on the expression of symptoms, regardless of the stress. Soil characteristics, most clearly nutrient relations and fertilizer practices, are vital. Unfavourable nutrient relations alone may cause distinct symptoms; but even when no such symptoms appear, they influence the development of all biotic pathogens and even abiotic stresses—including injury by air pollutants. And deficiencies alone may cause symptoms incorrectly attributed to air pollutants. Soil acidity can also be important, either by itself or in combination with other factors. So can the soil structure and texture that influence water relations and soil aeration.

Presence of a Pollutant Source

Naturally the presence or absence of a pollution source is the keystone to diagnosis as related to air pollution. But, as should be clear by now, the presence of such a source by no means assures that the emissions are responsible for all, or even any, of the symptoms observed, or otherwise influence plant health.

Where air pollution injury is suspected, proximity of the symptomatic plants to the source is a primary consideration. With few exceptions, symptoms will decrease with distance. The exceptions arise when no sensitive species occurs near the source, or cultural practices have altered the sensitivity in some local areas.

Proximity is most clearly relevant where point sources are involved— smelters, power plants, or other industries. However, urban pollution, involving automobiles and other vehicles, may have more far-reaching effects. Ozone and other photochemical pollutants may spread out in toxic concentrations one hundred miles (160 klm) or more (Treshow, 1981). Although industrial pollutants may well extend as far, their degree of toxicity in the diluted, low concentrations is more controversial.

CHEMICAL ANALYSIS

When all the diagnostic considerations and their interactions have been studied, and the cause of a condition is still in doubt, but when air pollutants are suspected, a further step is sometimes helpful. This is the chemical analysis of plant tissues. Leaves are generally used, as they take up the largest amounts of a pollutant; but other organs can also be analysed. Not all pollutants are accumulated by plants, and not all accumulate in them sufficient concentrations above background levels to be helpful, but analysis for certain chemicals can provide one more diagnostic tool.

The analysis itself is generally left to a reliable chemistry laboratory and will not be treated here. But collecting the samples for analysis is the domain of the diagnostician, so will be discussed briefly.

Sampling requires mostly common sense! What we want is a representative idea of how much of a chemical the leaf tissues contain over a given geographical area. Depending on the seriousness of a problem, samples would be collected at successive 1/2 to 1 km intervals from the source, starting as close as possible and extending therefrom for several kilometres—sufficiently to be well out of the affected area. At each sampling point, a composite, representative sample should be collected of uniformly aged leaves well away from dusty roads or other sources of contamination. Leaves of one or two of the most important species showing symptoms should be collected. The growing conditions of each should be as nearly identical as possible.

Fluoride

Chemical analysis probably has its greatest relevance and value in diagnosing fluoride injury. This is because background concentrations of fluoride are normally quite low, and must increase several-fold before injury occurs. Background fluoride concentrations—those occurring remotely from any sources of atmospheric fluoride—range approximately from 5 to 15 ppm. Analytic variation can increase indications of this spread, as can soil fluorides particularly under acid conditions. On alkaline soils, fluoride uptake is negligible. Even when fluorides may reach several thousand ppm in the soil, they are usually bound in chemical complexes that are not taken up by the plant.

There are a few classic exceptions to this, mostly in members of the tea family (*Theaceae*) including Camellia spp. These plants actually have an affinity for fluorides, and foliar contents normally reach or even exceed 100 ppm.

But as a general rule, if foliar fluoride concentrations exceed roughly 30 ppm, some atmospheric fluoride source is probably present in the area. A few of the most sensitive species may be injured when fluoride concentrations exceed 30–50 ppm, but most are unaffected at far higher concentrations of 100 ppm or more. Conversely, one must bear in mind that the presence of high fluoride concentrations, even in leaves of sensitive plants, does not necessarily mean that they are responsible for the injury in question. Other pathogens may well be responsible.

Low concentrations in the leaves do not conclusively mean that some other organ may not have been damaged. Fruits, in some instances, have been injured when there has been very little foliar uptake. Under experimental conditions, pollen development in cherries (*Prunus* spp.) has been impaired, with the implication of reduced fruit-set, when the fluoride content of leaves was not especially high (Facteau *et al.*, 1973).

Sulphur Dioxide

Chemical analysis is only slightly helpful in diagnosing SO_2 injury. As a natural, required element, the normal, background sulphur content is already rather high and quite variable. Sulphur is taken up from the soil to a varying degree, so that background concentrations typically range from 0.1 to 0.2% and sometimes higher. Most significantly, as the normal leaf-content of sulphur is already fairly high, it takes a substantial build-up before any increase is significant. Furthermore, injury from atmospheric SO_2 can occur when levels in leaves are no higher than the upper range of background.

If visual symptoms indicate that SO_2 may have caused a particular injury, then a high sulphur content in the leaf tissues helps to support the diagnosis, but does not necessarily establish it. The known sensitivity of the affected species, and other parameters must also be considered.

Other Pollutants

The common photochemical pollutants, ozone, PAN (peroxyacetyl nitrate), and nitrogen oxides, do not accumulate in plant tissues to a point where they can be helpful for diagnosis. The diagnosis of injury that is suspected to have been caused by such pollutants as ammonia and chlorine, could be aided by tissue analysis; but such injury is usually associated with a known, accidental spill, so the cause is already obvious.

Heavy-metal contamination from lead, mercury, etc., rarely causes visible symptoms in plants. Concern involves, rather, their accumulation and possible human consumption.

HISTOPATHOLOGY

Histopathology is the microscopic examination and study of unwell tissues. Pollutants injure specific tissues, cells, and organelles. Sometimes the specific, microscopic injury varies—depending on the pollutant or other stress present. It is then that histopathology may aid diagnosis. As the procedure is laborious, this technique is utilized only when all other steps have been inconclusive. This is infrequent when broad-leafed plants are involved, but may be useful for diagnosing injury in conifers.

The cause of needle-tip necrosis is perhaps the most difficult to establish, and histopathology may have its greatest application in diagnosing injury to pines. Stewart *et al.* (1973) compared the histological differences caused by various air pollutants with such other stress-factors as table salt (NaCl), boron, moisture-stress, winter injury, and suffocation. All of these can cause needle necrosis, but when the narrow zone between the necrotic and green tissue is examined under the microscope, using thin tissue sections, some differences appear that are characteristic of each stress.

Resin-duct occlusion is common regardless of the stress; mesophyll collapse occurs when plants are exposed to ozone or SO_2 but not when they are exposed to HF; phloem abnormalities are common except when ozone has been the pathogen, and transfusion tissue hypertrophy is often associated with fluoride. However, this last also occurs from moisture-stress, winter injury, and natural senescence. Combing all these responses fails to establish precisely which stress is responsible for needle necrosis, but can define which stresses are not.

Symptoms of ozone injury in broad-leafed species are far more definite, as the interveinal cells on the upper half of the leaf—the palisade layers—are particularly sensitive (Hill *et al.*, 1961). This sensitivity is fairly specific for ozone, although chlorides can sometimes cause a similar response. Injury from SO_2 and PAN is most prevalent in the lower, spongy mesophyll and tends to be associated with the stomata.

REMOTE-SENSING

Aerial photography (i.e. remote-sensing), using infrared and other special films, has been applied to detecting areas of plant stress. Very simplistically, when plants begin to be unwell, they respire more rapidly than formerly and their temperature increases, much as in humans. These higher temperatures appear on heat-sensitive infrared film. When the tissues are killed, respiration decreases, and this, too, appears on the infrared film. Large areas of diseased plants, such as may occur in forests, can be located in this way (Tingey *et al.*, 1979). Comparison of polluted and unpolluted areas is critical, and on-ground verification surveys are still essential to determining the cause of the disease or stress.

Much the same applies to satellite imagery. The large-scale images are helpful in locating areas of stress, but the cause of the stress cannot be distinguished. Diagnosis of any problems still requires on-ground study.

These techniques have also been tested to delimit different kinds of plant communities and relate them to the presence of smelters or other pollutant sources. The investigator must still complete the study on the site, in order to interpret the aerial reconnaissance. Consequently, while aerial photography and special sensitive films provide useful survey tools, they do not provide the complete story and have limited application to diagnosis.

MIMICKING SYMPTOMS

Symptoms caused by quite different and unrelated stresses can look very much the same. It would seem that the fundamental physiological mechanisms imposed on plants are sufficiently similar to cause rather similar responses. Thus, we see that various abiotic, and sometimes biotic, pathogens can cause injury that 'mimics' air pollution injury. This is one reason why the diagnostician must always be careful to consider all the facets of diagnosis, and carefully examine all the possible causes of a disorder (Lacasse & Treshow, 1976).

Abiotic Pathogens

Abiotic pathogens are stresses imposed by the physical environment. No organism is involved. In addition to air pollutants, they can include everything from soil relations to weather, pesticides, and lightning. The plant injuries which they cause are, however, often mistaken for air pollution effects.

Water: Plants, as with all forms of life, require water. But too much, or too little, can be harmful. Too much water has an indirect effect in that the water fills the air spaces in the soil, so there is not enough oxygen for the plant roots. Consequently, they cannot respire or produce the energy needed for metabolism. They suffocate! Movement of nutrients and water to the aerial parts of the plant is impaired, and leaf processes become arrested. As with so many stresses, chlorophyll synthesis is one of the first processes to be affected. New molecules are not synthesized, and as the old ones break down, the green leaf-colour fades. The leaves become yellowed and ultimately brown. This chlorosis tends to begin around the leaf margin producing a symptom reminiscent of SO_2 toxicity. However, SO_2 usually produces a sharper colour contrast between healthy and sick tissues, rather than the more diffuse fading that results from suffocation.

The effects of too little water are roughly comparable. Again, root function is impaired, this time from inadequate water to support metabolism. As water becomes deficient, the plant undergoes gradual, at first reversible, but finally permanent, wilting. Progressively, more and more plant processes become involved in this manner.

The first defence of a plant is often to roll up its leaves to minimize water loss. When this is not enough, the leaves gradually dry, mostly becoming necrotic from the margin inwards. A combination of too little water, suddenly alleviated by a good irrigation, causes the 'white spot' disease (Richards, 1929). The sudden hydration causes a symptom remarkably similar to SO_2 injury.

The Red Belt condition, producing a striking needle-burn symptom following winter desiccation often combined with sudden temperature changes (Hensen, 1923), bears a particularly close resemblance to air pollution injury.

Temperature: Proteins can begin to coagulate at about 35–40°C. As this temperature is approached, their activity declines—first one protein, often an enzyme that is vital to some physiological process, and then another. The initially lowered metabolism and reduced growth goes unnoticed but, as with adverse water relations, if it persists, then the reduced vitality can be important to the plant's health. Most noticeably, chlorophyll is broken down and the leaf-tips become yellowed or 'scorched' in a manner not unlike fluoride injury. Leaves that have matured in a cool, moist environment are most predisposed to heat stress. Several high-temperature disorders have rather widespread occurrence. Symptoms of needle blight of conifers, the name given to needle scorch of pine and fir, most closely resemble air pollution injury. Sunscald of vegetable crops and other plants may also be misinterpreted for air pollutant injury.

There is also a temperature below which protein stability is lost, and as temperatures drop below freezing during the growing season, enzymes may congeal and have their ability to function normally destroyed. Again, the visible expression most often is a 'burning', or more accurately, necrosis, beginning at the leaf margin. This symptom can be reminiscent of fluoride injury. On other occasions, where radiation frost is a factor, and the leaf tissue cools below freezing by losing heat into a clear, still atmosphere, a bleaching or bronzing expression over the leaf surface is more common. On conifers, a sudden temperature-drop is especially likely to cause needle necrosis.

Each year in temperate, fruit-growing areas of the world, flower buds or blossoms are killed by frosts. This is easy to recognize. Freezing temperatures are recorded, and pistils turn black during the next day or so. More subtle responses occur, though, when the pistil and ovary look normal but the ovules and sometimes associated tissues are injured. The fruits may continue to develop for a few days or weeks, but lack the integrity to cling. Under even a slight stress of normal higher temperatures and a paucity of water, the young developing fruits drop. Such dropping is a normal self-thinning process in many fruit crops, but, aggravated by low-temperature injury, it can be excessive and leave a light crop (personal observation). It is then that blame may be mistakenly placed on air pollutants.

Soil and nutrient relations: Adverse moisture-relations are especially common, but other soil parameters can be equally important in causing mimicking symptoms. Often the effects are indirect. The soil texture, for instance, influences the water-holding capacity of the soil—a heavy, clay soil often holds too much, whereas a sandy soil loses moisture too rapidly. The soil structure involves the soil particles, mostly the fine clay particles, to which water adheres and thus influences moisture availability. The salinity of the soil also influences water availability. Adverse relations in any of these parameters can produce a chlorotic expression on the leaves of affected plants that can be mistaken for air pollution injury.

Soil acidity or alkalinity is also critical, largely in influencing the availability of nutrients. Iron and phosphorus, for instance, become increasingly unavailable to plants in alkaline soils as the pH exceeds about 8.0. These and other nutrients also become less available as acidity increases below a pH of about 5.5. This results in symptoms that are characteristic of the nutrients which are in shortest supply. Often the symptom consists of leaf chlorosis. The leaf margin is most vividly yellowed, with chlorosis extending between the veins. There are many variations in this pattern, and in the age of the leaves most affected, but the general symptom resembles that caused by SO_2.

Nutrient deficiencies by themselves can cause symptoms resembling air pollution toxicity. When manganese or zinc are deficient, for instance, the leaf response can be mistaken for fluoride injury. Other deficiencies that cause chlorosis, such as nitrogen and magnesium deficiency, are also likely candidates for misinterpretation.

Pesticides: Pesticides are chemicals that are meant to kill pests. Herbicides, as they are designed specifically to kill plants, naturally have the greatest impact on vegetation, and in sublethal doses can cause symptoms in leaves and fruits that may be especially similar to symptoms caused by air pollutants. Leaf chlorosis is the most common expression. This may range from the sharply delimited bright- to pale-yellow border caused by some pre-emergence herbicides, to the more diffuse chlorosis encompassing much of the leaf surface and caused more often by post-emergence weed-killers. Such symptoms are really quite distinct from those caused by air pollutants, but nevertheless have been mistaken for them.

Even more distinct are the leaf twisting, distortion, and overgrowth, expressions caused by the phenoxyacetic acid chemicals—2,4-D and its relatives. The thick, rough, crinkled and sometimes cupped, leaf expression should not be mistaken for air pollution injury, but sometimes is. Low concentrations drifting over ripening fruits, on the other hand, cause symptoms virtually indistinguishable from fluoride effects on peaches. The premature, reddened suture area caused by the phenoxyacetic acids can only be distinguished because the suture area near the stem end is more affected than towards the other end of the fruit.

The greatest threat comes when pesticides are persistent, so that effects can appear several years after application. Then the crop history is most critical to diagnosis.

Insecticides are designed to kill insects, so the effects on plants are not necessarily injurious. But when they are, the symptoms very often consist of leaf chlorosis, which can be reminiscent of air pollution injury.

Biotic Relations

Biotic pathogens—fungi, bacteria, viruses, mycoplasmas, insects, and nematodes—also must all be considered as possible causal agents of many kinds of symptoms. While the organism itself may give away its identity, it is not always there, or is not apparent to the naked eye. Then other clues must be sought.

Fungi and Bacteria: Symptoms caused by fungi and bacteria are apt to be so very diverse that it is not surprising that some resemble those caused by air pollutants. However, this resemblance is not so great as to be indistinguishable. Furthermore, in the case of fungi, some reproductive structures are often present. Organisms that cause necrotic spots on the leaves are the most troublesome. When the spots are tiny, they can be mistaken for ozone symptoms.

The greatest similarity is indirect, and the primary pathogen affects the stem or roots rather than the leaves. When fungi infect stem, trunk, or root tissues, the cankers produced disrupt the nutrient, water, and food, movements in the plant. Leaves are 'starved' or desiccated, resulting in chlorosis or browning around the leaf margin, and often extending inwards between the veins.

Viruses and mycoplasmas: Symptoms caused by these organisms also are extremely varied. Viruses—sub-microscopic strands of nucleic acids and protein—can cause mosaics, leaf-cupping or twisting, and spotting symptoms that, while not closely resembling air pollution injuries, have nevertheless been mistaken for them by inexperienced observers.

Mycoplasmas—microscopic bodies reminiscent of bacteria but lacking a cell wall—often cause leaf-yellowing along the margin and extending inwards between the larger veins. This symptom can be mistaken for SO_2 injury when it occurs in sensitive plants. Generally, though, necrosis is associated with SO_2 injury.

Nematodes and insects: Once again, it is the secondary effects on the leaves, following stem or root injury, that are most likely to be mistaken for air pollution injury. Some insects, though, can cause a stippling or minute flecks on the leaves, that are reminiscent of ozone injury. Photochemical pollution symptoms are also simulated by mites, especially the eriophyd mites that cause a bronzing or silvering of the leaf surface. This type of injury is, however, not as delimited by the veins as when it is caused by ozone.

BIOINDICATORS

Plant species and varieties show a striking variation in their sensitivity to air pollutants. By far the greatest majority of all plants are relatively tolerant, and remain free from injury under all but the most severe air pollution episodes. A few, though, are highly sensitive and are injured by concentrations that may be scarcely above background levels. These unique, sensitive species or varieties can be used to monitor the presence of low concentrations of specific air pollutants. They are often called biological indicators or bioindicators (Heck, 1966; Feder & Manning, 1979), and they are most valuable for diagnosis.

In order to be useful, the indicator plant must be found in the area that is being studied. No one plant has such a universal distribution. Consequently, it is helpful to include several sensitive species in bioindicator lists. Bioindicators furthermore should, as well as being highly sensitive and ubiquitous species, provide symptoms that are definitive for a particular pollutant. To achieve these objectives, it may be necessary to set out plants of known response. The approach of placing plants of known sensitivity over the area under investigation has been used in many parts of the world.

Relatively few plant species and varieties are known to be sufficiently

Table I   Plant species most sensitive to major air pollutants.

Sulphur dioxide

Alfalfa (*Medicago sativa*)
Barley (*Hordeum vulgare*)
Cotton (*Gossypium hirsutum*)
Douglas Fir* (*Pseudotsuga menziesii*)

Ponderosa Pine* (*Pinus ponderosa*)
Soybean (*Glycine max*)
Wheat (*Triticum* sp.)
White Pine* (*Pinus strobus*)

Fluoride

Apricot (Chinese cultivar)
   (*Prunus armeniaca*)
Gladiolus (some cultivars)
Grape (some European cultivars)
   (*Vitis vinifera*)

Oregon Grape (*Mahonia repens*)
Peach (fruit) (*Prunus persica*)
Ponderosa Pine* (*Pinus ponderosa*)
St. John's-Wort (*Hypericum perforatum*)

Ozone

Alfalfa (*Medicago sativa*)
Barley (*Hordeum vulgare*)
Bean (*Phaseolus vulgaris*)
Green Ash (*Fraxinus pennsylvanica*)
Oats (*Avena sativa*)

Quaking Aspen (*Populus tremuloides*)
Spinach (*Spinacia oleracea*)
Tobacco (some cultivars)
   (*Nicotiana tabacum*)
Wheat (*Triticum* sp.)
White Pine (*Pinus strobus*)

* While these species are traditionally regarded as 'sensitive', they are far more tolerant than the herbaceous species listed in their category.

sensitive to be valuable as bioindicators. The Eastern White Pine (*Pinus strobus*), for instance, can be sensitive to SO_2, ozone, and fluoride, depending on the individual strain. Symptoms are discouragingly similar, but by using strains of known sensitivity, the particular pollutant or pollutants present can be detected. Coniferous species are often regarded to be among the more sensitive to a number of pollutants, and they are. But there are generally other species in any given area that are far more sensitive (Table I). The main limitation to observation is that possible growth or reproductive impairment at levels that are too low to cause visible symptoms may remain undetected.

Photochemical Pollutants

Ozone, PAN, and nitrogen oxides, are particularly common, and some excellent bioindicators have been used for many years to evaluate their presence. As early as the 1940s, sensitive varieties of lettuce, chard, tomato, bean, and other truck-crops, provided sensitive indicators for PAN pollution in the Los Angeles area of California although the chemical concerned had not yet been identified chemically, the injury being broadly attributed to 'smog'. To quantify better the distribution of photochemical pollutant injury, pinto beans (*Phaseolus* sp.) and Annual Bluegrass (*Poa annua*), grown under known, controlled conditions, were placed at numerous locations. Similarly the Bel W-3 tobacco, cultivar has been used as a bioindicator for ozone. The White Cascade *Petunia* cultivar (*Petunia hybrida*) is also highly sensitive, as are certain tomato and bean cultivars. Their very sensitivity, however, has led to their reduced use in commercial plantings. While field observations of these plants were valuable in their day, they have now been largely replaced by automated chemical methods.

Sulphur Dioxide

Lichens have provided an inadvertent bioindicator or urban pollution, associated for the most part with coal smoke and SO_2, for many decades (*see* Chapter 12). The scarcity of lichens in polluted areas is well-documented, beginning with the early work by the Finnish lichenologist W. Nylander, who made a list of lichens growing on the trunks of chestnut trees in the Luxembourg Gardens in Paris (1866). Subsequently, the demise of lichens near industrial and urban areas has been documented in over a hundred reports from around the world. LeBlanc and his colleagues have pursued this facet for many years, and described similar conditions in Montreal (LeBlanc & DeSlover, 1970). LeBlanc has developed an Index of Atmospheric Purity (IAP) based on the relative sensitivity of several lichen species that can be used to map air quality around urban centres or point sources of SO_2 (LeBlanc, 1972). There are, however, limitations to the widespread application of this approach, as the lichen species used for the IAP are not universally distributed, and are difficult to identify. Furthermore, more readily identifiable flowing plant species are at least equally sensitive.

Alfalfa (*Medicago sativa*) has been most commonly utilized as an SO_2 bioindicator. The species is among the most sensitive as well as being widely planted. Unfortunately the symptoms are not always the most definitive, so diagnosis must combine other criteria. Ragweed (*Ambrosia* spp.), Alfalfa, Barley (*Hordeum vulgare*), Soybean (*Glycine max*), and wheat (*Triticum* sp.), are useful in various parts of the USA. Other less well-known plants are occasionally valuable in local areas.

Fluoride

Gladiolus gandavensis, particularly the cultivars Snow Princess and Pink Prospector, provides the classic bioindicator for fluorides, but these plants are not always available. Thus native species, including Oregon Grape (*Mahonia repens*) in the Western US and St John's-Wort (*Hypericum perforatum*), of European origin and commonly naturalized over much of the USA, are used. Ponderosa Pine (*Pinus ponderosa*) is also considered sensitive, but is far more tolerant than any of the above, as is Douglas Fir (*Pseudotsuga menziesii*), another moderately sensitive species. Of the cultivated species, prunes (*Prunus* spp.), and the 'Chinese' cultivar of Apricot (*Prunus armeniaca*) are the most sensitive, with some cultivars of European Grape (*Vitis vinifera*) running a close third.

SUMMARY

Diagnosis is to recognize the cause of a disorder. As many unrelated stresses can cause symptoms that resemble those caused by air pollutants, special caution and methods must be applied in their diagnosis. Many elements of diagnosis must be integrated: symptoms, presence or absence of organisms, plant parts affected, distribution of affected plants, species affected, crop or ecosystem history, and presence of a pollutant source, must all be considered.

Chemical analysis may prove helpful in diagnosing fluoride, and sometimes SO_2, toxicity. Histopathology also can be helpful but may not warrant the time and cost. On a large scale, remote-sensing imagery may be useful. The numerous agents that can cause symptoms which incorrectly may be attributable to pollutants, include such abiotic factors as temperature or water-stress, and biotic factors involving viruses, mycoplasmas, bacteria, fungi, insects, and mites. Bioindicators—plants especially sensitive to a given pollutant—are valuable tools to aid diagnosis, as these are the first to show visible symptoms.

REFERENCES

Facteau, T. J., Wang, S. Y. & Rowe, K. E. (1973). The effect of hydrogen fluoride on pollen germination and pollen-tube growth in *Prunus avium* L. cv. 'Royal Anne'. *J. Amer. Soc. Hort Sci.,* **98**, pp. 234–6.

Feder, W. A. & Manning, W. J. (1979). Living plants as indicators and monitors. Ch. 9 of 14 pp. in *Methodology for the Assessment of Air Pollution Effects on Vegetation* (Ed. W. W. Heck, S. V. Krupa & S. N. Linzon). Air Poll. Contr. Assoc. Info. Rept No. 3, 380 pp., illustr. + append.

Heck, W. W. (1966). The use of plants as indicators of air pollution. *Internat. J. of Air and Water Poll.,* **10**, pp. 99–111.

Hensen, W. R. (1923). Chinook winds and Red Belt injury to Lodgepole Pine in the Rocky Mountains Park area of Canada. *Div. for Biol. Sci. Ser. Dept. Agr.,* **28**, pp. 62–4.

Hill, A. C., Pack, M. R., Treshow, M., Downs, R. J. & Transtrum, L. G. (1961). Plant injury induced by ozone. *Phytopathology,* **51**, pp. 356–63.

Jacobson, J. S. & Hill, A. C. (1970). *Recognition of Air Pollution Injury to Vegetation: A Pictorial Atlas.* Air Poll. Control Assoc. Pittsburgh, Pennsylvania, USA: vii + 45 pp., illustr.

Lacasse, N. L. & Treshow, M. (1976). *Diagnosing Vegetation Injury Caused by Air Pollution.* Environmental Protection Agency, Washington, DC, USA: 139 pp., illustr. and append.

LeBlanc, F. & Rao, D. N. (1972). The epiphytic study of *Populus balsamifera* and its significance as air pollution indicator in Sudbury, Ontario. *Can. J. Bot.,* **50**, pp. 519–28.

Le Blanc, F. & J. DeSloover (1970). Relation between industrialization and the distribution and growth of epiphytic lichens and mosses in Montreal. *Can. J. Bot.,* **48**, pp. 1485–96.

Nylander, W. (1866). Les lichens du jardin du Luxembourg. *Bull. Soc. Bot.,* **13**, pp. 364–72.

Richards, B. L. (1929). White spot of Alfalfa and its relation to irrigation. *Phytopathology,* **19**, pp. 125–41.

Stewart, D. M., Treshow, M. & Harner, F. M. (1973). Pathological anatomy of conifer needle necrosis. *Can. J. Bot.,* **51**, pp. 983–8.

Tingey, D. T., Wilhour, R. G. & Taylor, O. C. (1979). The measurement of plant responses. Ch. 7 of 35 pp. in *Methodology for the Assessment of Air Pollution Effects on Vegetation* (Ed. W. W. Heck, S. V. Krupa, & S. N. Linzon). Air Poll. Contr. Assoc. Info. Rept. No. 3, 380 pp., illustr. + append.

Treshow, M. (1970). *Environment and Plant Response.* McGraw-Hill, New York, NY, USA: xv + 422 pp., illustr.

Treshow, M. (1981). Pollution effects on plant distribution. *Environmental Conservation,* **7**, pp. 279–86, 5 figs.

Air Pollution and Plant Life
Edited by M. Treshow
© 1984 John Wiley & Sons Ltd

CHAPTER 7

Biochemical and Physiological Impact of Major Pollutants

S. S. MALHOTRA

Canadian Forestry Service, Environment Canada, Northern Forest Research Center, 5320 122nd Street, Edmonton, Alberta T6H 3S5, Canada

&

A. A. KHAN

Alberta Environmental Center, Vegreville, Alberta T0B 4L0, Canada

BACKGROUND

The harmful effects of air pollution on various components of vegetation such as forest trees, agricultural crops, ornamental plants, and lichens, are now well recognized. The major pollutants studied in this regard are sulphur dioxide, ozone, oxides of nitrogen, peroxyacetyl nitrate, and fluoride. These pollutants can have a deleterious effect on a variety of biochemical and physiological processes and on structural organization within the cells.

Following an episode it is often assumed that there has been no injury to vegetation unless visible symptoms of phototoxicity have developed. However, this can be misleading. In many controlled environment studies, air pollutants have been shown to reduce the growth and yield before any visible symptoms appeared. It is now commonly believed that injury initially takes place at the biochemical level (interference with photosynthesis, respiration, lipid and protein biosyntheses, etc.), subsequently progressing to the ultra-structural level (disorganization of cellular membranes), and then to the cellular level (cell-wall, mesophyll, and nuclear breakdown). Finally, visible symptoms develop (chlorosis and necrosis of foliar tissues).

Biochemical injury results when the concentration of the pollutant exceeds the capacity of the tissues to detoxify it through their normal metabolism. The subtle and varied nature of the biochemical and physiological effects produced by air pollutants suggest that reduction in plant growth and yield because of air pollution may be more widespread and serious than is generally suspected.

SULPHUR DIOXIDE

Stomatal Response

Plants absorb sulphur dioxide (SO_2) mainly by gaseous diffusion through the stomata. Some uptake of SO_2 also occurs from moist cuticular surfaces but is of minor significance. The number of stomata and size of aperture play major roles in the uptake of SO_2, as do regulating factors that can affect the turgidity of guard cells, such as humidity, wind velocity, light, and temperature. Thomas & Hill (1935) showed that absorption of SO_2 was correlated with humidity. The presence of SO_2 in the air has been shown to stimulate stomatal opening (Majernik & Mansfield, 1970, 1971; Mansfield & Majernik, 1970), or closing (Menser & Heggestad, 1966), both of which are regulated by relative humidity and the concentrations of SO_2 and CO_2 in the air (Majernik & Mansfield, 1972). Recently it was demonstrated in *Vicia faba* (Broad or Field Bean) that low concentrations of SO_2 stimulated stomatal conductance within 15 minutes of exposure, and that this persisted for several days (V. J. Black & Unsworth, 1980). This may have been due to extensive destruction of epidermal cells adjacent to the stomata (C. R. Black & V. J. Black, 1979*a*). High concentrations of SO_2, on the other hand, frequently caused severe ultracellular disorganization (C. R. Black & V. J. Black, 1979*b*).

The different tolerances of plant species to SO_2 under similar biophysical conditions suggest that fine differences in biochemical and physiological mechanisms, operative in different plants, could influence the sensitivity of a particular plant to SO_2. It has been suggested that SO_2 injury depends on the rate of SO_2 absorption (Thomas, 1951; Bressan *et al.*, 1978; Caput *et al.*, 1978). Recently, Furukawa *et al.* (1980*a*) found highly significant correlations between foliar injury and the amount of SO_2 absorbed; plants that are sensitive to SO_2 absorbed greater amounts of gas than did those which are resistant to it. In contrast, no correlation between sulphur uptake and foliar injury was observed in *Agropyron smithii* (Lauenroth *et al.*, 1979). Resistance of *Picea abies* (Norway Spruce) seedlings to SO_2 when stomata were open was attributed to cellular mechanisms for detoxifying SO_2 (Oku *et al.*, 1980).

The effects of SO_2 on the biochemical and physiological processes directly related to stomatal response have recently been studied in detail. An increased concentration of carbon dioxide (CO_2) induces stomatal closure. Abscisic acid (ABA) is also known to produce a similar response, and has the ability to control CO_2-induced stomatal closure (Raschke, 1975). Plant sensitivity to SO_2 could therefore also be an indication of changes in the ABA levels of the fumigated foliage. Kondo & Sugahara (1978) measured the effect of 2 ppm SO_2 on the transpiration rates (to some extent a measure of stomatal behaviour) of sensitive and resistant plants. They found a rapid decrease in the transpiration rate of SO_2-resistant plants, and either a gradual decrease or an increase in transpiration rates of sensitive plants, submitted to this level of SO_2. Analysis of the ABA content of the foliage revealed that the resistant plants contained higher levels of ABA than the sensitive ones. When ABA-treated *Raphanus sativus* (Radish) was exposed to SO_2, the transpiration rate began to drop immediately, thereby making the plant much more resistant to SO_2 than it was originally. Later experiments by Kondo *et al.* (1980*b*), involving a number of SO_2-resistant plant species, confirmed the relationship between ABA content and decreased transpiration rate following SO_2 'fumigation' (i.e. exposure).

Upon diffusion through the stomata, gaseous SO_2 dissolves in water on the moist cellular surfaces to form sulphite (SO_3^{2-}), bisulphite (HSO_3^-), and other ionic 'species' (depending on the pH of the surrounding cellular surfaces); in such transformation, cellular pH would also be influenced by the generation of protons. The influence of these species on the ABA-related stomatal response was studied recently by Kondo *et al.* (1980*b*), using epidermal strips of *Vicia faba*. In the absence of ABA, SO_3^{2-} slightly stimulated stomatal opening, but in the presence of ABA, SO_3^{2-} produced no effect. No additional decrease in aperture size occurred on adding SO_3^{2-} in the presence of ABA concentrations that, alone, reduced aperture size. Changing the pH (between 4 and 7) similarly did not affect aperture size in the absence of ABA, but during its presence the stomatal aperture size was markedly reduced at pH 4.

These results suggested that ABA-related stomatal closure as a result of SO_2 fumigation was an acidic effect on the surface or cytoplasm of guard-cells. Other studies, however, have reported an increase in stomatal opening at low pH (Squire & Mansfield, 1973; Dittrich *et al.*, 1979). Since in these studies the levels of ABA in the SO_2-fumigated plants were not examined before and after the fumigation (in either sensitive or resistant plant species), the role of ABA in regulating stomatal responses during fumigation remains speculative and contradictory. For example, the transpiration rate of *Zea mays* (Maize) leaves in response to SO_2 fumigation was similar to that of other resistant plant species, yet the ABA content of the leaves was the lowest of all the plant species examined. Even if the cellular content of ABA was a major factor in controlling the absorption of SO_2, the mechanisms involved in the regulation of ABA levels in guard- and other subsidiary cells following SO_2 fumigation remains to be determined.

It appears that stomatal regulation in plants not possessing the ABA-type mechanism occurs by other biochemical and physiological means. Cellular increase of H^+ (normally brought about by SO_2) could cause leakage of K^+,

Cl^-, and malate (F. A. Smith & Raven, 1979). In K^+-regulated stomatal opening, anions such as malate and Cl^- play important roles (Raschke, 1979). Malate has been shown to be synthesized in the guard-cells by carboxylation of phosphoenol pyruvate (PEP) (Outlaw & Kennedy, 1978). In *Zea mays* and *Spinaciea oleracea* (Spinach), phosphoenolpyruvate (PEP) carboxylase is inhibited by SO_3^{2-} (Ziegler, 1973*a*; Mukerji & Yang, 1974), which in turn reduces malate synthesis and leads to changes in stomatal opening or plant sensitivity towards SO_2. Another biochemical control of stomatal response occurs through metabolic regulation of glycollate content (Zelitch, 1971). It has been shown that SO_2 inhibits glycollate oxidase activity in needles of *Pinus banksiana* (Khan & Malhotra, 1982*a*); this enzyme is also inhibited by low concentrations of SO_3^{2-} (Zelitch, 1957; Khan & Malhotra, 1982*a*).

Biochemical Transformations of Absorbed SO$_2$

The phytotoxic effects of SO_2 are greatly influenced by the ability of plant tissues to convert dissolved SO_2 into relatively non-toxic forms. Sulphite and HSO_3^- are the major chemical species formed upon dissolution of SO_2 in aqueous solutions; their respective concentrations depend on the pH of the medium (Puckett *et al.*, 1973). Both SO_3^{2-} and HSO_3^- have been shown to be phytotoxic to many biochemical and physiological processes (Zeigler, 1975; Malhotra & Hocking, 1976). Plants can overcome these phytotoxic effects by converting SO_3^{2-} and HSO_3^- to less-toxic forms. Oxidation of SO_3^{2-} to sulphate (SO_4^{2-}) in plant cells can occur by both enzymic and non-enzymic mechanisms, and SO_4^{2-} thus accumulated is considerably less toxic than SO_3^{2-} (Thomas *et al.*, 1943).

Plants exposed to SO_2 can accumulate sulphur compounds. Accumulation of large amounts of sulphur takes place at low SO_2 concentrations (Guderian, 1977), while at high SO_2 concentrations the accumulation is impaired due to collapse of stomatal regulation. In general, metabolically active young leaves accumulate more sulphur than older leaves (Guderian, 1977); however, *Agropyron smithii* accumulates less sulphur in young leaves than in older ones on the same plant (Lauenroth *et al.*, 1979).

Experiments with isotopic SO_2 ($^{35}SO_2$) have shown that there are labelled SO_3^{2-} and SO_4^{2-} in the treated foliage (Garsed & Read, 1977*a*, 1977*b*). *Glycine max* (Soybean) leaves exposed to $^{35}SO_2$ incorporated five times as much radioactivity in light as in the dark (Garsed & Read, 1977*a*), the major product being SO_4^{2-} (Garsed & Read, 1977*b*). Studies on the residence time of SO_3^{2-} following SO_2 fumigation, showed that *Glycine max* cultivars resistant to SO_2 converted SO_3^{2-} more rapidly to SO_4^{2-} than the sensitive cultivars (J. E. Miller & Xerikos, 1979). It appears, therefore, that the presence of SO_3^{2-}-oxidizing mechanisms can influence plant resistance to SO_2.

Sulphite has been oxidized in the light by isolated chloroplasts in a reaction induced by the electron transport system (McCord & Fridovich, 1969; Asada & Kiso, 1973; Libera *et al.*, 1973; Khan & Malhotra, ms*a*). Sulphite-oxidizing activities have also been reported in isolated mitochondria (Tager & Rautanen, 1956; Arrigoni, 1959; Ballantyne, 1977). Recently, Kondo *et al.* (1980*a*) separated various SO_3^{2-}-oxidizing activities and characterized their natures. In a number of plants they found a cytochrome *c*-linked SO_3^{2-}-oxidizing substance that had a low molecular weight and was non-proteinaceous in nature. High-molecular-weight SO_3^{2-} oxidases, on the other hand, were not linked to cytochrome *c*.

Oxidation of SO_3^{2-} can also be stimulated by cellular enzymes such as peroxidase, cytochrome oxidase, and ferredoxin-NADP reductase, and by catalysts such as metals and ultraviolet light (Hällgren, 1978). Production of the superoxide radical (O_2^-) in the chloroplast during illumination also stimulates SO_3^{2-} oxidation. This is supported by the observation that, in the presence of superoxide dismutase (SOD), photooxidation of SO_3^{2-} was inhibited (McCord & Fridovich, 1969; Asada & Kiso, 1973); however, free-radical oxidation of SO_3^{2-} was accelerated by indoleacetic acid (tryptophan) and Mn^{2+}, but not by SOD (Yang & Saleh, 1973). The superoxide radical formed during illumination has been found *in vivo* (Radmer & Kok, 1976) and in isolated chloroplasts (Asada & Kiso, 1973; Epel & Neumann, 1973; Asada *et al.*, 1974); it could serve as a source of other active oxygens ($^1O_2$, OH', and H_2O_2).

In the presence of SO_3^{2-} and HSO_3^-, more O_2^- is formed by free-radical chain oxidation than otherwise. This process also generates OH' and SO_3^{2-} radicals. Together these oxidizing radicals can affect a number of cellular mechanisms. Apparently in order to arrest uncontrolled production of such free radicals, plants have developed natural scavengers (Asada, 1980) that protect them from injurious effects. For example, K. Tanaka & Sugahara (1980) found that leaves with a high content of superoxide dimutase (SOD) were resistant to SO_2. Similarly, increased peroxidase activity that often results from fumigation of plants with SO_2 (Horsman & Wellburn, 1975; Keller, 1976; Keller *et al.*, 1976; Khan & Malhotra, 1982*b*) could be the result of either cellular activity to metabolize H_2O_2 produced by a free radical mechanism or the oxidation of absorbed SO_2. Ascorbic acid, another scavenger of free radicals, has also been reported to protect plants from SO_2 injury (Keller & Schwager, 1977; Grill *et al.*, 1979*a*).

The $^{35}SO_2$ absorbed by plant leaves does not remain fixed at the site of absorption but has a substantial degree of mobility (Jensen & Kozlowski, 1975; Garsed & Read, 1977*c*); such translocations occur from leaves to roots, from old leaves to young leaves, and from roots to surrounding medium. Plants can, therefore, act as sinks for atmospheric SO_2.

Absorbed SO_2 can also be utilized by plants in the reductive sulphur cycle. Ziegler (1975) stated that this reduction seemed to involve reactions similar to SO_4^{2-} reduction. As details of the reduction mechanisms have been well reported elsewhere (Schiff & Hodson, 1973; Hallgren, 1978), only recent work on this topic will be dealt with here. Sulphite reduction in plant leaves occurs mainly in the chloroplasts (Schiff & Hodson, 1973). Recently it has been shown that, in plants with C_4-type photosynthesis, the assimilation of sulphur is initiated only in the bundle-sheath chloroplasts; the mesophyll chloroplasts are inactive (Gerwick *et al.*, 1980).

Sulphur dioxide can specifically inhibit the activity of adenosine phosphosulphate (APS) sulphotransferase in the leaves of *Phaseolus vulgaris* (Kidney Bean) seedlings, but the inhibition can be reversed upon the removal of SO_2 (Wyss & Brunold, 1980). This is a key enzyme in the intermediary SO_4^{2-}-reduction metabolism, and is involved in SO_4^{2-} transfer from APS to a carrier (Car-SH) to form a carrier-S-SO_3^{-} complex. Other sulphur compounds (SO_4^{2-}, H_2S, and cysteine) also affect the activity of this enzyme (Wyss & Brunold, 1979, 1980). It is possible that the high SO_2 concentration used in the above studies affected the enzyme owing to excessive accumulation of SO_4^{2-} and other sulphur compounds.

Recently it has been suggested that both SO_3^{2-} and SO_4^{2-} are transported to the inner chloroplast membranes by phosphate translocators (Hampp & Ziegler, 1977) and that light stimulates this process (Ziegler & Hampp, 1977). The light-activated uptake and reduction appear to be controlled by photosystem-dependent electron transport, as inhibition of the photoelectron transport system completely blocked both the uptake and reduction of SO_4^{2-} (Ziegler & Hampp, 1977). Sulphate reduction was dependent on ATP produced during photophosphorylation, because uncoupling of photophosphorylation inhibited SO_4^{2-} uptake and its reduction. Sulphite, on the other hand, was metabolized directly, without any ATP requirement; in fact SO_3^{2-} uptake and reduction were stimulated by the uncoupler (Ziegler & Hampp, 1977).

As SO_3^{2-} can be utilized directly in sulphur metabolism, these results are of considerable significance in SO_2 metabolism—especially in view of the inhibitory effects of SO_2 on chloroplast photoelectron transport and photophosphorylation systems (to be discussed later). In leaves of *Lemna minor* (Lesser Duckweed), however, it has been shown that absorbed SO_2 was assimilated only after oxidation to SO_4^{2-} (Brunold & Erismann, 1976). In support of direct reduction of SO_3, Ziegler (1977a) showed that sulphur from SO_2 and SO_3^{2-} was incorporated in the chloroplast lamellae to a much greater extent than sulphur from SO_4^{2-}. She suggested that SO_3^{2-} was either directly incorporated into the sulphonic groups of the sulphoquinovosyl moiety of the sulpholipids as reported by Benson (1963), or was taken up at the binding sites in the lamellae (Schwenn *et al.*, 1976).

Accumulation of SH-containing amino acids following fumigation with SO_2 has been reported for a number of plant species (Ziegler, 1975; Grill *et al.*, 1979b, 1980; Malhotra and Sarkar, 1979); similar accumulations have also been noticed upon continuous application of SO_3^{2-} to the cultures of *Chlorella vulgaris* (Soldatini *et al.*, 1978). Photoreduction of SO_2 to H_2S and its subsequent release from plants was reported by Cormis (1968). Similar results were reported with $^{35}SO_2$ in *Spinacia oleracea* leaves and isolated chloroplasts (Silvius *et al.*, 1976). Reduction of SO_2 to H_2S could be catalysed by sulphite reductase, found in many plants (Schiff & Hodson, 1973). It is interesting to note that short-time fumigation of *Spinacia oleracea* plants with SO_2 caused a marked increase in free and masked thiol in the chloroplasts (Miszalski & Ziegler, 1979). Such an increase in the concentration of SH compounds can influence the natural balance of sulphydryl (SH)/disulphide (SS) level in the cells. Plants exposed to injurious concentrations of SO_2 emit considerable amounts of H_2S; a positive correlation between such emissions and SO_2 resistance was shown in a recent report by Sekiya *et al.* (1980).

Photosynthesis

The response of photosynthetic processes to SO_2 depends on the duration of exposure and the concentration of SO_2. Short-time exposure to low SO_2 concentrations generally stimulates photosynthesis in a number of plants. In most cases, however, high doses of SO_2, or continued exposures at even low concentrations, are very inhibitory to photosynthesis. The sensitivity of plants to SO_2 at low concentrations can nevertheless be influenced by environmental factors such as wind-speed, light, and humidity. Ashenden & Mansfield (1977) showed that 0.11 ppm SO_2 significantly reduced the growth of *Lolium perenne* (Ryegrass) only at high wind-speed; the lack of effect at lower wind-speed was due to increased boundary-layer resistance. In *Vicia faba*, a decrease in net photosynthesis occurred at SO_2 concentrations exceeding 0.035 ppm, and the decrease was influenced by boundary-layer resistance and light intensity (V. J. Black & Unsworth, 1979*a*, 1979*b*). Hällgren (1978) has reviewed physiological conditions that influence the SO_2 effect on the photosynthetic response of intact leaves.

A rapid decrease in net photosynthesis in leaves of *Helianthus annuus* (Sunflower) upon exposure to 1.5 ppm SO_2 led Furukawa *et al.* (1980*b*) to suggest that the chloroplasts were the primary site of attack. Biochemical reactions associated with photosynthesis have been studied in order to ascertain more specifically the sites and mechanisms of SO_2 action. Ziegler (1972) observed that SO_3^{2-} inhibited ribulose bisphosphate (RuBP) carboxylase in *Spinacia oleracea* chloroplasts. The kinetics of SO_3^{2-}-inhibition indicated competition between HCO_3^- and SO_3^{2-} at the CO_2-binding sites of the enzyme, which suggests that the concentration of CO_2 (or HCO_3^-) at the site of carboxylation influences the degree of SO_3^{2-} inhibition.

A similar type of competitive inhibition by SO_3^{2-} with respect to HCO_3^- has been observed with isolated preparations of RuBP carboxylase from a lichen, *Pseudevernia furfuracea* (Ziegler, 1977*b*). Recently, however, Gezeilus & Hällgren (1980) showed that the SO_3^{2-}-inhibition of RuBP carboxylase from *Pinus sylvestris* (Scots Pine) and *Spinacia oleracea* was non-competitive with respect to HCO_3^-, and the nature of the inhibition was not affected by the presence of SO_3^{2-} during the activation. It is interesting to note that the enzyme from *Pinus sylvestris* was inhibited similarly by 10 mM SO_3^{2-} or SO_4^{2-}. As SO_4^{2-} was also shown to inhibit *Spinacia oleracea* RuBP carboxylase non-competitively with respect to HCO_3^- (Trown, 1965), the possible presence of SO_3^{2-}-oxidizing systems in crude enzyme extracts of *Pinus sylvestris* cannot be overruled. Gezeilus & Hallgren (1980) have minimized such an interference by assaying in N_2 atmosphere. Using purified preparations of RuBP carboxylase from *Pinus banksiana* (Jack Pine), Khan & Malhotra (1982*a*) found a competitive type of inhibition by SO_3^{2-}, with a $Ki(SO_3^{2-})$ value of 5 mM; the enzyme was also inhibited by SO_4^{2-}, but to a much lesser extent than by SO_3^{2-}.

It must be pointed out that crude extracts and isolated chloroplasts of *Pinus banksiana* needles contain very active SO_3^{2-}-oxidizing systems (Khan & Malhotra, ms*a*). It is therefore necessary that purified preparations of RuDP carboxylase be used for assessing the nature of SO_3^{2-} inhibition. Reduction in the activity of RuBP carboxylase was observed in *Alnus crispa* (Green Alder) and *Betula papyrifera* (Paper Birch) seedlings upon fumigation with 0.34 and 0.51 ppm SO_2 (Khan & Malhotra, ms*b*). Reduction in the activity of RuBP carboxylase upon exposure to gaseous SO_2 has also been reported in leaves of other plants (Horsman & Wellburn, 1975; Miszalski & Ziegler, 1980).

Sulphite also affects phosphoenolpyruvate (PEP) carboxylase (Ziegler, 1973*a*; Mukerji & Yang, 1974), which is involved in the C_4 pathway of photosynthesis. In *Zea mays* (Ziegler, 1973*a*), the SO_3^{2-} inhibition was competitive with respect to HCO_3^-, being similar to inhibition of RuBP carboxylase; however, the inhibitor constant Ki of SO_3^{2-} for PEP carboxylase, was much higher than for RuBP carboxylase. As PEP carboxylase has much higher affinity for CO_2 (Ziegler, 1973*a*) than has RuBP carboxylase, the replacement of CO_2 by SO_3^{2-} at the carboxylation site would be more difficult in the PEP carboxylase-catalysed reaction than in the RuBP carboxylase reaction.

The effect of SO_3^{2-} on PEP carboxylase from *Spinacia oleracea* (Mukerji & Yang, 1974) appeared to be different from that of the *Zea mays* enzyme. The *Spinacia oleracea* enzyme was stimulated by SO_3^{2-} at a low concentration (0.5 mM) and was inhibited at a high concentration (5 mM). Sulphite inhibited the enzyme activity, and the inhibition was of a mixed type with respect to HCO_3^-. Kinetic analysis of the *Spinacia oleracea* enzyme suggests that SO_3^{2-} affected PEP carboxylase activity, not by interfering with the substrate binding sites, but by unspecific binding with the enzyme protein (Mukerji &

Yang, 1974). As SO_3^{2-} has been shown to bind rapidly with PEP *in vitro* (Lehmann & Benson, 1964), such a reaction during PEP carboxylase assay would limit the availability of PEP for carboxylation. In the presence of PEP carboxylase, however, a SO_3^{2-} reaction with PEP did not appear to occur, because SO_3^{2-} either had no effect or acted non-competitively with respect to PEP (Ziegler, 1973*a*; Mukerji & Yang, 1974). Bisulphite compounds have also been shown to inhibit PEP carboxylase reaction (Osmond & Avadhani, 1970; Mukerji & Yang, 1974). The extent to which PEP carboxylase activity is influenced by gaseous SO_2 has not been determined.

In addition to influencing the carboxylation reactions, SO_2 can affect photosynthesis by attacking photosynthetic electron transport and photophosphorylation reactions. Chloroplasts isolated from needles of *Pinus contorta* (Lodgepole Pine) treated with varying concentrations of aqueous SO_2 showed that, at a low concentration (50 ppm), SO_2 stimulated Hill (evaluation of oxygen due to photolysis of water and the transfer of electrons to an electron acceptor, e.g. ferricyanide and dyes are generally used as artificial electron acceptors) reaction activity, but this activity was completely inhibited at high concentrations (500–1000 ppm) (Malhotra, 1976). The chloroplasts isolated from old tissues were more sensitive to SO_2 than were those from young, actively growing tissues. A decrease in the Hill reaction activity was accompanied by swelling and disintegration of chloroplast membranes. Such alterations in the membranes can cause disorganization of the two photosystems. Photosystems I and II are both localized in the membranes of chloroplasts, and have been separated by density-gradient centrifugation of digitonin-treated chloroplasts (Boardman, 1968).

Recently, Shimazaki & Sugahara (1980*a*, 1980*b*) studied in detail the effect of gaseous SO_2 on chloroplast photosystems in *Spinacea oleracea*. Fumigation with SO_2 at 1 and 2 ppm for 1 hour produced no effect on 2,6-dichloroindophenol (DCIP) photoreduction (Hill reaction); however, there was rapid inhibition following longer exposures. Shimazaki & Sugahara investigated the site of SO_2 attack in the electron transport systems by studying both photosystems.

Electron transport of both the whole chain and Photosystem II was inhibited to the same magnitude by SO_2, but SO_2 did not inhibit electron-flow from reduced DCIP to nicotinamide adenine dinucleotide phosphate (NADP) under uncoupled conditions, which suggests that the site of SO_2 action in the photosystems was associated with Photosystem II and not Photosystem I. A similar effect of SO_2 was observed in photosystems of *Lactuca sativa* (Garden Lettuce) chloroplasts (Shimazaki & Sugahara, 1980*b*). The work with chloroplasts isolated from SO_2-fumigated leaves of *Lactuca sativa* (Shimazaki & Sugahara, 1980*b*) demonstrated that the site of SO_2 action was located closer to the oxidizing side rather than the reducing side of Photosystem II. This was supported by the observation that the addition of an artificial electron-donor, diphenylcarbazide (DCP), did not change the rate of DCIP reduction in Photosystem II.

The work of Shimazaki & Sugahara (1980b) also suggests that SO_2 did not inactivate the electron-flow from the reductant to the primary electron acceptor (Q) of Photosystem II. Time-course analysis of fluorescence intensity in SO_2-treated plants indicated that SO_2 inhibited the accumulation of reduced Q. Furthermore, the addition of 3-(3′,4′dichlorophenyl)-1,1-dimethyl urea (DCMU), an inhibitor acting on the reducing side of Photosystem II (Bishop, 1958), caused a rapid increase in fluorescence in SO_2-inhibited chloroplasts, which suggests that Q was in the oxidized state. This could happen because of SO_2-inactivation of either the primary electron donor or the reaction centre itself in the electron transport chain.

The above inhibitory effects of gaseous SO_2 on the photosystems are different from those reported from earlier studies in which treatment of isolated chloroplasts with solutions of SO_3^{2-} either produced no overall effect on electron transfer (Asada *et al.*, 1965) or else stimulated a non-cyclic-type of electron transfer (Libera *et al.*, 1973). The effect of gaseous SO_2 seems to be specific and not associated with the acidity-related decrease in Photosystem II activity, as a decrease in Photosystem II activity due to low pH could be restored by adding electron donors of Photosystem II but not in chloroplasts from SO_2-treated plants (Shimazaki & Sugahara, 1980b). Inhibition of Photosystem II activity by exposure to SO_2 was accompanied by a similar inhibition in non-cyclic photophosphorylation, but not in cyclic photophosphorylation (Shimazaki & Sugahara, 1980a). On the other hand, treatment of isolated chloroplasts with solutions of SO_3^{2-}, HSO_3^{-}, and SO_2, inhibited both cyclic and non-cyclic photophosphorylations (Asada *et al.*, 1965; Libera *et al.*, 1973; Silvius *et al.*, 1975).

The differences between the *in vivo* effects of SO_2 on both photoelectron transport and phosphorylations, and the *in vitro* effects of treatment of isolated chloroplasts with aqueous SO_2 (HCO_3^{-}, SO_3^{2-}, and SO_2), are difficult to reconcile. Shimazaki & Sugahara (1980a) have attributed such differences to production of O_2^{-} and other radicals during photooxidation of SO_3^{2-}. As SO_4^{2-} is formed by the oxidation of SO_3^{2-}, it is possible that the effects observed *in vitro* are the effects of free radicals and SO_4^{2-}. It has been shown that SO_4^{2-} irreversibly inhibits both cyclic and non-cyclic photophosphorylations (Ryrie & Jagendorf, 1971).

Pigments

As chlorophyll and other plant pigments are necessary in harnessing light-energy by Photosystems I and II, the effect of SO_2 on these pigments would greatly influence the photosynthetic ability of plants. Rao & LeBlanc (1965) found that destruction of chlorophyll occurred in lichens following exposure to large doses (5 ppm for 24 hours) of gaseous SO_2. At this high concentration, chlorophyll molecules were degraded to phaeophytin and Mg^{2+}. A similar conversion of chlorophyll to phaeophytin can occur with acids or acidic substances. In this process Mg^{2+} in the chlorophyll molecule is replaced by two atoms of hydrogen, thereby changing the light-spectrum characteristic of the chlorophyll molecules.

A decrease in chlorophyll-content has often been suggested as an indicator of air pollution (mainly SO_2) injury. In sensitive lichens, chronic exposure to even a low concentration (0.01 ppm) of SO_2 resulted in a loss of chlorophyll (Gilbert, 1968). In *Evernia mesomorpha*, control fumigation with a low level of SO_2 caused a gradual decline in the chlorophyll content, which was accompanied by a decrease in photosynthesis; the effects on both chlorophyll destruction and photosynthesis became more rapid and pronounced at 0.34 ppm SO_2 concentration (Malhotra & Khan, unpublished results).

Sulphur dioxide can influence chlorophyll by various mechanisms. Malhotra (1977) showed that treatment of *Pinus contorta* needles with aqueous SO_2 markedly affected the chlorophyll content of the foliage, and that this was not due to increased acidity alone. As, *in vivo*, chlorophyll is stabilized by its organization as a complex with proteins, it is possible that SO_2 first attacks this complex before the actual breakdown of chlorophyll occurs. In *Pinus*, low aqueous SO_2 concentrations had very little effect on either chlorophyll *a* or *b*, but at high concentration chlorophyll *a* appeared to be much more sensitive to aqueous SO_2 than was chlorophyll *b*. Increased destruction of chlorophyll *a* was accompanied by a parallel increase in phaeophytin *a* and loss of Mg^{2+} from the tissues.

As the destruction of chlorophyll in pine needles was markedly greater in the presence of aqueous SO_2 than in the presence of HCl solutions of the same pH, it was suggested that SO_2 destroyed chlorophyll as a result of its strong redox properties. In needles of *Pinus contorta*, a decline in chlorophyll *b* content in the presence of SO_2 did not produce a corresponding increase in phaeophytin *b*, which suggests a different mechanism in chlorophyll *b* breakdown. In fact, an increase in chlorophyllide *b* content following SO_2 treatment suggested that chlorophyll *b* breakdown was probably a result of splitting of the phytol chain by chlorophyllase. The maximum increase in tissue chlorophyllide *b* content was observed at 50 ppm aqueous SO_2 concentration—the same concentration that produced the maximum stimulation in chlorophyllase activity.

Rapid *in vitro* chlorophyll destruction can also be caused by free radicals produced during the oxidation of HSO_3^{-}-catalysed decomposition of linoleic acid hydroperoxide (Peiser & Yang, 1977, 1978). Reently, Shimazaki *et al.* (1980) presented evidence that SO_2 fumigation of leaves increases the formation of O_2^{-} in chloroplasts that in turn destroys chlorophylls. Superoxide radical has been shown to influence chlorophyll at very low concentrations

(10^{-8} to 10^{-7} M) (Asada *et al.*, 1977). In *Spinacia oleracea* leaves, gaseous SO_2 destroyed chlorophyll *a* more rapidly than chlorophyll *b*, but the loss of chlorophyll *a* was not accompanied by a corresponding increase in phaeophytin *a* (Shimazaki *et al.*, 1980). As scavengers of free radicals inhibited chlorophyll breakdown in *Spinacia oleracea* leaves, it was suggested that SO_2 destroys chlorophyll mainly by a free-radical oxidation. This was further supported by the observation that chlorophyll *a* breakdown was inhibited by superoxide dismutase. Sulphur dioxide inhibits the superoxide dismutase activity in the fumigated tissues (Shimazaki *et al.*, 1980). Furthermore, accumulation of malonaldehyde, a lipid peroxidation product, and a decrease in chlorophyll *a* in SO_2-fumigated *Spinacia oleracea* leaves, was related to the free-radical oxidation of chlorophyll.

Chlorophylls and other pigments in the chloroplasts are stabilized by forming complexes with proteins, but little is known about the effect of SO_2 on these complexes. Recently, Sugahara *et al.* (1980) showed that, *in vitro*, water-soluble protein complexes of chlorophyll and chlorophyllide were stable and were not destroyed by even 40 mM SO_3^{2-}. The photoconversion of the dark form of the chlorophyll *a* and chlorophyllide *a* protein complex (CP 668) to the illuminated form (CP 743) was, however, inhibited by SO_3^{2-}. The inhibition was apparently due to irreversible denaturation of the protein component in the pigment protein complex, probably caused by destruction of disulphide bonds.

Respiration

Photorespiration has been found to be inhibited by SO_2 (Ziegler, 1975). Exposure of *Lolium perenne* to SO_2 (0.15 ppm) caused an inhibition in glycine and serine synthesis, even though CO_2 fixation was stimulated; this suggests inhibition of photorespiration (Koziol & Cowling, 1978). A similar reduction in these amino acids following exposure to SO_2 was reported by Tanaka *et al.* (1972*b*). Glycollate oxidase, an important enzyme for the synthesis of glycine and serine, was inhibited by low concentrations of SO_3^{2-} *in vitro* (Zelitch, 1957; Paul & Bassham, 1978; Khan & Malhotra, 1982*a*) and by gaseous SO_2 (Khan & Malhotra, 1982*a*). In leaves of *Nicotiana tabacum* (Tobacco), however, exposure to a high SO_2 concentration (1.3 ppm for 18 hours) induced an increased synthesis of glycollate oxidase (Soldatini & Ziegler, 1979).

It has been suggested that a decrease in photorespiration as a result of SO_2 or SO_3^{2-} exposure is due to the formation of glyoxylate bisulphite, which is a potent inhibitor of glycollate oxidase (Zelitch, 1957). Glyoxylate bisulphite was found to accumulate in the leaves of *Oryza sativa* (Rice) plants exposed to high concentrations of SO_2 (Tanaka *et al.*, 1972*a*). Similarly, *Pisum sativum* (Garden Pea) exposed to high SO_2 concentration, produced toxic bisulphite compounds of glyceraldelyde, α-ketoglutarate, pyruvate, and oxaloacetate (Jiracek *et al.*, 1972). The high concentration of SO_2 used in these studies leaves doubt as to the formation of such compounds in plants exposed to the more common, low levels of SO_2 experienced under field conditions.

Unlike photorespiration, dark respiration either showed no response or was stimulated by SO_2 (V. J. Black & Unsworth, 1979*b*; Furukawa *et al.*, 1980*b*). The significance of these results at the biochemical level was not ascertained. The *in vitro* addition of SO_3^{2-} to plant mitochondrial preparations, however, inhibited the formation of adenosine triphosphate (ATP) (Ballantyne, 1973). It is possible that, due to its strong reducing nature, SO_3^{2-} caused reduction of electron transport components and uncoupling, which resulted in subsequent decreased ATP synthesis.

Sulphur dioxide and its dissolved reactive 'species' can affect a variety of other cellular metabolites and enzyme systems through either non-specific or specific mechanisms. At high concentrations, SO_3^{2-} reacts with a number of metabolites (Mudd, 1975*a*; Petering & Shih, 1975; Malhotra & Hocking, 1976), but at the low concentrations that are generally present under field conditions, such reactions will be of rare occurrence. The following sections discuss the effects of low and medium concentrations of SO_2 on enzymes and cellular metabolites.

Amino Acids and Proteins

In a number of plant species, free amino acid content increases following SO_2 fumigation (Arndt, 1970; Jäger & Grill, 1975; Malhotra & Sarkar, 1979). However, in sulphur-deficient *Lolium perenne* (Ryegrass), SO_2 fumigation caused a decrease in total amino acids and amines, asparagine, and glutamine (Cowling & Bristow, 1979). In general, SO_2 fumigation results in an increase specifically in sulphur-containing amino acids (Ziegler, 1975).

Increased contents of glycine, alanine, thionine, lysine, and methionine, in needles of *Pinus banksiana* treated with SO_2, were thought to be due mainly to increased breakdown of needle proteins (Malhotra & Sarkar, 1979). Further work showed that SO_2 decreased the soluble cytoplasmic and chloroplast protein contents of the needles, and that the decrease was higher in the chloroplast than in the soluble cytoplasmic fraction (Khan & Malhotra, msc). A decrease in the total protein content upon SO_2 fumigation has been reported for a number of plants (Fischer, 1971; Godzik & Linskens, 1974; Constantinidou & Kozlowski, 1979). Such a decrease could be attributed to breakdown of the existing proteins and to reduced *de novo* synthesis. Exposure of an epiphytic lichen, *Evernia mesomorpha*, to 0.1 ppm SO_2 for 2 days, resulted in a considerable reduction in protein biosynthesis, and this effect became more severe after longer exposures. Fumigation for 3 days at 0.34

ppm SO_2 caused a very pronounced inhibition that was irreversible in clean air (Malhotra & Khan, ms). Similarly, SO_2 (0.34 ppm) markedly inhibited *de novo* biosynthesis of both cytoplasmic and chloroplast proteins in *Pinus banksiana* (Jack Pine) the effect being most marked on the chloroplast proteins (Khan & Malhotra, msc). These results indicate that membrane-associated processes are more sensitive to SO_2 than are the cytoplasmic-processes.

In *Pisum sativum*, SO_2 fumigation at 0.3 ppm concentration for 18 days resulted in a marked increase in free and bound polyamines such as putrescine and supermidine (Priebe *et al.*, 1978). Polyamines, which are metabolic products derived from amino-acids, play important roles in nucleic acid metabolism and in the regulation of cellular pH (Cohen, 1971). The increased formation of polyamines in *Pisum sativum* leaves upon exposure to SO_2, was accompanied by a marked increase in the contents of arginine and ornithine—the precursors of putrescine and supermidine (Preibe *et al.*, 1978). The arginine content of *Pinus banksiana* needles also increased upon fumigation with SO_2 (Malhotra & Sarkar, 1979); however, its relevance to polyamine synthesis was not studied. The accumulation of these metabolic products following SO_2 fumigation led Priebe *et al.* (1978) to suggest that increased polyamine synthesis was a regulatory process for binding excessive H^+ that was produced in the tissues as a result of SO_2 absorption.

It has been suggested that polyamines, whose basicity is comparable to that of NaOH, could form polyvalent cations by binding H^+ and thus act as buffering compounds in the cells (Priebe *et al.*, 1978). In seedlings of *Hordeum vulgare* (Barley), acidic conditions were shown to activate the enzymes arginine decarboxylase and N-carbamylputrescine amidohydrolase, which are involved in the synthesis of putrescine (T. A. Smith & Sinclair, 1967).

Several enzymes involved in amino-acid metabolism have also been shown to be affected by SO_2. In *Pisum sativum*, fumigation with SO_2 affected gluatamate dehydrogenase activity by stimulating reductive amination on one hand and inhibiting oxidative deamination on the other (Pahlich *et al.*, 1972). These effects were attributed to changes in izoenzyme pattern (Pahlich *et al.*, 1972) and to enzyme kinetics (Pahlich, 1971). Fumigation of *Pisum sativum* seedlings with SO_2 inhibited the mitochondrial glutamate oxaloacetate transaminase; the cytoplasmic enzyme showed no effect (Pahlich, 1973). The inhibition of the mitochondrial enzyme by SO_3^{2-}/HSO_3^- was of a mixed type. On the other hand, the transaminase(s) activity (glutamate-oxaloacetate and glutamate-pyruvate) of *Pisum sativum* seedlings was slightly stimulated by 0.2 to 2 ppm SO_2 fumigation (Horsman & Wellburn, 1975). Rabe & Kreeb (1980), however, showed that transaminases (alanine and aspartate) were stimulated only by low SO_2 concentrations; at high concentrations they were inhibited.

Lipids and Fatty Acids

Lipids are important constituents of biological membranes. In chloroplasts, glycerolipids constitute about 50% by weight of thylakoid membranes (James & Nichols, 1966). A major portion of these glycerolipids is present as monogalactosyl diglyceride, digalactosyl diglyceride, and sulphoquinovosyl diglyceride. In *Pinus contorta*, SO_2 caused a marked reduction in the concentration and composition of these glycolipids (Khan & Malhotra, 1977) in both young and fully developed needles. As galactolipids have been shown to be involved in the structure and function of chloroplasts (Shaw *et al.*, 1976), it was suggested that structural alterations in *Pinus contorta* chloroplasts (Malhotra, 1976) were due to changes in galactolipid concentration and composition (Khan & Malhotra, 1977). Isolated chloroplast preparations of *Pinus banksiana* needles contained enzyme systems responsible for the biosynthesis of these membrane lipids (Khan & Malhotra, 1978).

Lipid biosynthesis in the epiphytic lichen *Evernia mesomorpha* has been shown to be affected by SO_2 (Malhotra & Khan, ms). Inhibition caused by low-level SO_2 exposures (0.1 ppm) was followed by complete recovery 8 days after termination of fumigation. At 0.34 ppm SO_2, biosynthesis was inhibited even after 1 day of fumigation, and on day 3 this inhibition became very severe; there was very little recovery in biosynthetic activity upon removal of SO_2.

Decreases in the lipid content following SO_2 exposure may be brought about by either reduced synthesis, increased lipase activity, peroxidation of fatty-acid chains, or a combination of the above. Malhotra & Khan (1978) showed that exposure of *Pinus banksiana* and *P. contorta* seedlings to SO_2 caused a marked inhibition in the *de novo* synthesis of phospho-, glyco-, and neutral lipids. The magnitude of inhibition was dependent on the concentration and duration of exposure to SO_2. The effect was, however, transitory in nature, as the inhibition was partially or completely reversed upon removal of plants from the SO_2 atmosphere. As no accumulation of free fatty acids occurred upon SO_2 treatment, it was suggested that SO_2 did not cause stimulation of lipolytic activity but inhibited the synthetic and acylation processes.

Treatment of *Pinus banksiana* needles with SO_2 produced a marked reduction in the content of linolenic acid and an increase in the content of palmitic acid; the effect was more pronounced in young needles than in fully-developed ones (Khan & Malhotra, 1977). As SO_2 treatment in both types of needles caused reduction in linolenic acid content and an increase in palmitic acid content, it is suggested that SO_2 inhibited both the elongation and desaturation processes. The decline in the linolenic acid content may have occurred by SO_2-stimulated free-radical peroxidation reactions, indicated by SO_2-related malonaldehyde accumulation (Khan & Malhotra, 1977). The formation of malonaldehyde also occurs in leaves of other plants injured by SO_2 (Peiser & Yang, 1979; Shimazaki *et al.*, 1980).

Carbohydrates and Organic Acids

Plants exposed to SO_2 exhibit increasing amounts of soluble sugars (Khan & Malhotra, 1977; Koziol & Jordan, 1978; Malhotra & Sarkar, 1979). In *Pinus banksiana*, SO_2 fumigation (0.34 and 0.51 ppm) increased the content of the reducing sugars and reduced that of the non-reducing sugars (Malhotra & Sarkar, 1979). It was suggested that the increase was due to a breakdown of polysaccharides rich in reducing sugars. Koziol & Jordan (1978) showed that SO_2 exposure of *Phaseolus vulgaris* seedlings caused a reduction in starch content. Reduction in non-structural total carbohydrates has also been reported following SO_2 exposure of *Ulmus americana* (American Elm) seedlings (Constantinidou & Kozlowski, 1979). Recently, Bucher-Wallin *et al.* (1979) found significant increases in the activities of several glycosidases in the foliage of clonal forest trees following exposure to low levels of SO_2, which suggests that the enzymes were involved in the breakdown of polysaccharides. As polyhydric sugars are known to act as scavengers of free radicals (Asada, 1980), this could act as a mechanism to help to cope with increasing SO_2 pollution.

Sulphur dioxide also affects a number of enzymes and metabolites involved in organic acid metabolism. Organic acids act as intermediates of a number of metabolic end-products and help to maintain cellular pH. Any changes in their metabolism would therefore have an influence on plant growth and yield. In *Pinus banksiana* needles, SO_2 decreased the content of two major organic acids, namely quinic acid and shikimic acid, and increased the content of syringic acid, a minor component (Sarkar & Malhotra, 1979). Reduction of ^{14}C incorporation into organic acids following fumigation of *Pinus thunbergii* (Japanese Black Pine) and *P. densiflora* (Japanese Red Pine) with SO_2 has also been reported (Ishizaki & Hasegawa, 1974). Such changes may be brought about by modification of either the structure or the kinetics of the enzymes involved.

In *Spinacia oleracea*, SO_3^{2-} inhibited NADP-dependent glyceraldehyde-3-phosphate dehydrogenase (Ziegler *et al.*, 1976); however, fumigation with SO_2 (0.7 ppm for 1 hour) caused a significant stimulation of the light-activated enzyme. The stimulation was thought to be due to generation of increased SH groups during fumigation (Miszalski & Ziegler, 1979). Low concentrations of SO_2 similarly stimulated the activities of glucose-6-phosphate and isocitrate dehydrogenases, but high SO_2 concentrations appreciably inhibited the two enzymes (Rabe & Kreeb, 1980). A considerable decline has also been reported in nicotinamide adenine dinucleotide (NAD)-dependent malate dehydrogenase activity of *Pinus banksiana* seedlings exposed to 0.34 ppm SO_2; isoenzyme analysis showed one less isoenzyme in the SO_2-treated plants compared with the control (Sarker & Malhotra, 1979). Decreases in NAD and NADP-dependent malate dehydrogenase activity by SO_3^{2-} were shown to be due to the SO_3^{2-} effect on the kinetics and multi-molecular forms of the enzymes (Ziegler, 1974).

OZONE

Stomatal Response

Stomatal control could be responsible for the differential resistances to ozone (O_3) by seedlings (Treshow, 1970). Humidity, which is highly conducive to increase in stomatal opening, engendered considerable ozone injury to plants (Otto & Daines, 1969) though little or no such injury occurred when stomata were closed (Heath, 1975). In *Phaseolus vulgaris*, however, maximum sensitivity to O_3 was not related to stomatal number or their resistance to the gas at either surface of the leaf (Evans & Ting, 1974*a*).

Abscisic acid is known to effect stomatal responses, and its application reduces O_3 injury to plants (Fletcher *et al.*, 1972; Jeong *et al.*, 1980). Fumigation with O_3 increased ABA content in *Oryza sativa*; the plants resistant to O_3 had higher ABA content than the sensitive ones (Jeong *et al.*, 1980).

Absorption and Permeability Effects

Moist surfaces in the mesophyll tissues provide the media in which gaseous O_3 can dissolve; however, little is known about O_3 diffusion and its chemical transformation within the tissues. It has been suggested that O_3 diffuses passively, due to a concentration gradient similar to that of CO_2 (Heath, 1975). The concentration of O_3 in the plant tissues is affected by its solubility, rate of decomposition, and the pH at the site of absorption.

Similarly to SO_2, O_3 can also give rise to the superoxide radical (O_2^-), which can produce other radicals such as $OH^{\cdot}$, $^1O_2$, and H_2O_2. These radicals can oxidize various cellular metabolites (Asada, 1980). A number of membrane constituents such as SH groups, amino acids, proteins, and unsaturated fatty acids, are affected by O_3 (Heath, 1975), probably as a result of free-radical attack.

Permeability changes in O_3-fumigated tissues are thought to be due to leaky plasmalemma; these include changes in permeability to water (Evans & Ting, 1973), glucose (Dugger & Palmer, 1969; Perchorowicz & Ting, 1974), and ions (Evans & Ting, 1974*b*; Heath & Frederick, 1979). Permeability of mitochondrial (T. T. Lee, 1968) and chloroplast membranes (Nobel & Wang, 1973) is also influenced by O_3. Sutton & Ting (1977) demonstrated that O_3-induced membrane injury can be repaired by energy-dependent processes.

Photosynthesis

Fumigation of plants with O_3 inhibited photosynthesis, though this was reversed about 24 hours after the termination of exposure (Hill & Littlefield, 1969; Pell & Brennan, 1973). The direct effect of O_3 on carboxylases involved in photosynthetic CO_2 fixations are not well known. Thomson *et al.* (1966) speculated that the granulation of chloroplast stroma, seen in electron micrographs of O_3-treated tissues, was due to oxidation of SH groups in the Fraction 1 protein (RuBP carboxylase). In *Oryza sativa* (Rice) plants, fumigation with O_3 reduced the activity of RuBP carboxylase in both young and old leaves 12–14 hours after fumigation, but 48 hours after fumigation the enzyme activity had recovered to some extent only in the young leaves (Nakamura & Saka, 1978).

Treatment of isolated *Spinacia oleracea* chloroplast suspensions with O_3 inhibited electron transport in both photosystems; Photosystem I, however, was more sensitive than Photosystem II (Coulson & Heath, 1974). The treatment inhibited photophosphorylation by reducing electron transport and not by uncoupling. Chang & Heggestad (1974) have also demonstrated the effect of O_3 on Photosystem II in *Spinacia oleracea* chloroplasts. Inhibition by O_3 of Photosystem II was proportional to the extent of visible injury in relatively resistant (Bel-B) and sensitive (Bel-W$_3$) *Nicotiana tabacum* (Tobacco) leaves (Rhoads & Brennan, 1978). Ozone treatment of isolated chloroplasts from the two varieties, however, inhibited their electron transport activities equally, which suggests that either the O_3 effect was at sites other than the electron transport systems or the relatively resistant variety was more capable of repairing the injury than the sensitive one.

Schreiber *et al.* (1978) showed that O_3 (0.3 ppm) altered the fluorescence characteristics in *Phaseolus vulgaris* long before any visible symptoms of injury appeared. Their data also suggested that a low concentration of O_3 for a long exposure was more injurious than a high concentration for a short exposure. On the basis of specific change in fluorescence, they suggested that the site of O_3 action was within the photosynthetic apparatus. The initial damage was thought to be at the donor site of Photosystem II (H_2O-splitting enzyme system) and not directly on its reaction centre. Increasing O_3 exposures also resulted in the inhibition of electron transport from Photosystem II to Photosystem I.

In addition to influencing the activity of the electron transport system, O_3 affected the chlorophyll content of the treated plants (Leffler & Cherry, 1974). In *Phaseolus vulgaris*, a high correlation was found between chlorophyll loss and visible necrosis (Knudson *et al.*, 1977), and the ratio of Chlorophyll *a* to Chlorophyll *b* declined with increasing O_3 injury. Such changes were due either to more O_3 susceptibility of chlorophyll *a* than chlorophyll *b*, or to their altered biosynthesis. Beckerson & Hofstra (1979*c*) observed a decrease in chlorophyll *a* and *b* following exposure of *Phaseolus vulgaris* to O_3.

Respiration

The effect of O_3 on respiration is variable; it can either stimulate (Todd, 1958; Dugger & Palmer, 1969; Barnes, 1972) or inhibit (MacDowell, 1965) plant respiration. Ozone exposure inhibited respiration and phosphorylation in *Nicotiana tabacum* leaf mitochondria (T. T. Lee, 1967). In *Phaseolus vulgaris*, there was no effect on respiration immediatley after O_3 exposure, but significant stimulation occurred within 24 hours. The content of ATP and total adenylate, however, increased immediately following O_3 exposure (Pell & Brennan, 1973).

The effect of O_3 on photorespiration and related metabolic processes is not known at the present time.

Amino Acids and Protein Metabolism

Ozone can cause either an increase or a decrease in amino-acid content of plants (Ting & Mukerji, 1971; Tomlinson & Rich, 1967; Tingey *et al.*, 1973). Using $^{14}CO_2$, Wilkinson & Barnes (1973) demonstrated an increase in the synthesis of alanine and serine after exposure of *Pinus strobus* (White Pine) and *P. taeda* (Loblolly Pine) to O_3. Similarly, the protein content of *Glycine max* leaves increased 24 hours after exposure to O_3 (Tingey *et al.*, 1973), which suggests an increase in the biosynthesis of amino acids and proteins. On the other hand, the protein content of other plants was markedly reduced by O_3 treatment (Ting & Mukerji, 1971; Constantinidou & Kozlowski, 1979), thereby suggesting either a breakdown of the existing proteins or no synthesis of the new ones.

Chang (1971) reported that in *Phaseolus vulgaris* leaves, O_3 can cause dissociation of the chloroplast polysomes but not of the cytoplasmic polysomes. It has been suggested that such changes were induced by desiccation brought about by O_3 (Heath, 1975).

Ozone can also affect nitrogen metabolism by inhibiting nitrate reductase in *Glycine max* leaves; the inhibition of the enzyme, however, occurred only under *in vivo* and not *in vitro* conditions (Tingey *et al.*, 1973). This was explained as inhibition in the *in vivo* formation of reduced NADP (required for the enzyme activity) and not as a direct effect on the enzyme. In *Zea mays* (Maize) and *Glycine max* leaves, O_3 treatment inhibited the activities of both nitrite and nitrate reductases (Leffler & Cherry, 1974). Nitrite reductase, a chloroplast enzyme, was inhibited more severely by O_3 than was the cytoplasmic nitrate reductase.

Fatty Acid and Lipid Metabolism

Treshow *et al.* (1969) reported that exposure to O_3 resulted in a loss of pigmentation and neutral lipids in the fungus *Colletotrichum lindemuthianum*, though its fatty acid composition remained unaffected. The response in the lipids was attributed to an inhibitory effect on their biosynthesis, probably brought about as a result of SH oxidation. On the other hand, recent work with two sensitive cultivars of *Nicotiana tabacum* showed that fumigations of plants with O_3 (0.25 or 0.30 ppm for 6 hours) significantly increased the lipid content of the leaves but caused a decrease in triglyceride fatty acids (Trevathan *et al.*, 1979). It was speculated that the increase in lipids was an injury response, because rust infection causes a similar accumulation of lipids (Lösel, 1978). Variable effects of O_3 have been reported on fatty acid content in different plant species (Tomlinson & Rich, 1969, 1970, 1971; Swanson *et al.*, 1973). In *Nicotiana tabacum* leaves, O_3 caused a reduction in all fatty acids, the largest decreases being in palmitic (16 : 0) followed by linolenic (18 : 3) acids (Tomlinson & Rich, 1969). Such a reduction could occur by inhibition of fatty acid synthesizing and desaturating of enzymes.

Ozone can also affect polyunsaturated fatty acids by oxidative mechanisms. Such oxidations can change the properties of the membranes. Malonaldehyde, an oxidation product of unsaturated fatty acids, is formed upon treatment of unsaturated lipids with O_3 (Mudd *et al.*, 1971b); similarly, increase in malonaldehyde was accompanied by a decrease in unsaturated fatty acid (Frederick & Heath, 1970). Increased production of malonaldehyde has also been observed in leaves of *Oryza sativa* (Nakamura & Saka, 1978) and *Phaseolus vulgaris* (Tomlinson & Rich, 1970) after exposure to O_3.

The direct effect of O_3 on the lipid biosynthesis of isolated chloroplasts has been reported by Mudd *et al.* (1971a, 1971b). Ozone impaired the ability of chloroplasts to metabolize [1-^{14}C] acetate or acetyl-CoA in lipid biosynthesis. This was attributed to oxidation of essential SH groups (Mudd *et al.*, 1971a). The biosynthesis of galactolipids in chloroplasts was similarly inhibited by O_3 and by other agents that bind SH groups (Mudd *et al.*, 1971b). Ozone treatment, however, did not change the fatty acid composition, which led Mudd *et al.* (1971 a) to suggest that fatty acids of the chloroplast membranes were not very accessible to oxidation by O_3.

Ozone-treated plants also exhibit a decrease in free sterols (Tomlinson & Rich, 1971, 1973; Spotts *et al.*, 1975; Menser *et al.*, 1977; Trevathan *et al.*, 1979) and an increase in sterol glycosides and esterified sterol glycosides (Tomlinson & Rich, 1971). These changes may affect membrane permeability.

Carbohydrates and Related Metabolism

Ozone can either increase (Tingey *et al.*, 1973, 1976a) or decrease (P. R. Miller *et al.*, 1969; Constantinidou & Kozlowski, 1979) the level of soluble sugars and carbohydrates in the leaves of treated plants. Tingey *et al.* (1973) showed that in *Glycine max* leaves, a single acute exposure of O_3 (0.49 ppm) caused a significant initial decrease of reducing sugars, which was followed by a subsequent increase. The starch content did not change, however—which suggests that the initial decrease in soluble sugars could have resulted from a depression in the photosynthetic rate, and that the subsequent increase could be due to reduced sugar utilization and/or reduced sugar translocation for carbohydrate synthesis. In *Pinus ponderosa* (Ponderosa Pine), O_3 exposure resulted in an increase in the content of soluble sugars, starch, and phenols, in the foliage, and a decrease in their contents in the roots (Tingey *et al.*, 1976*b*).

Using $^{14}CO_2$, Wilkinson & Barnes (1973) demonstrated that, in *Pinus strobus* and *Pinus taeda*, O_3 decreased the incorporation of the label into simple soluble sugars while increasing its incorporation into sugar phosphates. Such alterations in the carbohydrate metabolism could be attributed to the effect of O_3 on enzyme activities. A decrease in the activity of glyceraldehyde-3-phosphate dehydrogenase and an increase in glucose-6-phosphate dehydrogenase in *Glycine max* leaves was attributed to O_3-induced inhibition of the glycolytic pathway and stimulation of the Pentose Phosphate Pathway (Tingey *et al.*, 1975, 1976*a*). Stimulation of the Pentose Phosphate Pathway also occurs in diseased and aged plant tissues (Goodman *et al.*, 1967; Gibbs, 1966).

As sugars also act as scavengers of free radicals (Asada, 1980), it is possible that an increase in their levels can partly overcome the phytotoxic effects of free radicals produced by O_3. This is a tentative speculation and awaits experimental evidence.

Phenols and Related Metabolism

Keen & Taylor (1975) found an accumulation of isoflavonoids in O_3-treated plants and suggested that plants under stress can trigger such metabolic responses. In *Medicago sativa* (Alfalfa or Lucerne), an accumulation of 4', 7-dihydroxyflavone occurred after exposure to 0.2 ppm O_3 for 2.5 hours, and its concentration markedly increased with increasing symptoms of visible injury (Hurwitz *et al.*, 1979). Increased production of isoflavonoids can also occur in response to pathogenic infections (Keen & Kennedy, 1974).

Enzymes involved in phenol metabolism, such as phenylalanine ammonia lipase, polyphenol oxidase, and peroxidase, are stimulated by O_3 (Tingey *et al.*, 1976*a*). Activation of these enzymes would stimulate oxidation of phenols to quinones and cause accumulation of their polymerization products. These products could be responsible for a necrotic appearance of injured leaves. Peroxidase activity has been reported to increase in plants after exposure to O_3 (Curtis & Howell, 1971; Dass & Weaver, 1972; Tingey *et al.*, 1975, 1976*a*; Curtis *et al.*, 1976). Isoenzyme analysis of peroxidase in two cultivars of *Glycine max* (Soybean) with varying sensitivities to O_3 showed that, in the resistant cultivar, O_3 caused an increase in the activity of a few isoenzymes whereas in the sensitive cultivar it increased the activity of all isoenzymes (Curtis *et al.*, 1976).

Plants subjected to stress produce elevated levels of ethylene (Abeles, 1973), which was also found in plants exposed to O_3 (Craker, 1971). Ethylene is also a product of normal plant metabolism and has been shown to stimulate the activities of phenylalanine ammonia lyase, polyphenol oxidase, and peroxidase (Abeles, 1973). It is therefore uncertain whether the changes in the activities of these enzymes occur as a direct effect of O_3-induced ethylene production.

OXIDES OF NITROGEN

Stomatal Response

Little is known about the effects of oxides of nitrogen (NO_x) on plant stomata and their biochemistry. Because of greater absorption by plants of labelled nitrogen dioxide ($^{15}NO_2$) during the day than at night, it was suggested that the gaseous uptake was dependent on stomatal aperture (Yoneyama *et al.*, 1979; Kaji *et al.*, 1980). High stomatal resistance to NO_2 absorption in the dark was considered to be the reason for decreased NO_2 uptake in *Phaseolus vulgaris*. NO_2 inhibited transpiration in illuminated leaves, apparently through causing partial stomatal closure (Srivastava *et al.*, 1975).

Absorption and Biochemical Transformation

Formation of nitrate (NO_3^-) and nitrite (NO_2^-) has been demonstrated in plants fumigated with NO_2^-; initially both NO_3^- and NO_2^- are formed in equal amounts, and this is followed by accumulation of only NO_2^- (Zeevaart, 1976). Nitrite is more toxic than NO_3^- (Mudd, 1973), and in many plants it is detoxified by enzymic mechanisms up to a certain concentration.

Plants absorb gaseous NO_2 more rapidly than NO (Bennett & Hill, 1975), mainly because NO_2 reacts rapidly with water, while NO is almost insoluble. Fumigation experiments using labelled NO_2 ($^{15}NO_2$) further demonstrated that absorbed NO_2 gas is easily converted to NO_3^- and NO_2^- before further utilization in plant metabolism (Yoneyama & Sasakawa, 1979; Kaji *et al.*, 1980). The NO_2 injury to plants occurs either as a result of acidification or because of a photooxidation process (Zeevaart, 1976).

Photosynthesis

Decreased photosynthesis has been demonstrated upon exposure of plants to gaseous NO and NO_2, even at concentrations that do not produce visible injury (Hill & Bennett, 1970; Bennett & Hill, 1975; Capron & Mansfield, 1976). The combined effect of the two gases was additive; the NO effect, however, was much more rapid than the effect of NO_2 (Hill & Bennett, 1970). Fumigated plants can recover their photosynthetic ability upon transfer to clean air (Hill & Bennett, 1970; Bennett & Hill, 1975). Srivastava *et al.* (1975) showed that a decrease in photosynthesis in *Phaseolus vulgaris* was related to NO_2 concentration and length of exposure.

Exposure of *Pisum sativum* to NO_2 produced little change in RuBP carboxylase activity (Horsman & Wellburn, 1975). Exposure of *Pinus banksiana* and *Picea glauca* (White Spruce) to low levels of NO_2 (0.2 to 1 ppm for 2 weeks), however, increased RuBP carboxylase activity; the extent of increase was dependent on NO_2 concentration (Khan & Malhotra, ms). Short exposures (48 hours) of *Pinus banksiana* to a relatively high concentration of 2 ppm NO_2 also stimulated the activity of RuBP carboxylase as well as that of glycollate oxidase; the stimulation was, however, greater in a solution of glycollate oxidase (174%) than in one of RuBP carboxylase (112%). As glycollate oxidase is involved in photorespiration, an increase in its activity to greater than that of RuBP carboxylase would imply a drop in photosynthesis. In contrast, the activities of both RuBP carboxylase and glycollate oxidase were markedly inhibited in *Betula papyrifera* and *Alnus crispa*, even at a low NO_2 concentration (0.2 ppm) (Khan & Malhotra, ms*b*).

Photosynthetic inhibition brought about by NO_x could be explained as competition for NADPH between the processes of nitrite reduction and carbon assimilation in chloroplasts. The acidity produced by NO_2 could also influence electron-flow and photophosphorylation. As photoelectron systems are associated with chloroplast membranes, any changes in their structures would influence activities of the photosystems. Nitrogen dioxide has been shown to cause swelling of chloroplast membranes (Wellburn *et al.*, 1972). Biochemical and membrane injury may result if ammonia, produced from NO_2, is not rapidly incorporated into amino-forms; ammonia has been shown to inhibit photosynthesis by uncoupling electron transport (Avron, 1960) and by inducing structural alterations (Puritch & Barker, 1967).

Little is known concerning NO_x effect on plant pigments involved in photosynthesis. Nash (1976) showed that NO_2 fumigation caused a significant reduction in the chlorophyll content of various lichen species. Unlike lichens, fumigation of *Pisum sativum* with NO_2 resulted in an increase (10%) in the chlorophyll content (Horsman & Wellburn, 1975). Zeevaart (1976) suggested, on the basis of visible injury, that NO_2 strongly inhibited pigment synthesis in the developing leaves; as development of chlorosis was generally light-dependent, it is possible that photooxidative processes may have affected the pigments.

Amino Acids and Proteins

NO_2 has been shown to stimulate amino acid synthesis from $^{14}CO_2$ (Matsushima, 1972). Plants have the ability to metabolize the dissolved NO_x through their NO_3^- assimilation pathway:

$$NO_x \rightarrow NO_3^- \rightarrow NO_2^- \rightarrow NH_4^+ \rightarrow \text{amino acids} \rightarrow \text{proteins}$$

Plants exposed to gaseous NO_2 can accumulate and metabolize various products of the above pathway (Zeevaart, 1976; Yoneyama & Sasakawa, 1979; Kaji *et al.*, 1980). Induction of nitrate reductase is known to occur by NO_3^- (Beevers & Hageman, 1969); and upon fumigation with NO_2 (Zeevaart, 1974). Stimulation of NO_3^- reductase in fumigated plants causes increased production of NO_2^-, which upon subsequent reduction can be utilized for amino acid synthesis. Reduction of NO_3^- and NO_2^- to NH_4^+ was greatly stimulated in light, which suggests that the process of photosynthesis was the major supplier of the reductant (Zeevaart, 1976).

Isolated chloroplasts with a high rate of CO_2 fixation contain an enzyme system that is capable of reducing NO_2^- in the presence of light (Magalhaes *et al.*, 1974; Miflin, 1974). The addition of ferredoxin stimulated such a reduction (Beevers & Hageman, 1969). A functional association of Photosystem I and ferredoxin with NO_2^--reduction in the chloroplasts has also been demonstrated (Neyra & Hageman, 1974). Plaut *et al.* (1977) showed that NO_2^- reduction in *Spinacia oleracea* chloroplasts was more rapid in the presence of HCO_3^- than in its absence, provided that the NO_2^- was added several minutes after the HCO_3^-. This led them to suggest that CO_2-fixation products may have a regulatory role in NO_2^- reduction.

Using labelled NO_2 ($^{15}NO_2$), it has been shown that most of the absorbed NO_2 was transformed into reduced organic nitrogen compounds; the incorporation of ^{15}N was predominantly in the amide nitrogen of glutamine, followed by gluatamate, aspartate, alanine, and amino-butyric acid (Yoneyama & Sasakawa, 1979; Kaji *et al.*, 1980). It was thought that, following NO_2^- reduction, the NH_4^+ was metabolized by glutamine synthetase (GS) and glutamate synthetase (GOGAT) systems. Differences in the distribution of ^{15}N occurred, depending on whether plants were fumigated during the day or at night (Kaji *et al.*, 1980). At night the label was present mainly in the amide nitrogen, with a small amount in amino acids, while during the day the amino acids synthesis was greatly stimulated. This led to the speculation that light stimulates GOGAT and not GS (Ito *et al.*, 1978; Kaji *et al.*, 1980).

In the manner of NO_2, NO also affected nitrogen metabolizing enzymes. For example, Wellburn *et al.* (1976) showed that NO fumigation caused an appreciable increase in the activity of nitrite reductase. Later work (Wellburn *et al.*, 1980) showed that NO not only stimulates nitrite reductase but also affects other related enzymes. Fumigation of a Tomato (*Lycopersicum esculentum*) cultivar (Ailsa Craig) that is sensitive to NO, resulted in a significant increase in the activities of glutamate dehydrogenese (GD), glutamate pyruvate transaminase, and glutamate oxaloacetate transaminase (GOT); fumigation of an NO-resistant cultivar (Santo) stimulated only GOT. These results suggest that nitrogen from NO_x could be assimilated by plants through different metabolic systems (GS and GOGAT or GD).

Fatty Acids and Lipids

In *Chlorella pyrenoidosa*, NO_2^- (25 mM) markedly inhibited lipid biosynthesis; the inhibition was greater in the dark than in the light (Yung & Mudd, 1966). It is possible that this was due to an increased supply of NADPH in the light-stimulated reduction of NO_2^- to NH_4^+, which does not affect lipid biosynthesis even in the dark (Yung & Mudd, 1966). Little is known about the effect of NO_x in lipid metabolism in vascular plants. Fumigation of *Pinus banksiana* seedlings with 2 ppm NO_2 for 48 hours inhibited the biosynthesis of lipids characteristic of chloroplast membranes (phospholipids and galactolipids) long before any visible symptoms of injury appeared (Khan & Malhotra, ms*d*).

Oxides of nitrogen have the potential to oxidize unsaturated fatty acids. It is, however, not yet known what NO_x concentration in the atmosphere would affect these fatty acids in plant membranes.

Other Metabolic Processes

Dark respiration in primary leaves of *Phaseolus vulgaris* was more severely inhibited by increasing doses of NO_2 than was photosynthesis, and the inhibition was not reversed quickly upon removal of NO_2 from the atmosphere (Srivastava *et al.*, 1975).

Peroxidase activity was found to be stimulated by fumigation with NO_2. This fumigation of *Pinus banksiana* at 2 ppm for 48 hours stimulated peroxidase activity by 25%, while in *Alnus crispa* the enzyme was stimulated (35%) at a much lower NO_2 dose (0.2 ppm for 24 hours) (Khan & Malhotra, ms*b*). No visible symptoms of injury were detected in either *Pinus banksiana* or *Alnus crispa* following these fumigations. Stimulation in peroxidase activity also occurred in plants fumigated with NO (Wellburn *et al.*, 1976).

PEROXYACYL NITRATES

Peroxyacetyl nitrate (PAN) is the most commonly studied homologue of peroxyacyl nitrates, and is a very phytotoxic component of photooxidative smog.

As with other gaseous pollutants, PAN is absorbed mainly through the stomata. In plant leaves, its inhibitory effect on photosynthesis is dependent on light (Taylor *et al.*, 1961; Dugger *et al.*, 1963).

Recent work has shown that fumigation conditions can influence the response of photosynthesis in lichens (Sigal & Taylor, 1979). Long exposures at low concentrations of PAN reduced photosynthesis in species such as *Parmelia sulcata* and *Hypogymnia enteromorpha*, but had no effect on *Collema nigrescens*. Short fumigations at high concentrations, on the other hand, produced a slight photosynthetic stimulation in all the species tried except *Hypogymnia enteromorpha*. Dugger *et al.* (1965), showed that fumigation at a low concentration of PAN for a short duration inhibited O_2 evolution in chloroplasts of treated plants, but had no effect on photophosphorylation. Treatment of isolated chloroplasts with PAN, however, inhibited electron transport, photophosphorylation, and CO_2-fixation. Coulson & Heath (1975) also showed that exposure of isolated *Spinacia oleracea* chloroplasts to PAN inhibited electron transport in both photosystems.

The inhibitory effect of PAN on enzymes has been attributed to its ability to oxidize SH groups in proteins and metabolites such as cysteine, reduced glutathione, CoA, lipoic acid, and methionine (Mudd, 1975*b*). The oxidizing nature of PAN is not limited to SH groups, because it can also oxidize NADH and NADPH, which would eventually interfere with metabolic reactions involving these reduced coenzymes. For more detailed accounts of biochemical effects of PAN, the reader is referred to reviews by Ziegler (1973*b*) and Mudd (1975*b*).

FLUORIDES

Uptake and Stomatal Response

Fluorides in the air occur in either gaseous or particulate form. The gaseous form (HF) is absorbed through the leaves, while the particulate form is generally adsorbed on the outer surfaces of the plant and is thus less injurious to plants (Davison & Blakemore, 1976). The accumulation of fluoride and development of injury in plants exposed to airborne fluorides depend on both environmental and biological factors. Gaseous fluoride absorption through stomata can influence stomatal responses. Poovaiah & Wiebe (1973) demonstrated a partial closure of *Glycine max* stomata after 1 hour of exposure to 0.15 ppm HF; the effect, however, became more severe after 4 hours of exposure. Stomatal opening did not recover completely in clean air until the following day. HF fumigation of *Glycine max* also severely depressed the transpiration rate and caused water-absorption potential and leaf temperature to increase. Plant uptake of fluoride was greater during daytime fumigations than at night. To date, nothing is known about the biochemical basis of such responses.

The site of fluoride accumulation inside the leaf cells is not known with certainty, although it has been suggested that, after passing through the cell-wall, fluoride attacks cytoplasmic membranes and is partially retained there and perhaps transferred to the vacuoles (Treshow, 1971). Chloroplasts have also been suggested as the site of fluoride accumulation in leaves of *Citrus sinensis* (Chang & Thompson, 1966).

Photosynthesis

Exposure of *Gladiolus* leaves to HF caused a decline in photosynthetic CO_2-fixation, but the effect was transient (Thomas & Hendricks, 1956). Little is known about the effect of fluoride on the enzymes involved in photosynthetic CO_2-fixation.

At concentrations that did not produce visible injury, HF had no effect on the activity of PEP carboxylase in leaves of *Phaseolus vulgaris* (McCune *et al.*, 1964). Yang & Miller (1963*b*), on the other hand, reported an increase in the activity of PEP carboxylase and dark CO_2-fixation in fluoride-injured leaves of *Glycine max*.

Potassium fluoride has been shown to inhibit the Hill reaction activity of *Phaseolus vulgaris* chloroplasts (Ballantyne, 1972). Ballantyne *et al.* (1979) also found, however, that application of fluoride through *Phaseolus vulgaris* petioles had little effect on Hill reaction activity of the leaf chloroplasts. Further work is needed to determine the site of fluoride action in these two cases.

The mechanism by which fluorides affect chlorophyll in plant leaves is not clear. Low concentrations of fluoride (1.3 to 12.4 ppb) caused slight depressions in the amounts of chlorophyll *a* and chlorophyll *b*; however, the chlorophylls returned to their normal levels after a recovery period (Weinstein, 1961). It has been suggested that fluoride causes a reduction in chlorophyll synthesis (McNulty & Newman, 1961).

Respiration

Fluorides can stimulate or inhibit plant respiration (Treshow, 1971; Chang, 1975). J. E. Miller & G. W. Miller (1974) observed that, depending on the duration of exposure, HF caused either a stimulation or a depression in intact tissue respiration. Respiratory rates and adenosine triphosphatase (ATPase) activity from isolated mitochondria showed the same pattern as intact tissue respiration, but mitochondrial phosphorylation was severely inhibited regardless of inhibition or stimulation of tissue respiration. Fluoride also stimulated mitochondrial swelling and caused leakage of proteins, which suggests that membranes were the primary site of fluoride action.

Psenak *et al.* (1977) showed that KF inhibited succinate oxidation more

severely than it inhibited malate and NADH oxidations in mitochondria of
Brassica oleracea (Cabbage) and *Glycine max*. The difference was suggested
as being due to membrane association of succinoxidase and matrix location of
malate oxidase. Respiration changes in fluoride-treated plants could also be
attributed to changes in the activities of the oxidative enzymes (C. Lee *et al.*,
1966).

Amino Acid and Protein Metabolism

The amino acid metabolism is markedly affected in plants exposed to
fluoride. In leaves of *Glycine max* fumigated with HF, necrosis was accompanied by an increase in free amino acids and asparagine content (Yang &
Miller, 1963*a*). Marked increases in free amino acids and amines
(asparagine/glutamine) were also reported in needles of *Picea abies* (Norway
Spruce) at all stages of HF-induced injury (Jäger & Grill, 1975). Among the
amino acids analysed, HF reduced the content of only alanine—perhaps due
to inhibition of enolase activity. It is, however, not certain whether the
observed increase in free amino acids was due to increased breakdown of
cellular proteins or to inhibition of protein synthesis, or to both.

Chang (1970*a*, 1970*b*) has shown that NaF caused a decrease in the size of
both free and bound ribosomes, as well as a decrease in their protein and
RNA content. The disorganization of the ribosomal systems was suggested to
be due to an increase in the activity of ribonuclease (RNAase) upon fluoride
treatment.

Fatty Acid and Lipid Metabolism

Not much is known about the effect of fluoride on plant fatty acid and lipid
metabolism. Yee-Meiler (1975) observed a significant increase in the activity
of non-specific esterase in foliage of *Betula verrucosa* (Silver Birch) and *Picea
abies* plants placed at various distances from a fluoride emission source.
Increased esterase activity may have an influence on the metabolism of lipids,
including those involved in plant membranes.

Recently, Simola & Koskimies-Soininen (1980) demonstrated the effect of
KF on the fatty acid composition of lipid fractions from *Sphagnum fimbriatum* gametophytes; the effect was more pronounced in gametophytes
grown at 25°C than at 15°C. Fluoride caused an increase in palmitic acid
(16 : 0) and a decrease in linoleic acid (18 : 2) in all lipid fractions, while
linolenic acid (18 : 3) decreased in glyco- and neutral lipids. The authors
suggested that fluoride inhibited elongation of the fatty acids; however, their
results showed that KF also caused an increase in stearic (18 : 0) and oleic
(18 : 1) acids in a number of lipid fractions from material grown at 25°C. It
therefore appears that inhibition of chain elongation was not the only event

affected by KF, and that a decrease in 18 : 2 and 18 : 3 fatty acids may also be due to inhibition of the desaturase(s) activity. Such changes in the composition of membrane lipids could influence metabolic functions associated with cellular membranes.

Carbohydrate and Organic Acid Metabolism

Plants fumigated with HF show changes in sugars, polysaccharides, and organic acid contents (Adams & Emerson, 1961; Weinstein, 1961; Yang & Miller, 1963*a*; McCune *et al.*, 1964). Some changes in the carbohydrates and organic acids can be explained as the direct effects of fluorides on various enzymes involved in the metabolism. Enzymes such as glucose-6-phosphate dehydrogenase (G. W. Miller, 1958; C. Lee *et al.*, 1966), enolase (McCune *et al.*, 1964; C. Lee *et al.*, 1966), and phosphoglucomutase (Ordin & Altman, 1965), have been shown to be affected by fluoride.

Psenak *et al.* (1977) tested the effects of various fluoride compounds on the activity of three isolated malate dehydrogenase isoenzymes and found that NaF and KF had no effect on the enzyme activity although these compounds inhibited malate oxidation in the mitochondria. Among other fluoride compounds, fluoropyruvic acid also inhibited isolated enzymes. It is possible that, in contrast to isolated enzymes, the sensitivity in the cells and organelles was influenced by other aspects of the metabolism.

Other Metabolic Functions

McCune *et al.* (1970) observed that HF had no effect on either the level of acid-soluble nucleotides or the distribution of ^{32}P into various nucleotide fractions from leaves of treated plants. Similarly, HF fumigation had no effect on acid-soluble phosphorus compounds (Pack & Wilson, 1967). Yee-Meiler (1975) studied the effects of various levels of fluoride on the activity of plant acid phosphatase, and found no correlation between the enzyme activity and the extent of plant exposure to fluoride. The acid phosphatase activity in *Pinus banksiana*, however, was severely inhibited by fluoride (Malhotra & Khan, 1980). The effects of fluoride on the activity of acid and alkaline phosphatases have been summarized by Chang (1975).

Keller & Schwager (1971) reported an increase in peroxidase activity of several forest tree species with decreasing distances from a fluoride emission source; the increase in peroxidase activity was related to fluoride content of the tissues. The investigators suggested that this increase was a direct response to high fluoride uptake by the foliage, rather than an artifact of fluoride dust interaction with the enzyme during extraction. This reasoning was based on the fact that *in vitro* addition of NaF and CaF_2 had no effect on the enzyme. Increased peroxidase activity is thought to be an indication of premature ageing, which could have an effect on the growth and yield of vegetation.

POLLUTANT MIXTURES

Stomatal Response

Very little is known about how stomatal regulation and associated biochemical and physiological processes are influenced by mixtures of pollutants. The results of some recent studies suggest that stomatal physiology may be affected differently by pollutant mixtures as compared with a single pollutant. Thus in *Phaseolus vulgaris* (Kidney Bean) the transpiration rates were stimulated by SO_2 and NO_2 individually, but were inhibited by a mixture of the two (Ashenden, 1979). Synergistic effects on stomatal closure and visible injury in *Helianthus annuus* (Sunflower) have been reported in response to an O_3–NO_2 mixture (Omasa *et al.*, 1980).

In a number of agricultural plant species, stomatal resistance increased more in response to an SO_2–O_3 mixture than to those pollutants singly (Beckerson & Hofstra, 1979*a*, 1979*b*). In spite of the similarities in stomatal response, these species exhibited considerable differences in the development of symptoms of visible injury, which suggests that the sensitivity was not dependent on only the stomatal behaviour. Changes in stomatal responses would, however, influence the absorption of each pollutant from the pollutant mixtures. Elkiey & Ormrod (1980) showed that in *Petunia hybrida*, the amount of SO_2 and O_3 absorbed from a mixture was generally less than that from a single pollutant. Similarly, SO_2 uptake in *Pinus banksiana* and *Betula papyrifera* was less in plants treated with a mixture of SO_2 and NO_2 than in those exposed to SO_2 alone (Khan & Malhotra, ms*b*).

Membrane Permeability

Membrane permeability has been shown to be affected by pollutants such as SO_2 and O_3 (*see* earlier sections). Beckerson & Hofstra (1980), using a number of different plant species, demonstrated that the effect of an SO_2–O_3 mixture on membrane permeability is different from that of the same pollutants individually.

In *Phaseolus vulgaris* and *Glycine max*, solute leakage increased significantly after O_3 treatment but not after an O_3–SO_2 mixture treatment, which suggests an antagonistic effect of SO_2 on membrane permeability. On the other hand, in *Cucumis sativus* (Cucumber) and *Raphanus sativus* (Radish), membrane permeability increased more after treatment with an O_3–SO_2 mixture than with O_3 alone.

Photosynthesis and Other Metabolic Processes

White *et al.* (1974) reported that, in *Medicago sativa* (Alfalfa or Lucerne), photosynthesis was inhibited synergistically at a low concentration of an SO_2–NO_2 mixture (0.15 ppm + 0.25 ppm, respectively), but not at a high concentration (0.5 ppm for both pollutants). Additive effects of an SO_2–NO_2 mixture have been reported on the reduction of photosynthesis in *Pisum sativum* (Bull & Mansfield, 1974).

Under field conditions, where emissions of gaseous pollutants are often accompanied by high levels of CO_2, the effect of a pollutant mixture on photosynthesis could be influenced by the CO_2 concentration in the atmosphere. Hou *et al.* (1977) observed that, at a high CO_2 concentration, photosynthesis in *Medicago sativa* was much less inhibited by a mixture of SO_2 and NO_2 than at a low CO_2 concentration. Other environmental conditions can also modify the photosynthetic response to pollutant mixtures; for example, plants exposed to a mixture of SO_2 and O_3 showed syngergistic reduction in photosynthesis, but the effect was greater at low light intensities and high humidity than at high light intensities and low humidity (Carlson, 1979).

Information about the effects of pollutant mixtures on activities of various enzymes is very limited. Horsman & Wellburn (1975) studied the effects of SO_2 and NO_2 mixtures on various enzymes in *Pisum sativum*, and observed synergistic effects on the activities of RuBP carboxylase and peroxidase. Recent work has also shown that various enzyme activities in several forest tree species exposed to SO_2 and NO_2 mixtures were affected differently from those exposed to single pollutants (Khan & Malhotra, ms*b*). These effects were additive (peroxidase and glycollate oxidase in *Pinus banksiana* and RuDP carboxylase in *Alnus crispa*), synergistic (peroxidase and glycollate oxidase in *Alnus crispa*), and antagonistic (peroxidase and glycollate oxidase in *Betula papyrifera*).

In *Ulmus americana* (American Elm), a mixture of SO_2 and O_3 had a different effect on cellular metabolites compared with the effects of these pollutants individually (Constantinidou & Kozlowski, 1979). These effects were either antagonistic, synergistic, or additive, depending upon the type of tissue examined and the length of time after fumigation. Antagonistic effects have also been reported on the destruction of chlorophylls upon treatment with a mixture of SO_2 and O_3 (Beckerson & Hofstra, 1979*c*). It is therefore obvious that many mixed pollutants can either prevent or accelerate the toxic responses produced by individual pollutants.

SUMMARY

Most plants evolve in a predominantly gaseous environment. When the composition of this environment exceeds the critical limits of adaptation and tolerance, stress is imposed and the most sensitive components of the system begin to malfunction. Any gas can cause such stress when a threshold concentration is reached. Regardless of the air pollutant, the impact invariably involves interactions with one or more biochemical metabolic processes. First exposed are the stomata and their guard-cells, which may respond first if they are sufficiently sensitive. The gas then passes into the intercellular spaces to become dissolved on the moist internal surfaces, characteristically contacting and influencing membranes and the cellular pH. Penetrating the cytoplasmic membrane, a pollutant is relatively free to attack the organelles within and the substances throughout. A pollutant may react with any number of metabolites along its course of migration through the cell. Consequently, numerous reaction-sites may be, and often are, involved.

The reactions affected depend on the properties and chemical form of the pollutant, but certain sites and reactions seem especially prone to disruption: these include membranes whose permeability may be altered; photosynthetic reactions, e.g. photophosphorylation and carboxylation; electron transport and respiration. Metabolic pools are affected: carbohydrates, organic acids and amino acids, proteins and lipids—are all involved, but the specifics vary with the pollutant.

REFERENCES

Abeles, F. B. (1973). *Ethylene in Plant Biology*. Academic Press, New York & London: xii + 392 pp., illustr.

Adams, D. F. & Emerson, M. T. (1961). Variations in starch and total polysaccharide content of *Pinus ponderosa* needles with fluoride fumigation. *Plant Physiol.,* **36**, pp. 261–5.

Arndt, U. (1970). Konzentrationsänderungen bei freien Aminosäuren im Pflanzen unter dem Einfluss von Fluorwasserstoff und Schwefeldioxid. *Staub Reinhalt. Luft,* **30**, pp. 256–9.

Arrigoni, O. (1959). The enzymic oxidation of sulphite in mitochondrial preparations of pea internodes. *Ital. J. Biochem.,* **8**, pp. 181–6.

Asada, K. (1980). Formation and scavenging of superoxide in chloroplasts, with relation to injury by sulphur dioxide. *Studies on the Effects of Air Pollutants on Plants and Mechanism of Phytotoxicity: Res. Rep. Natl. Inst. Environ. Stud.,* Japan, **11**, pp. 165–79.

Asada, K. & Kiso, K. (1973). Initiation of aerobic oxidation of sulphite by illuminated spinach chloroplasts. *Eur. J. Biochem.,* **33**, pp. 253–7.

Asada, K., Kiso, K. & Yoshikawa, K. (1974). Univalent reduction of molecular oxygen by Spinach chloroplasts on illumination. *J. Biol. Chem.,* **249**, pp. 2175–81.

Asada, K., Kitoh, S., Deura, R. & Kasai, Z. (1965). Effect of α-hydroxysulfonates on photochemical reactions of Spinach chloroplasts and participation of glycolate in photophosphorylation. *Plant and Cell Physiol.,* **6**, pp. 615–29.

Asada, K., Takahashi, M., Tanaka, K. & Nakano, Y. (1977). Formation of active oxygen and its fate in chloroplasts. Pp. 45–63 in *Biochemical and Medical Aspects of Active Oxygen* (Ed. O. Hayaishi & K. Asada). University of Tokyo Press, Tokyo, Japan: 313 pp., illustr.

Ashenden, T. W. (1979). Effects of SO$_2$ and NO$_2$ pollution on transpiration on *Phaseolus vulgaris* L. *Environ. Pollut.,* **18**, 45–9.

Ashenden, T. W. & Mansfield, T. A. (1977). Influence of wind speed on the sensitivity of Ryegrass to SO$_2$. *J. Exp. Bot.,* **281**, pp. 729–35.

Avron, M. (1960). Photophosphorylation by Swisschard chloroplasts. *Biochem. Biophys. Acta,* **40**, pp. 257–72.

Ballantyne, D. J. (1972). Fluoride inhibition of the Hill reaction in bean chloroplasts. *Atmos. Environ.,* **6**, pp. 267–73.

Ballantyne, D. J. (1973). Sulphite inhibition of ATP formation in plant mitochondria. *Phytochemistry,* **12**, pp. 1207–9.

Ballantyne, D. J. (1977). Sulfite oxidation by mitochondria from green and etiolated peas. *Phytochemistry,* **16**, pp. 49–50.

Ballantyne, D. J., Johnson, A. M. & Dijak, M. (1979). Physiological effects of systemic fluoride applications to Bean (*Phaseolus vulgaris*) leaves and Pea (*Pisum sativum*) epicotyls. *Fluoride,* **12**, pp. 155–62.

Barnes, R. L. (1972). Effects of chronic exposure to ozone on photosynthesis and respiration of pines. *Environ. Pollut.,* **3**, pp. 133–8.

Beckerson, D. W. & Hofstra, G. (1979*a*). Stomatal responses of white bean to O$_3$ and SO$_2$ singly or in combination. *Atmos. Environ.,* **13**, pp. 533–5.

Beckerson, D. W. & Hofstra, G. (1979*b*). Response of leaf diffusive resistance of Radish, Cucumber and Soybean to O$_3$ and SO$_2$ singly or in combination. *Atmos. Environ.,* **13**, pp. 1263–8.

Beckerson, D. W. & Hofstra, G. (1979*c*). Effect of sulphur dioxide and ozone, singly or in combination, on leaf chlorophyll, RNA and protein in white bean. *Can. J. Bot.,* **57**, pp. 1940–5.

Beckerson, D. W. & Hofstra, G. (1980). Effects of sulphur dioxide and ozone, singly or in combination, on membrane permeability. *Can. J. Bot.,* **58**, pp. 451–7.

Beevers, L. & Hageman, R. H. (1969). Nitrate reduction in higher plants. *Annu. Rev. Plant Physiol.,* **20**, pp. 495–522.

Bennett, J. H. & Hill, A. C. (1975). Interaction of air pollutants with canopies of vegetation. Pp. 273–306 in *Responses of Plants to Air Pollution* (Ed. J. B. Mudd & T. T. Kozlowski). Academic Press, New York–San Francisco–London: xii + 383 pp., illustr.

Benson, A. A. (1963). The plant sulfolipid. *Adv. Lipid Res.,* **1**, pp. 387–94.

Bishop, N. I. (1958). The influence of the herbicide, DCMU, on the oxygen-evolving system of photosynthesis. *Biochem. Biophys. Acta,* **27**, pp. 205–6.

Black, C. R. & Black, V. J. (1979*a*). The effect of low concentration of sulphur dioxide on stomatal conductance and epidermal cell survival in Field Bean (*Vicia faba* L.). *J. Exp. Bot.,* **30**, pp. 291–8.

Black, C. R. & Black, V. J. (1979*b*). Light and scanning electron microscopy of SO$_2$-induced injury to leaf surfaces of Field Bean (*Vicia faba* L.). *Plant, Cell and Environment,* **2**, pp. 329–33.

Black, V. J. & Unsworth, M. H. (1979*a*). A system for measuring effects of sulphur dioxide on gas exchange of plants. *J. Exp. Bot.,* **30**, pp. 81–8.

Black, V. J. & Unsworth, M. H. (1979*b*). Effects of low concentrations of sulphur dioxide on gas exchange of plants and dark respiration of *Vicia faba. J. Exp. Bot.,* **30**, pp. 473–83.

Black, V. J. & Unsworth, M. H. (1980). Stomatal responses to sulphur dioxide and vapour pressure deficit. *J. Exp. Bot.,* **31**, pp. 667–77.

Boardman, N. K. (1968). The photochemical systems of photosynthesis. *Adv. Enzymol.,* **30**, pp. 1–79.

Bressan, R. A., Wilson, L. G. & Filner, P. (1978). Mechanisms of resistance to sulfur dioxide in Cucurbitacea. *Plant Physiol.*, **61**, pp. 761–7.

Brunold, C. & Erismann, K. H. (1976). Sulfur dioxide as a sulfur source in duckweeds (*Lemna minor* L.). *Experientia*, **32**, pp. 296–7.

Bucher-Wallin, I. K., Bernhard, L. & Bucher, J. B. (1979). Einfluss niedriger SO_2-Konzentrationen auf die Aktivität einiger Glykosidasen der Assimilationsorgane verklonter Waldbäume. *Eur. J. For. Path.*, **9**, pp. 6–15.

Bull, J. N. & Mansfield, T. A. (1974). Photosynthesis in leaves exposed to SO_2 and NO_2. *Nature* (London), **250**, pp. 443–4.

Capron, T. M. & Mansfield, T. A. (1976). Inhibition of net photosynthesis in tomato in air polluted with NO and NO_2. *J. Exp. Bot.*, **27**, pp. 1181–6.

Caput, C., Belot, Y., Auclair, D. & Decourt, N. (1978). Absorption of sulphur dioxide by pine needles leading to acute injury. *Environ. Pollut.*, **16**, pp. 3–15.

Carlson, R. W. (1979). Reduction in the photosynthetic rate of *Acer, Quercus* and *Fraxinus* species caused by sulphur dioxide and ozone. *Environ. Pollut.*, **18**, pp. 159–70.

Chang, C. W. (1970*a*). Effect of fluoride on ribosomes from corn roots: Changes with growth retardation. *Physiol. Plant.*, **23**, pp. 536–43.

Chang, C. W. (1970*b*). Effect of fluoride on ribosomes and ribonuclease from corn roots. *Can J. Biochem.*, **48**, pp. 450–4.

Chang, C. W. (1971). Effect of ozone on ribosomes in pinto bean leaves. *Phytochemistry*, **10**, p. 2863–8.

Chang, C. W. (1975). Fluorides. Pp. 57–95 in *Responses of Plants to Air Pollution* (Ed. J. B. Mudd & T. T. Kozlowski), Academic Press, New York–San Francisco–London: xii + 383 pp., illustr.

Chang, C. W. & Thompson, C. R. (1966). Site of fluoride accumulation in navel orange leaves. *Plant Physiol.*, **41**, pp. 211–3.

Chang, C. W. & Heggestad, H. E. (1974). Effect of ozone on Photosystem II in *Spinacia oleracea chloroplasts*. *Phytochemistry*, **13**, pp. 871–3.

Cohen, S. S. (1971). *Introduction to Polyamines*, Prentice-Hall, Inc., Englewood Cliffs, New Jersey, USA: xii + 180 pp., illustr.

Constantinidou, H. A. & Kozlowski, T. T. (1979). Effects of sulfur dioxide and ozone on *Ulmus americana* seedlings, II: Carbohydrates, proteins, and lipids. *Can. J. Bot.*, **57**, pp. 176–84.

Cormis, L. D. (1968). Dégagement d'hydrogène sulfure par des plantes soumises à une atmosphère contenant de l'anhydride sulfureaux. *C.R. Acad. Sci. Ser. D.*, **266**, pp. 683–5.

Coulson, C. L. & Heath, R. L. (1974). Inhibition of the photosynthetic capacity of isolated chloroplasts by ozone. *Plant Physiol.*, **53**, pp. 32–8.

Coulson, C. L. & Heath, R. L. (1975). The interaction of peroxyacetyl nitrate (PAN) with the electron flow of isolated chloroplasts. *Atmos. Environ.*, **9**, pp. 231–8.

Cowling, D. W. & Bristow, A. W. (1979). Effects of SO_2 on sulfur and nitrogen fractions and on free amino acids in perennial ryegrass. *J. Sci. Food Agric.*, **30**, pp. 354–60.

Craker, L. E. (1971). Ethylene production from ozone-injured plants. *Environ. Pollut.*, **1**, pp. 299–304.

Curtis, C. R. & Howell, R. L. (1971). Increases in peroxidase isoenzyme activity in bean leaves exposed to low doses of ozone. *Phytopathology*, **61**, pp. 1306–7.

Curtis, C. R., Howell, R. L. & Kremer, D. F. (1976). Soybean peroxidases from ozone injury. *Environ. Pollut.*, **11**, pp. 189–94.

Dass, H. C. & Weaver, G. M. (1972). Enzymatic changes in intact leaves of *Phaseolus vulgaris* following ozone fumigation. *Atmos. Environ.*, **6**, pp. 759–63.

Davison, A. W. & Blakemore, J. (1976). Factors determining fluoride accumulation in forage. Pp. 17–30 in *Effects of Air Pollutants on Plants* (Ed. T. A. Mansfied). Cambridge University Press, Cambridge, UK: 209 pp., illustr.

Dittrich, P., Mayer, M. & Meusel, M. (1979). Proton-stimulated opening of stomata in relation to chloride uptake by guard-cells. *Planta,* **144**, pp. 305–9.

Dugger, W. M., Taylor, O. C., Klein, W. H. & Shropshire, W. (1963). Action spectrum of peroxyacetyl nitrate damage to bean plants. *Nature* (London), **198**, pp. 75–6.

Dugger, W. M., Mudd, J. B. & Koukol, J. (1965). Effect of PAN on certain photosynthetic reactions. *Arch. Environ. Health,* **10**, pp. 195–200.

Dugger, W. M. & Palmer, R. L. (1969). Carbohydrate metabolism in leaves of rough lemon as influenced by ozone. *Proc. First Intern. Citrus Symp.,* **2**, pp. 711–5.

Elkiey, T. & Ormrod, D. P. (1980). Sorption of ozone and sulfur dioxide by petunia leaves. *J. Environ. Qual.,* **9**, pp. 93–5.

Epel, B. L. & Neumann, J. (1973). The mechanism of the oxidation of ascorbate and Mn^{++} by chloroplasts: The role of radical superoxide. *Biochem. Biophys. Acta,* **325**, pp. 520–9.

Evans, L. S. & Ting, I. P. (1973). Ozone-induced membrane permeability changes. *Amer. J. Bot.,* **60**, pp. 155–62.

Evans, L. S. & Ting, I. P. (1974a). Ozone sensitivity of leaves: Relationship to leaf-water content, gas transfer resistance, and anatomical characteristic. *Amer. J. Bot.,* **61**, pp. 592–7.

Evans, L. S. & Ting, I. P. (1974b). Effect of ozone on ^{86}Rb-labeled potassium transport in leaves of *Phaseolus vulgaris* L. *Atmos. Environ.,* **8**, pp. 855–61.

Fischer, K. (1971). Methoden zur Erkennung und Beurteilung forstschädlicher Luftverunreinigungen: Chemische und physikalische Reaktionen SO_2-begaster Pflauzen und Blatter. *Mitt. Forstl. Bandes-Versuchsanst. Wien,* **92**, pp. 209–31.

Fletcher, R. A., Adedipe, N. O. & Ormrod, D. P. (1972). Abscissic acid protects bean leaves from ozone-induced phytotoxicity. *Can. J. Bot.,* **50**, 2389–91.

Frederick, P. E. & Heath, R. L. (1970). Ozone-induced fatty acid and viability changes in *Chlorella. Plant Physiol.,* **55**, pp. 15–9.

Furukawa, A., Isoda, O., Iwaki, H. & Totsuka, T. (1980a). Interspecific differences in resistance to sulfur dioxide. *Studies on the Effects of Air Pollutants on Plants and Mechanisms of Phytotoxicity: Res. Rep. Natl Inst. Environ. Stud.,* Japan, **11**, pp. 113–26.

Furukawa, A. ,Natori, T. & Totsuka, T. (1980b). The effect of SO_2 on net photosynthesis in sunflower leaf. *Studies on the Effects of Air Pollutants on Plants and Mechanisms of Phytotoxicity: Res. Rep. Natl Inst. Environ. Stud.,* Japan, **11**, pp. 1–8.

Garsed, S. G. & Read, D. J. (1977a). Sulphur dioxide metabolism in soybean, *Glycine max* var. Biloxi, 1: The effects of light and dark on the uptake and translocation of $^{35}SO_2$. *New Phytol.,* **78**, pp. 111–9.

Garsed, S. G. & Read, D. J. (1977b). Sulphur dioxide metabolism in soybean, *Glycine max* var. Biloxi, II: Biochemical distribution of $^{35}SO_2$ products. *New Phytol.,* **79**, pp. 538–92.

Garsed, S. G. & Read, D. J. (1977c). The uptake and metabolism of $^{35}SO_2$ in plants of differing sensitivity to sulphur dioxide. *Environ. Pollut.,* **13**, pp. 173–86.

Gerwick, B. C., Ku, S. B. & Black, C. C. (1980). Initiation of sulfate activation: A variation in C_4 photosynthesis plants. *Science,* **209**, pp. 513–5.

Gezeilus, K. & Hallgren, J. E. (1980). Effect of SO_3^{2-} on the activity of ribulose biphosphate carboxylase from seedlings of *Pinus sylvestris. Physiol. Plant.,* **49**, pp. 354–8.

148 *Air Pollution and Plant Life*

Gibbs, M. (1966). Carbohydrates: Their role in plant metabolism and nutrition. Pp. 3–115 in *Plant Physiology* (Ed. F. C. Steward) Vol. IVB. Academic Press, New York & London: xiv + 599 pp., illustr.

Gilbert, O. L. (1968). *Biological Indicators of Air Pollution.* Ph.D thesis, University of Newcastle upon Tyne, Newcastle upon Tyne, England, UK: [not available for checking].

Godzik, S. & Linskens, H. F. (1974). Concentration changes of free amino-acids in primary leaves after continuous and interrupted SO_2 fumigation and recovery. *Environ. Pollut.,* 7, pp. 25–38.

Goodman, R. N., Kiraly, Z. & Zaitlin, M. (1967). *The Biochemistry and Physiology of Infectious Plant Disease.* D. Van Nostrand, Princeton, New Jersey, USA: 354 pp.

Grill, D., Esterbauer, H. & Welt, R. (1979a). Einfluss von SO_2 auf das Ascorbinsäuresystem der Fichtennadeln. *Phytopath. Z.,* 93, pp. 361–8.

Grill, D., Esterbauer, H. & Klosch (1979b). Effect of sulphur dioxide on glutathione in leaves of plants. *Environ. Pollut.,* 19, pp. 187–94.

Grill, D., Esterbauer, H., Scharner, M. & Felgitsch, C. (1980). Effect of sulfur dioxide on protein-SH in needles of *Picea abies. Eur. J. For. Pathol.,* 10, pp. 263–7.

Guderian, R. (1977). Air pollution: Phytotoxicity of acid gases and its significance in air pollution control. In *Ecological Studies* (Ed. W. D. Billings, F. Golley, O. L. Lange & J. S. Olson), Vol. 22. Springer Verlag, Berlin–Heidelberg–New York: viii + 127 pp.

Hällgren, J. E. (1978). Physiological and biochemical effects of sulphur dioxide on plants. Pp. 163–209 in *Sulphur in the Environment: Part II, Ecological Impact* (Ed. J. O. Nriagu). John Wiley & Sons, New York–Chichester–Brisbane–Toronto: xii + 482 pp., illustr.

Hampp, R. & Ziegler, I. (1977). Sulfate and sulfite translocation *via* the phosphate translocator of the inner envelope membrane of chloroplasts. *Planta,* 137, pp. 309–12.

Heath, R. L. (1975). Ozone. Pp. 23–55 in *Responses of Plants to Air Pollution* (Ed. J. B. Mudd & T. T. Kozowski). Academic Press, New York–San Francisco–London: vii + 383 pp., illustr.

Heath, R. L. & Frederick, P. E. (1979). Ozone alteration of membrane permeability in *Chlorella. Plant Physiol.,* 64, pp. 455–9.

Hill, A. C. & Bennett, J. H. (1970). Inhibition of apparent photosynthesis by nitrogen oxides. *Atmos. Environ.,* 4, pp. 341–8.

Hill, A. C. & Littlefield, N. (1969). Ozone effect on apparent photosynthesis, rate of transpiration, and stomatal closure, in plants. *Environ. Sci. Technol.,* 3, pp. 52–6.

Horsman, D. C. & Wellburn, A. R. (1975). Synergistic effect of SO_2- and NO_2-polluted air upon enzyme activity in pea seedlings. *Environ. Pollut.,* 8, pp. 123–33.

Hou, L. Y., Hill, A. C. & Soleimani, A. (1977). Influence of CO_2 on the effects of SO_2 and NO_2 on alfalfa. *Environ. Pollut.,* 12, pp. 7–16.

Hurwitz, B., Pell, E. J. & Sherwood, R. T. (1979). Status of coumestrol and 4′,7-dihydroxyflavone in alfalfa foliage exposed to ozone. *Phytopathology,* 69, pp. 810–3.

Ishizaki, A. & Hasegawa, M. (1974). Physiological and morphological properties of the Japanese Black and Red Pine trees (*Pinus thunbergii* Parl. and *P. densiflora* Sieb. et Zucc.) on the resistance to air pollution [in Japanese with English summary]. *Bull. Fac. Agric.,* Tamagawa University, 14, pp. 40–50.

Ito, O., Yoneyama, T. & Kumazawa, K. (1978). Amino acid metabolism in plant leaf, IV: The effect of light on ammonium assimilation and glutamine metabolism in the cells isolated from Spinach leaves. *Plant and Cell Physiol.,* 19, pp. 1109–19.

Jäger, H. J. & Grill, D. (1975). Einfluss von SO_2 und HF auf freie Aminosauren der Fichte (*Picea abies* [L.] Karsten). *Eur. J. For. Path.,* 5, pp. 279–86.

James, A. T. & Nichols, B. W. (1966). Lipids of photosynthetic systems. *Nature* (London), **210**, pp. 372–5.

Jensen, K. F. & Kozlowski, T. T. (1975). Absorption and translocation of sulfur dioxide by seedlings of four forest tree species. *J. Environ. Qual.,* **4**, 379–82.

Jeong, Y. H., Nakamura, H. & Ota, Y. (1980). Physiological studies on photochemical oxidants injury in rice plants, 1: Varietal difference of abscisic acid content and its relation to the resistance to ozone [in Japanese with English summary]. *Jap. J. Crop Sci.,* **49**, pp. 456–60.

Jiracek, V., Machackova, I. & Kastir, J. (1972). Nachweis der Bisulfite-Addukte (α-Oxysulfonäuren) von Carbonylverbindungen in den mit SO_2 behandelten Erbesenkeimlingen. *Experientia,* **28**, pp. 1007–9.

Kaji, M., Yoneyama, T., Tostuka, T. & Iwaki, H. (1980). Absorption of atmospheric NO_2 by plants and soils, VI: Transformation of NO_2 absorbed in the leaves and transfer of the nitrogen through the plants. *Studies on the Effects of Air Pollutants on Plants and Mechanisms of Phytotoxicity: Res. Rep. Natl Inst. Environ. Stud.,* Japan, **11**, pp. 51–8.

Keen, N. T. & Kennedy, B. W. (1974). Hydroxyphaseolin and related isoflavonoids in the hypersensitive resistant response of soybeans against *Pseudomonas glycinea.* *Physiol. Plant Pathol.,* **4**, 173–85.

Keen, N. T. & Taylor, O. C. (1975). Ozone injury in soybeans: Isoflavonoid accumulation is related to necrosis. *Plant Physiol.,* **55**, 731–3.

Keller, T. (1976). Auswirkungen niedriger SO_2-Konzentrationen auf jung Fichten. *Schweiz. Z. Forstwes.,* **127**, pp. 237–51.

Keller, T. & Schwager, H. (1971). Der Nachweis unsichtbarer ('Physiologischer') Fluor—Immissionsschädigungen an Waldbäumen durch eine einfache kolorimetrische Bestimmung der Peroxidase-Aktivität. *Eur. J. For. Path.,* **1**, pp. 6–18.

Keller, T., Schwager, H. & Yee-Meiler, D. (1976). Der Nachweis Winterlicher SO_2-Immissionen an jungen Fichten: Ein vergleich dreier Methoden. *Eur. J. For. Pathol.,* **6**, pp. 244–9.

Keller, T. & Schwager, H. (1977). Air pollution and ascorbic acid. *Eur. J. For Pathol.,* **7**, pp. 338–50.

Khan, A. A. & Malhotra, S. S. (1977). Effects of aqueous sulphur dioxide on pine needle glycolipids. *Phytochemistry,* **16**, pp. 539–43.

Khan, A. A. & Malhotra, S. S. (1978). Biosynthesis of lipds in chloroplasts isolated from Jack Pine needles. *Phytochemistry,* **17**, pp. 1107–10.

Khan, A. A. & Malhotra, S. S. (1982*a*). *Studies on Ribulose Diphosphate Carboxylase and Glycollate Oxidase from Needles of* Pinus banksiana: *Effects of Sulphur Dioxide and Metals on Enzyme Activities.* [Manuscript in preparation.]

Khan, A. A. & Malhotra, S. S. (1982*b*). Peroxidase activity as an indicator of SO_2 injury in Jack Pine and White Birch. *Biochem. Physiol. Pflanzen* (in press).

Khan, A. A., & Malhotra, S. S. (ms*a*). *Sulphite Oxidation by Chloroplast and Cytoplasmic Fractions of* Pinus banksiana *Needles.* [Manuscript in preparation.]

Khan, A. A., & Malhotra, S. S. (ms*b*). *Effects of SO_2, NO_2, and Their Mixture, on the Enzymic Activities of Forest Plant Species.* [Manuscript in preparation.]

Khan, A. A. & Malhotra, S. S. (ms*c*). *Protein Biosynthesis in Pine Needles and Its Inhibition by Sulphur Dioxide.* [Manuscript in preparation.]

Khan, A. A. & Malhotra, S. S. (ms*d*). *Response of NO_2 Fumigation on Lipid Biosynthesis in Pine Needles.* [Manuscript in preparation.]

Knudson, L. L., Tibbitts, T. W. & Edwards, G. E. (1977). Measurement of ozone injury by determination of leaf chlorophyll concentration. *Plant Physiol.,* **60**, pp. 606–8.

Kondo, N. & Sugahara, K. (1978). Changes in transpiration rate of SO_2-resistant and -sensitive plants with SO_2 fumigation and the participation of abscisic acid. *Plant and Cell Physiol.,* **19**, 365–73.

Kondo, N., Akiyama, Y., Fujiwara, M. & Sugahara, K. (1980a). Sulfite oxidizing acitivities in plants. *Studies on the Effects of Air Pollutants in Plants and Mechanisms of Phytotoxicity: Res. Rep. Natl Inst. Environ. Stud.* Japan, **11**, pp. 137–50.

Kondo, N., Maruta, I. & Sugahara, K. (1980b). Abscisic acid-dependent changes in transpiration rate with SO_2 fumigation and the effect of sulfite and pH on stomatal aperture. *Studies on the Effects of Air Pollutants on Plants and Mechanisms of Phytotoxicity: Res. Rep. Natl Inst. Environ. Stud.,* Japan, **11**, pp. 127–36.

Koziol, M. J. & Jordan, C. F. (1978). Changes in carbohydrate levels in Red Kidney Bean (*Phaseolus vulgaris* L.) exposed to sulphur dioxide. *J. Exp. Bot.* **29**, pp. 1037–43.

Koziol, M. J. & Cowling, D. W. (1978). Growth of Ryegrass (*Lolium pernne* L.) exposed to SO_2. *J. Exp. Bot.,* **29**, pp. 1431–9.

Lauenroth, W. K., Black, C. J. & Dodd, J. L. (1979). Sulfur accumulation in Western Wheatgrass exposed to three controlled SO_2 concentrations. *Plant and Soil,* **53**, pp. 131–6.

Lee, C., Miller, G. W. & Welkie, G. W. (1966). The effects of hydrogen fluoride and wounding on respiratory enzymes in Soybean leaves. *Air Water Pollut. Int. J.,* **10**, pp. 169–81.

Lee, T. T. (1967). Inhibition of oxidative phosphorylation and respiration by ozone in tobacco mitochondria. *Plant Physiol.,* **42**, pp. 691–6.

Lee, T. T. (1968). Effect of ozone on swelling of tobacco mitochondria. *Plant Physiol.,* **43**, pp. 133–9.

Leffler, H. R. & Cherry, J. H. (1974). Destruction of enzymatic activities of corn and soybean leaves exposed to ozone. *Can. J. Bot.,* **43**, pp. 677–85.

Lehmann, J. & Benson, A. A. (1964). The plant sulfolipid, IX: Sulfosugar synthesis from methyl hexosenides. *J. Amer. Chem. Soc.,* **86**, 4449–72.

Libera, W., Ziegler, H. & Ziegler, I. (1973). Forderung der Hill Reaktion und der CO_2-Fixierung in isolierten Spinatchloroplasten durch niedere Sulfitkonzentratinen. *Planta,* **109**, pp. 269–79.

Lösel, D. M. (1978). Lipid metabolism of leaves of *Poa pratensis* during infection by *Puccinia poarum. New Phytol.,* **80**, pp. 2167–74.

McCord, J. M. & Fridovich, I. (1969). Superoxide dismutase: An enzymic function of erythrocuprein. *J. Biol. Chem.,* **244**, pp. 6049–55.

McCune, D. C., Weinstein, L. H., Jacobson, J. S. & Hitchcock, A. E. (1964). Some effects of atmospheric fluoride on plant metabolism. *J. Air Pollut. Contr. Assoc.,* **14**, pp. 465–8.

McCune, D. C., Weinstein, L. H. & Mancini, J. F. (1970). Effects of hydrogen fluoride on the acid-soluble nucleotide metabolism of plants. *Contrib. Boyce Thompson Inst.,* **24**, pp. 213–26.

MacDowell, F. D. H. (1965). Stages of ozone damage to respiration of tobacco leaves. *Can. J. Bot.,* **43**, pp. 419–27.

McNulty, I. B. & Newman, D. W. (1961). Mechanisms of fluoride-induced chlorosis. *Plant Physiol.,* **36**, pp. 385–8.

Magalhaes, A. C., Neyra, C. A. & Hageman, R. H. (1974). Nitrite assimilation and amino nitrogen synthesis in isolated spinach chloroplasts. *Plant Physiol.,* **53**, pp. 411–5.

Majernik, O. & Mansfield, T. A. (1970). Direct effect of SO_2 pollution on the degree of opening of stomata. *Nature* (London), **227**, pp. 377–8.

Majernik, O. & Mansfield, T. A. (1971). Effects of SO_2 pollution on stomatal movements in *Vicia faba. Phytopathol. Z.,* **71**, pp. 123–218.

Majernik, O. & Mansfield, T. A. (1972). Stomatal responses to raised atmospheric CO_2 concentrations during exposure of plants to SO_2 pollution. *Environ. Pollut.,* **3**, pp. 1–7.

Malhotra, S. S. (1976). Effects of sulphur dioxide on biochemical activity and ultrastructural organization of pine needle chloroplasts. *New Phytol.,* **76**, pp. 239–45.

Malhotra, S. S. (1977). Effects of aqueous sulphur dioxide on chlorophyll destruction in *Pinus contorta. New Phytol.,* **78**, pp. 101–9.

Malhotra, S. S. & Hocking, D. (1976). Biochemical and cytological effects of sulphur dioxide on plant metabolism. *New Phytol.,* **76**, pp. 227–37.

Malhotra, S. S. & Khan, A. A. (1978). Effects of sulphur dioxide fumigation on lipid biosynthesis in pine needles. *Phytochemistry,* **17**, 241–4.

Malhotra, S. S. & Khan, A. A. (1980). Effects of sulphur dioxide and other air pollutants on acid phosphatase activity in pine seedlings. *Biochem. Physiol. Pflanzen,* **175**, pp. 228–36.

Malhotra, S. S. & Khan, A. A. (ms). *Sensitivity of Various Metabolic Processes to Sulphur Dioxide in an Epiphytic Lichen,* Evernia mesomorpha. [Manuscript in prepatation.]

Malhotra, S. S. & Sarkar, S. K. (1979). Effects of sulphur dioxide on sugar and free amino-acid content of pine seedlings. *Physiol. Plant.,* **47**, pp. 223–8.

Mansfield, T. A. & Majernik, O. (1970). Can stomata play a part in protecting plants against air pollutants? *Environ. Pollut.,* **1**, pp. 149–54.

Matsushima, J. (1972). Influence of SO_2 and NO_2 on assimilation of amino-acids, organic acids and saccharoid in *Citrus natsudaidai* seedlings. *Bull. Fac. Agr., Mie University* (Japan), **44**, pp. 131–9.

Menser, H. A. & Heggestad, H. E. (1966). Ozone and sulfur dioxide synergism: Injury to tobacco plants. *Science,* **153**, pp. 424–5.

Menser, H. A., Chaplin, J. F., Cheng, A. L. S. & Sorokin, T. (1977). Polyphenols, polysterols, and reducing sugars in air-cured tobacco leaves injured by ozone air pollution. *Tobacco Sci.,* **21**, pp. 35–8.

Miflin, B. J. (1974). Nitrite reduction in leaves; studies in isolated chloroplasts. *Planta,* **116**, pp. 187–96.

Miller, G. W. (1958). Properties of enolase in extracts from pea seed. *Plant Physiol.,* **33**, pp. 199–206.

Miller, J. E. & Miller, G. W. (1974). Effects of fluoride on mitochondrial activity in higher plants. *Physiol. Plant.,* **32**, pp. 115–21.

Miller, J. E. & Xerikos, P. (1979). Residence time of sulfite in SO_2 'sensitive' anu 'tolerant' Soybean cultivars. *Environ. Pollut.,* **18**, pp. 259–64.

Miller, P. R., Parmeter, J. R., Flick, B. H. & Martinez, C. W. (1969). Ozone dose response of Ponderosa Pine seedlings. *J. Air Pollut. Cont. Assoc.,* **19**, pp. 435–8.

Miszalski, Z. & Ziegler, I. (1979). Increase in chloroplastic thiol groups by SO_2 and its effect on light modulation of NADP-dependent glyceraldehyde 3-phosphate dehydrogenase. *Planta,* **145**, pp. 383–7.

Miszalski, Z. & Ziegler, H. (1980). 'Available SO_2'—a parameter for SO_2 toxicity. *Phytopath. Z.,* **97**, pp. 144–7.

Mudd, J. B. (1973). Biochemical effects of some air pollutants on plants. *Adv. Chem. Ser.,* **122**, pp. 31–47.

Mudd, J. B. (1975a). Sulfur dioxide. Pp. 9–22 in *Responses of Plants to Air Pollution* (Ed. J. B. Mudd & T. T. Kozlowski). Academic Press, New York–San Francisco–London: xii + 383 pp., illustr.

Mudd, J. B. (1975*b*). Peroxyacyl nitrates. Pp. 97–119 in *Responses of Plants to Air Pollution* (Ed. J. B. Mudd & T. T. Kozlowski). Academic Press, New York–San Francisco–London: xii + 383 pp., illustr.

Mudd, J. B., McManus, T. T. & Ongun, A. (1971*a*). Inhibition of lipid metabolism in chloroplasts by ozone. Pp. 256–60 in *Proc. 2nd Int. Clean Air Congr., 1970.* (Ed. H. M. England & W. T. Beery). Academic Press, New York, NY, USA: xxv + 1354 pp., illustr.

Mudd, J. B., McManus, T. T., Ongun, A. & McCollogh, T. E. (1971*b*). Inhibition of glycolipid biosynthesis in chloroplasts by ozone and sulfhydryl reagent. *Plant Physiol., 48*, pp. 335–9.

Mukerji, S. K. & Yang, S. F. (1974). Phosphoenolpyruvate carboxylase from spinach leaf tissue: Inhibition by sulfite ion. *Plant Physiol., 53*, pp. 829–34.

Nakamura, H. & Saka, H. (1978). Photochemical oxidants injury in rice plants, III: Effect of ozone on physiological activities in rice plants [in Japanese with English summary]. *Jap. J. Crop Sci., 47*, pp. 707–14.

Nash, T. H. (1976). Sensitivity of lichens to nitrogen dioxide fumigations. *Bryologist, 79*, pp. 103–6.

Neyra, C. A. & Hageman, R. H. (1974). Dependence of nitrite reduction on electron transport in chloroplasts. *Plant Physiol., 54*, pp. 480–3.

Nobel, P. S. & Wang, C. T. (1973). Ozone increases the permeability of isolated pea chloroplasts. *Arch. Biochem. Biophys., 157*, pp. 388–94.

Oku, T., Shimazaki, K. & Sugahara, K. (1980). Resistance of spruce seedlings to sulfur dioxide fumigation. *Studies on the Effects of Air Pollutants on Plants and Mechanisms of Phytotoxicity: Res. Rep. Natl Inst. Environ. Stud.,* Japan, **11**, pp. 151–4.

Omasa, K., Abo, F., Natori, T. & Totsuka, T. (1980). Analysis of air pollutant sorption by plants. *Studies on the Effects of Air Pollutants on Plants and Mechanisms of Phytotoxicity: Res. Rep. Natl Inst. Environ. Stud.,* Japan, **11**, pp. 213–24.

Ordin, L. & Altman, A. (1965). Inhibition of phosphoglucomutase activity in oat coleoptiles by air pollutants. *Physiol. Plant., 18*, pp. 790–7.

Osmond, C. B. & Avadhani, P. N. (1970). Inhibition of the β-carboxylation pathway of CO_2 fixation by bisulfite compounds. *Plant Physiol., 45*, pp. 228–30.

Otto, H. W. & Daines, R. H. (1969). Plant injury by air pollutants: Influence of humidity on stomatal apertures and plant response to ozone. *Science, 163*, pp. 1209–10.

Outlaw, W. H., jr & Kennedy, J. (1978). Enzymic and substrate basis for the anaplerotic step in guard-cells. *Plant Physiol., 62*, pp. 648–52.

Pack, M. R. & Wilson, A. M. (1967). Influence of hydrogen fluoride fumigation on acid-soluble phosphorus compounds in bean seedlings. *Environ. Sci. Technol., 1*, pp. 1011–13.

Pahlich, E. (1971). Allosterische Regulation der Aktivität der Glutamate-hydrogenase aus Erbsenkeimlingen durch das Substrat α-Ketoglutarsäure. *Planta, 100*, pp. 222–7.

Pahlich, E. (1973). Uber den Hemm-Mechanismus mitochondrialer Glutamat-Oxalacetat-Transaminase in SO_2 begasten Erbsen. *Planta, 110*, pp. 267–78.

Pahlich, E., Jäger, H. J. & Steubing, L. (1972). Beeinflussung der Aktivitäten von Glutamatdehydrogenase und Glutaminsynthetase aus Erbsenkeimlingen durch SO_2. *Angew. Botanik., 46*, pp. 183–97.

Paul, J. S. & Bassham, J. A. (1978). Effects of sulfite on metabolism in isolated mesophyll cells from *Papaver somniferum. Plant Physiol., 62*, pp. 210–4.

Peiser, G. D. & Yang, S. F. (1977). Chlorophyll destruction by the bisulfite-oxygen system. *Plant Physiol., 60*, pp. 277–81.

Peiser, G. D. & Yang, S. F. (1978). Chlorophyll destruction in the presence of bisulfite and linoleic acid hydroperoxide. *Phytochemistry, 17*, pp. 79–84.

Peiser, G. D. & Yang, S. F. (1979). Ethylene and ethane production from sulfur dioxide injured plants. *Plant Physiol.*, **63**, pp. 142–5.

Pell, E. J. & Brennan, E. (1973). Changes in respiration, photosynthesis, adenosine *5'-triphosphate*, and total adenylate content of ozonated Pinto Bean foliage as they relate to symptom expression. *Plant Physiol.*, **51**, pp. 378–81.

Perchorowicz, J. T. & Ting, I. P. (1974). Ozone effects on plant cell permeability. *Amer. J. Bot.*, **61**, pp. 787–93.

Petering, D. H. & Shih, N. T. (1975). Biochemistry of bisulfite-sulfur dioxide. *Environ. Res.*, **9**, pp. 55–65.

Plaut, Z., Lendzian, K. & Bassham, J. A. (1977). Nitrite reduction in reconstituted and whole Spinach chloroplasts during carbon dioxide reduction. *Plant Physiol.*, **59**, pp. 184–8.

Poovaiah, B. W. & Wiebe, H. H. (1973). Influence of hydrogen fluoride fumigation on the water economy of Soybean plants. *Plant Physiol.*, **51**, pp. 396–9.

Priebe, A., Klein, W. H. & Jäger, H. J. (1978). Role of polyamines in SO_2-polluted pea plants. *J. Exp. Bot.*, **29**, pp. 1045–50.

Psenak, M., Miller, G. W., Yu, M. H. & Lovelace, C. J. (1977). Separation of malic dehydrogenase isoenzymes from soybean tissue in relation to fluoride treatment. *Fluoride*, **10**, pp. 63–72.

Puckett, K. J., Nieboer, E., Flora, W. P. & Richardson, D. H. S. (1973). Sulphur dioxide: Its effects on photosynthetic ^{14}C fixation in lichens and suggested mechanisms of phytotoxicity. *New Phytol.*, **73**, pp. 141-54.

Puritch, G. S. & Barker, A. V. (1967). Structure and function of Tomato leaf chloroplasts during ammonium toxicity. *Plant Physiol.*, **42**, pp. 1229–38.

Rabe, R. & Kreeb, K. H. (1980). Wirkungen von SO_2 auf die Enzymaktivität in Pflauzenblättern. *Z. Pflanzenphysiol.*, **97**, pp. 215–26.

Radmer, R. J. & Kok, B. (1976). Photoreduction of O_2 primes and replaces CO_2 assimilation. *Plant Physiol.*, **58**, pp. 336–40.

Rao, D. N. & LeBlanc, B. F. (1965). Effects of sulfur dioxide on the lichen Alga, with special reference to chlorophyll. *Bryologist*, **69**, pp. 69–75.

Raschke, K. (1975). Stomatal action. *Annu. Rev. Plant Physiol.*, **26**, pp. 309–40.

Raschke, K. (1979). Movements of stomata. Pp. 383–441 in *Physiology of Movements: Encyclopedia of Plant Physiology*, Vol. 7 (Ed. W. Haupt & M. E. Feinleib). Springer-Verlag, Berlin, W. Germany: xvii + 731 pp., illustr.

Rhoads, A. & Brennan, E. (1978). The effect of ozone on chloroplast lamellae and isolated mesophyll cells of sensitive and resistant Tobacco selections. *Phytopathology*, **68**, pp. 883–6.

Ryrie, I. J. & Jagendorf, A. T. (1971). Inhibition of photophosphorylation in chloroplasts by inorganic sulfate. *J. Biol. Chem.*, **246**, pp. 582–8.

Sarkar, S. K. & Malhotra, S. S. (1979). Effects of SO_2 on organic acid content and malate dehydrogenase activity in Jack Pine needles. *Biochem. Physiol. Pflanzen.*, **174**, pp. 438–45.

Schiff, J. A. & Hodson, R. C. (1973). The metabolism of sulfate. *Annu. Rev. Plant Physiol.*, **24**, pp. 381–414.

Schreiber, U., Vidaver, W., Runeckles, V. C. & Rosen, P. (1978). Chlorophyll fluorescence assay for ozone injury in intact plants. *Plant Physiol.*, **61**, pp. 80–4.

Schwenn, J. D., Depka, B. & Hennies, H. H. (1976). Assimilatory sulfate reduction in chloroplasts: Evidence for the participation of both stromal and membrane-bound enzymes. *Plant & Cell Physiol.*, **17**, pp. 165–71.

Sekiya, J., Wilson, L. G. & Filner, P. (1980). Positive correlation between H_2S emission and SO_2 resistance in Cucumber. *Plant Physiol.*, *Suppl.*, **65**, Abs. 407.

Shaw, A. B., Anderson, M. M. & McCarty, R. E. (1976). Role of galactolipids in Spinach chloroplast lamellar membranes, II: Effects of galactolipid depletion on phosphorylation and electron flow. *Plant Physiol.,* **57**, pp. 724–9.

Shimazaki, K. & Sugahara, K. (1980a). Specific inhibition of photosystem II activity in chloroplasts by fumigation of Spinach leaves with SO_2. *Studies on the Effects of Air Pollutants on Plants and Mechanisms of Phytotoxicity: Res. Rep. Natl Inst. Environ. Stud.,* Japan, **11**, pp. 69–77.

Shimazaki, K. & Sugahara, K. (1980b). Inhibition site in electron transport system in chloroplasts by fumigation of Lettuce leaves with SO_2. *Studies on the Effects of Air Pollutants on Plants and Mechanisms of Phytotoxicity: Res. Rep. Natl Inst. Environ. Stud.,* Japan, **11**, pp. 79–89.

Shimazaki, K., Sakaki, T. & Sugahara, K. (1980). Active oxygen participation in chlorophyll destruction and lipid peroxidation in SO_2-fumigated leaves of Spinach. *Studies on the Effects of Air Pollutants on Plants and Mechanisms of Phytotoxicity: Res. Rep. Natl Inst. Environ. Stud.,* Japan, **11**, pp. 91–101.

Sigal, L. L. & Taylor, O. C. (1979). Preliminary studies of the gross photosynthetic response of lichens to peroxyacetyl nitrate fumigations. *Bryologist,* **82**, 564–75.

Silvius, J. E., Ingle, M. & Baer, C. H. (1975). Sulfur dioxide inhibition of photosynthesis in isolated Spinach chloroplasts. *Plant Physiol.,* **56**, pp. 434–7.

Silvius, J. E., Baer, C. H., Dodrill, S. & Patrick, H. (1976). Photoreduction of sulfur dioxide by Spinach leaves and isolated Spinach chloroplasts. *Plant Physiol.,* **57**, pp. 799–801.

Simola, L. K. & Koskimies-Soininen, K. (1980). The effect of fluoride on the growth and fatty acid composition of *Sphagnum fimbriatum* at two temperatures. *Physiol. Plant.,* **50**, pp. 74–7.

Smith, F. A. & Raven, J. A. (1979). Intracellular pH and its regulation. *Annu. Rev. Plant Physiol.,* **30**, pp. 289–311.

Smith, T. A. & Sinclair, C. (1967). The effect of acid feeding on amine formation in Barley. *Ann. Bot.,* **31**, pp. 103–11.

Soldatini, G. F., Ziegler, I. & Ziegler, H. (1978). Sulfite: Preferential sulfur source and modifier of CO_2 fixation in *Chlorella vulgaris. Planta,* 143, pp. 225–31.

Soldatini, G. F. & Ziegler, I. (1979). Induction of glycolate oxidase by SO_2 in *Nicotiana tabacum. Phytochemistry,* **18**, pp. 21–2.

Spotts, R. A., Lukezic, F. L. & LaCasse, N. L. (1975). The effect of benzimidazole, cholesterol, and a steroid inhibitor on leaf sterols and ozone resistance of bean. *Phytopathology,* **65**, pp. 45–9.

Squire, G. R. & Mansfield, T. A. (1973). A simple method of isolating stomata on detached epidermis by low pH treatment: observation of the importance of subsidiary cells. *New Phytol.,* **71**, pp. 1033–43.

Srivastava, H. S., Jolliffe, P. A. & Runeckles, V. C. (1975). Inhibition of gas exchange in bean leaves by NO_2. *Can. J. Bot.,* **53**, pp. 466–74.

Sugahara, K., Uchida, S. & Takimoto, M., (1980). Effects of sulfite ions on water-soluble chlorophyll proteins. *Studies on Effects of Air Pollutants on Plants and Mechanisms of Phytotoxicity: Res. Rep. Natl Inst. Environ. Stud.,* Japan, **11**, pp. 103–12.

Sutton, R. & Ting, I. (1977). Evidence for the repair of ozone-induced membrane injury. *Amer. J. Bot.,* **64**, pp. 404–11.

Swanson, E. S., Thomson, W. W. & Mudd, J. B. (1973). The effect of ozone on leaf cell membranes. *Can. J. Bot.,* **51**, pp. 1213–9.

Tager, J. M. & Rautanen, N. (1956). Sulfite oxidation by plant mitochondrial system: Enzymic and non-enzymic oxidation. *Physiol. Plant.,* **9**, 665–73.

Tanaka, H., Takanashi, T. & Yatazawa, M. (1972*a*). Experimental studies on sulfur dioxide injuries in higher plants, I: Formation of glyoxylate bisulfite in plant leaves exposed to sulfur dioxide. *Water, Air, and Soil Pollution,* 1, pp. 205–11.

Tanaka, H., Takanashi, T., Kadota, M. & Yatazawa, M. (1972*b*). Experimental studies on sulfur dioxide injuries in higher plants, II: Disturbance of amino-acid metabolism in plants exposed to sulfur dioxide. *Water, Air, and Soil Pollution,* 1, pp. 343–6.

Tanaka, K. & Sugahara, K. (1980). Role of superoxide dismutase in the defense against SO_2 toxicity and induction of superoxide dismutase with SO_2 fumigation. *Studies on the Effects of Air Pollutants on Plants and Mechanisms of Phytotoxicity: Res. Rep. Natl Inst. Environ. Stud.,* Japan, 11, pp. 155–64.

Taylor, O. C., Dugger, W. M., Cardiff, E. A. & Darley, E. F. (1961). Interaction of light and atmospheric photochemical products (SMOG) within plants. *Nature* (London), 192, pp. 814–6.

Thomas, M. D. (1951). Gas damage to plants. *Annu. Rev. Plant Physiol.,* 2, pp. 293–322.

Thomas, M. D. & Hill, G. R. (1935). Absorption of sulfur dioxide by Alfalfa and its relation to leaf injury. *Plant Physiol.,* 10, pp. 291–307

Thomas, M. D., Hendricks, R. H., Collier, T. R. & Hill, G. R. (1943). The utilization of sulfate and sulfur dioxide for the nutrition of Alfalfa. *Plant Physiol.,* 18, pp. 343–71.

Thomas, M. D. & Hendricks, R. H. (1956). Effect of air pollution on plants. P. 44 in *Air Pollution Handbook* (Ed. P. L. Magill, F. R. Holden & C. Ackley). McGraw-Hill, New York, NY, USA: xi + 670 pp., illustr.

Thomson, W. W., Dugger, W. M., jr & Palmer, R. L. (1966). Effect of ozone on the fine structure of the palisade parenchyma cells of bean leaves. *Can. J. Bot.,* 44, pp. 1677–82.

Ting, I. P. & Mukerji, S. K. (1971). Leaf ontogeny as a factor in susceptibility to ozone: Amino-acid and carbohydrate changes during expansion. *Amer. J. Bot.,* 58, pp. 497–504.

Tingey, D. T., Fites, R. C. & Wickliff, C. (1973). Ozone alteration of nitrate reduction in Soybean. *Physiol. Plant.,* 29, pp. 33–8.

Tingey, D. T., Fites, R. C. & Wickliff, C. (1975). Activity changes in selected enzymes from Soybean leaves following ozone exposure. *Physiol. Plant.,* 33, pp. 316–20.

Tingey, D. T., Fites, R. C. & Wickliff, C. (1976*a*). Differential foliar sensitivity of Soyabean cultivars to ozone associated with differential enzyme activities. *Physiol. Plant.,* 37, pp. 69–72.

Tingey, D. T., Wilhour, R. G. & Standley, C. (1976*b*). The effect of chronic ozone exposures on the metabolite content of Ponderosa Pine seedlings. *Forest Sci.,* 22, pp. 234–41.

Todd, G. W. (1958). Effect of ozone and ozonated 1-hexene on respiration and photosynthesis of leaves. *Plant Physiol.,* 33, pp. 416–20.

Tomlinson, H. & Rich, S. (1967). Metabolic changes in free amino-acids of bean leaves exposed to ozone. *Phytopathology,* 57, pp. 972–4.

Tomlinson, H. & Rich, S. (1969). Relating lipid content and fatty acid synthesis to ozone injury of Tobacco leaves. *Phytopathology,* 59, pp. 1284–6.

Tomlinson, H. & Rich, S. (1970). Lipid peroxidation, a result of injury in bean leaves exposed to ozone. *Phytopathology,* 61, pp. 1531–2.

Tomlinson, H. & Rich, S. (1971). Effect of ozone on sterols and sterol derivatives in bean leaves. *Phytopathology,* 61, pp. 1404–5.

Tomlinson, H. & Rich, S. (1973). Anti-senescent compounds reduce injury and steroid changes in ozonated leaves and their chloroplasts. *Phytopathology*, **63**, pp. 903–6.

Treshow, M. (1970). Ozone damage to plants. *Environ. Pollut.*, **1**, pp. 155–61.

Treshow, M. (1971). Fluorides as air pollutants affecting plants. *Annu. Rev. Phytopathol.*, **9**, pp. 21–44.

Treshow, M., Harner, F. M., Price, H. E. & Kormelink, J. R. (1969). Effects of ozone on growth, lipid metabolism, and sporulation of Fungi. *Phytopathology*, **59**, pp. 1223–5.

Trevathan, L. E., Moor, L. D. & Orcutt, D. M. (1979). Symptom expression and free sterol and fatty acid composition of flue-cured Tobacco plants exposed to ozone. *Phytopathology*, **69**, pp. 582–5.

Trown, P. W. (1965). An improved method for the isolation of carboxydismutase: Probable identity with Fraction-1 protein and the protein moiety of protochlorophyll holochrome. *Biochemistry*, **4**, pp. 908–18.

Weinstein, L. H. (1961). Effects of atmospheric fluoride on metabolic constituents of tomato and bean leaves. *Contrib. Boyce Thompson Inst.*, **21**, pp. 215–31.

Wellburn, A. R., Majernik, O. & Wellburn, F. A. M. (1972). Effects of SO_2- and NO_2-polluted air upon the ultrastructure of chloroplasts. *Environ. Pollut.*, **3**, pp. 37–49.

Wellburn, A. R., Capron, T. M., Chan, H. S. & Horsman, D. C. (1976). Biochemical effects of atmospheric pollutants on plants. Pp. 106–14 in *Effects of Air Pollutants on Plants* (Ed. T. A. Mansfield). Cambridge University Press, Cambridge, England, UK: 209 pp., illustr.

Wellburn, A. R., Wilson, J. & Aldridge, P. H. (1980). Biochemical responses of plants to nitric-oxide-polluted atmospheres. *Environ. Pollut.*, **22**, pp. 219–28.

White, K. L., Hill, A. C. & Bennett, J. H. (1974). Synergistic inhibition of apparent photosynthesis rate of Alfalfa by combinations of sulfur dioxide and nitrogen dioxide. *Environ. Sci. Tech.*, **8**, pp. 575–6.

Wilkinson, T. G. & Barnes, R. L. (1973). Effects of ozone on $^{14}CO_2$ fixation patterns in pine. *Can. J. Bot.*, **51**, pp. 1573–8.

Wyss, H. R. & Brunold, C. (1979). Regulation of adenosine 5′-phosphosulfate sulfotransferase activity by H_2S and cyst(e)ine in primary leaves of *Phaseolus vulgaris* L. *Planta*, **147**, pp. 37–42.

Wyss, H. R. & Brunold, C. (1980). Regulation of adenosine 5′-phosphosulfate sulfotransferase by sulfur dioxide in primary leaves of beans (*Phaseolus vulgaris*). *Physiol. Plant.*, **50**, pp. 161–5.

Yang, S. F. & Miller, G. W. (1963*a*). Biochemical studies on the effect of fluoride on higher plants. 1. Metabolism of carbohydrates, organic acids and amino acids. *Biochem. J.*, **88**, 509–16.

Yang, S. F. & Miller, G. W. (1963*b*). Biochemical studies on the effect of fluoride on higher plants, 3: The effect of fluorine on dark carbon dioxide fixation. *Biochem. J.*, **88**, 517–22.

Yang, S. F. & Saleh, M. A. (1973). Destruction of indole-3-acetic acid during the aerobic oxidation of sulfite. *Phytochemistry*, **12**, 1463–6.

Yee-Meiler, D. (1975). Über die Eignung von Phosphotase- und Esteraseaktivitätsbestimmungen an Fichtennadeln und Birkenblättern Zum Nachweis, 'Unsichtbarer Physiologischer' Fluorimissionsschädigungen. *Eur. J. For Path.*, **5**, pp. 329–38.

Yoneyama, T. & Sasakawa, H. (1979). Transformation of atmospheric NO_2 absorbed in Spinach leaves. *Plant and Cell Physiol.*, **20**, pp. 263–6.

Yoneyama, T., Sasakawa, H. & Ishizuka, S. (1979). Absorption of atmospheric NO_2 by plants and soils, II: Nitrite accumulation, nitrite reductase activity, and diurnal change, of NO_2 absorption in leaves. *Soil Sci. Plant Nutr.*, **25**, pp. 267–75.

Yung, K. H. & Mudd, J. B. (1966). Lipid synthesis in the presence of nitrogenous compounds in *Chlorella pyrenoidosa*. *Plant Physiol.,* **41**, pp. 506–9.

Zeevaart, A. J. (1974). Induction of nitrate reductase by NO_2. *Acta Bot. Neerl.,* **23**, pp. 345–6.

Zeevaart, A. J. (1976). Some effects of fumigating plants for short periods with NO_2. *Environ. Pollut.,* **11**, pp. 97–108.

Zelitch, I. (1957). α-Hydroxysulfonates as inhibitors of the enzymatic oxidation of glycolic and lactic acids. *J. Biol. Chem.,* **224**, pp. 251–60.

Zelitch, I. (1971). *Photosynthesis, Photorespiration and Plant Productivity.* Academic Press, New York–San Francisco–London: xiv + 347 pp., illustr.

Ziegler, I. (1972). The effect of SO_3 on the activity of ribulose-1,5-diphosphate carboxylase in isolated Spinach chloroplasts. *Planta,* **103**, pp. 155–63.

Ziegler, I. (1973*a*). Effect of sulphite on phosphoenolpyruvate carboxylase and malate formation in extracts of *Zea mays. Phytochemistry,* **12**, pp. 1027–30.

Ziegler, I. (1973*b*). The effect of air polluting gases on plant metabolism. Pp. 182–208 in *Environmental Quality and Safety: Global Aspects of Chemistry, Toxicology and Technology as Applied to the Environment* (Ed. F. Coulson & F. Korte), Vol. II. Georg Thime, Stuttgart, West Germany and Academic Press, New York & London: xviii + 333 pp., illustr.

Ziegler, I. (1974). Action of sulphite on plant malate dehydrogenase. *Photochemistry,* **13**, pp. 2411–6.

Ziegler, I. (1975). The effect of SO_2 pollution on plant metabolism. *Residue Rev.,* **56**, pp. 79–105.

Ziegler, I. (1977*a*). Subcellular distribution of ^{35}S-sufur in Spinach leaves after application of [35]SO_3^{2-}, and [35]SO_2. *Planta,* **135**, pp. 25–32.

Ziegler, I. (1977*b*). Sulfite action on ribulose diphosphate carboxylase in the lichen *Pseudevernia furfuracea. Oecologia,* **29**, pp. 63–6.

Ziegler, I. Marewa, A. & Schoepe, E. (1976). Action of sulphite on the substrate kinetics of chloroplastic NADP-dependent glyceraldehyde 3-phosphate dehydrogenase. *Phytochemistry,* **15**, pp. 1627–32.

Ziegler, I. & Hampp, R. (1977). Control of [35]SO_4^{2-} and [35]SO_3^{2-} incorporation into Spinach chloroplasts during photosynthetic CO_2 fixation. *Planta,* **137**, pp. 303–7.

Air Pollution and Plant Life
Edited by M. Treshow
© 1984 John Wiley & Sons Ltd.

CHAPTER 8

Cellular and Ultrastructure Effects

SIRKKA SOIKKELI & LAURI KARENLAMPI

Ecological Laboratory, Department of Environmental Hygiene, University of Kuopio, P.O. Box 138, SF-70101 Kuopio 10, Finland

INTRODUCTION

Light- and electron-microscope studies concerning the effects of air pollution on plants are rather limited compared with those made at the biochemical or physiological level, not to mention those concerning visible injuries. Thus light-microscopical methods have not been used very effectively in past or recent years, and whereas electron-microscopical studies were conducted in the 1960s, they only became prominent in the 1970s. Some reviews on the ultrastructural changes caused by air pollution were also written in the 1970s (e.g. Mudd & Kozlowski, 1975; Karenlampi & Soikkeli, 1980). The studies were restricted mainly to pollutants which were thought to be most harmful to plants—such as SO_2, NO_2, O_3 and HF. Most studies have been concerned with the effects of a single pollutant; combined effects or effects under field conditions were rarely reported.

The aims of structural research on the effects of pollutants can be superficially classified as: (1) to clarify the mechanisms of the influence of the pollutants, and (2) to develop and evaluate methods of practical management of the environment. The problems addressed by different authors have included: can the injury be detected before visible symptoms appear?, what kind of injury develops?, what is the primary site of action?, and are there differences in reactions between different ages? This chapter deals with current information about the structural effects of air pollutants on higher plants as learned from light- and transmission-electron-microscope studies.

STRUCTURAL EFFECTS

Effects on Cell Level of Organization

Regardless of the kind of pollutant or pollutants responsible for cell injury, the effects on both broad-leafed plants and conifers are relatively similar.

Most often the microscopic damage is reported from material that also has macroscopic or visible symptoms, such as tip-burn. Disorganization of chloroplasts is described as one of the first events to appear. This can be distinguished by a change in the staining properties of the cell or tissue (Bursche, 1955; Solberg & Adams, 1956). When injury is more severe, the cell contents, including the chloroplasts, aggregate into one or more dark masses in the middle of the cell or near the cell wall (Solberg & Adams, 1956; Wei & Miller, 1972). Finally the cell walls collapse. This is the most common description of injury that is given in several studies (Solberg & Adams, 1956; Evans & Miller, 1973; Wei & Miller, 1972; Stewart *et al.*, 1973; Smith & Davis, 1978).

More recently, Soikkeli & Tuovinen (1979) have observed badly damaged cells in apparently healthy Norway Spruce (*Picea abies*) needles. Soikkeli (1981) reported in more detail the stages of cell injury in green Norway Spruce and Scots Pine (*Pinus sylvestris*) needles collected from industrial areas chronically polluted by SO_2, NO_x, and/or fluorides. Three different

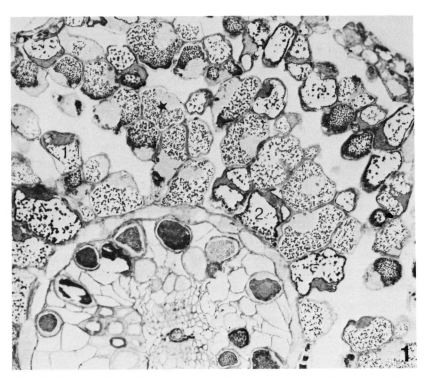

Fig. 1   Cross-section of part of a green Norway Spruce needle collected from area polluted mainly by S-compounds. Healthy cells (star) and cells at the first (1) and second (2) stages of cell injury can be seen. × 250.

stages of cell injury were described. In the first stage the tannin appeared as a thin ribbon and/or scattered particles in the central vacuole, whereas normally it appeared granular (Soikkeli, 1978, 1980). Simultaneously the cytoplasmic constituents became darker than formerly. In the second stage the tannin mainly appeared as a thick ribbon, and cytoplasm could be seen as a dark, rather thin mass between the tannin and cell walls. In the third stage of cell injury, tannin appeared as a ribbon or disappeared entirely, and no cytoplasmic constituents could be resolved. Collapsed cell walls could rarely be found. Different stages of cell injuries are apparent in the same mesophyll tissue (Fig. 1).

It is obvious that the final disorganization and collapse of cell walls, accompanied by the widening of intercellular spaces, occurs to a great extent only when macroscopic symptoms also exist. The collapse of cells is likely to be an acute injury caused by the high concentration of pollutants used in experiments. In the field, the long-term action of low concentrations during a longer period is more usual and significant than the short-term action of high concentrations. When the injury is chronic, it is possible to follow and determine the stages of cell injury as described by Soikkeli (1981).

Effects on Tissue Level

Pell & Weissberger (1976) reported that cells along the veins of the primary leaves of Soybean (*Glycine max*) are the most ozone-sensitive cells. They suggested that, because these cells have a function in the transport of photosynthates, injury here may be significant to other functions of the Soybean plant. Ledbetter *et al.* (1959) and Hill *et al.* (1961) reported that the palisade parenchyma cells are preferentially injured by ozone. Pell & Weissberger (1976) found injury in both types of mesophyll—the greater amount of injury in palisade parenchyma than in spongy parenchyma being, according to them, directly related to the numbers of cells of each type that were present.

The histological responses caused by several stresses (e.g. salt, boron, moisture stress, winter injury, HF, O_3, SO_2, SO_2 + HF, and O_3 + HF) on each of the major tissue types of conifer needles were studied by Stewart *et al.* (1973). This was done in order to evaluate the use of pathological anatomy as a diagnostic tool to distinguish the causal agents. They found that all stresses caused a general hypertrophy of the epithelial tissue of the resin canals, with granulation of transfusion and mesophyll parenchyma. Collapse of the mesophyll cells, according to these authors, best characterized necrosis that was caused by ozone, sulphur dioxide, salt, or boron toxicity. Smith & Davis (1978) described very similar changes in tip-burned cotyledons, and in primary and secondary needles of Scots Pine seedlings. Despite a few specific variations, the basic response of Pine needles to most stresses is fundamentally the same. Accordingly, Stewart *et al.* (1973) reported that the application of pathological anatomy to field diagnosis appears to be limited, though elimination of certain pathogens as causal agents is possible.

Soikkeli (1981) reported the development of damage in mesophyll of Norway Spruce and Scots Pine needles collected from different polluted environments in Finland. The first cells to be injured were either immediately beneath the epidermis, near to the stomata, or else close to the endodermis and near the intercellular spaces. Injury later spread into the inner mesophyll. The last healthy-looking cells could be found only in the corners of the needle cross-section. Thus the development of injury as viewed in cross-section indicates that pollutants enter needles through the stomata and spread from there *via* the intercellular spaces. The extent of damage can be classified from slight to average, severe, or very severe, and it develops similarly in both of the conifer species studied. The stage of cell injury is minor when the damage seen in cross-section is slight. Cell injury intensifies as the damage in cross-section spreads. Near pollutant sources, the advancement of damage leads more rapidly to the very severe category than farther away.

The value of light-microscopical studies for diagnostic use in relation to air pollutants, according to current knowledge, is questionable. In some respects, however, they render many valuable possibilities. For instance, the relationship between pollutant concentrations of the ambient air and the extent of damaged tissue could be worth studying. Dose–response curves thus obtained could be more informative than those obtained from concentrations and visible symptoms.

Soikkeli (1981) studied the relationship between the extent of damage at different pollutant concentrations in the ambient air and chemical analyses of conifer needles. Together with the electron-microscopical studies, the light-microscope observations provide a most valuable tool in localizing the damaged tissues or cells, which are then studied ultrastructurally. In this way, the detailed ultrastructural observations can be better understood than otherwise.

ULTRASTRUCTURAL EFFECTS

Plants exposed to low concentrations of pollutants can show a reduction of growth and yield without any visible symptoms. Many previous biochemical and physiological studies have shown that various aspects of photosynthesis are inhibited by pollutants. Consequently, previous ultrastructural studies have strongly concentrated on the photosynthesizing cells and especially on their chloroplasts.

Shape, Size, etc., of the Chloroplasts

Needles of *Larix leptolepis* (Japanese Larch) fumigated with SO_2 (2.5 ppm for 8 hr during three days) by Mlodzianowski & Bialobok (1977) showed

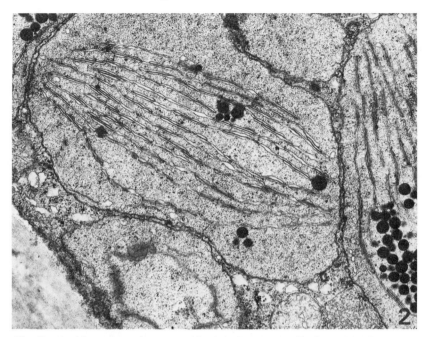

Fig. 2　A chloroplast of a green Norway Spruce needle from S-polluted environment. The grana are reduced, forming 2 or 3 lamellae. × 24,000.

gradual changes in the chloroplast profile, from ellipsoidal to oval and then spherical. The observations were made from mesophyll cells of the middle portion of needles having clear, visible signs of degradation as well. The same phenomenon was reported by Soikkeli & Tuovinen (1979) in *Picea abies* and by Soikkeli (1981) in *Pinus sylvestris* from mesophyll cells of needles collected from areas polluted mainly by S-compounds (Fig. 2).

The rounded appearance of chloroplasts has also been reported in Soybean plants fumigated with 40–50 ppb of HF for 1 to 6 days by Wei & Miller (1972). The overall decrease in chloroplast size was found in Sugar-cane (*Saccharum officinarum*) leaves during the development of chlorotic fluoride injury symptoms (Engelbrecht & Louw, 1973). Bligny *et al.* (1973) found that *Abies alba* (Silver Fir) plants affected by fluoride also had smaller chloroplasts than the controls.

The first apparent incipient changes induced by ozone in *Nicotiana tabacum* (Tobacco) (Swanson *et al.*, 1973) and in Kidney Bean (*Phaseolus vulgaris*) consisted of an irregular chloroplast shape, particularly at the surface adjacent to the plasmalemma (Thomson *et al.*, 1974).

Chloroplast Envelope

Godzik & Knabe (1973) have reported invaginations from the inner membrane of chloroplast envelopes, and doubling of the envelopes in needles of some *Pinus* species collected from industrial areas polluted by S-compounds. The same also was observed in *Phaseolus vulgaris* fumigated with 0.7 ppm SO_2 for 72 hours by Godzik & Sassen (1974). The observations were made from material having no symptoms that were visible to the naked eye.

Chloroplast envelopes of green needles of conifers growing in areas polluted by S-compounds were ruptured at the later stage of cell injury (Soikkeli & Tuovinen, 1979; Soikkeli, 1981). In the first stage of cell injury, the envelope became stretched and later ruptured in apparently healthy conifer needles collected from areas polluted by fluorides (Figs. 4 and 5). Envelope disruption was also found in Sugar-cane leaves fumigated with HF by Engelbrecht & Louw (1973). However, in this species the rupturing was observed only in the visibly necrotic leaves.

Several authors have reported small but noticeable alterations in the ultrastructure of the envelope as one of the first signs of injury caused by ozone (Thomson *et al.*, 1966, 1974; Swanson *et al.*, 1973; Pell & Weissberger, 1976). These changes showed primarily an increase in the staining density and accumulation of electron-dense material between the two membranes of the envelope.

Dolzmann & Ullrich (1966) and Lopata & Ullrich (1975) observed tubular protrusions extending from the envelopes of chloroplasts in Kidney Bean leaves fumigated with NO_2. The protrusions were associated with mitochondria and appeared to enclose the mitochondria either partially or entirely.

Stroma

Godzik & Sassen (1974) described two types of vesicles in the periphery of the chloroplast stroma in SO_2-treated Kidney Bean leaves. They also found rod-like bundles that may have a connection with crystalline bodies found in stroma of some chloroplasts in a later stage of damage. Granulation of stroma was observed as the first change induced by SO_2 by Fischer *et al.* (1973). The granulation has been reported also in *Larix leptolepis* polluted by SO_2 (Mlodzianowski & Bialobok, 1977), and in *Picea abies* and *Pinus sylvestris* growing in S-polluted, or S- and F-polluted, areas in Finland (Soikkeli & Tuovinen, 1979; Soikkeli, 1981). These injuries, however, were found only at the later stage of cell injury.

Several authors have indicated a significant change in the stroma after ozone fumigation. Thomson *et al.* (1966) described an increase in the granulation and electron density of the stroma to be among the first changes before the appearance of any visible symptoms. They also reported the appearance of clusters and ordered arrays of fibrils in some chloroplasts.

Later Thomson & Swanson (1972) found that crystalline structures were present in the stroma after ozone treatment. Swanson *et al.* (1973) reported an increase in the density of stroma but no crystalloids.

Thomson *et al.* (1974), studying the effects of O_3 on *Phaseolus vulgaris*, also found fibrils and crystalloids. These were noted in the mild stage of injury frequently associated with the envelopes, especially where the envelopes had symptoms of injury. The crystalloids in moderately injured cells were extremely large and appeared not only near the envelopes but also in the vicinity of grana and stroma thylakoids. The crystalloids were also seen in severely damaged cells where they were distributed throughout the aggregated mass (Thomson *et al.*, 1974).

The crystalline bodies have also been observed in NO_2-fumigated material by Dolzmann & Ullrich (1966). Lopata & Ullrich (1975) reported a dense layer of filaments in the stroma of *Phaseolus vulgaris* after NO_2 treatment.

Thylakoids

Changes described in the chloroplast disks, or thylakoids, of plants fumigated with SO_2, include swelling of the lamellae and reduction of the grana. Wellburn *et al.* (1972) in *Vicia faba* (Broad Bean), Godzik & Knabe (1973) in some *Pinus* species, Malhotra (1976) in *P. contorta* (Lodgepole Pine), and Wong *et al.* (1977) in *Pisum*, all described slight swelling of stroma lamellae in the first stage of cell injury caused by SO_2. Later the swelling increased and could be detected in the granum thylakoids, particularly in those at the 'top' and 'bottom' of the granum stacks. In severe cell-injury, all thylakoids were swollen. Wellburn *et al.* (1972) also described this kind of swelling in NO_2-treated material. They further found that the swelling appeared to be reversible when the samples, after fumigation, were treated with unpolluted air.

Mlodzianowski & Bialobok (1977) described two types of injury caused by SO_2 in thylakoids of *Larix leptolepis* (Japanese Larch). One involved the disappearance of thylakoids and the other their swelling. These authors suggested that the first type prevails in plants which are more resistant to SO_2.

Godzik & Knabe (1973) in some *Pinus* species, and Godzik & Sassen (1974) in *Phaseolus vulgaris*, reported reduction of grana in otherwise apparently healthy material affected by SO_2. Later, Soikkeli & Tuovinen (1979) in *Picea abies*, and Soikkeli (1981) in *Pinus sylvestris*, found that the grana were often reduced, consisting of 2–3 lamellae in apparently healthy needles collected from areas polluted by S-compounds (Fig. 2 and contrast Fig. 3). The reduced lamellae were found to swell only at a later stage of injury, after the envelopes had shown disintegration.

Reduction of grana has also been established in HF-fumigated *Phaseolus vulgaris* by Wei & Miller (1972), and in Sugar-cane by Engelbrecht & Louw

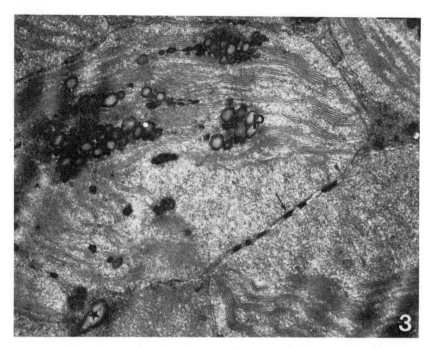

Fig. 3 Lightening of plastoglobuli (P) and small electron-dense particles (arrow) and lipid-like material (star) in cytoplasm of green Norway Spruce needle from S-polluted environment in Finland. × 21,000.

(1973). Horvath *et al.* (1978) described a lack of grana and dilation of the thylakoids of *Vicia faba* 24 hours after HF fumigation.

Soikkeli & Tuovinen (1979) in *Picea abies*, and Soikkeli (1981) in *Pinus sylvestris*, have reported two types of lamellar injures in green needles collected from F-polluted areas. One was the 'smooth' swelling of thylakoids, which began from stroma lamellae. At a later stage of injury, this spread to the grana lamellae (Fig. 4). (The swelling of the granal-fretwork system has been reported earlier in *Abies alba* treated with fluoride (Bligny *et al.*, 1973).) The other injury described by Soikkeli & Tuovinen (1979) in F-polluted conifer needles was the strong curling of the thylakoids without any apparent swelling (Fig. 5). At a very late stage of this type of injury, the thylakoids were found as undulating structures in entirely damaged cytoplasm.

Fumigation of *Spinacia oleracea* (Spinach) with HCl revealed that low concentrations accelerated the development of grana in young leaves (Masuch *et al.*, 1973). In fully developed chloroplasts, 1.6 mg HCl/m^3 of air caused a greater average thickness of grana when compared with the controls.

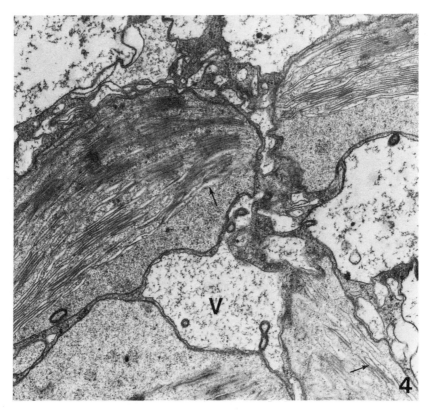

Fig. 4   Part of a cell of green Norway Spruce needle from S- and/or F-polluted area. Note the stretched envelopes and 'smooth' swelling of the thylakoids (arrow). The cytoplasm has plenty of vacuoles (V). × 22,000.

Plastoglobuli

An increase in size and number of plastoglobuli was described in *Spinacia oleracea* fumigated with SO_2 or HCl by Masuch *et al.* (1973). They also reported the appearance of many osmiophilic granules in close contact with the thylakoids in SO_2-treated material. The increase in size and number of plastoglobuli also has been observed in Soybean (Wei & Miller, 1972) and Sugar-cane (Engelbrecht & Louw, 1973) fumigated with HF. In visibly healthy conifer needles exposed to S-compounds, Soikkeli & Tuovinen (1979) and Soikkeli (1981) reported that the lightening of plastoglobuli as the first signs of injury (cf. Fig. 3). At a later stage, the shape of plastoglobuli changed, and their number increased.

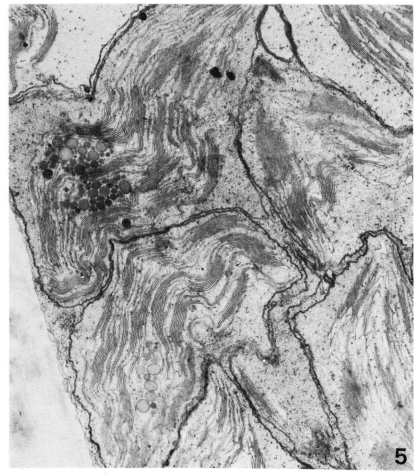

Fig. 5   Chloroplasts with stretched envelopes and curled thylakoids. Green
Norway Spruce needle from S- and/or F-polluted environment. × 22,000.

Mitochondria

Observations on the ultrastructure of mitochondria from polluted material
are scanty. Wellburn *et al.* (1972) and Wong *et al.* (1977) did not observe any
alterations in the extra-chloroplastic cytoplasm in material fumigated with
SO_2 or NO_2. Soikkeli & Tuovinen (1979), in *Picea abies* growing either in
SO_2- or SO_2 and /or F-polluted areas, reported the mitochondria to stay
mainly healthy although the chloroplasts showed severe injury. The portion
of injured mitochondria increased when the general injury progressed.

Engelbrech & Louw (1973) did not observe injury in mitochondria in

HF-fumigated Soybean when many other injuries were evident. When the damage became worse after 4 days from fumigation, deterioration of mitochondria was more evident.

Swelling of mitochondria has also been reported on Tobacco leaves fumigated with ozone (Swanson *et al.*, 1973). Later, Thomson *et al.* (1974) found mitochondria which had electron-dense accumulations in association with the bounding envelopes in moderately ozone-damaged bean cells.

Cytoplasm, Ribosomes, and Endoplasmic Reticulum

Results concerning the cytoplasm are very scanty. Mlodzianoswki & Bialobok (1977) reported granulation of the cytoplasm in evidently injured needles of *Larix leptolepis*. Soikkeli & Tuovinen (1979) found that in Norway Spruce needles polluted by S-compounds, small, electron-dense particles and/or lipid-like material appeared in the cytoplasm in connection with lightened plastoglobuli (cf. Fig. 3). The disturbance was light at first, but later, as the cell injury became worse, the amount of lipid-like material increased. Also, the cytoplasm appeared granulated in severely injured cells.

Wei & Miller (1972) found that the first changes in HF-fumigated Soybean leaves involved an increase in the amount of endoplasmic reticulum (ER) and its aggregation before any macroscopically visible injuries appeared. Subsequently, when the injury became visible, the amount of ER decreased and small vacuoles as well as lipid droplets appeared in the cytoplasm. A decrease in polysomes and a detachment of ribosomes from the rough endoplasmic reticulum were also observed. Soikkeli & Tuovinen (1979) have reported that the polysomes in conifer needles polluted by S- and/or F-compounds were scarce in spring and summer, when normally they are numerous. The vacuolization was described by Soikkeli & Tuovinen (1979) in *Picea abies* exposed to fluorides (cf. Fig. 4), and by Lopata & Ullrich (1975) in *Phaseolus vulgaris* exposed to NO_2. Engelbrecht & Louw (1973) reported the cytoplasm to be very granular in Sugar-cane only when leaves fumigated with HF became necrotic.

The effects of ozone on cytoplasm are mostly described only in severely damaged cells. The cell contents are then aggregated into a dense, collapsed mass around the periphery of the cells (Swanson *et al.*, 1973; Thomson *et al.*, 1974; Pell & Weissberger, 1976).

Plasmalemma, Tonoplast, Cell-walls, etc.

Observations of membranes other than those of chloroplasts are very limited. When they are mentioned, their disturbance is reported to occur at a later stage of injury. Following exposures to ozone, neither the tonoplast nor plasmalemma were damaged even in moderately injured cells (Thomson *et al.*,

1974). Wei & Miller (1972) reported that in HF-treated Soybean leaves the tonoplast began to rupture in an early phase of the injury. Plasmalemma, instead, appeared to be intact although cell disintegration was severe (in severely chlorotic leaves 5 days after fumigation).

Engelbrecht & Louw (1973) described the swelling and dense staining of the middle lamellae that accompanied the development of chlorotic fluoride-injury systems. During persistence of the dark-brown condition, the tono-plast disappeared whereas the plasmalemma was reported to disappear only when the leaf was in the necrotic condition.

Functional Connections of the Ultrastructure Changes

Many researchers believe that the membranes, especially those of chloro-plasts, are the sites most sensitive to pollutants. Current results often indicate that SO_2 causes the swelling of the thylakoids or reduction in grana lamellae. Whatever level or type the injury is, its appearance in chloroplast structure is obviously harmful to photosynthesis. The injuries in thylakoids are likely to be connected with the decline in the amount of chlorophyll and reduced CO_2 uptake observed, for example in pine needles, before other changes became evident (Malhotra & Hocking, 1976; Malhotra, 1977).

Swelling and reduction of grana has been described in HF-fumigated mater-ial in some species. Changes in thylakoids, differing from those found in S-polluted material, have been reported in spruce and pine needles exposed to fluoride. These include the stretching of chloroplast envelopes and simul-taneous curling or swelling of thylakoids without reduction of grana. These changes are thought to be caused by fluorides.

Many results concerning the effects of ozone indicate that membranes are very sensitive to this pollutant. Pell & Weissberger (1976) reported that O_3 does not react with the same membranes. They suggested that an attack on the plasmalemma and/or tonoplast may lead to death, whereas injury to other membranes only limits cell function. Thomson *et al*. (1974) described the first changes induced by O_3 in chloroplast envelopes, explaining that the per-meability of the envelopes changes, after which the stroma looses water. The dehydration of stroma could then be observed as the formation of crystalloids.

Disturbances in the formation of polysomes, or the decrease in ribosomes induced by pollutants, suggest the inhibition of protein synthesis. The mitochondria, when observed, often do not show injury or show it only at a later stage of cell injury. These observations agree with those made by Furukawa *et al*. (1980), where mitochondrial activity was probably not in-hibited by SO_2.

The same ultrastructural changes have often been observed to be associated with different pollutants either in the same or in different plant species. For instance, crystalloids have been reported not only in O_3 -treated material but also in plants fumigated with SO_2 and NO_2; swelling of thylakoids and changes in plastoglobuli have been observed both in materials fumigated with SO_2 and in those fumigated with HF. However, many authors have shown that ultrastructural studies can be used for diagnostic purposes, especially when no obvious injury exists (e.g. Godzik & Knabe, 1973; Masuch *et al.*, 1973; Soikkeli & Tuovinen, 1979).

When applying the ultrastructural methods for resolving the mechanisms of phytotoxicity or for diagnostic purposes, several factors must be noticed. Results obtained on one species cannot be extrapolated directly to others. Also, the same species can react to the same pollutant differently on different occasions—depending on varying climatic or edaphic factors, seasons, age of plants, nutritional status, or other environmental variables.

Whatever the species or climatic or other factors are, electron micro-scopical studies should be directed to resolve the effects of low pollutant concentrations over a long period. In this way it would be possible to distinguish the very first injury symptoms and identify the specific changes caused by different pollutants. High concentrations usually cause acute injury (collapse of cells) and cell death. Then, and at the very last stages of the gradually proceeding cell injury, the changes caused by different pollutants—such as the granulation of cytoplasm and stroma often described in severely damaged cells—and the final disruption of membranes and plasmolysis, resemble each other. In these cases it is much more difficult than earlier on to distinguish or suspect the causal agents, although sometimes it may be possible.

The final destruction of cytoplasm and cell organelles could be due to the denaturing effect of tannin after tonoplast rupture, or to releasing of lytic enzymes after the vacuole tonoplast breaks down (Butler & Simon, 1971). The causes of all structural changes, such as those in the structure of tannin which were seen at both the light- and electron-microscope levels, cannot yet be understood from the functional point of view.

Further, it would be useful to study the same species both in controlled experiments and in the field, in order to clarify the role of different pollutants. Other methods such as light microscopy, and biochemical, physiological, and chemical, measurements should be included in the ultrastructural observations more effectively than has been done hitherto. All these factors are important in improving our understanding of the mechanisms of action, as well as in the application of fine-structural methods to practical management of environmental problems.

CONCLUSIONS

In structural studies, light-microscopy is a good method for evaluating the extent of damage as well as in resolving dose–response questions. Electron microscopy provides a tool to study the mechanisms of pollutant phytotoxicity. It also makes possible the study of injuries before the appearance of damage that is visible to the naked eye, and also the distinguishing of probable causal agents. These two methods should be used together when both the quantity and quality of the effects of pollutants on plants are to be studied.

Structural studies should be directed more strongly than hitherto to resolve the effects of low and combined pollutant concentrations over long periods of time. Other biological studies may be required for both theoretical considerations and practical applications.

SUMMARY

Studies of the effects of pollutants on the ultrastructure of cells have helped to clarify the mechanisms of their influence. The electron-microscope can reveal injuries before any are apparent to the naked eye or can even be seen through the light-microscope. The chloroplast membrane becomes stretched and then ruptured while leaves still appear healthy.

This appears to be the typical response to sulphur compounds, fluorides, and ozone—regardless of the plant species. Following this the chloroplast stroma becomes granulated, the thylakoids begin to swell, and plastoglobuli increase in size and number. Other organelles, membranes, and the cytoplasm, subsequently are effected as symptoms become apparent through light-microscopy.

REFERENCES

Bligny, R., Bisch, A.-M., Garrec, J. P. & Fourcy, A. (1973). Observations morphologiques et structurales des effets du fluor sur les cires épicuticulaires et sur les chloroplastes des ai-guilles de sapin (*Abies alba* Mill.) *J. Microscopie,* **17**, pp. 207–14.

Bursche, E. M. (1955). Vegetationsschäden durch Fluor. *Schr.-R. Ver. Wasser-, Boden- u. Lufthyg.* **10**.

Butler, R. D. & Simon, E. W. (1971). Ultrastructural aspects of senescence in plants. Pp. 73–129 in *Advances in Gerontological Research* (Ed. B. L. Strehler). Academic Press, New York & London: xiv + 262 pp., illustr.

Dolzmann, P. & Ullrich. H. (1966). Einige Beobachtungen über Beziehungen zwischen Chloroplasten und Mitochondria im Palisadenparenchym von *Phaseolus vulgaris. Z. Pflanzenphysiol.,* **55**, pp. 165–80.

Engelbrecht, A. H. P. & Louw, C. W. (1973). Hydrogen fluoride injury in sugar-cane: Some ultrastructural changes. Pp. A157–159 in *Proceedings of the Third International Clean Air Congress.* VDI-Verlag GmbH., Düsseldorf, West Germany: xxii + A170 pp. + B108 pp. + C139 pp. + D71 pp. + E151 pp. + F36 pp., illustr.

Evans, L. S. & Miller, P. R. (1973). Ozone damage to Ponderosa Pine: a histological and histochemical appraisal. *Am. J. Bot.,* **59**, pp. 297–304.

Fischer, F., Kramer, D. & Ziegler, H. (1973). Elektronenmikroskopische Untersuchungen SO$_2$-begaster Blätter von *Vicia faba*. I. Beobachtungen am Chloroplasten mit akuter Schädigung. *Protoplasma,* **76**, pp. 83–96.

Furukawa, A., Natori, T. & Totsuka, T. (1980). The effect of SO$_2$ on net photosynthesis in sunflower leaf. Pp. 1–8 in *Studies on the Effects of Air Pollutants on Plants and Mechanisms of Phytotoxicity. Res. Rep. Nat. Inst. Environ. Stud. No. 11.* Yatabe-machi, Tsukuba, Ibraki 305, Japan: 265 pp., illustr.

Godzik, S. & Knabe, W. (1973). Vergleichende elektronenmikroskopische Untersuchungen der Feinstruktur von Chloroplasten einiger Pinus-Arten aus den Industriegebieten an der Ruhr und in Oberschlesien. Pp. A164–170 in *Proceedings of the Third International Clean Air Congress.* VDI-Verlag GmbH., Düsseldorf, West Germany xxii + A170 pp. + B108 pp. + C139 pp. + D71 pp. + E151 pp. + F36 pp., illustr.

Godzik, S. Sassen, M. M. A. (1974). Einwirkung von SO$_2$ auf die Feinstruktur der Chloroplasten von *Phaseolus vulgaris. Phytopath. Z.,* **79**, pp. 155–9.

Hill, A. C., Pack, M. R. Treshow, M., Downs, R. S. & Transtrum, L. G. (1961). Plant injury induced by ozone. *Phytopathology,* **51**, pp. 356–63.

Horvath, I., Klasova, A. & Navara, J. (1978). Some physiological and ultrastructural changes of *Vicia faba* L. after fumigation with hydrogen fluoride. *Fluoride,* **11**, pp. 89–99.

Karenlampi, L. & Soikkeli, S. (1980). Morphological and fine structural effects of different pollutants on plants: Development and problems of research. Pp. 92–9 in *Papers Presented to the Symposium on the Effects of Air-borne Pollution on Vegetation.* United Nations, Economic Commission for Europe, Warsaw, Poland: xx + 410 pp., illustr.

Ledbetter, M. C., Zimmerman, P. W. & Hitchcock, A. E. (1959). The histological effects of ozone on plant foliage. *Contrib. Boyce Thomson Inst.,* **20**, pp. 225–82.

Lopata, W. D. & Ullrich, H. (1975). Untersuchungen zu stofflichen und strukturellen Veränderungen an Pflanzen unter NO$_2$-Einfluss. *Staub-Reinhalt. Luft,* **35**(5), pp. 196–200.

Malhotra, S. S. (1976). Effects of sulphur dioxide on biochemical activity and ultrastructural organization of pine needle chloroplasts. *New Phytol.,* **76**, pp. 239–245.

Malhotra, S. S. (1977). Effects of aqueous sulphur dioxide on chlorophyll destruction in *Pinus contorta. New Phytol.,* **78**, pp. 101–9.

Malhotra, ·S. S. & Hocking, D. (1976). Biochemical and cytological effects of sulphur dioxide on plant metabolism. *New Phytol.,* **76**, pp. 227–37.

Masuch, G., Weinert, H. & Guderian, R. (1973). Wirkungen von Chlorwasserstoff und Schwefeldioxid auf die Ultrastruktur der Chloroplasten von *Spinacia oleracea* L. Pp. A160–163 in *Proceedings of the Third International Clean Air Congress.* VDI-Verlag GmbH., Düsseldorf: xxii + A170 pp. + B108 pp. + C139 pp. + D71 pp. + E151 pp. + F36 pp., illustr.

Mlodzianowski, F. & Bialobok, S. (1977). The effect of sulphur dioxide on ultrastructural organization of larch needles. *Acta Soc. Bot. Poloniae,* **46**(4), pp. 629–34.

Mudd, J. B. & Kozlowski, T . T. (1975). *Responses of Plants to Air Pollution.* Academic Press, New York–San Francisco–London: xii + 383 pp., illustr.

Pell, E. J. & Weissberger, W. C. (1976). Histopathological characterization of ozone injury to Soybean foliage. *Phytopathology,* **66**, pp. 856–61.

Smith, H. J. & Davis, D. D. (1978). Histological changes induced in Scotch Pine needles by sulfur dioxide. *Phytopathology,* **68**, pp. 1711–6.

Soikkeli, S. (1978). Seasonal changes in mesophyll ultrastructure of needles of Norway Spruce (*Picea abies* (L.) Karst.). *Can. J. Bot.,* **56**(16), pp. 1932–40.

Soikkeli, S. (1980). Ultrastructure of the mesophyll in Scots Pine and Norway Spruce: Seasonal variation and molarity of the fixative buffer. *Protoplasma,* **103**, pp. 241–52.

Soikkeli, S. (1981). Comparison of cytological injuries in conifer needles from several polluted industrial environments in Finland. *Ann. Bot. Fennici,* **18**, pp. 47–61.

Soikkeli, S. & Tuovinen, T. (1979). Damage in mesophyll ultrastructure of needles of Norway spruce in two industrial environments in central Finland. *Ann. Bot. Fennici,* **16**, pp. 50–64.

Solberg, A. & Adams, D. F. (1956). Histological responses of some plant leaves to hydrogen fluoride and sulfur dioxide. *Am. J. Bot.,* **43**, pp. 755–760.

Stewart, D., Treshow, M. & Harner, F. M. (1973). Pathological anatomy of conifer needle necrosis. *Can. J. Bot.,* **51**, pp. 983–8.

Swanson, E. S., Thomson, W. W. & Mudd, J. B. (1973). The effect of ozone on leaf cell membranes. *Can. J. Bot.,* **51**, pp. 1213–9.

Thomson, W. W. & Swanson, E. S. (1972). Some effects of oxidant air pollutants (ozone and peroxyacetyl nitrate) on the ultrastructure of leaf tissues. *Proc. Electron. Microsc. Soc. Amer.,* **30**, pp. 360–1.

Thomson, W. W., Dugger, W. M., jr. & Palmer, R. L. (1966). Effects of ozone on the fine structure of the palisade parenchyma cells of bean leaves. *Can. J. Bot.,* **44**, pp. 1677–82.

Thomson, W. W., Nagahashi, J. & Platt, K. (1974). Further observation on the effects of ozone on the ultrastructure of leaf tissue. Pp. 83–93 in *Air Pollution Effects on Plant Growth* (Ed. W. M. Dugger, jr) (ACS symposium series 3). Am. Chem. Soc., Washington, DC: x + 150 pp., illustr.

Wei, L.-L. & Miller, G. W. (1972). Effect of HF on the fine structure of mesophyll cells from *Glycine max,* Merr. *Fluoride,* **5**(2), pp. 67–73.

Wellburn, A. R., Majernik, O. & Wellburn, F. A. M. (1972). Effects of SO_2 and NO_2 polluted air upon the ultrastructure of chloroplasts. *Environ. Pollut.,* **3**, pp. 37–49.

Wong, C. H., Klein, H. & Jäger, H.-J. (1977). The effect of SO_2 on the ultrastructure of *Pisum* and *Zea* chloroplasts. *Angew. Botanik.,* **51**, pp. 311–9.

Air Pollution and Plant Life
Edited by M. Treshow
© 1984 John Wiley & Sons Ltd.

CHAPTER 9

Organismal Responses of Higher Plants to Atmospheric Pollutants: Sulphur Dioxide and Fluoride

GOTTFRIED HALBWACHS

Botanisches Institut der Universität für Bodenkultur, Gregor Mendel-Strasse 33, A-1180 Wien, Austria

INTRODUCTION

Sulphur Dioxide (SO_2)

Any attempt to rate the most important air pollutants with regard to their potential hazard for vegetation will find SO_2 in the leading position due to its wide distribution over the world and its considerable phytotoxicity. It should be noted, however, that sulphur is one of the essential nutrients of plants. Under certain circumstances, e.g. on sulphur-deficient soils, SO_2 occurring in low concentrations may become a sulphur source as confirmed by numerous investigations (e.g. Thomas *et al.*, 1943; Faller, 1972). Cowling *et al.* (1973) have reported increasing yields of Ryegrass (*Lolium perenne*) under SO_2-emission in sulphur-deficient soils. But even when slight SO_2-injury occurred on the leaves of Oats (*Avena sativa*), there was often an increase in biomass (Heck & Dunning, 1978).

Although this chapter deals with SO_2, it should be noted that during the combustion of fuels there is always emitted a small quantity of SO_3. For quantitative data on SO_3-levels in the air, for instance, in the vicinity of conventional power-plants, *see* Gartrell *et al.* (1963). Investigations by Commins (1967) further show that high atmospheric concentrations of SO_2 are accompanied by high concentrations of free sulphuric acid and sulphates as aerosols.

Fluoride

Among the fluorine compounds in the atmosphere, the gaseous hydrogen fluoride (HF) is of particular interest because of its adverse influence on vegetation. Although the amounts of gaseous HF that are being emitted are lower than those of SO_2, by some orders of magnitude, HF ranks second to SO_2 in importance as a phytotoxicant where the pollutants are listed with respect to their relevance for causing adverse effects on vegetation (Knabe, 1979). Among air pollutants in the United States, fluorine is ranked fifth in importance with respect to the amount of plant damage produced. But there is no doubt that HF is the most phytotoxic of these pollutants (Weinstein, 1979).

In contrast to sulphur, fluorine is no macronutrient for plants. Its role as a micronutrient is still unresolved for most plant species (Garber *et al.*, 1967). Only with Maize (*Zea mays*) has fluorine been verified as a trace element (Bertrand, 1969). In fluoride-polluted regions the fluoride content of plants increases considerably, particularly in the foliage. Numerous analyses of plants and soils have shown that high fluoride contents in plants are caused primarily by the impact from the atmosphere and cannot be attributed to uptake from the soil. No close relationship exists between fluoride levels in the plants and natural fluorine content of the soils.

Although gaseous HF is the most dangerous fluoride compound for vegetation, fluoride-containing dusts should not be ignored, especially in connection with secondary injuries. Particulate fluorides are much less dangerous to vegetation than is gaseous HF, because very little particulate matter can actually enter the plant. Most of it is washed off the plant surface rather easily by rainfall (McCune *et al.*, 1965). Indeed, particulates are only phytotoxic if they are dissolved in moisture on the surfaces of the foliage and then penetrate the leaves.

SYMPTOMATOLOGY

The visible responses caused by SO_2- or HF-emissions in higher plants—e.g. necrosis, discolorations, etchings, or chlorosis of the foliage, early abscission of needles or leaves, changes in growth-habit, inhibition of growth, decrease in yield, premature ageing—are not specific for a certain pollutant. A diagnosis of emission effects which is based solely on symptomatology therefore seems problematical. However, combined with other methods, symptomatology may contribute to assess emission situations. As foliar symptoms are the best documented responses of plants to air pollutants, symptoms of flowers and fruits are not taken into close consideration here.

Two types of visible symptoms on foliage are commonly considered: acute and chronic (both of which may occur at the same time). Acute injury of leaves results from the absorption of high pollutant concentrations from short-term exposures, often happening but once. It is characterized by encircled necrotic or discoloured leaf regions. In contrast, chronic injury is caused by repeated exposures to sublethal concentrations of a pollutant and most commonly results in chlorosis.

Literature dealing with symptoms on foliage is extensive, and so only a few of those papers that include informative pictorial presentation (partially in colour) of SO_2- and HF-caused symptoms are noted (Barrett & Benedict, 1970; Haut & Stratmann, 1970; Treshow & Pack, 1970).

When evaluating visible symptoms of injury, it should be remembered that these represent a final stage of a process that begins with the entrance of the pollutant into the leaf and consists of various reactions at the cellular level. Thereby the main interest is focused on changes of the assimilation organs; they are the preferred pollutant-absorbing plant parts because of their intensive gas exchange. However, the scope of visible responses would be incomplete—especially in the case of perennials, particularly trees—if only leaf symptoms were considered. Observation of tree responses during longer periods provides information on the resistance of various tree species in polluted areas.

Sulphur Dioxide

Characteristics of acute injury of the foliage—Considering first dicotyledonous herbaceous plants and deciduous trees, localized necrotic areas that are predominantly intercostal but sometimes—as with narrow leaves—on the tips and margins, are typical of acute injury. The necrotic lesions are visible from both sides of the leaf. The absorption of high concentrations during short exposure-times obviously causes a rapid accumulation of intracellular sulphite (SO_3^{2-}), which interferes with metabolic processes taking place in the mesophyll cells. The destroyed parts of the tissue at first appear greyish-green and water-soaked, but become dry later on and change their colour to reddish-brown. The intensity of the colour is greatest on the upper surfaces of the leaves. Additionally, pale ivory-coloured spots may occur. The large and extended necrotic areas are often accompanied by numerous point-like necrotic spots that give the leaves a stippled appearance. Necrosis of the leaf-margin is found rather rarely, as it develops only within a narrow range of concentrations. The larger necrotic spots and areas often merge to form intercostal stripes. Because of their brittleness, the necrotic areas tear and fall out of the surrounding tissue, giving the leaves a perforated appearance and, indeed, form.

Considering next monocotyledonous plants, the most common visible sign of acute injury is a light, yellowish-white or ivory-coloured necrosis beginning at the leaf tips, which then extends down the blade. Besides this type of symptom, necrotic leaf-margins and point-like spots or a definite pattern of stripes between the veins on the blade may be observed.

Considering thirdly conifers: because of their anatomy, which differs from that of broad-leafed plants, a separate discussion about their needles seems necessary. Except in the larches (*Larix* spp.), most coniferous needles show xeromorphic features: a thick cuticle, stomata sunk below the surface, a sclerenchymatous hypodermis, and a minor intercellular volume. These morphological factors are regarded as being responsible for the relatively low SO_2- susceptibility of the older needles of some species, and has been termed 'physical resistance' (Vogl & Börtitz, 1965). This type of resistance is based on a limited uptake of pollutants, termed as 'avoidance' by Levitt (1980).

After exposure to higher SO_2-concentrations, the tips of the needles seem to be water-soaked; however, this soon turns to a red-brown to light-brown tip necrosis that may extend to the base of the needle. Often, a well-defined dividing line between the necrotic tip and the unaffected green tissue occurs, e.g. in *Pinus* and *Abies*. This boundary may be deeper-coloured than the adjacent leaf parts. Occasionally, the injury may be observed as a transverse banding. An appearance resembling that of 'semimature-tissue needle-blight' (Linzon, 1966), i.e. a necrosis beginning below the needle-tip, has been described by Wentzel (1963) as a consequence of SO_2-impact on (Pine) *Pinus*. This progresses towards the tips.

The colour of the injured tissues varies with the age of the needle: the younger the needle, the lighter the colour. Likewise, there is a relation between the location of the injury pattern and the age of a needle. While young needles of *Pinus* show symptoms at the ultimate tip, those needles that are expanding are injured just below the tip. With increasing age, localized necrotic regions may appear most frequently in the middle and basal sections of the needle (Haut & Stratmann, 1970). The colours of necrotic tissue are usually red-brown, foxy-red, yellow-brown, and rarely even pink.

Characteristics of chronic injury: Chronic injury may result from two mechanisms (Barrett and Benedict, 1970): After the entrance of SO_2 into the leaf, it reacts with water to produce the sulphite ion (SO_3^{2-}) which is then oxidized to sulphate. When the accumulation rate is low, it will not exceed the capacity of the cells for removal of SO_3^{2-} by oxidation. Injury of the tissue could then be due to increasing salt concentration after prolonged sulphate accumulation and is displayed as chlorosis. On the other hand, sulphite, which is regarded to be 30 times more toxic than sulphate (Linzon, 1969), could bleach chlorophyll even at sub-lethal concentrations occurring after short-term fumigation with SO_2.

Both mechanisms generate very similar chlorotic symptoms, but the leaves remain turgid and exert, at least partially, their function. An influence upon the buffer capacity, as described by Grill and Härtel (1972) seems very likely. Chronic SO_2-emissions do not cause irreversible injury to cells of coniferous needles, but growth is inhibited. As a consequence, the needles remain short not only because of smaller cells but also due to a decrease in cell numbers. Chlorosis also is frequently observed. Pale green or yellow-green areas appear at the tip of the needle, this extends further as longitudinal stripes to the base and thereafter over the entire needle. The next step is the appearance of pale-yellow colours followed by a colour change to brown and needletip dieback. Changes in the colour of the needles of chronic injured *Picea*-trees have been taken as criterion for assessing emission-caused stress to the vegetation (Pelz *et al.*, 1966). However, chronic injury often resembles natural senescence of leaves and needles.

Hydrogen Fluoride

Characteristics of acute and chronic types of injury on foliage—Differentiation between acute and chronic injury is of only limited interest in the case of fluorides (Brandt, 1967), as the fluoride-concentration within the tissue which is responsible for the injury is not necessarily correlated with the HF-concentration in the atmosphere. When the accumulated fluoride exceeds a certain value, acute symptoms become apparent. Similar results have been obtained by Treshow & Pack (1970), who found no strict dose–response relationship. The reasons originate partially in leaching of fluorides from the leaves by rain, and partially in the deposition of fluorides on the foliage in the form of harmless chemical compounds, e.g. CaF_2.

Although Guderian *et al.* (1969) describe chlorotic discoloration as chronic leaf injury, they regard this type of injury to be less valuable *per se* for evaluation. As chlorosis often precedes necrosis, or appears at the same time as necrosis, both chlorosis (chronic injury) and necrosis (acute injury) will be discussed together here in this section.

The symptomatology of HF—primarily necrotic markings at the tips and margins of the foliage (Halbwachs, 1963)—may be well understood in relation to the fact that the element fluorine is, in contrast to sulphur, not an essential nutrient of the plant. HF enters the foliage mainly *via* the stomata, spreads within the intercellular system and dissolves in the water of the cell-walls. The plasmalemma, as the outermost biomembrane of the protoplasm, is then capable of excluding fluorine which is not required—provided the concentration is not too high. As a consequence, fluorine remains in the cell wall and follows the transpiration stream to the tip and margin of the leaf (Woltz, 1964; Jacobson *et al.*, 1966). In these leaf areas, the fluoride concentration increases to a level that causes injury of the plasmalemma and therefore damage of the cells.

This transport mechanism and the consequent fluoride accumulation, have been investigated using various methods—including imbibition of solutions containing low fluoride concentrations into twigs cut from various woody plants. Uptake in the transpiration stream caused the same symptoms as appeared after fumigation with HF (Halbwachs, 1963). The connection between water transport and localization of fluoride is evident, as the typical symptoms occurred more rapidly than otherwise in cases of increased transpiration, while inhibition of transpiration has prevented the formation of tip and edge necrosis. Numerous analytical results obtained with various plant species proved that, in the usually injured parts of the foliage (tips and margins), the tissue contained from two- to a hundred-fold more fluoride than the middle or basal regions of leaves and needles (Zimmerman & Hitchcock, 1956; Kronberger & Halbwachs, 1974).

   Transport directed to the tips and edges of leaves and needles has also been observed for sulphur by means of autoradiograms which were obtained with $^{35}SO_2$-fumigated plants (Kohout & Materna, 1966). However, the results are not as striking as with fluorides because of the relatively small range of SO_2-concentrations which can produce marginal injury.

   Turning to groups, let us first consider dicotyledorous herbaceous plants and broad-leafed trees: Treshow & Pack (1970), and Weinstein & McCune (1970), agree in describing necrosis at the tips and margins of leaves as characteristic symptoms of HF-injury of broad-leafed plants. The tissue first attains a water-soaked appearance, its colour changing to dull grey-green and later to various tones of brown. The necrotic tissue may be separated from the intact part of the leaf by a narrow, reddish-brown band resulting from deposition of resin and tannins. The necrotic parts of the tissue frequently break and fall out. From the margins, the necrosis spreads mainly into the intercostal areas and to the midrib of the leaf. Chlorosis and necrosis may appear simultaneously, the colours being different with various kinds of plants.

   Monocotyledonous and coniferous plants: Within this group, leaf tip injury clearly predominates. Necrosis frequently proceeds at the margin of one side of the leaf farther down than on the other side. In cereals, light-brown to white injury of the tips frequently occurs, although in Maize (*Zea mays*) chlorotic spots are located towards the leaf tips and leaf-margins, while sometimes chlorotic stripes develop between the veins. In *Gladiolus*, *Iris*, and some cereals, this narrow band separating necrotic from adjacent healthy tissue is often chlorotic (Thomas, 1961).

   In conifers, the first symptoms of injury appear as chlorotic fading and the development of greenish-yellow spots mainly at the needle-tips. Rarely does the chlorotic colour-change spread all over the needle. With progress of the injury, the needle-tip becomes necrotic and turns to brown, thus accomplishing the step to acute injury. Acute symptoms consist of initially ashy-grey to yellow, later on red-brown to brown, necrosis which begins at the tip and progresses various distances towards the base, depending upon the concentration of HF. A characteristic feature of fluoride effects is that the acute symptoms often do not cover all needles of a young shoot uniformly, so that some unaffected ones may always be observed among heavily injured needles.

Symptoms in Perennial Plants (Trees)

Especially with perennial plants, e.g. trees, marked changes of their growth-habit may occur in addition to the leaf injury in SO_2 and HF-polluted areas following exposures of several years. These phenotypical changes originate primarily in a depression in the normal growth rate. The general reduction of growth can be documented by measurement of alterations in diameter increment as consequences of disordered annual ring formation (Keller, 1979), height increment, shoot growth, and size and number of needles or leaves. Owing to thorough investigations, the measurement of the radial increment-loss has become a standard method in forestry to assess the effects of emissions upon forests from the economic point of view (Vinš, 1965; Vinš & Ludera, 1967; Pollanschütz, 1971). Decrease in height increment is also considered to be a suitable criterion for assessing the effects of emission (Wentzel, 1971). The reduction of shoot and foliage growth very often merely affects the upper parts of the crown, where annual shoots tend in any case to be shorter and the needles smaller than normal.

Under extreme conditions of repeated high concentration peaks, the development of *Picea*, *Betula*, and *Salix*, may be disturbed in such a way that dwarfing occurs—as observed in the HF-polluted vicinity of an aluminium plant (Halbwachs & Kisser, 1967). Continued impact for several years caused biometrically detectable deviations in the anatomical structure of the wood of dwarfed trees, by changing the number and dimension of the different types of wood cells. Histometrical examinations on changes in the wood of trees affected by pollution have also been reported for *Picea* by Liese *et al.* (1975), and for several conifers and deciduous trees by Höster (1977) and Grill *et al.* (1979).

Pelz & Materna (1964) made an exact analysis of the process of *Picea* perishing in SO_2-polluted areas, which is also valid for fluoride-polluted areas. The process of dying off, which may be regarded as proceeding in 5 steps, begins with a loss of needles in the upper tree-crown, close to the top, the young shoot at the top remaining unaffected (phase 1). This is in accordance with the view that the most vital region of a plant—in this case the top shoot—is supplied with mineral and organic nutrients at the expense of the somewhat older twigs below (Mathes, 1960). In the second phase the loss of needles from the middle crown continues, which results in an increasing isolation of the uppermost crown (phase 3). Phase 4 is characterized by a perishing of the whole tree-top, while in the final stage (phase 5) the loss of needles extends to the lower parts of the crown. An increasing number of withering twigs and the continuing loss of needles, leads to a striking thinning out of the top of individual trees and finally to a disintegration of the whole stand.

The relations between the visible changes in growth habit of trees and their water-balance have been studied repeatedly (Halbwachs, 1970, 1971, 1973). That the upper parts of a tree-crown are particularly affected by pollutant emissions is certainly not a function only of their physiology, but may also be attributed to their exposure. Trees are open systems that go on growing until their death, and therefore can reach considerable heights. Compared with herbs, trees occur in zones of more vigorous air-flow. Therefore, more pollutants arrive per unit of time at the tree tops than at the less exposed lower parts of the crown. These findings are verified by some experiments. Exposure of plants at various heights above ground over several months in HF-polluted areas revealed a higher and higher impact of pollutants with increasing height (Knabe, 1968). Furthermore, fumigation experiments have demonstrated the significant influence of wind-speed on plant responses (Ashenden & Mansfield, 1977; Ashenden *et al.*, 1978). The reason for this is the magnitude of the resistance of the diffusion pathway outside the leaf, which in still air is much greater than that of open stomata. With increasing wind-speed the boundary-layer resistance becomes small, and the sensitivity of the plant depends mainly on its stomatal resistance.

Besides these changes in the growth habit caused primarily by the loss of needles, those changes should be mentioned that lead to deformations in the shape of trunks, branches, and crowns, of trees as a consequence of chronic emissions over several years. Striking changes in the shape of tree-crowns—especially inhibition of the normal development of the crown at that side which is turned towards the pollutant source—have been demonstrated during the last century, and were recently designated as 'flag-crown (Fahnen-krone)' by Wentzel (1971). In this connection, the different capability of tree species to change the shape of their crowns is worth mentioning. While species of *Pinus*, for instance, show a high flexibility (e.g. to a pinon-like, flattened crown), *Picea* continues to grow in its vertical way. Among others, these qualities may be responsible for the different resistance of species of these two important genera.

Summarizing this section on 'Symptomatology', it may be concluded that there are striking symptoms displayed by the foliage of herbaceous plants, of deciduous trees, and of conifers, in response to exposure to SO_2 and HF; yet they are not very specific, as similar symptoms may be caused by other factors such as frost, herbicides, deficiencies of macro- or microelements, parasites, viral infections, and yet other circumstances (*see* Chapter 6). As both changes in the growth-habit of trees and symptoms on their foliage are rather non-specific, they are of less value for differential diagnosis of emission effects. However, in connection with other methods (e.g. chemical analysis of air and plants), symptomatology may serve as a valuable criterion for gaining a swift survey of pollution impact in a polluted area.

INFLUENCE ON GROWTH, DEVELOPMENT, AND REPRODUCTION, OF PLANTS

During the more than 100-years-old history of emission research in Europe and the United States, evidence appeared repeatedly of an influence on the physiological activity of plants in the absence of the above-discussed visible symptoms on the foliage. The terms for this kind of injury vary, and range from invisible (Sorauer & Ramann, 1899; Wieler, 1903), hidden (Hill *et al.*, 1958), physiological (Vogl *et al.*, 1965), invisible physiological (Keller & Schwager, 1971; Dässler, 1976), invisible chronic (Reckendorfer, 1952), and latent (Keller, 1977b), to subtle (Linzon, 1978). Nevertheless these expressions have essentially the same meaning: an influence of pollutant-concentrations under a certain threshold that causes transient local functional injuries of the protoplasm and of some enzyme systems. Because of the rapid regeneration, disorders of metabolism may only be recognized by a small decline in the total productivity—which, however, cannot be classified as a decrease in yield, because of difficulties of quantification (Dässler, 1976). Börtitz (1974) therefore avoided the expression 'injury' and set up the term 'invisible effect'—including contamination of the plant material, responses of the plant metabolism, and changes in the structure inside the cells. Guderian (1977) believes that it is HF rather than SO_2 that may have an 'invisible' effect.

Few authors, if any, still deny the existence of pollutant effects in the absence of visible symptoms. The various influences upon physiological and metabolic processes discussed in Chapter 7 are generally accepted to be the beginning of emission-induced changes involving growth, development, yield, and reproduction.

The plant responses to SO_2 and HF-emissions are dependent on pollutant concentration, environmental factors, and stage of development, as well as on the species-resistance, and may range from phenomena of stimulation to marked decrease in productivity. This has been clearly formulated by Guderian (1978) for SO_2 and by Treshow (1969) for HF. The same concentrations may induce acute injuries (necrosis) with very sensitive plants, inhibitions of growth with less sensitive plants, and in certain cases stimulation of growth with resistant plants. In the following, some typical effects of SO_2 and HF on growth and yield will be summarized according to the recent literature.

Sulphur Dioxide

Among Gramineae, the cereals *Secale, Triticum*, and *Hordeum*, react with reductions in grain-yield and reduction of the thousand-grain-weight, even in cases where only the beard-hairs are affected by SO_2 (Haut & Stratmann, 1970). Inhibition of growth, reduction of dry-weight of the shoots, and a significant loss of yield, were reported with *Lolium perenne* (Ryegrass) by several authors (e.g. Bell & Clough, 1973; Bell *et al.*, 1979; Crittenden & Read, 1978). Bell *et al.* (1979) described a decrease in yield of up to 51% of the control value without observable necrosis.

Similarly, Crittenden & Read (1979) found decreases in the dry weight of the shoots of up to 40% with *Dactylis glomerata* (Cocksfoot). Marchesani & Leone (1980) developed a bioassay technique to measure the effects of SO_2 on *Avena* seedlings and found that the growth-rate of the seedlings was retarded without any visible symptoms when they were exposed to various concentrations of SO_2. During the post-fumigation intervals, only about 15% of the fumigated seedlings recovered to the normal growth-rates of the prefumigation phase. These findings again support the theory of subtle or invisible injury in the form of decreased growth-rates. The observation of considerable losses of yield in the absence of necrosis disproves the previously adopted view that the reduction in yield is a function of the percentage of damaged leaf area (Thomas, 1961).

Among herbaceous dicotyledons, *Nicotiana* (tobacco), *Medicago* (alfalfa), and *Phaseolus* (bean), have been especially favoured objects of research. Tingey & Reinert (1975) observed inhibitions of growth with *Nicotiana* and *Medicago* in the presence of SO_2 both singly and in combination with ozone. Guderian (1966) examined numerous species of forage plants not only in relation to the susceptibility of their leaves, but also with assessment of their productivity after SO_2-impact. He could observe reductions in yields of *Phaseolus*, *Pisum* (bean), *Vicia* (vetch), and *Lupinus* (lupine), which did not correlate with the percentage of the necrotic leaf area. While the loss of yield was larger than the percentage of injured leaf area with *Phaseolus*, *Pisum*, and *Vicia*, *Lupinus* showed an exactly inverse response. These findings again contradict the former view that the reduction in yield quantitatively agrees with the percentage of the injured leaf areas.

On the influence of SO_2 upon growth and production, the most information is undoubtedly available for forest trees. There is apparently a close relationship between restricted growth and decreased photosynthesis. Keller (1978*a*) periodically measured CO_2 uptake of *Abies alba* (silver fir) and *Picea abies* (Norway spruce) grafts during a 10-week's fumigation with low SO_2 concentrations. SO_2 at 0.05 ppm* did not cause visible injury, but the uptake of CO_2 was decreased. With increasing concentration and duration of exposure to SO_2, CO_2 uptake decreased. A stimulation of photosynthesis was not detected in any case. Concentration up to 0.1 ppm SO_2 caused injury in needle tissue of sensitive clones of *Pinus strobus* (Eastern White Pine) (Dochinger *et al.*, 1970). Garsed *et al.* (1979) reported different responses of seedlings of *Pinus sylvestris* (Scots Pine) and deciduous trees (*Betula* (birch), *Quercus* (oak), *Acer* (maple) to continuous fumigation with low SO_2-concentrations. While the growth of *Pinus* was decreased significantly, the dry weight of roots and trunks of deciduous tree species remained unaffected; only leaf weight and leaf area of *Acer* were reduced. *Quercus* and *Acer* reacted with a stimulation of height growth and with premature ageing of leaves when compared with controls.

With young trees, the reduction in biomass production results in smaller leaves and particularly striking reductions in shoot elongation (Haut & Stratmann, 1970). When biomass production culminates in *Picea*, the negative effects upon height- and girth-increment, shoot growth, and the number and size of needles, are the greatest. With trees in maturing stands, effects on

---

*1 ppm SO_2 (v/v) equals 2.6 mg SO_2/m^3 at 20°C and 760 torr. 1 mg SO_2/m^3 equals 0.35 ppm SO_2 (v/v) at 20°C and 760 torr.

seed production also should be taken into account. Length and weight of cones, weight per thousand seeds, and germination, may all be reduced in *Picea*. Similarly with *Fagus*, an influence on seed quality has been experimentally determined (Haut & Stratmann, 1970). Interference with fructification of *Picea* and *Pinus*, respectively, has been observed by Pelz (1963), Mrkva (1969), and Mamajew & Shakrlet (1972). Pospišil & Richtar (1970) described anomalies of *Larix* pollen after impact of SO_2 upon the floral organs.

With *Abies*, a very significant negative linear relation between the SO_2 concentration during winter fumigation and the germination of pollen in the following spring was found (Keller, 1977*b*). Matsuoka *et al.* (1969) showed experimentally with *Pyrus* that fumigation with SO_2 induces an inhibition of pollen germination. The pear trees cultivated in greenhouses under optimum conditions of growth and flowering were fumigated for some hours with SO_2 before, during, and after, their artificial impregnation with viable pollen. As a result, the number of fruits was drastically reduced.

Significant effects on the reproductive organs of *Pinus sylvestris* occurred when medium SO_2 concentrations were accompanied by fluoride pollution. As a consequence, decrease in the number of seeds per cone, increase in the abortion rate, and reduction in dimensions and weight of cones, were observed (Roques *et al.*, 1980). SO_2 pollution alone caused only slight modifications.

Hydrogen Fluoride

A consideration of the effects of HF emission on the rate of growth and yield of different plant species covers but one side of the problem of fluoride impact. In the case of fluoride uptake by food or forage plants, the fluoride content accumulated in the plant tissues attains considerable importance as a criterion for the quality of these plants. In contrast to SO_2, the plant responses to HF-emissions are often related not so much to the concentration in the atmosphere, as to the fluoride concentration in the plant tissue. The primary hazard for vegetation, particularly for perennial plants, arises from long-term exposures to very low concentrations. Guderian (1971) postulated that, for forage plants in general, the effects of HF on growth and yield are of minor importance, but that the fluoride concentration in plant organs is of great relevance for animals. He could show with 10 grass and clover species that very low HF concentrations caused fluoride accumulation up to levels which were hazardous for animals but did not interfere significantly with the plants' growth rate. Occasionally, however, some decrease in yield from HF could be observed even in the absence of chlorosis or other symptoms in the leaves with *Zea* and *Sorghum* (Hitchcock *et al.*, 1963, 1964).

There are numerous examples of interference by fluoride in the growth of Gramineae, herbaceous dicotyledons, and woody plants, from field observations and from fumigation experiments, the criteria being either the dry-weight or the change of dimension, i.e. length, surface area, or volume, compared with the controls. The difficulties in making general statements on the effects on growth, yield, and reproduction, especially after fumigation experiments, have been pointed out by Weinstein & McCune (1970), who gave the following reasons: the relatively short exposure-times, the limited number of species and varieties so far examined, incomplete knowledge of the influences of climatic and edaphic factors upon growth and reproduction, and finally the difficulty of comparing and interpreting the results obtained with different experimental arrangements. Although the following statement is made especially for assessing the effects of fluoride, its principle may validly be extended to other pollutants.

Among the Gramineae, yield reductions connected with HF emissions were observed in the form of decreases in weight of the epigeal plant parts in *Zea* and *Sorghum* (Hitchcock *et al.*, 1963, 1964), diminution of growth in *Hordeum* and *Avena* (Guderian, 1971), and reduction of the thousand-grain-weight in *Hordeum* (Szalonek & Wateresiewicz, 1979).

Among herbaceous dicotyledons, *Medicago* and *Lactuca sativa* ssp. *longifolia* responded with reductions in the dry weight of young shoots as well as of roots (Benedict *et al.*, 1964), *Vicia sativa* (Common Vetch), *Lupinus* sp., *Pisum sativum* (Garden Pea), and *Allium cepa* (Onion), with reduction in yield, *Beta vulgaris* with affected leaf growth (Guderian, 1971), *Solanum tuberosum* (Potato), *Beta* sp., *Trifolium* sp. and *Phaseolus* in the vicinity of an aluminium plant with reduction in yield up to 40%–60% (Szalonek & Warteresiewicz, 1979), while in *Solanum tuberosum* (Potato) the starch content also was diminished. Inhibition of growth of *Phaseolus* occurred only when the fluoride content of the leaves exceeded 200 ppm (Treshow & Harner, 1968). With concentrations in the leaf tissues of between 300 and 400 ppm, the dry weight of the plants dropped to half of the controls without the appearance of visible necrosis.

Similar results were obtained with *Medicago*: reductions in dry-weight to 63% of controls did not occur before the fluoride content of the leaves exceeded 500 ppm. After fumigation with low HF-concentrations for 22 weeks, *Lycopersicum* (Tomato) showed, besides marginal injuries of the young leaves, a diminution of the plant weight and a reduction in the number of fruits as a result of the destruction of young shoots and a decrease in the average fruit's weight (Pack, 1966). MacLean *et al.* (1977) exposed *Phaseolus* and *Lycopersicum* (Tomato) for 43 and 99 days, respectively, to HF at a mean concentration of 0.6 μgF/m^3.[*] The results obtained with bean plants led to the conclusion that HF effects on leaf injury and fruiting were independent.

[*]1 ppb F (V/V) equals 0.800 μgF/m^3 at 15°C and 760 torr.
1 ppb F (V/V) equals 0.777 μgF/m^3 at 25°C and 760 torr.
1 ppb HF (V/V) equals 0.846 μgF/m^3 at 15°C and 760 torr.
1 ppb HF (V/V) equals 0.818 μF/m^3 at 25°C and 760 torr.

HF did not affect growth or induce foliar injury in bean, but the fresh mass of pods was reduced by almost 25%. Although Tomato plants usually were reported to be more susceptible than bean plants, no effects of HF were detected on their growth or fruiting.

These findings corroborate earlier results with *Phaseolus* reported by Pack (1971) and are also consistent with the investigations on fruiting of some *Fragaria* varieties which suggested that HF affected fruiting by interfering with fertilization or seed development (Pack, 1972). The effects on fruiting seemed to be independent of visible injury to the foliage. Pack & Sulzbach (1976) conducted numerous investigations on the response of plant fruiting to HF exposure, e.g. with *Pisum sativum* (Pea), *Lycopersicum esculentum* (Tomato), *Cucumis sativus* (Cucumber), *Zea mays* (Maize), *Avena sativa* (Oats), and *Triticum aestivum* (Wheat). The response found most frequently was a reduction in seed production due to pollen germination or pollen-tube growth being affected by fluorine (Sulzbach & Pack, 1972). Similar effects of HF on pollen germination and pollen-tube growth in varieties of *Prunus avium* (Sweet Cherry) and *P. armeniaca* (Apricot), respectively, have been reported (Facteau *et al.*, 1973; Facteau & Rowe, 1977). But there is also one publication pointing to a stimulating effect of fluorides to pollen-tube growth in *Malus* sp., *Pyrus* sp., and *Prunus avium* (Lai Dinh *et al.*, 1973).

A smaller leaf-size associated with fluorides has been described repeatedly in woody plants (e.g. Leonard & Graves, 1970). This coincides with the findings of Halbwachs & Kisser (1967) for *Betula* and *Picea*. The fluoride content of needles of dwarfed *Picea* was up to 1400 ppm. Conifer shoot-elongation also may occur, but mainly in younger trees, while a decrease of girth was detected in older trees (Keller, 1973). While growth interference of *Pinus* and *Picea* species (Wentzel, 1965) is often accompanied by the needle necrosis, growth of *Pseudotsuga menziesii* (Douglas Fir) can be inhibited even in the absence of chlorosis and necrosis (Treshow *et al.*, 1967). The latter finding is supported by Keller (1977a), who reported on a drastic decrease of photosynthesis in *Pinus nigra* when fluoride content was below 40 ppm in dry matter. A decrease in girth increment has been reported for *Pseudotsuga* (Treshow *et al.*, 1967) and *Citrus* (Brewer *et al.*, 1969). With *Pseudotsuga*, increasing fluoride content of the needles caused, on the one hand, a reduction in girth, and, on the other, an elongation of the needles. When the fluoride content reached several hundred ppm, needle tip-burn and loss of needles occurred.

Leonard & Graves (1970) observed a significant negative correlation between the fruit yield of *Citrus* varieties and the mean fluoride concentration in the air as well as the fluoride content of spring leaves. The fluoride content of these leaves, measured in May and in October, was 20–40 ppm and 50–100 ppm, respectively. The yield of *Citrus* cultures decreased by 27% when the fluoride content of 5- to 6-months-old leaves increased by 50 ppm. Visible chlorosis appeared at fluoride contents above 20–30 ppm.

These examples on yield reductions originated in field observations and fumigation experiments with various HF-concentrations and exposure-times. There are also investigations that do not show significant losses of yield or growth with certain species of plants under the given conditions. A few cases even indicate growth stimulation (e.g. Adams & Sulzbach, 1961). But there are some observations which point (Weinstein, 1977) to the fact that growth occurring as a result of fluoride stimulation is often abnormal. Both HF and SO_2 may induce a premature ageing of plants, as is commonly assumed among physiologists. This has been repeatedly confirmed with different methodology, i.e. by determining the activity of the ageing enzyme peroxidase (e.g. Keller & Schwager, 1971; Bucher-Wallin, 1976), by observation of the abnormal premature ripening of cereals (Haut & Stratmann, 1970) and of premature casting of leaves or needles from trees (Garsed *et al.*, 1979).

FACTORS AFFECTING PLANT RESPONSES TO SO_2 AND HF

The intensity of a plant's reponse is influenced, not only by the concentration, duration, and frequency of exposure, but also by its resistance to the pollutant, which moreover depends on internal and external factors. According to Vogl & Börtitz (1965) and Levitt (1980), there are two types of resistance: the resistance against pollutant uptake or stress avoidance, and the resistance against the absorbed pollutant or stress tolerance.

While resistance to uptake acts unspecifically with respect to all substances, the tolerance may differ distinctly with various pollutants. Both types of resistance within a genetically determined and species-dependent variability of resistance are subjected to environmental influences by climatic and edaphic factors, and undergo changes with the stage of development and age of the plant concerned. A broad discussion of causes and criteria of resistance to air pollution by *Picea abies* was offered by Braun (1977, 1978), who concluded that there is an important relationship between resistance against air pollution and drought. Moreover, the deposition of SO_2 on vegetation is affected by several environmental factors, many of which undergo diurnal, seasonal, and spatial variations (Wesely & Hicks, 1977); stomatal resistance is of special interest, although other external and internal factors affecting resistance may play a more important role under certain conditions.

Temperature

Temperature is most important in the ranges below 5°C and above 30°C. However, while SO_2-resistance may appear greater at temperatures below 5°C, because of the lower physiological activity, Materna (1974) demonstrated that *Picea* trees absorb SO_2 during winter, and that this exerts its effect later on under more favourable external conditions.

Huttunen (1976) referred to the fact that different ecological conditions (north–south) can affect the responses of plants to air pollution. The relative resistance of *Pinus sylvestris* to pollutants in Northern Europe seems to be distinctly connected with its general resistance to the winter, though this could not be detected in *Picea abies* (Norway Spruce). Huttunen (1978) showed that greater injury to Spruce saplings was caused during winter than during the summer, though the chronic damage occurring regularly every winter became visible only in the spring. This result corroborates Materna's experiments and suggests the existence of combined effects of SO_2 and frost. Feiler *et al.* (1981) analysed the complex effect of SO_2 and frost on *Picea abies* and assumed that SO_2-treated plants are unable to regulate their membrane permeability and therefore react more sensitively to frost. Keller (1978*b*) observed that frost injury became apparent after fumigation of forest trees (*Fagus sylvatica* [Beech], *Pinus nigra* [Austrian pine] and *Picea abies*) with low SO_2-concentrations. This made a so-called latent injury evident: SO_2-pollution had weakened the vitality of the plants and this weakness continued for a longer period than was expected. Yet, the seasonal differences in susceptibility may not be explained merely by the different temperatures.

The influence of four temperatures, ranging from 18°C to 30°C, during the growth period, and the influence of relative humidity (RH) at constant temperatures on the SO_2-resistance of *Avena sativa* varieties, were studied by Heck & Dunning (1978). The effects of the temperature of exposure were unclear, but lower temperatures for a given time during the growth-period increased the SO_2-resistance of the plants. This resistance was further increased by the lowered humidity and decreased by high humidity at the time of exposure. Rist & Davis (1979) investigated the influence of three different temperatures (13°, 21°, and 32°C) and three different relative humidities (40%, 60%, and 80%) on leaves of *Phaseolus vulgaris* var. Pinto III treated with SO_2. Injury induced by SO_2 was generally greatest at RH 80% and 32°C. The foliar content of sulphate also was greater at higher than at lower temperatures. This may be due to the increased stomatal conductance with increased temperature and RH. The observation in this experiment that the stomatal conductance of leaves on exposed plants was less than that of those on unexposed controls, is in contrast to many other reports on SO_2-stimulated stomatal opening found with other plant species.

A relationship also exists between temperature and the development of necrosis with fluorine compounds. The higher the temperature is, the more pronounced are the necrotic symptoms (Treshow, 1969). A distinct sensitivity can be observed at temperatures above 35°C. With *Gladiolus* species under conditions of constant vapour-pressure deficit, MacLean & Schneider (1971) observed increasing tip-burn with increasing temperature between 16°C and 23°C. An influence of higher temperatures upon fluorine accumulation within the tissues of roots was observed by Benedict *et al.* (1965). This observation presumes that fluoride can be translocated from the shoot to the roots, as has been experimentally proved in *Picea* by Kronberger *et al.* (1978), although it is in contrast to previously stated views (e.g. Treshow, 1969).

One aspect still has to be taken into consideration: the metabolic rate of plants is highly dependent upon the air temperature and solar radiation. If these are high, the response-time of plants to pollutant exposure will be of the order of 10–15 minutes. This points to the necessity for establishing air quality standards for short periods (Shinn, 1979).

Relative Humidity (RH) and Soil Moisture

As the role of the stomata for entrance of gaseous pollutants into the plant has been stressed repeatedly, doubtless all factors which induce stomatal opening will increase the uptake of pollutants. In this way, SO_2 uptake has been described as a function of the SO_2-gradient between the air and the mesophyll, and also of the diffusion resistance, involving aerodynamic, cuticular, stomatal, and internal, resistances (Jäger & Klein, 1980). Obviously, high RH leads to decrease in stomatal resistance provided neither illumination nor soil moisture is limiting.

Beckerson & Hofstra (1979) revealed that it is important, when measuring effects of air pollutants on stomatal resistance, that both upper and lower leaf surfaces of amphistomatic leaves be considered, as stomatal responses vary. SO_2 generally decreased stomatal resistance of bean leaves, but the resistance of the upper surface was about double that of the lower surface—possibly owing to the difference in stomatal number of the two surfaces. In this connection the influence of wind-speed has to be mentioned, as this may surpass all other factors affecting stomatal resistance—especially under the particular conditions that often prevail in fumigation experiments (Ashenden & Mansfield, 1977).

A distinct correlation between SO_2-susceptibility and increasing RH was established with *Medicago* (Thomas, 1961). McLaughlin & Taylor (1981) emphasized the strong effects of RH on the foliar uptake of SO_2 by *Phaseolus vulgaris* (Bush Blue Lake var. 274). An increase of RH from 35% to 75% enhanced the foliar uptake of SO_2 by two to three times at all given concentrations. The mechanism which was thought to be responsible was an altered internal leaf-resistance to uptake, rather than stomatal regulation. Humidity-induced changes in plant susceptibility to pollutants may be a function of humidity or water-flux through the leaf (Barton *et al.*, 1980). SO_2 penetrates the leaf and is absorbed in the water of mesophyll cell walls. It is assumed that at low humidity the enhanced outward flow of transpired water through cell-walls would provide a resistance to inward, liquid phase diffusion. Also, with conifers and hardwood trees, the sensitivity was increased by RH above 70% (Linzon, 1969). However, according to investigations with *Vicia faba* (Majernik & Mansfield, 1971), the effect of RH should always be considered in connection with a direct action of SO_2 upon the stomata. When RH is high,

0.25 ppm SO_2 indices opening, whereas at a lower RH, the same SO_2-concentration induces closing of stomata.

A generalization of these experimental results seems at least problematical, according to results obtained with *Nicotiana tabacum* (Menser & Heggestad, 1966), which showed that a distinct decrease of the aperture was caused by SO_2-fumigation at 80%–100% RH, and results with *Petunia* cultivars, which showed only slight effects on the leaf's diffusion-resistance after SO_2 fumigation at RH 50% and 90% (Elkiey & Ormrod, 1979). Furthermore, differences not only with respect to resistance but also concerning the reaction of the stomata have been observed in three *Betula* species, i.e. *B. pendula*, *B. lutea* (Yellow Birch), and *B. populifolia* (White Birch), at a constant RH of 75% but varying SO_2 concentrations. A concentration of 0.3 ppm SO_2 for 1 or 2 hours induced an opening (higher conductivity) with the two first-mentioned species but a closing with the other one (Biggs & Davis, 1980). The frequently observed tendency for opening of stomata and reduction of stomatal resistance in response to relatively small SO_2 concentrations, which is further increased by higher RH, has been taken as an explanation of SO_2-mediated growth-stimulations (Roberts, 1975).

Fumigation experiments with hydrogen fluoride revealed that the incidence of injury was not affected by differing RH (50%, 65%, and 80%), but the severity of HF-induced injury on leaves of *Gladiolus* was greatest on plants exposed at RH 80%. Furthermore, the injury occurred sooner and was more pronounced at RH 80% than at 65% and 50% RH (MacLean *et al.*, 1973). The authors concluded that the effect of RH must not be seen only in the direct influence on stomata, but in the alteration of the relationships between foliar necrosis and fluoride accumulation. The indirect effects of pollutant-induced injury on absorption and translocation of fluoride are suggested as being most important.

The influence of soil moisture upon SO_2-sensitivity has been investigated with *Hordeum* (Markowski & Grzesiak, 1974). At a soil moisture slightly above the wilting point, which induced a temporary water-loss from the tissues, the susceptibility decreased. Analytical determination of the sulphur content in different plant tissues showed that water-stress is accompanied by decreased SO_2-uptake. Again, the water stress-induced stomatal closure seems responsible for this decreased uptake-rate.

Similar results have been reported repeatedly for fluoride. In principle, the results with woody plants, which showed that resistance decreases with increasing RH, fit well into this pattern (Rohmeder & Schönborn, 1968). RH at values above 90% drastically increase injury. However, in the case of fluoride emissions, the relationships between water-supply and injury are not as clear as with SO_2, because the results of controlled greenhouse fumigation experiments do not agree with results observed in field experiments. Greenhouse plants, well supplied with water and nutrients, were extremely sensitive to fluorides (Zimmerman & Hitchcock, 1956). In contrast, extensive field observations (Treshow & Pack, 1970) gave the impression that trees growing under unfavourable conditions (of aridity, gravelly soil, exposure to the south, etc.) are liable to be injured more severely than irrigated ones.

Light

As all environmental factors that act upon stomatal movement are apt to influence the uptake of gaseous pollutants, light should also be included when we consider the resistance of plants. Generally, plants show a high degree of resistance to SO_2 in the dark, and become more sensitive with increasing irradiance (Mukammal, 1976). However, SO_2 uptake in the dark can be demonstrated (Zahn, 1963). This seems less surprising when one recalls that various plant species keep their stomata at least partially open during the night (Hübl, 1963). Rohmeder & Schönborn (1968) noted that, with increasing irradiance (and increasing rate of photosynthesis as a consequence), sensitivity to fluoride increased. Benedict *et al.* (1965) measured a fluoride-uptake in the dark of about 40% compared with light-conditions; however, fluoride-uptake occurred when stomata were closed. This aspect points to fluoride-uptake *via* the epidermis, which has been discussed in addition to stomatal uptake by Guderian (1977), and for soluble particulate forms of fluoride by Weinstein (1979). The resulting hazard of fluoride emission even during night is thereby stressed. However, correlation of foliar content resulting in visible markings in daylight and darkness at equivalent fumigation levels, revealed that about 50% as much foliar fluoride was required in darkness to produce equivalent injury (Adams *et al.*, 1957).

The influence of light quality upon sensitivity to air pollutants has been given considerable importance in explaining the differences in the response of outdoor and greenhouse-cultivated plants (Brandt & Heck, 1968). However, this might vary for different air pollutants. Changes in the response of plants to the same concentration of SO_2 during the course of a day may not be due solely to changes of light quality and irradiance. Fumigation experiments with SO_2 under constant conditions of light, temperature, and humidity, showed changes in sensitivity during the course of a day (Haut, 1961): the plants were most resistant in the afternoon. Low sugar-content in the early morning may explain the plants' relative sensitivity at that time (Thomas, 1961).

Nutrient Supply

Optimum supply with nutrients, especially nitrogen, increases the resistance of plants to pollutants or at least improves their recovery and state of health after exposure. Conversely, nutrient deficiency increases the sensitivity of plants (Haut & Stratmann, 1970). Forage plants showed a higher resistance than controls when sufficiently supplied with nitrogen and potassium. The positive action of nitrogen fertilizing has been attributed to nitrogen's role in protein metabolism, and that of potassium to its function in regulating permeability of protoplasm. Jäger & Klein (1976) analysed the relation between nutrient supply and SO_2-susceptibility, and showed that potassium content and buffer capacity of Garden Pea plants (*Pisum sativum*) were of special importance in the detoxification of SO_2-borne products. Plants grown on substrates with ammonia as the main source of nitrogen, deficiency of potassium, and low pH values, showed a reduction in buffer capacity and therefore an increase in SO_2 effects. Decreased sulphur nutrition resulted in a considerable delay of the beginning of injury. The most important result derived from these investigations was undoubtedly that resistance of SO_2-treated plants depended less on the nitrogen quantity than on its source. Nitrogen in the form of nitrate (NO_3^-) was able to increase the SO_2 tolerance of plants, whereas nitrogen in the form of ammonia (NH_4^+) exerted adverse effects.

Increasing complete fertilization increased the resistance of some dicotyledons, such as *Brassica napus* (Turnip), *Spinacia oleracea* (Spinach), and *Raphanus sativus* (Radish) var. *nanus*. Nitrogen was the critical component for this effect (Haut & Stratmann, 1970). In contrast, a positive effect was absent with monocotyledons, young plants becoming even more sensitive than formerly when fertilized. However, findings have often conflicted. Linzon (1978) correlates all external factors that favour plant growth (light, moderate temperature, high RH, soil moisture and, among others, also nutrition) with increased SO_2 injury.

The greater sensitivity sometimes found under optimum nutrient supply might be explained by an increased physiological activity, which may for instance promote pollutant uptake. Because an abundant nutrient supply favours water balance, the stomata may remain open for a longer time than otherwise.

Some studies from field investigations show that pollutant effects can be minimized by fertilizing with nitrogen (Edelbauer & Halbwachs, 1982). Results with fertilizing as a therapy against pollutant injury are very spectacular in agriculture for the reason that the effect of fertilizing appears rapidly.

However, as for fertilizing as a therapy in forests influenced by pollutants, the findings in the literature are less definite. Kisser (1965) has discussed the different aspects thoroughly. In forests, 'time' is also a factor of considerable influence because of the long periods of wood production. The impact of acid substances such as SO_2 or HF over several years, affects not only the assimilating foliage or the entire trees but can also exert a harmful effect on soil. Although the acid concentrations are low, they can cause an unfavourable shift in the ratio of ions in time, particularly in soils with a low buffer capacity. This can alter the availability of nutrients and even pollutants (e.g. heavy metals) within the soil solution (Ulrich *et al.*, 1979).

To summarize, fertilization appears, in spite of some contradictory conclusions, to be favourable on the whole. The best results can be expected in regions with a chronic type of injury when the induced growth-increment improves the whole situation of the stand. In areas with an acute type of injury, no influence of fertilization upon the rate at which trees die has been observed. This suggests that fertilizing measures primarily act upon growth in a positive way, but are unable to increase significantly the resistance of trees to SO_2 and HF.

Fluoride injury appeared intensively in various experiments provided certain elements were present in only small amounts. For example, deficiency of Ca induced increased tip-burn of leaves and affected fruit size and seed production (Pack, 1966; Pack & Sulzbach, 1976), and leaf damage was most severe in *Pisum sativum* when calcium levels in the leaves were low (Holub & Navara, 1966). These obvious connections between Ca supply and fluoride injury have already been made use of in agriculture by improving the Ca supply of plants with Ca sprays or adding lime to the soil. The transformation into sparingly soluble CaF_2 of any fluoride taken up from the atmosphere should thereby be accelerated.

In contrast, McCune *et al.* (1966) reported that, with the extremely sensitive 'Snow Princess' *Gladiolus*, Ca- and N-deficiency led to decreasing tipburn. Potassium- and phosphorus-deficiency increased tip necrosis. Absence of Mn and Fe was without effect.

The connection between HF and Mg was stressed by Garrec *et al.* (1977), who showed that fluoride accumulation led to a depletion in the magnesium content of *Abies alba* (Silver Fir) needles.

Optimum nutrient supply increased the resistance of forest trees to HF—with similar restrictions as discussed with SO_2 (Rohmeder & Schönborn, 1968).

Age and Stage of Development of Plants

Sensitivity is influenced by the age and developmental stage of the various plant organs. Haut & Stratmann (1970) have shown that sensitivity of cereals to SO_2 is especially high in certain critical developmental stages: the 3-leaf stage and the period before inflorescence. The sensitivity of the entire plant was determined by the sensitivity of successive leaves, each additional leaf being more resistant than the preceding one.

With herbaceous dicotyledons, e.g. *Phaseolus*, there is a critical stage for the whole plant between flowering and the beginning of ripening (Guderian, 1977). As for leaf sensitivity to SO_2, several authors regard very young, not yet fully expanded leaves to be relatively resistant, the fully expanded leaves to be very sensitive, and the older leaves to be less sensitive again. However, these reactions depend largely on the concentration of pollutant. In cases of long-term exposure to low SO_2-concentrations, the older leaves of *Beta vulgaris* (Beet), *Vicia fava*, *Malus*, and *Pyrus*, were damaged usually earlier than the younger ones, although they showed a lower physiological activity and a lower SO_2-uptake. A three-days' fumigation of *Alnus glutinosa* (Alder) with

3 mg SO_2/m^3 injured the older leaves before the younger leaves; a case of fumigation with 6 mg SO_2/m^3 revealed a correlation between the extent of injury of the young but fully expanded leaves, their highest sulphur accumulation, and their highest rate of photosynthesis (Guderian, 1977).

Perennial plants also show periods of sensitivity to SO_2 pollution in the course of their development. Besides the seasonal variations of sensitivity, which are paralleled by the variations in physiological activity of the foliage of deciduous trees and conifers, there are also variations of sensitivity depending on age. Young leaves of younger *Helianthus* (Sunflower) plants were physiologically more active, accumulated more sulphur, but were more resistant to it, than young leaves of older plants (Guderian, 1977).

Furthermore, throughout shoot development, the sensitivity of the leaves changes with age, as one leaf follows another and the younger ones are progressively more and more sensitive. With conifers, the previous year's needles are less sensitive than the current year's needles when fully developed. Smith & Davis (1978) found differences in sensitivity related to the needle types at various stages of development in *Pinus sylvestris*. A comparison of *P. sylvestris* seedlings in the cotyledon, primary, and secondary needle, stages treated with 2.6 mg SO_2/m^3 for 1, 3, or 5 hours, revealed a generally higher susceptibility of the secondary needles than of the primary ones. Cotyledons were generally more tolerant than the other two needle types. The susceptibility for all three needle types was lowest at the time of needle emergence, increased through 3 to 5 weeks, and then remained constant or decreased from 5 to 7 weeks. All needle types exhibited tip-necrosis.

Age-dependent differences in sensitivity may also be found with perennial woody plants, dissimilarities between deciduous trees and conifers becoming obvious. Deciduous trees are relatively sensitive in their early growth-period and become more resistant with increasing age (Guderian & Stratmann, 1968), whereas conifers are relatively resistant in their early growth-period and become very sensitive during the peak of woody growth, remaining so until the early stage of maturity (Wentzel, 1968).

Numerous observations with SO_2 have shown that differences in susceptibility are often interconnected with uptake of pollutants. Differences in susceptibility between *Pisum sativum* (sensitive) and *Zea mays* (tolerant), for instance, are partly related to the leaf's diffusion resistance (Klein *et al.*, 1978).

The situation with fluorides is quite different. Following fluoride exposure, very young leaves of *Beta, Populus, Alnus glutinosa, Acer platanoides* (Norway Maple), *Betula, Dahlia, Dianthus caryophyllus* (Carnation), and *Pinus ponderosa* (Ponderosa Pine),showed the first symptoms (Adams *et al.*, 1956), although their fluoride content was lower than that of older leaves or needles. In contrast to SO_2, long-term intermittent exposure to HF produced a higher fluoride concentration in the older leaves. It is important, however, to take into account the biomass production during the pollutant-free periods, as a high growth-rate dilutes fluoride within the tissue. In this way fertilization may act positively in reducing the pollutant effect and increasing quality (lowering of the fluoride content hazardous for animals). Kronberger & Halbwachs (1978) showed, in field investigations with *Zea mays*, that the fluoride content in plant tissue varied markedly in different stages of development—due to the variable rate of dry-matter production, although the plant continuously accumulated HF from the atmosphere, which caused a steady increase in the absolute fluoride content of the entire plant. Bligny *et al.* (1972) also used *Zea mays* for their experiments and found that resistance of leaves to necrosis varied during leaf development. They related these variations to the changes in the amount of calcium in the leaves during their ageing. There seem to be close similarities at the level of absorption mechanism of calcium and fluorine ions, suggesting a close relationship between fluoride and endogenous Ca concentrations in different parts of plants.

Needles of conifers are regarded as being most sensitive during their elongation in spring and early summer. In the following months, they become more resistant, and older needles are rarely injured (Treshow & Pack, 1970). This observation might be a consequence of elevated Ca levels in older needles, as shown by Garrec *et al.* (1974) in needles of *Abies alba.*

As fluorine translocation occurs from older to newly formed organs, it is necessary to include this possibility when assessing fluoride contents in plant tissues. The circuit of a small fraction of the fluoride within a plant, suggested by Kronberger *et al.* (1978), i.e. xylem, radial diffusion, phloem, radial diffusion, xylem, and so on, could explain the small part of fluoride that is translocated from two-years-old needles into new growth in *Picea* (Guderian, 1977).

Genetic Resistance

Genetically determined, specific individual resistance to SO_2 and HF has been established by Rohmeder *et al.* (1962) with *Picea abies* and *Pinus sylvestris*, and by Schönbach *et al.* (1964) with *Larix decidua* (European Larch) and *L. leptolepis* (Japanese Larch). Also, Treshow (1969), and Treshow & Pack (1970), pointed out that numerous plant species, and even among those certain varieties and clones, show considerable variance of sensitivity to fluorides. Although some of the differences might be attributed to environmental factors, most of them are genetically determined. But because of the influence of environmental factors, conclusions on sensitivity and resistance are not of general validity, being restricted to the individual experimental conditions. Results from fumigation experiments under controlled environmental conditions often disagree with field observations. Therefore, 'resistance series' established by numerous authors, e.g. recently for SO_2 (Davis & Wilhour, 1976; Zahn, 1978); for HF (Treshow & Pack, 1970; Guderian, 1971; Weinstein, 1977); and for both SO_2 and HF (Dässler, 1976; Prinz & Brandt, 1980), display considerable variations. For they depend first on the opinion of the author, but also on particular plant species under incomparable environmental conditions, and the application of different criteria for plant responses. Also, the position that a plant species occupies on the scale of pollutant resistance depends upon what response is measured (Weinstein & McCune, 1979). When the susceptibility of leaves, the dry-matter production rate, and the capability to survive in polluted areas, are taken for criteria, three different resistance series may be obtained with the same plant species. Therefore, clear declaration of the chosen criterion and of the experimental conditions is indispensable.

Usually in extensive tables a large number of plants is listed according to one or two series of three categories of sensitivity: either very sensitive, sensitive, and less sensitive, or sensitive, intermediate, and relatively resistant. As foliage injury is often taken for the criterion of sensitivity, *Abies alba* and *Picea abies* are not included, although these trees as a whole are regarded as being extremely sensitive to SO_2 and HF. The sensitivity of conifers is indeed commonly accepted; however, it may not be explained by the susceptibility of the needles. This was confirmed by Smith & Davis (1978), who demonstrated experimentally the relatively high resistance of coniferous needles to various SO_2 concentrations.

To summarize the findings from the various resistance- or sensitivity-series seems to be almost impossible, because of the difficulties of comparing the variable criteria for assessing the effects and of the different test conditions on which the results are based.

The sensitive plants that respond to SO_2-impact with visible symptoms on the foliage much earlier than other plants are termed 'indicator plants'. A compilation of those higher plants that are especially suitable for biomonitoring of SO_2 and HF has been presented by Manning & Feder (1980). The use of higher plants as indicators of different components of air pollution, including SO_2 and HF, is discussed in Chapter 5. Sensitive plants such as *Medicago sativa* var. Du Puits or *Fagopyrum esculentum* (Buckwheat) for SO_2, and *Gladiolus gandavensis* var. Snow Princess or cultivar Flowersong for HF, show macroscopical effects (chlorosis or necrosis, abortion of flowers and fruits, leaf-drop, and inhibition of growth) which indicate the presence of pollution. Other plants act as mere receivers or absorbers of pollutants, without any injury, and the accumulated pollutant can be determined by chemical leaf analysis. It was suggested some years ago to differentiate between visibly injured indicator-plants and pollutant-absorbing test plants (Halbwachs & Richter, 1971). Posthumus (1980), moreover, has elaborated standardized methods for using plants as indicators for oxidized sulphur components, and for the use of plants as indicators and accumulators for fluoride-containing air pollution (cf. Chapter 5).

Using the grass species *Lolium multiflorum* (Italian Ryegrass), Scholl (1971, 1976) has developed a standardized method for determination of the fluorine emission rate. The accumulation of fluorine within the leaves is analytically measured and is taken as the effective emission dose ('Immissionswirkdosis'). It is the result of the emission-flux and the working of the mechanisms of excretion, metabolism, and translocation, within the acceptor plant. From the effective emission dose established with *Lolium*, the hazards to other plant species and animals in a pasture can be established.

DOSE–RESPONSE RELATIONSHIPS

The scientific basis for establishing 'Air Quality Standards' for the protection of vegetation can be seen in the 'Air Quality Criteria', which describe the relationship between air pollutants and their adverse effects on vegetation, man, other animals, and property. Scientific research in this respect should provide comprehensive and unequivocal information by means of experimental and epidemiological determination of dose–response relationships on which to judge injurious responses of plants and plant communities in the case of a particular pollutant dose (Prinz & Brandt, 1980). Furthermore, a basis should be provided to estimate the risks of injury upon change of the emissions.

In the case of SO_2, it is useful to define the pollutant dose by a combination of concentration and time of exposure, but with HF it seems desirable also to focus on the quantity of fluoride absorbed by plants per unit time.

Early work by O'Gara (1922) suggested that concentration and exposure-time are directly proportional, but recent work has demonstrated that this is over-simplistic. Haut (1961) found that, with *Raphanus sativus* ssp. *radicola*, injury increased progressively with increasing SO_2 concentrations. Reduction in yield of the plant community also correlated better with the concentration than with exposure-time (Guderian, 1966). The simple relationship, i.e. concentration × exposure-time = constant degree of injury, seems to be limited to certain higher SO_2 concentrations and time-periods of a few hours (Caput *et al.*, 1978). During longer time-periods, the degree of injury is influenced by the SO_2 concentration as well as by chemical processes within the cells, e.g. the oxidation of sulphite to sulphate.

In SO_2 fumigation experiments conducted to determine the dose–response relationships for four species of urban trees, Temple (1972) found differences in species' response to duration of exposure and concentration. While concentration and duration of exposure were of equal importance in producing injury in *Ulmus parvifolia*, and probably in *Quercus palustris* (Spanish or Pin Oak), SO_2 concentration was of greater importance in injuring *Acer platanoides* and *Ginkgo biloba* (Ginkgo) than the duration of exposure.

Furthermore, the exposure intervals are crucial to causing injurious effects. Comparing continuous *versus* intermittent fumigation with intervals of different time-length, Guderian (1977) found that when pollutant-free periods increased above a certain minimum, injury decreased although the amount of SO_2 absorbed did not decrease. In general, S-accumulation increases with the frequency of pollutant exposure (Guderian, 1970).

With fluorine, the degree of injury and the accumulation of this element varies considerably among different plant species and therefore no general relationship between concentration, length of exposure, and degree of injury, can be established. With some plant species, e.g. of *Pinus*, injury increased progressively with increasing concentration, but with species of *Gladiolus* this was not the case. Also, some tree species (*Abies balsamea* [Balsam Fir], *Picea mariana* [Black Spruce], *Larix* sp., *Betula pendula*, and *Alnus glutinosa*) displayed considerable differences in fluoride accumulation in the current year's foliage at the first appearance of symptoms within 3–4 weeks of exposure. At the end of the growing season damage seemed to be higher, but was not proportional to the increase in fluoride levels in the leaves or needles (Sidhu, 1979).These data indicate that the fluoride content of the foliage is indeed an indicator of fluoride exposure, but is not directly suited for evaluating the injurious effects. MacLean & Schneider (1973) found that fluoride accumulation was greater after continuous fumigations than when the same HF dose was provided in alternate 48-hours' exposures.

From the above results, the difficulties in extrapolating data derived from continuous fumigation experiments to field conditions become obvious. The causes of these different responses might be seen in the differences in secondary fluoride-translocation mechanisms among several plant species (Jacobson *et al.*, 1966). But the washing-out and washing-off of fluorides from plant tissues during emission-free periods, as well as dilution by rapid growth, also play an important role for the reduction in fluoride content.

Pollutant effects are described at every level of organization—from cellular to community. In the case of economically important forest trees and forests, Knabe (1971) provides some useful criteria of injury. Owing to the various criteria of injury and the previously discussed numerous influences of environmental and endogenous factors, it becomes evident that, for each plant species and even variety, several curves of the dose–response relationships may be established according to the chosen criterion and the environmental parameters. This is the reason why the overwhelming majority of results come from fumigation studies performed under well-defined conditions with relatively high concentrations over relatively short exposure-periods.

The impossibility of providing for each plant a relation between pollutant exposure and response, leads to establishing rather broad resistance-groups. In regions where coniferous forests are a major component of the landscape, proposals for limitation of SO_2- and HF-pollution may be based upon responses of *Picea*. It is of advantage, especially for Central and Northern Europe, to use *Picea abies* as a guiding species because it is most common, and its sensitivity to air pollution provides an indicator to protect most of the other species of higher plants as well.

A particular problem in forestry, exceeding the air pollution aspect, is the decline of *Abies alba* on many sites. Although *Abies alba* is regarded as the most sensitive tree species to SO_2- and HF-pollution in Europe (Wentzel, 1980), pollution is but one of several factors causing this decline.

A working group on air pollution, of IUFRO (1979), has established effective limiting values of pollutant concentrations that produce no injury to forest stands on the basis of the latest scientific knowledge and silvicultural experiences. For three different time-periods (1 year, 24 hours, and 30 minutes), two SO_2 concentration limits have been established to guarantee, respectively,

(a) full production on most sites, and
(b) the maintenance of full production and environmental protection, e.g. against erosion and avalanches in higher regions of mountains, in boreal zones, and on 'extreme' sites (Table I).

If these thresholds are exceeded, a reduction in vitality, growth performance, and resistance to biotic and abiotic influences, must be expected. The strict limitation of allowed concentrations, as proposed by the IUFRO (1979) Resolution, is based essentially upon the investigations of Materna (1973), who described the responses of *Picea* growing on poor sites to long-term exposure (several years) to average annual concentrations of $25-40 \mu$g SO_2/m^3. Although a decrease in yield, reduction in dry-weight, and inhibition of growth, has also been observed in several other plant species and varieties (Table II) under the influence of low SO_2 concentrations (av. conc. $<200 \mu$g) during extended periods, these other plants are not considered to be as sensitive as conifers.

The Commission of the European communities, dealing with environment and quality of life, proposes annual mean concentration limits of 50μg SO_2/m^3 for very sensitive, 80μg SO_2/m^3 for sensitive, and 120μg SO_2/m^3 for less-sensitive plants, to protect vegetation sufficiently to fulfil its main economical and ecological functions (Prinz & Brandt, 1980). Knabe (1970) has pointed out that there is a close correlation between the distribution range of *Pinus sylvestris* in polluted areas and the arithmetic mean of the annual SO_2 concentration.

Table I SO_2 concentration limits for the protection of *Picea abies* (SO_2 in micrograms per cubic metre of air).

	Annual average	Average of 24 hours	97.5 percentile of 30 minutes values in period of vegetation
(a)	50	100	150
(b)	25	50	75

To guarantee the full physiological efficiency and productivity of vegetation, exceeding mere survival, it is necessary to establish concentration limits for short time-periods. An annual mean concentration will not be exceeded even though high concentration peaks during short periods may occur. Such peaks are compensated by pollutant-free periods, but the physiological and metabolic processes may be influenced in a way that excludes complete recovery. Larsen & Heck (1976) tried to find out which exposure durations were most suited to setting ambient air quality standards to prevent acute responses. The experimental work was conducted with 19 plant species or groups, including trees. The authors recommended a non-overlapping ambient air quality measurement for averaging times of 1, 3 and 8 hours. These monitoring times were found to be best related to patterns of ambient pollutant concentration, even though SO_2 is emitted by isolated point-sources. From the point of view of a plant physiologist, a 30-minutes' averaging value is appropriate when functional disorders of plants are involved.

When discussing the relations between a certain HF concentration in the air and the response of plants, it is important to consider another factor: Fluorides are accumulative toxicants and injury usually occurs with long-term exposure. So even extremely low concentrations can cause injury to sensitive species. As fluorine, in contrast to sulphur, is not metabolized by the plant, increasing accumulation of it within the tissues bears hazards for people and other animals consuming plants that have been exposed to it. For assessing the risks to plants and animals, both fluoride concentration in the air and the fluoride content of the plants must be taken into consideration. The IUFRO (1979) Resolution claims that an annual average of 0.3 μg HF/m^3, and a 97.5 percentile of 30 minutes' value of 0.9 μg HF/m^3 during the growing-season, is necessary to keep full production of *Picea abies* on most sites. The proposal for a limitation of atmospheric HF concentration, put forward by the European Economic Community (Prinz & Brandt, 1980) distinguishes between three classes of sensitivity: very sensitive, sensitive, and less sensitive, and contains limits for 24 hours, for 30 days, and for the whole growing-season from April to October (Table III).

However, average atmospheric concentrations of a pollutant that accumulates within the plant tissue do not reflect the real stress upon plants. Therefore a second criterion is necessary, namely the fluoride accumulation within the plant. Under field conditions it must be assumed that the relationship between the considerable varying HF-concentration in the air and the fluoride content of the plant tissue is a non-linear function, as those three processes that mediate the concentration of fluorine in plants (i.e. uptake, distribution of penetrated fluoride and of fluoride on the exterior surface, and translocation and elimination, e.g. by leaching or abscission) are far more influenced by environmental factors than is the case in fumigation experiments.

Table II Effects of low concentrations of SO_2 on several species of higher plants.

Species	Conc. $\mu g\ SO_2/m^3$	Exposure time	Effect	Reference
Picea abies	25–40	several years	Slight to medium needle injury; loss of older needles; reduction in photosynthesis	Materna (1973)
Lolium perenne	43	96 days (winter)	Reduction in dry-weight of shoots up to 21%	Bell et al. (1979)
Lolium perenne	43	126 days (winter)	Reduction in dry-weight of shoots up to 51%	Bell et al. (1979)
Pinus strobus	45	7 months	Growth reduction	Linzon (1971)
Several agricultural crops and conifers	50	90 days	Influence on physiological processes	Steubing (1978)
Lolium multiflorum and Dactylis glomerata	50–90	8–10 weeks	Reduction in dry-weight of shoots	Crittenden & Read (1979)
Lolium perenne	69	12 weeks	Reduction in dry-weight; reduction in number of shoots	Crittenden & Read (1978)

Species		Duration	Effect	Reference
Pinus strobus (sensitive clones)	70	6 hours	Needle necrosis	cited after Auclair (1976)
Lolium perenne	106	194 days	Reduction in dry-weight of shoots up to 24%	Bell *et al.* (1979)
Lolium perenne	122	44 days	Reduction in yield up to 25%	Bell *et al.* (1979)
Raphanus sativus var. Cherry Belle	130	40 hrs/week 5 weeks	Reduction in: fresh-weight of the entire plant, leaf fresh-weight, root fresh- and dry-weight, root length and width	Tingey *et al.* (1971)
Nicotiana tabacum (BEL W-3)	131	40 hrs/week 4 weeks	Growth reduction; reduction in dry-weight of leaves up to 22%	Tingey & Reinert (1975)
Medicago sativa	131	40 hrs/week 4 weeks	Reduction in: dry-weight of foliage up to 26%, dry-weight of root up to 49%	Tingey & Reinert (1975)
Pinus sylvestris	170	> 1 year	Growth reduction of seedlings	Farrar *et al.* (1977)
Lolium perenne S 23	191	26 weeks	Decrease in dry-weight of leaves	Bell & Clough (1973)
Hordeum vulgare cv. Dickenson and *Zea mays* cvs	160–210	27 days	Lesions on foliage	Mandl *et al.* (1975)

Table III   Concentration limits established by the Commission of the European Communities for the protection of vegetation from atmospheric HF (in microgram per cubic metre; $\mu g/m^3$.

Sensitivity category of plants	Average over 24 hours	Average over 30 days	Average over growing-season
Very sensitive	2.0	0.4	0.3
Sensitive	3.0	0.8	0.5
Less sensitive	4.0	2.0	1.4

Recent proposals concerning the experimental investigation of the action of airborne fluoride upon plants, have been put forward by McCune *et al.* (1976). Using eight different patterns of exposure, these authors showed with *Sorghum* that the effect of a change in concentration of HF depended on the concentration of HF in a previous or subsequent exposure. These data indicate that several exposures to an accumulating pollutant such as HF, resulting in the same mean concentration, are not equivalent in causing effects.

Brandt (1971) correlated the fluoride level of the plant tissue with recognizable responses, and distinguished between three groups of plants with different sensitivity: very sensitive plants responded at a fluoride level below 50 ppm (μg F/g dry weight), sensitive plants at levels from 50 to 200 ppm, and relatively resistant plants at levels above 200 ppm. The average fluoride concentration at which no visible symptoms are expected ranges from 0.2 μg F/m^3 with conifers to 0.4 μg F/m^3 with deciduous trees (Sidhu, 1980). The Canadian Air Quality Standards limit the fluoride concentration to 0.2 μg HF/m^3 for 70 days, and are comparable in their order of magnitude with the IUFRO-proposed level of 0.3 μg HF/m^3 for 1 year (IUFRO, 1979). According to the calculations of Sidhu (1980), an atmospheric concentration of 0.3 μg HF/m^3 would lead to an accumulation of up to 20 ppm fluoride in the foliage after a two-years' exposure.

Therefore, with the present state of knowledge, observance of these limits guarantees health and growth of plants and excludes hazards for animals by fluoride accumulation in the food chain.

SUMMARY

Although foliar symptoms are the best-documented response of plants to air pollutants, it must be borne in mind that these are often preceded by growth and yield effects. Visible symptoms are characterized by various patterns of chlorosis and necrosis, depending on the pollutant and the plant species. Fluorides typically cause necrosis around the margins of leaves—or at the tip of the needles if conifers are involved. When chlorosis appears it may extend interveinally. Sulphur dioxide tends to cause chlorosis, also extending interveinally—or, with conifers, needle-tip necrosis. The classic symptom of ozone toxicity is a chlorotic or necrotic flecking.

The intensity of plant responses is influenced strongly, not only by the concentration, duration, and frequency, of exposure, but also by the relative genetic tolerance to the pollutant and to every environmental factor. Temperature, humidity, soil moisture, light, nutrient conditions, and the age of tissues, are all critical. Under the environmental conditions that are most conducive to injury, the highly sensitive plants might be injured by HF concentrations as low as 2.0 $\mu g/m^3$ for a 24 hours' period, or by SO_2 concentrations as low as 50 $\mu g/m^3$ for 24 hours.

REFERENCES

Adams, D. F., Shaw, C. G., Gnagy, R. M., Koppe, R. K., Mayhew, D. J. & Yerkes, W. D. (1956), Relationship of atmospheric fluoride levels and injury indexes on *Gladiolus* and Ponderosa Pine. *J. Agr. Food Chem.,* **4**, pp. 64–66.

Adams, D. F., Hendrix, J. W. & Applegate, H. G. (1957). Relationship among exposure periods, foliar burn, and fluorine content of plants exposed to hydrogen fluoride. *J. Agr. Food Chem.,* **5**, pp. 108–16.

Adams, D. F. & Sulzbach, C. W. (1961). Nitrogen deficiency and fluoride susceptibility of bean seedlings. *Science,* **133**, pp. 1425–6.

Ashenden, T. W. & Mansfield, T. A. (1977). Influence of wind-speed on the sensitivity of Ryegrass to SO_2. *J. Exp. Bot.,* **28**, pp. 729–35.

Ashenden, T. W., Mansfield, T. A. & Wellburn, A. R. (1978). Influence of wind on the sensitivity of plants to SO_2. *VDI-Berichte,* **314**, pp. 231–5.

Auclair, D. (1976). Effets du dioxyde de soufre sur les arbres et la forêt (Étude bibliographique). *Centre de Recherches Forestieres d'Orleans,* Doc. 76/15, 50 pp.

Barrett, T. W. & Benedict, H. M. (1970). Sulfur dioxide. In *Recognition of Air Pollution and Injury to Vegetation. A Pictorial Atlas* (Ed. J. Jacobson & A. C. Hill). Air Pollution Control Association, Pittsburgh, Pennsylvania, USA: pp. C1–C17.

Barton, J. R., McLaughlin, S. B. & McConathy, R. K. (1980). The effects of SO_2 on components of leaf resistance to gas exchange. *Environ. Pollut.,* **21**, pp. 255–65.

Beckerson, D. W. & Hofstra, G. (1979). Stomatal responses of white bean to O_3 and SO_2 singly or in combination. *Atmospheric Environment,* **13**, pp. 533–5.

Bell, J. N. B. & Clough, W. S. (1973). Depression of yield in Ryegrass exposed to sulphur dioxide. *Nature* (London), **241**, p. 47.

Bell, J. N. B., Rutter, A. J. & Relton, J. (1979). Studies in the effects of low levels of sulphur dioxide on the growth of *Lolium perenne* L. *New Phytol.,* **83**, pp. 627–43.

Benedict, H. M., Ross, J. M. Wade, R. H. (1964). The disposition of atmospheric fluorides by vegetation. *Int. J. Air Water Pollut.,* **8**, pp. 279–89.

Benedict, H. M., Ross, J. M. & Wade, R. H. (1965). Some responses of vegetation to atmospheric fluorides. *J. Air Poll. Contr. Assoc.,* **15**, pp. 253–5.

Bertrand, D. (1969). Le fluor, oligo-element dynamique pour le mais. *C. R. H. Acad. Sci.,* Ser. D 269, pp. 1767–9.

Biggs, A. R. & Davis, D. D. (1980). Stomatal response of three birch species exposed to varying acute doses of SO_2. *J. Amer. Soc. Hort. Sci.,* **105**(4), pp. 514–6.

Bligny, R., Garrec, J. P. & Fourcy, A. (1972). Migration et accumulation du fluor chez *Zea mays. C. R. Acad. Sci.,* **275**, pp. 755–8.

Börtitz, S. (1974). Bedeutung 'unsichtbarer' Einflüsse industrieller Immissionen auf die Vegetation. *Biol. Zbl.,* **93**, pp. 341–9.

Brandt, C. S. (1967). Effects of air pollution on crop production. *Ontario Pollution Control Conf. 1967.* Cit. after *NAPCA Abstr. Bulletin* **1**(7), p. 121.

Brandt, C. S. (1971). Ambient air quality criteria for hydrofluorine and fluorides. *VDI-Berichte,* **164**, pp. 23–7.

Brandt, C. S. & Heck, W. W. (1968). Effects of air pollutants on vegetation. Pp. 401–28 in *Air Pollution* Vol. 1 (Ed. A. C. Stern). Academic Press, New York & London: 694 pp.

Braun, G. (1977). Über die Ursachen und Kriterien der Immissions resistenz bei Fichte. Teil I–III. *Eur. J. For. Path.,* **7**, pp. 23–43, pp. 129–52, pp. 303–19.

Braun, G. (1978). Über die Ursachen und Kriterien der Immissionsresistenz bei Fichte; Teil IV. *Eur. J. For. Path.,* **8**, pp. 83–96.

Brewer, R. F., Sutherland, F. H. & Guillemet, F. B. (1969). Effects of various fluoride sources on citrus growth and fruit production. *Environ. Sci. Technol.,* **3**, pp. 378–81.

Bucher-Wallin, I. (1976). Zur Beeinflussung des physiologischen Blattalters von Waldbäumen durch Fluor-Immissionen. *Mitt. Eidg. Anst. forstl. Vers'wes.,* **52**, pp. 101–155.

Caput, C., Belot, Y., Auclair, D. & Decourt, N. (1978). Absorption of sulphur dioxide by pine needles leading to acute injury. *Environ. Pollut.,* **16**, pp. 3–15.

Commins, B. T. (1967). Some studies on the synthesis of particulate acid sulfate from the products of combustion of fuels and measurement of the acid in polluted atmospheres. *Proc. of the Symposium on the physico-chemical transformation of sulfur compounds in the atmosphere and the formation of acid smogs.* OECD, Mainz, West Germany: pp. 178.

Cowling, D. W., Jones, L. H. P. & Lockyer, D. R. (1973). Increased yield through correction of sulphur deficiency in Ryegrass exposed to sulphur dioxide. *Nature* (London), **243**, pp. 479–80.

Crittenden, P. D. & Read, D. J. (1978). The effects of air pollution on plant growth with special references to sulphur dioxide, II. Growth studies with *Lolium perenne* L. *New Phytol.,* **80**, pp. 49–62.

Crittenden, P. D. & Read, D. J. (1979). The effects of air pollution on plant growth with special reference to sulphur dioxide, III: Growth studies with *Lolium multiflorum* Lam. and *Dactylis glomerata* L. *New Phytol.,* **83**, pp. 645–51.

Dässler, H.-G. (1976). *Einfluss von Luftverunreinigungen auf die Vegetation.* VEB Gustav Fischer, Jena, E. Germany: 189 pp.

Davis, D. D., & Wilhour, R. G. (1976). Susceptibility of woody plants to sulfur dioxide and photochemical oxidants. *U.S. Environ. Protec. Agency, Ecol. Res. Ser.* EPA-600/3-76–102, Corvallis, Oregon, USA: 71 pp.

Dochinger, L. S., Bender, F. W., Fox, F. L. & Heck, W. W. (1970). Chlorotic dwarf of Eastern White Pine caused by an ozone and sulphur dioxide interaction. *Nature* (London), **225**, p. 476.

Edelbauer, A. & Halbwachs, G. (1982). Mineralstoff- und Fluoridgehalt von Grünfutter unter Rauchgaseinfluss in Abhängigkeit von der Entfernung zur Emissionsquelle und der Bewirtschaftungsform. *Carinthia II,* **39**, Sonderheft, pp. 233–63.

Elkiey, T. & Ormrod, D. P. (1979). Leaf diffusion resistance responses of three *Petunia* cultivars to ozone and/or sulphur dioxide. *J. Air Poll. Control Assoc.,* **29**, pp. 622–5.

Facteau, T. J., Wang, S. Y. & Rowe, K. E. (1973). The effect of hydrogen fluoride on pollen germination and pollen tube growth in *Prunus avium* L. cv.

'Royal Ann'. *J. Amer. Soc. Hort. Sci.,* **98**, pp. 234–6.

Facteau, T. J. & Rowe, K. E. (1977). Effect of hydrogen fluoride and hydrogen chloride on pollen-tube growth and sodium fluoride on pollen germination in 'Tilton' apricot. *J. Amer. Soc. Hort. Sci.,* **102**, pp. 95–6.

Faller, N. (1972). Schwefeldioxid, Schwefelwasserstoff, nitrose Gase und Ammoniak als ausschliessliche S-bzw. N-Quelle der höheren Pflanze. *Z. Pflanzenernährung und Bodenkunde,* **131**, pp. 120–30.

Farrar, J. F., Relton, J. & Rutter, A. J. (1977). Sulphur dioxide and the growth of *Pinus sylvestris. J. of Applied Ecology,* **14**, pp. 861–75.

Feiler, S., Michael, G., Ranft, H., Tesche, M. & Bellmann, C. (1981). Zur Komplexwirkung von SO_2 und Frost auf Fichte (*Picea abies* (L.) Karst.). *Biol. Rdsch.,* **19**, pp. 98–100.

Garber, K., Guderian, R. & Stratmann, H. (1967). Untersuchungen über die Aufnahme von Fluor aus dem Boden durch Pflanzen. *Qual. Plant. et. Mat. Veget.,* **14**(3), pp. 223–36.

Garrec, J. P., Oberlin, J. C., Ligeon, E., Bisch, A. M. & Fourcy, A. (1974). Fluoride–calcium interaction in polluted fir needles. *Fluoride,* **7**, pp. 78–84.

Garrec, J. P., Plebin, R. & Lhoste, A. M. (1977). Influence du fluor sur la composition minérale d'aiguilles poluées de sapin. *Environ. Pollut.,* **13**, pp. 159–67.

Garsed, S. G., Farrer, J. F. & Rutter, A. J. (1979). The effects of low concentrations of sulphur dioxide on the growth of four broadleaved tree species. *J. Appl. Ecology,* **16**, pp. 217–26.

Gartrell, F. E., Thomas, F. W. & Carpenter, S. B. (1963). Atmospheric oxidation of SO_2 in coal-burning power-plant plumes. *Am. Ind. Hyg. Assoc. J.,* **24**, pp. 113–20.

Grill, D. & Härtel, O. (1972). Zellphysiologische und biochemische Untersuchungen an SO_2-begasten Fichtennadeln: Resistenz und Pufferkapazität. *Mitt. Forstl. Bundesversuchsanstalt Wien,* **97**, pp. 367–85.

Grill, D., Liegl, E. & Windisch, E. (1979). Holzanatomische Untersuchungen an abgasbelasteten Bäumen. *Phytopath. Z.,* **94**, pp. 335–42.

Guderian, R. (1966). Reaktionen von Pflanzengemeinschaften der Feldfutterbaues auf Schwefeldioxideinwirkungen. Essen, West Germany: Girardet-Verlag, Schriftenreihe Landesanst. Immissions-Bodennutzungsschutz d. Landes Nordrhein-Westfalen 4, pp. 80–100.

Guderian, R. (1970). Untersuchungen über quantitative Beziehungen zwischen dem Schwefelgehalt von Pflanzen und dem Schwefeldioxidgehalt der Luft. *Z. Pflanzenkrankh. Pflanzenschutz,* **77**, I. Teil, pp. 200–220, II. Teil, pp. 289–308. III. Teil, pp. 387–399.

Guderian, R. (1971). Ergebnisse aus Begasungsexperimenten zur Emitlung pflanzenschädigender HF-Konzentrationen. *VDI-Berichte,* **164**, pp. 33–7.

Guderian, R. (1977). Air Pollution. (*Ecological Studies 22.*) Springer-Verlag, Berlin–Heidelberg–New York: 127 pp.

Guderian, R. (1978). Wirkungen sauerstoffhaltiger Schwefelverbindungen. *VDI-Berichte,* **314**, pp. 207–17.

Guderian, R., Haut, H. van & Stratmann, H. (1969). Experimentelle Untersuchungen über pflanzenschädigende Fluorwasserstoff-Konzentrationen. *Forschungsbericht des Landes Nordrhein-Westfalen* 2017. Westdeusccher Verlag, Köln & Opladen, West Germany: 54 pp.

Guderian, R. & Stratmann, H. (1968). Freilandversuche zur Ermittlung von Schwefeldioxidwirkungen auf die Vegetation. III. Teil: Grenzwerte schädlicher SO_2-Immissionen für Obst und Forstkulturen sowie für landwirtschaftliche und gärtnerische

Pflanzenarten. *Forsch. Ber. d. Landes Nordrhein-Westfalen* 1920. Köln & Opladen, West Germany: 114 pp.

Halbwachs, G. (1963). Untersuchungen über gerichtete aktive Strömungen und Stofftransporte im Blatt. *Flora,* **153**, pp. 333–57.

Halbwachs, G. (1970). Vergleichende Untersuchungen über die Wasserbewegung in gesunden und fluorgeschädigten Holzgewächsen. *Cbl. f. d. gesamte Forstwesen,* **87**, pp. 1–22.

Halbwachs, G. (1971). Physiologische Probleme der Vegetationsschädigung durch gasförmige Immissionen. *Ber. Dtsch. Bot. Ges.,* **84**, pp. 507–14.

Halbwachs, G. (1973). Besondere Umwelteinflüsse und Baumwachstum. *100-Jahre Hochschule f. Bodenkultur,* Studienrichtung Forst-u. Holzwirtschaft, Wien, **4**(1), pp. 167–77.

Halbwachs, G. & Kisser, J. (1967). Durch Rauchimmissionen bedingter Zwergwuchs bei Fichte und Birke. *Centralbl. f. d. ges. Forstwesen,* **84**, pp. 156–73.

Halbwachs, G. & Richter, H. (1971). Pflanzen als Anzeiger für Luftverunreinigungen. *Proc. First Europ. Biophysic. Congr.,* Vienna, pp. 319–24.

Haut, H. van (1961). Die Analyse von Schwefeldioxidwirkungen auf Pflanzen im Laboratoriumsversuch. *Staub,* **21**, pp. 52–6.

Haut, H. van & Stratmann, H. (1970). *Farbtafelatlas über Schwefeldioxid-Wirkungen an Pflanzen.* Verlag W. Girardet, Essen, West Germany: 206 pp.

Heck, W. W. & Dunning, J. A. (1978). Response of oats to sulphur dioxide: Interactions of growth temperature with exposure temperature or humidity. *J. Air Poll. Contr. Assoc.,* **28**, pp. 241–6.

Hill, A. C., Transtrum, L. G., Pack, M. R. & Winters, W. S. (1958). Air pollution with relation to agronomic crops, VI: An investigation of the 'hidden injury' theory of fluoride damage to plants. *Agron. J.,* **50**, pp. 562–5.

Hitchcock, A. E., Zimmermann, P. W. & Coe, R. R. (1963). The effect of fluorides on milo maize (*Sorghum* sp.). *Contrib. Boyce Thomps. Inst.,* **22**, pp. 175–206.

Hitchcock, A. E., Weinstein, L. H., McCune, D. C. & Jacobson, J. S. (1964). Effects of fluorine compounds on vegetation, with special reference to sweet corn. *J. Air Poll. Contr. Assoc.,* **14**, pp. 503–8.

Holub, Z. & Navara, J. (1966). The protecting effect and influence of calcium, magnesium and phosphorus on the uptake of fluoride by pea seeds. *Biologia Bratislava,* **21**, pp. 177–82.

Höster, H. R. (1977). Veränderungen der Holzstruktur als Indikator für Umweltbelastungen bei Bäumen. *Ber. Dtsch. Bot. Ges.,* **90**, pp. 253–60.

Hübl, E. (1963). Über das stomatäre Verhalten von Pflanzen verschiedener Standorte im Alpengebiet und auf Sumpfwiesen der Ebene. *Sitzungsber. d. Österr. Akad. d. Wissensch., Mathem.-naturw. Kl.,* Abt. I., Bd. **172**, H. 1u. 2, pp. 1–84.

Huttunen, S. (1976). The influence of air pollution on the northern forest vegetation. Pp. 97–101 in *Proc. of the Kuopio Meeting on Plant Damages Caused by Air Pollution* (Ed. L. Kärenlampi). Savon Sanomain Kirjapaino Oy, Kuopio, Finland: 160 pp.

Huttunen, S. (1978). The effects of air pollution on provenances of Scots Pine and Norway Spruce in northern Finland. *Silva fennica,* **12**, pp. 1–16.

IUFRO (Subject group 'Air Pollution') (1979). Resolution über maximale Immissionswerte zum Schutze der Wälder. *Supplement to IUFRO News No. 25* (3/1979), pp. 2.

Jacobson, J. S., Weinstein, L. H., McCune, D. C. & Hitchcock, A. E. (1966). The accumulation of fluorine by plants. *J. Air Poll. Contr. Assoc.,* **16**, pp. 412–6.

Jäger, H.-J. & Klein, H. (1976). Modellversuche zum Einfluss der Nährstoffversorgung auf die SO_2-Empfindlichkeit von Pflanzen. *Eur. J. For. Path.,* **6**, pp. 347–54.

Jäger, H.-J. & Klein, H. (1980). Biochemical and physiological effects of SO₂ on plants. *Angew Botanik,* **54,** pp. 337–48.

Keller, T. (1973). Über die schädigende Wirkung des Fluors. *Schweizerische Zeitschrift f. Forstwesen,* **124,** pp. 700–6.

Keller, T. (1977*a*). Der Einfluss von Fluorimmissionen auf die Nettoassimilation von Waldbäumen. *Mitt. d. Eidg. Anstalt f. d. forstl. Versuchsw.,* **53,** pp. 161–98.

Keller, T. (1977*b*). Begriff und Bedeutung der 'latenten Immissionsschädigung'. *Allg. Forst-u. Jagd-Zeitung,* **148,** p. 115–20.

Keller, T. (1978*a*). Einfluss niedriger SO₂-Konzentrationen auf die CO₂-Aufnahme von Fichte und Tanne. *Photosynthetica,* **12,** pp. 316–22.

Keller, T. (1978*b*). Frostschäden als Folge einer 'latenten' Immissionsschädigung. *Staub,* **38,** pp. 24–6.

Keller, T. (1979). Der Einfluss mehrwöchiger niedriger SO₂-Konzentrationen auf CO₂-Aufnahme und Jahrringbau der Fichte. *Ber. d. X. Tagung d. IUFRO-Fachgruppe 'Air Pollution'* Ljubljana, Yugoslavia, pp. 73–87.

Keller, T. & Schwager, H. (1971). Der Nachweis unsichtbarer ('physiologischer') Fluor-Immissionsschädigungen an Waldbäumen durch eine einfache kolorimetrische Bestimmung der Peroxidase-Aktivität. *Europ. J. Forest Path.,* **1,** pp. 6–18.

Kisser, J. (1965). Forstliche Rauchschäden aus der Sicht des Biologen. *Mitt. forstl. Bundesversuchsanstalt Mariabrunn,* **73,** pp. 7–48.

Klein, H., Jäger, H.-J., Domes, W. & Wong, C. H. (1978). Mechanisms contributing to differential sensitivities of plants to SO₂. *Oecologia,* **33,** pp. 203–8.

Knabe, W. (1968). Experimentelle Prüfung der Fluoranreicherung in Nadeln und Blättern von Pflanzen in Abhängigkeit von deren Expositionshöhe über Grund. *Int. Tagung forstl. Rauchschadensachverständiger.,* Referate d. VI, Katowice, Poland, pp. 101–16.

Knabe, W. (1970). Kiefernwaldverbreitung und Schwefeldioxid-Imissionen im Ruhrgebiet. *Staub,* **30,** pp. 32–5.

Knabe, W. (1971). Air quality criteria and their importance for forest. *Mitt. forstl. Bundesversuchsanstalt,* Wien, **92,** pp. 129–50.

Knabe, W. (1978). Luftverunreinigung und Waldwirtschaft. Pp. 697–709 in *Natur-u. Umweltschutz in der Bundesrepublik Deutschland* (Ed. G. Olschowy). Verlag Paul Parey, Hamburg & Berlin, West Germany: 926 pp.

Kohout, R. & Materna, J. (1966). Die Ergebnisse der Untersuchungen über die Sorption des SO₂ und seines Transportes in den Blättern einiger Holzarten. *Referate d. V. Int. Tagung 'Wirkungen industrieller Emissionen auf die Forstwirtschaft'.* Janské Lázně, Czechoslovakia, pp. 167–80.

Kronberger, W. & Halbwachs, G. (1974). Über eine einfache Methode zur Bestimmung des Fluorgehaltes von Pflanzen mittels ionenspezifischer Elektrode. *Referate d. IX. Int. Tagung 'Luftverunreinigung und Forstwirtschaft'.* Mariánské Lázně, Czechoslovakia, pp. 121–9.

Kronberger, W. & Halbwachs, G. (1978). Distribution of fluoride in *Zea mays* grown near an aluminum plant. *Fluoride,* **12,** pp. 129–35.

Kronberger, W., Halbwachs, G. & Richter, H. (1978). Fluortranslokation in *Picea abies* (L.) Karst. *Angew. Bot.,* **53,** pp. 149–54.

Lai Dinh, D., Buchloh, G. & Oelschläger, W. (1973). Auswirkung von Fluorverbindungen auf die Pollenkeimung und den Fruchtansatz von Obstgewächsen. *Der Erwerbsobstbau,* **15,** pp. 154–157.

Larsen, R. I. & Heck, W. W. (1976). An air quality data analysis system for interrelating effects, standards, and needed source reductions, Part 3: Vegetation injury. *J. Air Poll. Contr. Assoc.,* **26,** pp. 325–33.

Leonard, C. D. & Graves, H. B., jr (1970). Some effects of airborne fluorine on growth and yield of six *Citrus* varieties. *Proc. Fla. State Hort. Soc.,* **83**, pp. 34–41.

Levitt, J. (1980). *Responses of Plants to Environmental Stresses.* Academic Press, New York & London: ix + 697 pp., illustr.

Liese, W., Schneider, M. & Eckstein, D. (1975). Histometrische Untersuchungen an Holz einer rauchgeschädigten Fichte. *Eur. J. For. Path.,* **5**, pp. 152–61.

Linzon, S. N. (1966). Damage to Eastern White Pine by sulphur dioxide, semi-mature-tissue needle blight, and ozone. *J. Air Poll. Contr. Assoc.,* **16**, pp. 140–4.

Linzon, S. N. (1969). Symptomatology of sulphur-dioxide injury on vegetation. Pp. 1–12 in *Handbook of Effects, Assessment, Vegetation Damage: Section VIII* (Ed. N. Lacasse & W. Moroz). Pennsylvania State University, University Park, Pennsylvania, USA: 183 pp. illust.

Linzon, S. N. (1971). Economic effects of sulfur dioxide on forest growth. *J. Air Poll. Contr. Assoc.,* **21**, pp. 81–86.

Linzon, S. N. (1978). Effects of airborne sulphur pollutants on plants. Pp. 109–62 in *Sulphur in the Environment, Part II: Ecological Impacts* (Ed. J. O. Nriagu). John Wiley & Sons, New York–Chichester–Brisbane–Toronto: 482 pp.

McCune, D. C., Hitchcock, A. E., Jacobson, J. S. & Weinstein, L. H. (1965). Fluoride accumulation and growth of plants exposed to particulate cryolite in the atmosphere. *Contr. Boyce Thompson Inst.,* **23**, pp. 1–12.

McCune, D. C., Hitchcock, A. E. & Weinstein, L. H. (1966). Effect of mineral nutrition on the growth and sensitivity of *Gladiolus* to hydrogen fluoride. *Contr. Boyce Thompson Inst.* **23**, pp. 295–9.

McCune, D. C., MacLean, D. C. & Schneider, R. E. (1976). Experimental approaches to the effects of airborne fluoride on plants. Pp. 31–46 in *Effects of Air Pollutants on Plants* (Ed. T. A. Mansfield). Cambridge University Press, Cambridge–London–New York–Melbourne: 209 pp.

McLaughlin, S. B. & Taylor, G. E. (1981). Relative humidity: Important modifier of pollutant uptake by plants. *Science,* **211**, pp. 167–9.

MacLean, D. C. & Schneider, R. E. (1971). Fluoride phytotoxicity: Its alteration by temperature. Pp. 292–5 in *Second International Clean Air Congress* (Ed. H. M. Englund & W. T. Beery). Academic Press, New York, NY, USA: 188 pp.

MacLean, D. C. & Schneider, R. E. (1973). Fluoride accumulation by forage: Continuous vs. intermittent exposures to hydrogen fluoride. *J. Environ. Quality,* **2**, pp. 501–3.

MacLean, D. C., Schneider, R. E. & McCune, D. C. (1973). Fluoride phytotoxicity as affected by relative humidity. *Proc. Third International Clean Air Congress,* VDI-Verlag, Düsseldorf, West Germany: pp. A143–5.

MacLean, D. C., Schneider, R. E. & McCune, D. C. (1977). Effects of chronic exposure to gaseous fluoride on yield of field-grown bean and tomato plants. *J. Amer. Soc. Hort. Sci.,* **102**, pp. 297–9.

Majernik, O. & Mansfield, T. A. (1971). Effects of SO_2 pollution on stomatal movements in *Vicia faba. Phytopath Zeitschr.,* **71**, pp. 123–8.

Mamajew, S. A. & Skarkolet, O. D. (1972). Effects of air and soil pollution by industrial waste on the fructification of Scotch Pine in the Urals. *Mitt. Forstl. Bundesversuchsanstalt,* Wien, **97**, pp. 443–50.

Mandl, R. H., Weinstein, L. H. & Keveny, M. (1975). Effects of hydrogen fluoride and sulphur dioxide alone and in combination on several species of plants. *Environ. Pollut.,* **9**, pp. 133–43.

Manning, W. J. & Feder, W. A. (1980). *Biomonitoring air pollutants with plants.* Applied Science Publishers, London, England: 142 pp.

Marchesani, V. J. & Leone, I. A. (1980). A bioassay for assessing the effect of sulphur dioxide on oat seedlings. *J. Air Poll. Contr. Assoc.,* **30**, pp. 163–5.

Markowski, A. & Grzesiak, S. (1974). Influence of sulphur dioxide and ozone on vegetation of bean and barley plants under different soil moisture conditions. *Bull. Acad. Polon. Sci. Ser. Sci. Biol.,* **22**, pp. 875–88.

Materna, J. (1973). Kriterien zur Kennzeichnung einer Immissionswirkung auf Waldbestände. *Proc. Third International Clean Air Congress,* VDI-Verlag, Düsseldorf, West Germany: pp. A121–3.

Materna, J. (1974). Einfluss der SO_2-Immissionen auf Fichtenpflanzen in Wintermonaten. *XI Int. Tagung 'Luftverunreinigung und Forstwirtschaft'.* Mariánské Lázně, Czechoslovakia, pp. 107–14.

Mathes, K. (1960). Über das Altern der Blätter und die Möglichkeit ihrer Wiederverjüngung. *Die Naturwissenschaften,* **15**, pp. 337–51.

Matsuoka, Y., Udagawa, O. & Ono, T. (1969). Effects of sulphur dioxide on pear pollination. *J. Japan Soc. Air Pollution,* **4**, p. 130.

Menser, H. A. & Heggestad, H. E. (1966). Ozone and sulphur dioxide synergism: Injury to tobacco plants. *Science,* **153**, pp. 424–5.

Mrkva, R. (1969). Einfluss der Immissionen auf die Saatgutgüte der Kiefer (*Pinus silvestris* L.) im Gebiet des Forstbetriebes Břeclav (Südmähren). *Acta Univ. Agriculturae,* Brno, Czechoslovakia, **38**, pp. 345–60. 　　　　　　　．

Mukammal, E. I. (1976). Review of present knowledge of plant injury by air pollution. *WMO Nr. 431, Techn. Note No. 147,* pp. 1–27.

O'Gara, P. J. (1922). Sulphur dioxide and fume problems and their solutions. *Ind. Eng. Chem.,* **14**, pp. 744–5.

Pack, M. R. (1966). Response of tomato fruiting to hydrogen fluoride as influenced by calcium nutrition. *J. Air Poll. Contr. Assoc.,* **16**, pp. 541–4.

Pack, M. R. (1971). Effects of hydrogen fluoride on production and organic reserves of bean seed. *Environ. Sci. Technol.,* **5**, pp. 1128–32.

Pack, M. R. (1972). Responses of strawberry fruiting to hydrogen fluoride fumigation. *J. Air Poll. Contr. Assoc.,* **22**, pp. 714–7.

Pack, M. R. & Sulzbach, C. W. (1976). Response of plant fruiting to hydrogen fluoride fumigation. *Atmosph. Environ.,* **10**, pp. 73–81.

Pelz, E. (1963). Untersuchungen über die Fruktifikation rauchgeschädigter Fichtenbestände. *Arch. Forstwes.,* **12**, pp. 1066–77.

Pelz, E. & Materna, J. (1964). Beiträge zum Problem der individuellen Rauchhärte von Fichte. *Arch. Forstwes.,* **13**, pp. 177–210.

Pelz, E., Temmlová, B. & Tesař, V. (1966). Die Nadelfarbe von Fichten als Hilfsmittel bei Rauchschadenuntersuchungen. *Biol. Zentralbl.,* **85**, pp. 325–44.

Pollanschütz, J. (1971). Die ertragskundlichen Messmethoden zur Erkennung und Beurteilung von forstlichen Rauchschäden. Methoden zur Erkennung und Beurteilung forstschädlicher Luftverunreinigungen, *Mitt. Forst. Bundesversuchsanstalt,* Wien, **92**, pp. 153–206.

Pospišil, J. & Richtar, V. (1970). Microsporogenesis in *Larix decidua* Mill. in stands affected by industrial air pollutants in the Ostrava region. *Lesnictvi,* **16**(XLIII), pp. 263–71.

Posthumus, A. C. (1980). *Elaboration of a Communitive Methodology for the Biological Surveillance of the Air Quality by Evaluation of the Effects on Plants.* Comm. of the Europ. Communities, EUR 6642 EN, 40 pp.

Prinz, B. & Brandt, C. J. (1980). *Study on the Impact of the Principal Atmospheric Pollutants on the Vegetation.* Comm. of the Europ. Communities, EUR 6644 EN, 110 pp.

Reckendorfer, P. (1952). Ein Beitrag zur Mikrochemie des Rauchschadens durch Fluor: Die Wanderung des Fluors im pflanzlichen Gewebe, I: Die unsichtbaren Schäden. *Pflanzenschutz-Berichte,* **9,** 3/4, pp. 33–55.

Rist, D. L. & Davis, D. D. (1979). The influence of exposure temperature and relative humidity on the response of Pinto Bean foliage to sulphur dioxide. *Phytopathology,* **69,** pp. 231–5.

Roberts, B. R. (1975). The influence of sulphur dioxide concentration on growth of potted White Birch and Pin Oak seedlings in the field. *J. Amer. Soc. Hort. Sci.,* **100,** pp. 640–2.

Rohmeder, E., Merz, W. & Schönborn, A. v. (1962). Züchtung von gegen Industrieabgase relative resistenten Fichten- und Kiefernsorten. *Forstw. Cbl.,* **81,** pp. 321–32.

Rohmeder, E. & Schönborn, A. v. (1968). Untersuchungen an phänotypisch relativ fluorresistenten Waldbäumen. *Forschungsberichte d. Deutsch. Forschungsgemeinschaft,* **14,** pp. 49–67.

Roques, A., Kerjean, M. & Auclair, D. (1980). Effets de la pollution atmospherique par le fluor et le dioxyde de soufre sur l'appareil reproducteure femelle de *Pinus sylvestris* en foret de roumare (Seine-Maritime, France). *Environ. Pollut. Ser. A.,* **21,** pp. 191–201.

Scholl, G. (1971). Ein biologisches Verfahren zur Bestimmung der Herkunft und Verbreitung von Fluorverbindungen in der Luft. *Landw. Forsch.,* **26,** pp. 29–35.

Scholl, G. (1976). Vorschläge für die Begrenzung der Aufnahmerate von Fluorid in standardisierter Graskultur zum Schutz von Pflanzen und Weidetieren. *Schriftenreihe Landesanst. Immissions-Bodennutzungsschutz d. Landes Nordrhein-Westfalen,* **37,** Girardet-Verlag, Essen, West Germany: pp. 129–32.

Schönbach, H., Dässler, H. G., Enderlein, H., Bellmann, E. & Kästner, W. (1964). Über den unterschiedlichen Einfluss von Schwefeldioxid auf die Nadeln verschiedener 2-jähriger Lärchenkreuzungen. *Der. Züchter,* **34,** pp. 312–6.

Shinn, J. H. (1979). Problems in the assessment of air-pollution effects on vegetation. Pp. 88–105 in *Advances in Environmental Science and Engineering,* Vol. 2 (ed. J. R. Pafflin & E. N. Ziegler). Gordon & Breach, New York–London–Paris.

Sidhu, S. S. (1979). Fluoride levels in air, vegetation and soil in the vicinity of a phosphorus plant. *J. Air Poll. Contr. Assoc.,* **29,** pp. 1069–1072.

Sidhu, S. S. (1980). Patterns of fluoride accumulation in boreal forest species under perennial exposure to emissions from a phosphorus plant. *Atmospheric Pollution 1980, Studies in Environmental Science,* **8,** pp. 425–432.

Smith, H. J. & Davis, D. D. (1978). Susceptibility of conifer cotyledons and primary needles to acute doses of SO_2. *Hort. Sci.,* **13**(6), pp. 703–4.

Sorauer, P. & Ramann, E. (1899). Sogenannte unsichtbare Rauchbeschädigung. *Bot. Cbl.,* **80,** pp. 50–6, 106–16, 156–68, 205–16, 251–62.

Steubing, L. (1978). Statement zu 'Schadstoff Sauerstoffhaltige Schwefelverbindungen'. Sachverständigenanhörung in Berlin. Pp. 35–6. in: *Medizinische, biologische und ökologische Grundlagen zur Bewertung schädlicher Luftverunreinigungen.* Umweltbundesamt, Berlin, West Germany: pp. 321.

Sulzbach, C. W. & Pack, M. R. (1972). Effects of fluoride on pollen germination, pollen-tube growth, and fruit development, in Tomato and Cucumber. *Phytopathology,* **62,** pp. 1247–53.

Szalonek, I. & Warteresiewicz, M. (1979). Yielding of cultivated plants in the vicinity of aluminium works. *Arch. Ochrony Środowiska,* **3–4,** pp. 39–62.

Temple, P. J. (1972). Dose-response of urban trees to sulphur dioxide. *J. Air Poll. Contr. Assoc.,* **22,** pp. 271–4.

Thomas, M. D. (1961). Auswirkungen der Luftverunreinigung auf Pflanzen. In *Die Verunreinigung der Luft,* WHO Monograph Series No. 46. Verlag Chemie, Weinheim–New York, NY, USA: 478 pp.

Thomas, M. D., Hendricks, R. H., Collier, T. R. & Hill, G. R. (1943). The utilization of sulphate and sulphur dioxide for the sulphur nutrition of Alfalfa. *Plant Physiol.,* **18,** pp. 345–71.

Tingey, D. T. & Reinert, R. A. (1975). The effect of ozone and sulphur dioxide singly and in combination on plant growth. *Environ. Pollut.,* **9,** pp. 117–25.

Tingey, D. T., Heck, W. W. & Reinert, R. A. (1971). Effect of low concentrations of ozone and sulphur dioxide on foliage, growth and yield of radish. *J. Amer. Soc. Hort. Sci.,* **96,** pp. 369–71.

Treshow, M. (1969). Symptomatology of fluoride injury on vegetation. Section VII, pp. 1–41 in *Handbook of Effects Assessment; Vegetation Damage* (Ed. N. Lacasse & W. Moroz). CAES, The Pennsylvania State University, USA.

Treshow, M., Anderson, F. K. & Harner, F. M. (1967). Responses of Douglas Fir to elevated atmospheric fluoride. *Forest Sci.,* **13,** pp. 114–20.

Treshow, M. & Harner, F. M. (1968). Growth responses of bean and alfalfa to sub-lethal fluoride concentrations. *Can. J. Botany,* **46,** pp. 1207–10.

Treshow, M. & Pack, M. R. (1970). Fluoride. Pp. D1–17 in *Recognition of Air pollution Injury to Vegetation: A Pictorial Atlas* (Ed. J. S. Jacobson & A. C. Hill). Air Pollution Control Association, Pittsburgh, Pennsylvania, USA: iii + 102 pp.

Ulrich, B., Mayer, R. & Khanna, P. K. (1979). Deposition von Luftverunreinigungen und ihre Auswirkungen in Waldökosystemen im Solling. *Schriften aus der Forstlichen Fakultät der Universität Göttingen und der Niedersächsischen Forstlichen Versuchsanstalt,* **58,** 291 pp.

Vinš, B. (1965). A method of smoke injury evaluation—determination of increment decrease. *Comm. Inst. For. Čechosloveniae,* pp. 235–45.

Vinš, B., & Ludera, J. (1967). Anwendung der Jahresringanalysen zum Nachweis der Rauchschäden. *Lesnický Časopis,* **13,** pp. 409–44.

Vogl, M. & Börtitz, S. (1965). Physiologische und biochemische Beiträge zur Rauchschadenforschung, 4 Mitt.: Zur Frage der physiologischen und physikalisch bedingten SO_2-Resistenz von Koniferen. *Flora,* **155,** pp. 347–52.

Vogl, M., Börtitz, S. & Polster, H. (1965). Physiologische und biochemische Beiträge zur Rauchschadenforschung, 6. Mitt.: Definition von Schädigungsstufen und Resistenzformen gegenüber der Schadgaskomponente SO_2. *Biol. Zbl.,* **84,** pp. 763–77.

Weinstein, L. H. (1977). Fluoride and plant life. *J. of Occupat. Med.,* **19,** pp. 49–78.

Weinstein, L. H. (1979). The effects of airborne fluorides on agriculture and forestry. *Proc. of the Ninth Conference on Environmental Toxicology,* March 1979, Paper No. 23, pp. 252–82.

Weinstein, L. H. & McCune, D. C. (1970). Effects of fluorides on vegetation. Pp. 81–106 in *Impact of Air Pollution on Vegetation Conference* (Ed. S. N. Linzon). Air Pollution Control Association, Pittsburgh, USA: 122 pp.

Weinstein, L. H. & McCune, D. C. (1979). Air pollution stress. Pp. 327–42 in *Stress Physiology in Crop Plants* (Ed. H. Mussel, R. C. Staples). John Wiley & Sons, New York, NY, USA: xiii + 510 pp. illustr.

Wentzel, K. F. (1963). Kennzeichen akuter Rauchgas-Einwirkung an Nadeln und Blättern. *Allg. Forstzeitschrift, 7*, pp. 107–8.

Wentzel, K. F. (1965). Fluorhaltige Immissionen in der Umgebung von Ziegeleien. *Staub, 25*, pp. 121–5.

Wentzel, K. F. (1966). Empfindlichkeit und Resistenzunterschiede der pflanzen gegenüber Luftverunreinigung. *Forstarchiv, 39*, pp. 189–94.

Wentzel, K. F. (1971). Habitus-Änderung der Waldbäume durch Luftverunreinigung. *Forstarchiv, 41*, pp. 165–72.

Wentzel, K. F. (1980). Weistanne—immissionsempfindlichste einheimische Baumart. *Allg. Forstzeitschrift, 14*(2), pp. 373–4.

Wesely, M. L. & Hicks, B. B. (1977). Some factors that affect the deposition rate of sulphur dioxide and similar gases on vegetation. *J. Air Poll. Contr. Assoc., 27*, pp. 1110–6.

Wieler, A. (1903). Über unsichtbare Rauchschäden. *Z. Forst- und Jagdwesen, 35*, pp. 204–35.

Woltz, S. S. (1964). Distinctive effects of root- *versus* leaf-acquired fluorides. *Proc. Florida State Hort. Soc., 77*, pp. 516–7.

Zahn, R. (1963). Über den Einfluss verschiedener Umweltfaktoren auf die Pflanzenempfindlichkeit gegenüber Schwefeldioxid. *Z. Pflanzenkrankh. Pflanzenschutz, 70*, pp. 81–95.

Zahn, R. (1978). Begründung der MIK-Werte für SO_2 zum Schutze der Vegetation. *VDI-Berichte, 314*, pp. 275–9.

Zimmerman, P. W. & Hitchcock, A. E. (1956). Susceptibility of plants to hydrofluoric acid and sulphur dioxide gases. *Contrib. Boyce Thomp. Inst., 18*, pp. 263–79.

Air Pollution and Plant Life
Edited by M. Treshow
© 1984 John Wiley & Sons Ltd.

CHAPTER 10

Organismal Responses of Higher Plants to Atmospheric Pollutants: Photochemical and Other

O. Clifton Taylor

Department of Botany and Plant Sciences, University of California, Riverside, California 92502, USA; Acting Director, Statewide Air Pollution Research Center

Background

Photochemical processes induce the formation and decomposition of many gaseous components of the atmosphere. Under certain conditions, ozone and the peroxyacetyl nitrates accumulate in sufficient quantities to be detrimental to vegetation. Nitrogen dioxide (NO_2) is sometimes included as a phytotoxic photochemical oxidant. While the formation of NO_2 from nitric oxide (NO), emitted during high-temperature combustion, is enhanced by photochemical processes, it is usually not considered an end-product of the process. NO_2 is a key component and is consumed in the photochemical processes that produce ozone and peroxyacetyl nitrates. However, NO_2 may occur in sufficient concentrations in close proximity to certain types of industrial operations to be phytotoxic. There is also experimental evidence that NO_2 may react synergistically to enhance the phytotoxicity of other gaseous air pollutants. At present, the potential for synergistic responses from mixtures of NO_2 and sulphur dioxide are considered to be the most important way that NO_2 in the atmosphere reacts with vegetation.

The effects of sulphur dioxide and fluorides on vegetation were discussed in the preceding chapter. However, it should be recognized that industrial operations, metallurgy, and the massive combustion of fossil fuels, have resulted in a complex mixture of air pollutants in the atmosphere. Some components of this mixture are known to be toxic to plants, but the potential phytotoxicity of the bulk of these airborne compounds and intermediates has not been investigated. Additional gaseous air pollutants, although of minor importance nationally and internationally, will be discussed in this chapter: ethylene, chlorine, hydrogen chloride, and hydrogen sulphide. These pollutants may produce serious injury and damage to vegetation in localized regions, but the quantity emitted is insufficient after dilution and dispersion to produce such injury over large regions.

OZONE

Symptoms

In 1958, Richards *et al.* showed that ozone was an important phytotoxicant in the Los Angeles-type smog when they identified 'oxidant' injury to Grapevine (*Vitis vinifera*) foliage in southern California. The next year Heggestad & Middleton (1959) reported that ozone was the cause of injury to foliage of Tobacco (*Nicotiana tabacum*). This symptom had been previously referred to as 'weather fleck', as the causal factor had not been identified. Soon after the reports on injury to Grape leaves and Tobacco were published, many investigators reported ozone injury to foliage of ornamentals, crop plants, trees, and native vegetation. In addition to numerous articles which describe the injury symptoms on specific plants, colour photographs of representative ozone injury symptoms have been published in many articles in journals and chapters of books (A. C. Hill *et al.*, 1961, 1970; Japanese Society of Air Pollution, 1973; Hicks, 1976).

Visible injury from ozone (O_3) is normally confined almost exclusively to the green foliage of plants, but in some instances it is suspected of inducing rind discoloration and corkiness of the skin on some fruits. Excessive external cork formation on some varieties of Avocado (*Persea americana*) growing in the South Coast Air Basin, which encompasses Los Angeles and neighbouring cities sharing a coastal valley in Southern California, while the same varieties grown in less polluted regions have little or no corkiness, suggests that air pollutants may be a causal factor. Similarly, ozone air pollution has been suggested as a causal agent of some rind-blemishes on *Citrus* fruits; but the implication of ozone in directly injuring fruit has not been conclusively proved experimentally.

A wide variety of visible injury symptoms on plant foliage can be associated with exposure to ozone. Some of the factors that influence the injury-symptom syndrome are: type of vegetation involved; physical characteristics of the foliage; vigour and succulence of the plants; age of plant and of leaf-tissue attacked; duration of ozone exposure; maximum ozone concentrations during exposure; and environmental conditions during plant development and during the pollutant exposure period. A few of the injury symptoms are sufficiently distinctive to be used with considerable confidence for diagnostic purposes. However, many of the injury symptoms produced by ozone can be mimicked by other pathogenic and physiological conditions (*see* Chapter 6). Thus, careful diagnostic procedures should be followed when the assessment of ozone injury is attempted. Conversely, the same care must be exercised when diagnosing other pathogenic and physiological problems, because ozone air pollution may be involved.

Open stomata provide the principal route for penetration of gaseous O_3 into the leaf tissues. When once ozone has entered the substomatal cavity, it first encounters the cellulose cell walls of surrounding tissues. No evidence has been presented to indicate that cellulose is affected by this strong oxidizing compound. However, there is evidence that it does attack the plasmalemma lining the inner walls of the cells. There is also evidence that the permeability of the plasmalemma is disrupted, allowing leakage of cell contents into the intercellular spaces (Wedding & Erickson, 1955; Perchorowicz & Ting, 1974). When this occurs, the injured leaf may develop dark-green or water-soaked-appearing areas. These represent some of the earliest evidence of the development of O_3 injury. After several hours, the water-soaked-appearing areas usually develop chlorosis, or the tissue may collapse and become necrotic. Under some conditions the water-soaked-appearing areas may disappear without leaving evidence of permanent injury, so indicating the presence of a repair mechanism.

J. B. Mudd (pers. comm., 1981) postulated that the protein component of the plasmalemma may be attacked by ozone, and that this may provide a pathway of O_3 to enter and reach organelles of the cell. Earlier research workers had suggested that the lipid component of the membrane was probably the site of O_3 attack, but no conclusive evidence to support this suggestion has been presented. It is not known for certain how the O_3 molecule is able to penetrate the cell membrane, or what form this chemically reactive, strongly oxidizing molecule is in as it travels from the intercellular space to organelles in the cell matrix. However, one of the first components to show changes after the leaf is exposed to ozone is the chloroplast (Thomson *et al.*, 1966). Electron-micrographic examination of bean leaf tissue immediately after exposure to O_3 revealed a disruption of the chloroplast. Crystal formations appeared in the stroma, and other disruptions associated with stress conditions were observed. The granulation or crystallization and general disorganization of the plastid was similar to that observed when plants were under stress from various environmental factors such as desiccation, and suggested that permeability of the plastid membrane was disrupted, leading to plasmolysis. Other research workers have reported changes in mitochondria when tissues were exposed to phytotoxic levels of O_3 (Lee, 1968).

The disruption and destruction of chloroplasts by O_3 exposure leads to a reduction in photosynthetic potential and also results in a foliage injury symptomatology known as chlorosis. Characteristics of the chlorotic symptoms are influenced by type of plant, concentration of O_3, duration of exposure, and other environmental factors. Consequently the appearance and extent of chlorosis range widely. Under some conditions the intensity of green fades gradually and more or less uniformly, so that the effects of O_3 exposure are difficult to detect unless the foliage can be compared with that of plants that have grown in ozone-free atmospheres. In other instances the chlorosis is much more intensive and extensive. The symptom syndrome ranges from the indistinct, incipient fading of green colour to various patterns of mottle, fleck, streaks, and general yellowing of the entire leaf. The last symptom closely resembles normal senescence of the leaf and usually results in early leaf-drop.

Herbaceous plants: On broad-leafed woody species and some herbaceous plants with well-developed palisade parenchyma, the primary lesion of injury is often limited to a cluster of only a few palisade cells at any one spot. As the injured lesion develops, it becomes visible on the upper or adaxial surface and is referred to as fleck or as stipple-like lesions. The size of the lesions can range from one that involves only a few cells to one that may be up to two millimetres in diameter, and the colour may range from light-green to white or dark-brown if the cells have been killed. Normally the spongy parenchyma and epidermis are not injured unless the foliage is exposed to very high concentrations of O_3 and the adjacent chlorophyll-containing cells are killed.

In monocotyledonous species such as those of the grass family, where the mesophyll tissues are undifferentiated, the injury may appear as chlorotic streaks or stippling between the parallel veins, and the lesions are usually visible on both sides of the leaf. The injury is often most severe at the bend in the maize, allium, or other grass leaf, and may cause almost complete collapse of the interveinal tissue in this region.

A prominent type of injury in many kinds of herbaceous plants as well as woody species is a localized pigmentation of groups of cells to form sharply defined, stipple-like lesions (Ledbetter *et al.,* 1959). On Tobacco (*Nicotiana tabacum*) leaves, the injury may appear as small, light-grey to tan or dark-brown to black, flecks on the upper surface of the leaf (Fig. 1). It has been suggested that tannins may be deposited in the cells as they are killed by ozone, and that the variation in coloration is a result of variations in tannin concentration. Some plant species have the characteristic of developing red to purple coloration of the leaf and stem tissues when stressed by low temperature or physical injury. Koukol & Dugger (1967) suggested that anthocyanin may be one of the important ozone-induced pigments formed when Curled or Yellow Dock (*Rumex crispus*) was exposed for long periods to photochemical smog. This pigmentation frequently resulted in a general reddening of the leaf tissue; the injury was not confined to the small spots but was generally spread over a significant amount of the leaf area.

Woody plants: Disruption of cells and tissues by ozone is essentially the same in woody as in herbaceous plants, and the symptoms produced are

Fig. 1 O_3 fleck and bifacial injury on Bel w_3 tobacco (left) and on Bel c tobacco (right). (Photograph courtesy of Statewide Air Pollution Research Center, University of California at Riverside.)

similar. The symptoms characteristically consist of minute chlorotic flecks developing on the upper surface of sensitive leaves as the chloroplast of the palisade parenchyma become disrupted. Epidermal cells overlaying the injured area usually remain uninjured. At first isolated, discrete, and pale-green in colour, as more cells become affected the lesions increase in size and coalesce, while the colour becomes chlorotic, bleached, or brownish. Pigmentation of the lesions from anthocyanins, and the resulting dark-green, purplish or reddish flecks, are particularly common in affected leaves of woody plants. Unpigmented 'bleaching' of the upper leaf surface can also occur but, is less frequent than in herbaceous plants. When severe, lesions may extend through the leaf, so that symptoms appear on both surfaces.

Islets of cells between the smallest veins are the first to be affected, but the vein tissues may also be involved. The delimitation makes the lesions angular in shape. The larger veins and then adjacent tissues, however, tend to be most tolerant, and often remain green even though half or more of the leaf is affected. The gradual colouring of the flecks gives ozone-affected leaves of many woody species a characteristic bronzed appearance, particularly over the upper leaf surface (A. C. Hill *et al.*, 1970).

Grape stipple is among the better known of the ozone-induced diseases of woody plants, and was the first to be described (Fig. 2). Tannins may form upon death of the cells, causing darker flecks (Fig. 3). Leaf chlorosis, stipple,

Fig. 2   Ozonic injury on Carignamic grape.

Fig. 3   Ozonic injury on grape. Stipple-like lesions on upper leaf surface.

subsequent bronzing, premature senescence, and defoliation, all contribute to the decline and reduced yields (*see also* Chapter 15). Symptoms in such polluted areas as the Los Angeles Basin may appear by midsummer (Thompson *et al.*, 1969). Even before discrete lesions appear, the Grape vines have begun to show senescence, dropping leaves and fruit. Midsummer senescence and defoliation have long been recognized as a response to smog in the Los Angeles area. Sensitive street trees such as Sycamores (*Platanus occidentalis*) and Carob (*Ceratonia silique*) are especially hard-hit.

Pine (*Pinus* spp.) trees generally tend to be among the more sensitive to photochemical pollution, and some species are especially sensitive. Ponderosa Pine (*P. ponderosa*) provides the best known example in the West. Chlorotic decline of this species was described in the San Bernadino Mountains of California in the 1950s, and the cause was well documented by 1968 (Parmeter & Miller, 1968). A chlorotic flecking or mottle appeared on the older needles, followed by a tan necrosis extending back from the tip. Needles began to drop in their second or third year rather than the usual fourth. The reduced photosynthate caused a stress that led to terminal dieback.

The response of Eastern White Pine (*Pinus strobus*) is quite similar. The earliest expression of injury is the chlorotic silver, pinkish, or reddish, flecks appearing on needles from one to six weeks of age. This progresses to a tip-burn that is followed by defoliation that gives the tree a thinned appearance. Death characteristically ensues. This disease, earlier known as 'chlorotic dwarf', was described first in 1908 (Dana, 1908), but it was not until the 1960s that the cause was discovered (Berry & Ripperton, 1963; Dochinger *et al.*, 1965; Dochinger & Seliskar, 1970).

Other coniferous species also are affected, and the geographic area of impact increases each year. Injury symptoms have been described in the mountains east of the Central Valley of California in White Fir (*Abies concolor*) and Sugar Pine (*Pinus lambertiana*). Less definitive chlorotic mottle has been observed on Giant Sequoia (*Sequoiadendron gigantea*), Incense Cedar (*Calocedrus decurrens*), and Lodgepole Pine (*P. contorta*) (Williams *et al.*, 1977). Symptoms were most severe at elevations of 6,000–8,000 feet (1,800–2,400 m). Broad-leafed species affected included Black Oak (*Quercus kelloggii*), but PAN (peroxyacetyl nitrate) may also have been a factor.

Hourly ozone concentrations in the areas where trees showed symptoms were between 8 and 11 pphm (parts per hundred millions) during the late summer. The polluted air was transported from urban areas 30 to 40 miles (48 to 64 km) to the west. Transport in the eastern United States may be much farther. Skelly (1980) has recognized significant injury to plants in the Blue Ridge and Southern Appalachia Mountains where average hourly-assessed concentrations exceeded 8 to 10 pphm during the summer months. The problem in Europe appears to be similar—even in Norway and Sweden.

Sensitivity

Sensitivity of *Pinus strobus* (Eastern White Pine) has been studied extensively (Skelly, 1980). Of 315 trees of this graceful species surveyed in the Blue Ridge Mountains of Virginia, 17%, 80%, and 3%, were considered tolerant, intermediate, and sensitive, respectively, to ozone. Ten of the 315 trees tagged in 1977 were dead from oxidants by 1980, following the continued appearance of symptoms. Growth increments declined between 1955 and 1978 even on the tolerant trees.

Eastern White Pine is especially sensitive to ozone, some individuals being injured by as little as 5 pphm. However, other trees may grow normally at concentrations above 50 pphm, thus suggesting a tremendous genetic diversity (Dochinger *et al.*, 1965). Fumigation studies showed *Pinus nigra* (Austrian Pine), *P. banksiana* (Jack Pine), and *P. virginiana* (Jersey or Scrub Pine), to be among the other more sensitive eastern pines, with *P. ponderosa* (Ponderosa Pine) and *P. jeffreyii* (Jeffrey Pine) the most sensitive in the West. Miller (1973) found that Western White Pine (*P. monticola*) Jeffrey × Coulter Pine (*P. coulteri*) hybrids and Red Fir (*Abies magnifica*) were next in sensitivity to Ponderosa and Jeffrey Pines. Seedlings of these species showed symptoms following 22 weeks of 6 hours per day exposures to 10 pphm O_3. Seedling growth was significantly reduced although there was no consistent association between growth and foliar injury.

The more sensitive broad-leafed species are only slightly more tolerant than the conifers. Leaves of Green Ash (*Fraxinus pennsylvannica* var. *lanceolata*), White Ash (*F. velutina*), and Tulip poplar (*Liriodendron tulipifera*), were severely damaged following a single 8-hours' exposure to 25 pphm. Kress (1980) found that growth in height was reduced in seedlings of ash, Sweetgum (*Liquidambar styraciflua*) and Sycamore (*Platanus occidentalis*) exposed to 10 pphm ozone for 28 days. Loblolly Pine (*P. taeda*) was similarly affected at 5 pphm. Berry (1974) considered Virginia Pine, Short-leaf Pine (*P. echinata*) and Slash Pine (*P. caribaea*), to be as sensitive as the Loblolly, all being injured at 477.5 $\mu g/m^3$ (0.2 ppm) for 2 hours. Hibben (1969) injured Sugar Maple (*Acer saccharum*) with a 2-hours' exposure to 20–30 pphm ozone or intermittent doses down to 10 pphm for 2 hours daily for 14 days. He noted that there was great variation in sensitivity from tree to tree.

Davis & Wilhour (1979) summarized the relative sensitivity of a number of woody plants to photochemical pollution. The list is based on the appearance of visible, or clinical, symptoms. It does not necessarily reflect the sensitivity in terms of the more subtle growth or physiological responses alone.

One must always bear in mind that any sensitive categories are tenuous, owing to genetic variation among individuals in a population and to environmental influences. Also, any listing of species sensitivity is bound to be somewhat arbitrary, being subject to what the observer considers to be 'sensitive' and how broad this category should be. Regardless of such drawbacks, however, as well as of the influence of environmental factors, the accompanying list (Table I) of sensitivity provides a general guide.

One of the studies that demonstrated genetic variation (Karnosky, 1977) showed that, of 11 different clones of Quaking Aspen (*Populus tremuloides*) fumigated with ozone at 20 pphm for 3 hours, leaf injury ranged from 6.6% to 56.2% of the leaves injured. Five to 10 of the 10 plants in each group showed the characteristic symptoms of bifacial chlorosis, sometimes together with dark necrotic lesions appearing within 20 to 48 hours after fumigation.

Variation in tolerance also occurs with the geographic area from which populations of a species grow (Karnosky, 1979). Thus Green Ash (*Fraxinus pennsylvanica*) seedlings grown from seeds collected in the eastern United States were more tolerant of ozone than those collected from trees in Manitoba. This suggests a natural selection of tolerance in response to the higher ozone concentrations in the eastern United States.

Table I Sensitivity of woody plants to ozone (based on Treshow, 1970).

Visible symptoms appearing at concentrations below 10 pphm.

Chinese Lilac	(*Syringa chinensis*)
Gambel Oak	(*Quercus gambelii*)
Incense Cedar	(*Calocedrus decurrens*)
Lodgepole Pine	(*Pinus contorta*)
Western White Pine	(*P. monticola*)
Red Fir	(*Abies magnifica*)
Eastern White Pine	(*Pinus strobus*)
Ponderosa Pine	(*P. ponderosa*)
Jeffrey Pine	(*P. jeffreyi*)
Sugar Pine	(*P. lambertiana*)
White Fir	(*Abies concolor*)
Giant Sequoia	(*Sequoiadendron gigantea*)

Visible symptoms appearing at concentrations below 25 pphm.

Snowberry	(*Symphoricarpus alba*)
Sumac	(*Rhus canadensis*)
Quaking Aspen	(*Populus tremuloides*)
Grape	(*Vitis vinifera*)
Sweet Cherry	(*Prunus avium*)
Privet	(*Ligustrum vulgare*)
Green Ash	(*Fraxinus pennsylvanica* var. *lanceolata*)
Honey locust	(*Gleditsia triacanthos*)
English Walnut	(*Juglans regia*)
White Ash	(*Fraxinus velutina*)
Tulip-poplar	(*Liriodendron tulipifera*)

Production Losses

Classical ozone injury symptoms in the foliage of plants provide a useful indicator that there have been phytotoxic concentrations in an area, and even more serious crop loss may have occurred than is readily visible. When the market price of a crop is directly associated with aesthetics of the commodity, as is the case with leafy vegetables, tobacco, flowers, and ornamentals, its value may be destroyed or drastically reduced as a result of ozone injury. The aesthetic effect of chlorosis, bifacial necrosis, mottle, and premature defoliation on such crops, can result in serious losses particularly when the injury affects only a portion of the total crop-area. In addition to the aesthetic effects, loss of photosynthetically active tissue to the injury lesions can reduce the amount of photosynthate available for optimum growth and development (G. R. Hill & Thomas, 1933). However, some studies have indicated that plants may produce more leaf area than is needed for maximum yield of fruit and seed, and suggest that a significant amount of the leaf area may be lost without affecting yield (Katz *et al.*, 1939; Thomas, 1956; Davis, 1972).

With the exception of those crops which are grown and marketed for their foliage, the yield of fruit, seed, fibre, or forage, is of prime interest. Foliar injury may in some instances correlate with reduced biomass and yield, but for the most part its relationship to yield of fruits and seeds is not understood. Studies conducted during the past 20 years have focused on the long-term (full growing season) effects of several air pollutants on crop yield and biomass production. Several of these studies have revealed that production is significantly reduced by ozone, and that the effects are usually enhanced by the presence of sulphur dioxide. One of the earliest such studies showed that ambient ozone (oxidant) in the South Coast Air Basin in California resulted in 30% to 40% reduction in Lemon (*Citrus limon*) and Navel Orange (*C. sinensis* var. Washington Navel) production (Thompson, 1968; Thompson & Taylor, 1969). Yield of grapes (*Vitis vinifera* cv. Zinfandel) was reduced by 50% to 70% with little evidence of foliage injury (Thompson *et al.*, 1969).

Brewer & Ferry (1974) found that ambient ozone in the polluted air in the San Joaquin valley of California resulted in 20% to 30% reduction in lint and seed production by cotton (*Gossypium hirsutum*). The studies by Thompson & Brewer using citrus, grapes, and cotton were all conducted by enclosing one to several plants in fully enclosed field chambers that were equipped with blowers to circulate the ambient and filtered air through the chamber. The citrus study by Thompson *et al.* (1969) included treatments designed to measure the responses to hydrogen fluoride and peroxyacetyl nitrate (PAN). The grape and cotton studies only compared the effects of ambient and filtered air. No provisions were made to monitor continuously the ozone concentrations at the sites where these experiments were conducted.

It is well known that environmental conditions such as light intensity, relative humidity, and air temperature, will influence plant response to air pollutants. Consequently, the fully enclosed chambers used in the grape, cotton, and citrus, studies did not accurately simulate field conditions, and the responses measured may have been greater or less than those of plants growing in the open field. However, these experiments did demonstrate that long-term exposure to ambient concentrations or dosages of oxidant air pollutants resulted in a pronounced reduction of yield, even though the percentage reduction reported may not be entirely accurate.

Yield of Spinach (*Spinacea oleracea*), Winter Wheat (*Triticum vulgare*), Field Corn (*Zea mays*), and Soybean (*Glycine max*), was reduced by exposure to ozone concentrations between 10 and 15 pphm for 7 hours a day during the growing season (Heagle, 1972; Heagle *et al.*, 1974, 1979*a*, 1979*b*, 1979*c* and 1980). The studies with these crops were designed to measure yield by using open-top field-chambers. The chambers were cylinders (3 m in diameter by 2.5 m high) constructed with aluminium frames and covered with a clear plastic film. The lower, 4-foot (*ca.* 1.3 m) panel of the covers is double-walled, with holes in the inner wall to distribute air and pollutants evenly through the vegetation canopy. With this structure it is possible to expose plants to the desired concentration and/or dosage of air pollutant throughout the growing season under essentially natural growing conditions.

Cameron *et al.* (1970) reported that, in portions of the Los Angeles Basin (South Coast Air Basin), several commercial Sweet Corn (*Zea Mays*) hybrids were severely damaged by ambient oxidant air pollutants in 1969 and 1970, while others were virtually unaffected. High concentrations of ozone during periods of high temperature (above 32°C) were considered to be responsible for the extensive necrosis and chlorotic streaks observed on foliage of the susceptible hybrids.

Heagle *et al.* (1972) selected Golden Midget and White Midget to represent a susceptible and moderately tolerant variety of Sweet Corn, from the group which Cameron *et al.* (1970) had observed, to determine the long-term effects of O_3 on yield. They used 5 and 10 pphm of O_3 for 6 hours per day from the time the plants emerged until harvest, and exposed them in completely closed field-chambers. As a result of the 10 pphm treatment, fresh weight of ears, number of kernels, and dry-weight of kernels were significantly reduced.

Thompson *et al.* (1976) used a susceptible and a tolerant hybrid from the group which Cameron *et al.* (1970) had observed, to determine the effect of ambient oxidant (primarily ozone) on yield in the Riverside, California, area. Open-top chambers constructed from opaque sheets of rigid plastic were used in the experiment, and the Corn was grown in large pots. In these experiments, plant growth was suppressed and kernel yield was significantly reduced as a result of the exposure.

Root Interactions

Growth and yield are suppressed by chronic exposures to ozone, and at least in some plants the reduction of root growth is greater than that of top growth. The air pollutant inhibits the assimilation of carbon and may in some way inhibit translocation of the metabolite to root systems. Hanson & Stewart (1970) observed that translocation of starch out of the leaves in the dark was inhibited when the plants were exposed to ozone or Los Angeles-type smog. This phenomenon was suggested as a possible explanation of the reduced population of mycorrhiza on *Citrus* roots which received chronic exposures of ozone (McCool *et al.,* 1979). Mycorrhiza live in close symbiotic association with roots of higher plants, which supply carbohydrates and other organic materials to the mycorrhiza and receive water and mineral nutrients from the fungi.

Neely *et al.* (1977) reported that chronic exposure of Alfalfa (*Medicago sativa*) to ozone, sulphur dioxide, and an ozone–sulphur dioxide mixture, caused an approximately 40% to 60% decrease in total symbiotically fixed nitrogen in the forage, stubble, and roots, of Alfalfa at the last harvest. They also reported that, although the pollutant-stressed forage tended to have a higher protein content than non-stressed forage, the total protein produced by stressed plants showed a net loss because of reduced growth. As the bacterial genus *Rhizobium,* which is associated with leguminous plants and the fixation of nitrogen, depends upon carbohydrates and other organic nutrients from the plant roots for its energy source, it may be postulated that the synthesis and/or transport of carbon compounds is probably an important factor in the success of the symbiotic relationship.

Manning *et al.* (1971) reported that Pinto bean roots developed poorly on plants exposed to 10 to 15 pphm O_3, and a larger number of colonies of fungi were found on the roots and hypocotyls compared with those of plants growing in activated charcoal-filtered air. The increased fungal population may have been due to increased exudates from the roots, or to a general decrease in root vigour. The authors also mentioned that the control bean plants developed nodules, but those exposed to 10 to 15 pphm O_3 developed no nodules. Tingey & Blum (1973) found that three-weeks-old Soybean plants developed fewer nodules, less nodule weight per plant, and less leghemoglobin content, than others, when exposed for 1 hour to 75 pphm O_3. They also found that top fresh weight and root growth were consistently reduced by exposure to O_3.

Blum & Letchworth (1976) exposed White Clover (*Trifolium repens*) plants of different ages, for one or two 2-hour periods, to 30 and 60 pphm O_3. The effect of O_3 on growth, nitrogen fixation, percentage nitrogen, total nitrogen, and nodulation, varied with age of the plant and ozone concentration and number of exposures. The reduction in nodulation compared with comparable-aged control plants was greatest in younger plants, and the difference diminished after the plants were about 40 days old.

There is considerable experimental evidence to support the concept that root response is as great as, or perhaps in some instances greater than, the response of plant tops when the plant is exposed to O_3 (Bennett & Oshima, 1976; Oshima *et al.*, 1978). At this point it can only be hypothesized that a reduction in available carbohydrates or other energy-producing organic compounds in the roots inhibits the symbiotic relationship between microorganisms such as mycorrhiza and *Rhizobium,* and the host plants. Similarly, these conditions may encourage the development of pathogenic fungi and increase the population on root surfaces of O_3-injured plants.

PEROXYACETYL NITRATE (PAN)

Middleton *et al.* (1950*a*) described a syndrome of injury symptoms that were first observed on several species of herbaceous plants in 1944. These authors concluded that the injury was not induced by sulphur dioxide but that it was related to the 'unpleasant, murky atmosphere' that developed along with the rapid population growth and industrial expansion of the Los Angeles area during and following World War II. The injury occurred only during periods of 'aggravated' air pollution, and no specific pollutant was associated with it.

The initial symptom, which developed 24 to 72 hours after exposure to the gaseous air pollution, was a glazed appearance on the undersurface of affected leaves (Middleton *et al.*, 1950*a*, 1950*b*) (Fig. 4). On crops such as Spinach, garden Beets (*Beta vulgaruis*), Romaine Lettuce (*Lactuca sativa*) and Chard (*Beta chilensis*), the glazing was silvery in the manner of that produced by freeze injury. Endive (*Cichorium endivia*) and Turnips (*Brassica rapa*) initially developed a bleaching of the lower leaf surface rather than a glazing; this often developed into light-tan, necrotic areas. Microscopic examination of affected areas revealed that the protoplasts of the mesophyll tissue, especially in the region of stomata, collapsed and large air pockets took their place. These air-filled spaces were responsible for the glazed or bleached appearance of the leaves. The epidermal layer was not initially affected. Usually progressive dehydration of affected leaf-tissue developed scorched areas that extended through the leaf's entire thickness, leaving necrotic areas with glazed margins. This leaf-scorching developed across veins and was not limited in area by any anatomical situation in the leaf.

From the field surveys conducted by Middleton *et al.* (1950*a*, 1950*b*), they concluded that Romaine Lettuce, Endive, and Spinach, were extremely susceptible to the air pollutants, and that Beets, Celery (*Apium graveolens*), Oats (*Avena sativa*), Chard, and Alfalfa (*Medicago sativa*) were moderately susceptible. This listing corresponds relatively well with the sensitivity rankings determined by later field observations, and by fumigations with synthesized

Fig. 4   PAN injury on lettuce. Silvering and glaze in diffuse transverse bands.

PAN (Taylor & MacLean, 1970). Notable differences were that these later studies moved Oats and Tomato (*Lycopersicum esculentum*) to the extremely susceptible group and listed Spinach as moderately susceptible. However, the detailed symptom descriptions and the susceptibility listings given by Middleton *et al.* (1950*b*) so closely matched the injury reproduced when plants were exposed to synthesized PAN, that it seemed clear that both groups of workers were dealing with the same type of air pollutant.

Noble (1955) found the location of damage produced by photochemical 'smog' to be related to the progressive development of the leaf. On many plants, damage from a single exposure was frequently found on only three mature leaves of any one stalk—the immature and senescent leaves showed no injury. Of the three that were damaged, the youngest would be injured at the tip, the next-older would be injured about one-third of the way down from the tip, and the next would be injured only near its base (Fig. 5). These locations correspond to the known progression of regions of maturation of

Fig. 5   PAN injury on Petunia.

leaves (Avery, 1933). On most plants a diffuse transverse band of injury develops across the central portion of the intermediate-aged leaf and at the base of the oldest susceptible leaf (Taylor *et al.*, 1960; Darley *et al.*, 1963).

When plants are exposed on successive days to damaging levels of PAN or photochemical 'smog', a series of bands of injury separated by relatively healthy tissue may occur. The injured bands were first associated with 'smog' by Noble (1955). Later, symptoms were reproduced by fumigations using irradiated nitrogen dioxide and 1-hexene (Taylor *et al.*, 1960), and, still later, with synthesized PAN (Darley *et al.*, 1963). PAN injury symptoms, formerly referred to as oxidant or 'smog' injury, are clearly distinct from those produced by ozone, which appear as necrotic stipple or flecks on the upper leaf-surface, various patterns of chlorosis (Fig. 6), or a bifacial collapse of tissue extending through the thickness of the leaf.

Fig. 6   PAN and O₃ injury on *Poa annua*. Centre leaf, PAN injury;
right leaf, O₃ injury.

It should be pointed out that other pollutants and environmental factors can induce glazing, silvering, and bronzing, that may be difficult to distinguish from PAN injury. Frost or cold temperatures sometimes produce a type of glazing of the lower leaf-surface. Ozone has been reported to induce bronzing, and exposure to chlorine or hydrogen chloride may produce the silvering or glazing symptom. Symptoms of ozone and PAN injury may be especially similar (cf. Fig. 6). Under some conditions, lower-surface glaze and bronzing may result from feeding by mites and other insects. Care should be exercised when one is diagnosing the causal agent responsible for leaf-injury on the basis of visible symptoms alone.

CHLORINE AND HYDROGEN CHLORIDE

The effects of hydrogen chloride (HCl) gas on vegetation were first noted in Europe during the mid-nineteenth century (Haselhoff & Lindau, 1903). More recent investigations of HCl phytotoxicity emphasized the response to chronic and acute exposures (Shriner & Lacasse, 1969; Guderian, 1977). Some of the more recent work has been concerned with responses to acute exposures characteristic of mass releases, e.g. spills, equipment failure during transport, and exhaust from solid-fueled rockets (Endress *et al.*, 1979; Swiecki *et al.*, 1982).

Chlorine is released to the atmosphere less frequently than HCl, but accidental spills and leaks do occur occasionally around industrial operations or during transport. Such spills usually result in acute exposures in localized regions and induce the formation of necrotic lesions or severe bleaching and chlorosis. Acute exposures to chlorine will stimulate the rapid development of the abscission zone on many tree species and result in heavy to almost complete defoliation within a few hours following exposure. Field observations following a known spill of chlorine have revealed that Eucalyptus (*Eucalyptus globulus*), Elm (*Ulmus parvifolia*), and Orange (*Citrus sinensus*) trees, exposed during the early part of a day, were heavily defoliated by the evening of the same day. Most of the foliage dropped without the formation of necrotic or chlorotic injury symptoms. Following exposure to chlorine, leaf-drop has been reported on Buckwheat (*Fagopyrum sagittatum*) and Peach (*Prunus persica*) seedlings (Zimmerman, 1955), on Peach, Apricot (*Prunus armeniaca*), Apple (*Malus pumila*), and Quince (*Cydonia oblonga*) (Schmidt, 1951), and on one species of *Chenopodium*.

The injury-symptoms produced by chlorine and HCl are variable, ranging from marginal necrosis to bifacial intercostal necrotic lesions or necrotic flecks, stippling, or spots, and a variety of chlorosis symptoms. Stout (1932) reported that the lower concentrations of chlorine produced a superficial glossy-grey or bronze discoloration to the undersurface of Lettuce leaves. A similar type of glazing of the abaxial leaf surface was observed when bean plants were exposed to HCl gas (Endress *et al.,* 1979).

Disposal of industrial wastes often involves combustion of chloride-containing gases, liquids, and solids. Such practices release relatively low concentrations of HCl to the atmosphere for long periods, and subject plants to sublethal concentrations. Under such conditions foliage continues to absorb HCl and deposit chloride salts at the apex and margin of the leaves. Injury symptoms, characteristic of those associated with salinity in soils or irrigation water, gradually develop. Substantial marginal necrosis and/or chlorosis may develop before the leaf is dropped. Leaf analyses for total chloride ion content is an effective diagnostic tool for chronic exposures to both HCl and chlorine, but it is not so effective for identifying the causal factor when an acute exposure produces injury. Exposure of susceptible plants to high concentrations of HCl (15 ppm or more), or of chlorine (7–10 ppm or more), for a few minutes may produce severe injury symptoms with only insignificant increase in chloride content of the leaves.

Injury symptoms from both HCl and chlorine may resemble those induced by sulphur dioxide, ozone, and other oxidants, hydrogen fluoride, drought, and salinity. For diagnostic purposes under field conditions, all of these factors must be carefully considered by an observer who is familiar with relative susceptibility of many plant species to oxidant, sulphur dioxide, and chloride air pollutants.

HYDROGEN SULPHIDE

Hydrogen sulphide (H_2S) is only slightly phytotoxic (Thornton & Setterstrom, 1940), and H_2S near oil-refineries, however obnoxious to our sense of smell, is probably harmless to plants. Thomas (1961), and Thompson and Kats (1978) continuously fumigated Alfalfa, Grapes, Lettuce, Sugar beet (*Beta vulgaris*), California Buckeye (*Aesculus californica*), Ponderosa Pine, and Douglas Fir (*Pseudotsuga menziesii*), with 30, 100, 300, and 3,000 ppb (parts per billion) H_2S. They found that above about 100 ppb, foliage on most species developed necrotic lesions or marginal leaf and needle tip burn. At these concentrations, growth was also suppressed. However, the 30 ppb and sometimes the 100 ppb exposures actually stimulated growth. It is unlikely that concentrations of H_2S that can be tolerated by Man will injure any significant amount of vegetation.

ETHYLENE

Ethylene is the only gaseous hydrocarbon of any significance as a phytotoxic air pollutant. It is a by-product of combustion processes and of several types of industrial operations. It is a contaminant in synthesized illuminating gas, and large amounts are produced by the internal combustion engine such as that used in the automobile. As so much is released by the highly developed society, ethylene is probably present in significant amounts in many areas of the United States.

Ethylene acts as a plant growth hormone, causes distortion of plant parts, decreases apical dominance, stimulates the growth of lateral buds and the development of abscission zones, and hastens the ageing process in plants. It is produced endogenously in sufficient quantity to control certain plant processes, and excessive amounts are released in response to stress and physical injury to plants' leaves. Ethylene is produced in significant quantity by ripening fruits. Abscission of leaves, buds, and other plant parts, and failure of flowers to open properly, has been reported to result (Abeles, 1973). Most of the descriptions of acute injury symptoms have been based on observations of injury produced during laboratory or greenhouse fumigations with ethylene (Crocker, 1948; Davidson, 1949; Heck & Pires, 1962). These experiments were conducted with concentrations of 10 pphm or more in fumigations lasting for only a few hours. In the field, such concentrations would be rare except in instances when accidental spills occur. Characteristically, plants are exposed to fractions of a part per billion (= thousand millions) (ppb) or to only a few ppb, but the exposure duration is for the full growing cycle of the

Fig. 7   Dry sepal on Orchid. Ethylene damage.

plant. Very little research has been done in the field or with experiments designed to simulate field conditions.

Chronic injury symptoms attributed to ethylene exposures include: various degrees of epinasty; excessive curvature and distortion of foliage; general chlorosis of leaf tissue associated with senescence; excessive drop of flower-buds, blossoms, fruits and leaves; and failure of blossoms to open properly, 'sleepiness' of Rosa and Carnation (*Dianthus caryophyllos*) blossoms. The sepals on Orchid (*Cattleya* sp.) blossoms are often injured by ethylene prior to opening, developing the necrotic lesions known as 'dry sepal' (Fig. 7), and making the blossoms unsaleable.

The chronic effects of ethylene in the field may be responsible for many of the symptoms of leaf senescence, chlorosis, defoliation, and premature drop of fruits and blossoms; but its involvement in such problems is difficult to assess.

SUMMARY

The principal phytotoxic components of photochemical air pollution are ozone and peroxyacetyl nitrate (PAN). As ozone first affects local areas of palisade leaf cells, the characteristic symptoms of injury consist of minute flecks of damaged cells on the upper leaf surface. This produces the classic light- to dark-brown 'stippling' symptom on sensitive broad-leafed plants. As damaged areas coalesce, and larger areas become involved, the leaf may appear chlorotic or bronzed. This is also characteristic of conifers, beginning at the needle-tip. Some of the more sensitive plant species include Alfalfa, Potato, Oats, Cotton, Soybean, Grape, Tobacco, and several species of *Pinus*. Citrus and many other crops also may suffer serious production losses. It is important to note that there may be considerable variation in sensitivity among individuals in a population.

PAN causes more of a glazing symptom, most prominently on the lower leaf surface. Leafy crops such as Lettuce and Chard are among the most sensitive of species. Tomatoes, Spinach, Beets, Oats, and Petunias are also highly sensitive.

Other pollutants are occasionally harmful, especially in local situations. These include hydrogen sulphide, chlorine, hydrogen chloride, and ethylene. Leaf chlorosis is the most general symptom of toxicity, but ethylene is notably important because it stimulates senescence and premature abscission of the buds and flowers

REFERENCES

Abeles, F. B. (1973). *Ethylene in Plant Biology*. Academic Press, New York, NY, USA: xii + 302 pp., illustr.

Avery, G. S., jr. (1933). Structure and development of the Tobacco leaf. *Amer. Jour. Bot.,* **20**, pp. 565–92.

Bennett, J. P. & Oshima, R. J. (1976). Carrot response to ozone. *J. Amer. Soc. Hort. Sci.,* **101**(6), pp. 638–9.

Berry, C. R. (1974). Age of pine seedlings with primary needles affects sensitivity to ozone and sulfur dioxide. *Phytopath.,* **64**, pp. 207–9.

Berry, C. R. & Ripperton, L. A. (1963). Ozone, a possible cause of White Pine emergence tipburn. *Phytopath.,* **53**, pp. 552–7.

Berry, C. R. & Hepting, G. H. (1964). Injury to Eastern White Pine by unidentified atmospheric contaminants. *For. Sci.,* **10**, pp. 2–13.

Blum, U. & Letchworth, M. (1976). Effects of ozone on nitrogen fixation in Ladino Clover. *Ecol. Res. Series* EPA-600/3-76-031. Nat. Tech. Info. Service, Springfield, Virginia, USA: ii + 21 pp.

Brewer, R. F. & Ferry G. (1974). Effects of air pollution on Cotton in the San Joaquin valley. *Calif. Agric.,* **28**(6), pp. 6–7.

Cameron, J. W., Johnson, H., Jr., Taylor, O. C. & Otto, H. W. (1970). Differential susceptibility of Sweet Corn hybrids to field injury by air pollution. *Hort. Science,* **5**(4), pp. 217–9.

Crocker, W. (1948). Physiological effects of ethylene and other carbon containing gases. Pp. 139–71, Chapter 4, in *Growth of Plants.* Reinhold, New York, USA: iii + 459 pp., illustr.

Dana, S. T. (1908). Extent and importance of White Pine blight. *U.S. Forest Serv. Leaflet,* Washington, D.C.: 4 pp., illustr.

Darley, E. F., Dugger, W. M., Mudd, J. B., Ordin, L., Taylor, O. C. & Stephens, E. R. (1963). Plant damage by pollution derived from automobiles. *Arch. Environ. Health,* **6**, pp. 761–70.

Davidson, O. W. (1949). Effects of ethylene on Orchid flowers. *Proc. Amer. Soc. Hort. Sci.,* **53**, pp. 440–6.

Davis, C. R. (1972). Sulfur dioxide fumigation of Soybeans: Effect on yield. *J. Air Pollut. Cont. Assoc.,* **22**, pp. 964–6.

Davis, D. D. & Wilhour, R. G. (1979). Susceptibility of woody plants to sulfur dioxide and photochemical oxidants. *EPA 600/3-76-102.* Corvallis, Oregon. Nat. Tech. Info. Serv., Springfield, Virginia 22161, USA: iii + 71 pp.

Dochinger, L. S., Seliskar, C. E. & Bender, F. W. (1965). Etiology of chlorotic dwarf of Eastern White Pine. *Phytopath.,* **55**, p. 1055.

Dochinger, L. S. & Seliskar, C. E. (1970). Air pollution and the chlorotic dwarf disease of Eastern White Pine. *For. Sci.,* **16**, pp. 46–55.

Endress, A. G., Oshima, R. J. & Taylor, O. C. (1979). Age-dependent growth and injury responses of Pinto bean leaves to gaseous hydrogen chloride. *J. Environ. Qual.,* **8**(2), pp. 260–4.

Guderian, R. (1977). *Air Pollution Phytotoxicity of Acidic Gases and its Significance in Air Pollution Control* (Translated from the German by C. Jeffrey Brandt). Springer-Verlag, Berlin–Heidelberg–New York: viii + 127 pp., illustr.

Hanson, G. P. & Stewart, W. S. (1970). Photochemical oxidants: Effect on starch hydrolysis in leaves. *Science,* **168**, pp. 1223–4.

Haselhoff, E. & Lindau, G. (1903). Chlor und Salzsaure, pp. 230–56 in *Die Beschadigang der Vegetation durch Rauch. Handbuch Erkennung und Beurteilung von Rauschschaden.* Verlag von Gebruder Brontraeger, Leipzig, East Germany.

Heagle, A. S. (1972). Effect of ozone and sulfur dioxide on injury, growth, and yield of Soybean. *Phytopath.,* **62**, p. 763, Abst.

Heagle, A. S., Body, D. E. & Heck, W. W. (1972). Effect of ozone on yield of Sweet Corn. *Phytopath.,* **62**, pp. 683–7.

Heagle, A. S., Body, D. E. & Neely, G. E. (1974). Growth and yield response of Soybean to chronic doses of ozone and sulfur dioxide in the field. *Phytopath.*, **64**, pp. 1372–6.

Heagle, A. S., Riordan, A. J. & Heck, W. W. (1979*a*). Field methods to assess the impact of air pollutants on crop yields. *72nd Annual Meeting of Air Pollut. Cont. Assoc., Cincinnati, Ohio, Paper 79-46.6,* pp. 3–24.

Heagle, A. S., Philbeck, R. B. & Letchworth, M. B. (1979*b*). Injury and yield responses of Spinach cultivars to chronic doses of ozone in open-top field chambers. *J. Environ. Qual.,* **8**(3), pp. 368–73.

Heagle, A. S., Philbeck, R. B. & Knott, W. M. (1979*c*). Thresholds for injury, growth, and yield loss caused by ozone on Field Corn hybrids. *Phytopath.*, **69**, pp. 21–6.

Heagle, A. S., Spencer, S. & Letchworth, M. B. (1980). Yield response of Winter Wheat to chronic doses of ozone. *Can. J. Bot.,* **57**(19), pp. 1999–2005.

Heck, W. W. & Pires, E. G. (1962). Growth of plants fumigated with saturated and unsaturated hydrocarbon gases and their derivatives. *Texas Agric. Expt. Sta. Rept. MP-603.*

Heggestad, H. E. & Middleton, J. T. (1959). Ozone in high concentrations as a cause of Tobacco leaf injury. *Science,* **129**, pp. 208–210.

Hibben, C. R. (1969). Ozone toxicity to Sugar Maple. *Phytopath.*, **59**, 1423–8.

Hicks, D. R. (1976). Diagnosing vegetation injury caused by air pollution. (Air Pollution Training Inst. EPA Contract 68-02-1344). US Env. Pollution Admin. Office of Air and Waste Management, Research Triangle Park, North Carolina, USA: iii + 278 pp.

Hill, A. C., Treshow, M., Downs, R. J. & Transtrum, L. G. (1961). Plant injury induced by ozone. *Phytopath.*, **51**(6), pp. 356–63.

Hill, A. C., Heggestad, H. E. & Linzon, S. N. (1970). Ozone. 22 pp. in *Recognition of Air Pollution Injury to Vegetation: A Pictorial Atlas.* Air Poll. Contr. Assoc. and Nat. Air Poll. Cont. Admin. Inf. Rept. No. 1. iii + 102 pp.

Hill, G. R. & Thomas, M. D. (1933). Influence of leaf destruction by sulfur dioxide and by clipping on yield of Alfalfa. *Plant Physiol.,* **8**, 223– 45.

Japanese Society of Air Pollution (1973). Photographs of air pollution injury to plants. Assoc. of Public Health, Tokyo, Japan: 211 pp.

Karnosky, D. F. (1977). Evidence for genetic control of response to sulfur dioxide and ozone in *Populus tremuloides. Can. J. Forest Res.,* 7, pp. 437–40.

Karnosky, D. F. (1979). Consistency from year to year in the response of *Fraxinus pennsylvanica,* provences to ozone. *INFRO Air Poll. Mtg.* Zabre, Poland, Aug. 27–29, 1979.

Katz, M., Ledingham, G. A. & Harris, A. E. (1939). Carbon dioxide assimilation and respiration of Alfalfa under influence of sulfur dioxide, Effect of sulfur dioxide on vegetation, *Nat. Research Council, Canada, Pub. No. 815,* pp. 393–428.

Koukol, J. & Dugger, W. M., jr. (1967). Anthocyanin formation as a response to ozone and smog treatment in *Rumex crispus* L. *Plant Physiol.,* **42**(7), pp. 1023–4.

Kress, L. W. (1980). Effect of O_3 and O_3 + NO_2 on growth of tree seedlings. Pp. 239–40 in *Symp. on Effects of Air Pollutants on Mediterranean and Temperate Forest Ecosystems,* Riverside, California, 1980. Pacific Southwest Forest and Range Exp. Sta., Berkeley, California, USA: 256 pp.

Ledbetter, M. C., Zimmerman, P. W. & Hitchcock, A. E. (1959). The histological effects of ozone on plant foliage. *Contrib. Boyce Thompson Inst.,* **20**(4), pp. 275–82.

Lee, T. T. (1968). Effects of ozone on swelling of mitochondria. *Plant Physiol.,* **43**, pp. 133–9.

McCool, P. M., Menge, J. A. & Taylor, O. C. (1979). Effects of ozone and HCl gas on development of the mycorrhizal fungus *Glomus fasciculatus* and growth of 'Troyer' citrange. *J. Amer. Soc. Hort. Sci.,* **104**, pp. 151–4.

Manning, W. J., Feder, W. A., Papia, P. M. & Perkins, I. (1971). Influence of foliar ozone injury on root development and root surface fungi of Pinto bean. *Environ. Pollut.,* **1**, pp. 305–12.

Middleton, J. R., Kendrick, J. B., jr. & Schwalm, H. W. (1950*a*). Injury to herbaceous plants by smog or air pollution. *Plant Disease Rept.,* **34**(9), pp. 245–52.

Middleton, J. T., Kendrick, J. B., jr. & Schwalm, H. W. (1950*b*). Smog in the south coastal area: Injury to herbaceous plants in the affected area found to be result of air pollution by gases and aerosols. *Calif. Agric.,* **4**(11), pp. 7–10.

Miller, P. R. (1973). Oxidant-induced community change in a mixed conifer forest. Pp. 101–17 in *Air Pollution Damage to Vegetation* (Ed. J. A. Naegel). Adv. Chem. Ser. 122, Amer. Chem. Soc., Washington DC, USA: vii + 137 pp.

Neely, G. E., Tingey, D. T. & Wilhour, R. G. (1977). Effects of ozone and sulfur dioxide singly and in combination on yield, quality, and N-fixation of alfalfa. *Proceedings International Conference on Photochemical Oxidant Pollution and its Control.* Vol. II: Ecolog. Res. Ser. EPA-600/3-77-001b US Env. Prot. Agency, Office of Research & Develop. Research Triangle Park, North Carolina, USA, pp. 633–73.

Noble, W. M. (1955). Air pollution effects: Pattern of damage produced on vegetation by smog. *Agric. and Food Chem.,* **3**(4), pp. 330–2.

Oshima, R. J., Bennett, J. P. & Braegelmann, P. K. (1978). Effect of ozone on growth and assimilate partitioning in parsley. *J. Amer. Soc. Hort. Sci.,* **103**(3), pp. 348–50.

Parmeter, J. R. jr. & Miller, P. R. (1968). Studies relating to the cause of decline and death of Ponderosa Pine in Southern California. *Plant Dis. Reptr.,* **52**, pp. 707–11.

Perchorowicz, J. R. & Ting, I. P. (1974). Ozone effect on plant cell permeability. *Amer. J. Bot.,* **61**(7), pp. 787–93.

Richards, B. L., Middleton, J. T. & Hewitt, W. B. (1958). Air pollution with relation to agronomic crops: V. Oxidant stipple to grape. *Agron. J.,* **50**, pp. 559–61.

Schmidt, H. (1951). Beobochtung uber Gosschaden an Obstabaumen. *Deut. Baumsch.,* **3**, pp. 10–2.

Shriner, D. S. & Lacasse, N. L. (1969). Histological response of Tomato to hydrogen chloride gas. *Phytopathology,* **59**, p. 1050, Abst.

Skelly, J. M. (1980). Photochemical oxidant impact on Mediterranean and temperate forest ecosystems: Real and potential effects. Pp. 38–50 in *Symposium on Effects of Air Pollutants on Mediterranean and Temperate Forest Ecosystems.* June, 1980, Riverside, California. Pacific Southwest Forest and Range Exp. Sta., Berkeley, California, USA: 256 pp.

Stout, G. L. (1932). Chlorine injury to Lettuce and other vegetation. *Calif. Dept. Agr. Bull.,* **21**, pp. 340–4.

Swiecki, T. J., Endress, A. G. & Taylor, O. C. (1982). Histological effects of aqueous acids and gaseous hydrogen chloride on bean leaves. *Amer. J. Bot.,* **69**(1), pp. 141–9.

Taylor, O. C., Stephens, E. R., Darley, E. F. & Cardiff, E. A. (1960). Effect of air-borne oxidants on leaves of Pinto bean and *Petunia. Proc. Am. Soc. Hort. Sci.,* **75**, pp. 435–44.

Taylor, O. C. & MacLean, D. C. (1970). Nitrogen oxides and the peroxyacyl nitrates. 14 pp. in *Recognition of Air Pollution Injury to Vegetation: A Pictorial Atlas.* Air Pollution Cont. Assoc., and Nat. Air Pollut. Cont. Adm., Pittsburgh., Pennsylvania, USA: iii + 102 pp., illustr.

Thomas, M. D. (1956). Invisible injury theory of plant damage. *Jour. Air Pollut. Contr. Assoc.,* **5**, pp. 205–8.

Thomas, M. D. (1961). Effects of air pollution on plants. Pp. 233–78 in *Air Pollution. World Health Organization, Geneva, Switzerland, Monograph Ser.*, **46**, 478 pp.

Thompson, C. R. (1968). Effects of air pollutants on Lemons and Navel Oranges. *Calif. Agric.,* **22**(9), 2–3.

Thompson, C. R. & Taylor, O. C. (1969). Effects of air pollutants on growth, leaf drop, fruit drop and yield of citrus trees. *Environ. Sci. Technol.,* **3**, pp. 934–40.

Thompson, C. R., Hensel, E. & Kats, G. (1969). Effects of photochemical air pollutants on Zinfandel grapes. *Hort. Sci.,* **4**(3), pp. 222–4.

Thompson, C. R., Kats, G. & Cameron, J. W. (1976). Effects of ambient photochemical air pollutants on growth, yield and ear characters of two Sweet Corn hybrids. *J. Environ. Qual.,* **5**, pp. 410–2.

Thompson, C. R. & Kats, G. (1978). Effects of continuous H_2S fumigation on crop and forest plants. *Environ. Sci. Technol.,* **12**, pp. 550–3.

Thomson, W. W., Dugger, W. M. jr. & Palmer, R. L. (1966). Effects of ozone on the fine structure of the palisade parenchyma cells of bean leaves. *Can. J. Bot.,* **44**, pp. 1677–82.

Thornton, N. C. & Setterstrom, C. (1940). Toxicity of ammonia, chlorine, hydrogen cyanide, hydrogen sulfide, and sulfur dioxide gases: III, Green plants. *Contrib. Boyce Thompson Inst.,* **11**, pp. 343–56.

Tingey, D. R. & Blum, U. (1973). Effects of ozone on Soybean nodules. *J. Environ. Qual.,* **2**, pp. 341–3.

Treshow, M. (1970). Ozone damage to plants. *Environ. Pollut.,* **1**, pp. 155–61.

Wedding, R. T. & Erickson, L. C. (1955). Changes in the permeability of plant cells to $P^{32}O_4$ and water as a result of exposure to ozonated hexene (smog). *Amer. J. Bot.,* **42**, pp. 570, 575.

Williams, W. T., Brady, M., & Wilson, S. C. (1977). Air pollution damage to the forests of Sierra Nevada Mountains of California. *Jour. Air Pollut. Contr. Assoc.,* **27**, pp. 230–4.

Zimmerman, P. W. (1955). Chemicals involved in air pollution and their effects upon vegetation. *Prof. Papers Boyce Thompson Inst.,* **2**, pp. 124–45.

Air Pollution and Plant Life
Edited by M. Treshow
© 1984 John Wiley & Sons Ltd.

CHAPTER 11

Impact of Air Pollutant Combinations on Plants

VICTOR C. RUNECKLES

Department of Plant Science, University of British Columbia, 2074 Wesbrook Mall, Vancouver, BC, V6T 2A2, Canada

INTRODUCTION

Normal, clean air consists of complex mixtures of gases and suspended matter; polluted air likewise consists of mixtures, but of still greater complexity. A single pollutant gas, such as sulphur dioxide (SO_2), may be the predominant pollutant; but in almost all situations other pollutant gases will be present, because almost all occur naturally, though at very low concentrations (Valley, 1965). Frequently, greater-than-background concentrations of more than one pollutant are introduced into an air-mass as a result of atmospheric mixing, the emission of a pollutant into already polluted air, or the simultaneous emission of more than one pollutant. Such combinations may be simultaneous or sequential over time, and may result in types of impact on vegetation that differ appreciably from those of the single pollutant.

The last decade has witnessed a growing recognition of the importance of understanding the nature of these interactive effects between pollutants and plants. These data are necessary for the realistic assessment of impacts of air emissions or other pollutants on vegetation, and for the development of standards or objectives of air quality.

The subject of the effects of combinations of pollutants on plants has previously been reviewed by Reinert *et al.* (1975), and more recently by Ormrod (1982). It has been discussed more briefly in other works on the effects of air pollution in general on plants, such as those of Guderian (1977) and Ormrod (1978), and in reviews of the effects of specific pollutants (National Academy of Sciences, 1977a, 1977b). Recently, McCune (in press) has critically analysed the effects on plants of interactions of hydrogen fluoride (HF) with several other gaseous pollutants.

TERMINOLOGY

The various factors and plant variables that affect plant response to air pollutants are discussed in other chapters of this book. When pollutant combinations are considered, additional factors come into play, namely (1) the effects of a concentration of a gas in a combination in relation to the effects of that concentration when administered separately; (2) the ratio of gas concentrations; and (3) the nature of the combination. This last term refers not only to the chemical composition of the combination but also to whether the gases are combined into a mixture or impinge upon vegetation separately and sequentially. In either case, the effects of the individual gases may be merely additive, or they may be truly interactive. Tingey & Reinert (1975) have discussed the terminology used to describe the interactive effects of combinations: *synergism* or *potentiation* occurs when the effect of the combination is greater than the sum of the effects of the individual pollutants, and *antagonism* or *interference* when the effect of the combination is less than this sum. In the case of sequential exposures, one pollutant may *predispose* a plant to the impact of exposure to a subsequent pollutant. In addition, there should be added the terms *desensitize* or *harden*, when exposure to one pollutant reduces the impact of a subsequent pollutant. As is the case with single pollutants, exposures to mixtures or sequences may be *continuous* or *intermittent*.

Most investigations have been concerned with binary combinations, with relatively few studies reported of the effects of combinations of three or more pollutants. Regardless of the nature of the combination, however, studies have ranged from those concerned solely with foliar injury, to investigations of physiological responses and plant growth and yield, to biochemical effects and the mechanism of plant response, and to secondary ecological effects.

It will be immediately obvious that the choices with regard to pollutant combinations, concentrations, exposure régimes, species, and types of response, give rise to a near-infinite number of possibilities for experimentation. As a result, in spite of the considerable efforts expended particularly over the last decade, our knowledge of the complexities of the effects of pollutant combinations is still rudimentary, and our understanding of the mechanisms involved is only now beginning to emerge.

EARLY STUDIES OF POLLUTANT COMBINATIONS

Several reports from the 1950s drew attention to the potential for one gaseous pollutant to modify the plant responses to a second pollutant. For example, Thomas *et al.* (1952) observed that the removal of SO_2 from ambient greenhouse air by water-spray scrubbers appeared to increase the amount of injury caused by oxidants to bean leaves (*Phaseolus vulgaris* L. cultivar Pinto). Subsequent studies by Middleton *et al.* (1958) were inconclusive, and it was Menser & Heggestad (1966) who first clearly demonstrated an interaction between ozone (O_3) and SO_2 on a sensitive cultivar of Tobacco (*Nicotiana tabacum* L. cultivar Bel W_3). This observation, together with a growing awareness of the existence and nature of pollutant combinations in ambient air, led to further studies of the effects of such combinations.

In the early 1970s, several investigations were carried out on the effects of mixtures of SO_2 and O_3, SO_2 and nitrogen dioxide (NO_2), and SO_2 and hydrogen fluoride (HF), on a range of herbaceous (crop) and tree species. These studies have been summarized in tabular form in Reinert *et al.* (1975), and illustrate a bewildering array of additive, more than additive (synergistic), and less than additive (antagonistic), effects.

With regard to SO_2 + O_3 mixtures covering ranges of concentrations varying between 0.025 and 1.7 ppm SO_2 and 0.03 and 0.5 ppm O_3, simple additive effects predominated for most species, but several studies of *Nicotiana tabacum* L. (Tobacco) consistently revealed synergisms. Of note is the frequent dissimilarity between effects on foliar injury and on growth responses. For example, several studies summarized in Reinert *et al.* (1975) demonstrated synergistic effects in terms of injury to tobacco leaves, while effects on foliar and root growth were simply additive. In the case of Alfalfa (*Medicago sativa* L.), generally greater than additive foliar injury resulted from SO_2 + O_3 mixtures, whereas on growth the effects were less than additive (Tingey *et al.*, 1973; Tingey & Reinert, 1975).

In the case of SO_2 + NO_2 mixtures, early studies summarized in Reinert *et al.* (1975) showed a preponderance of synergistic effects on leaf injury to several species exposed to concentrations of the two pollutants in the ranges of those experienced in industrial and metropolitan areas. Studies on specific physiological and overall growth responses, however, have only been conducted more recently, and will be discussed in a later section of this chapter.

Up to 1975, few studies had been conducted on SO_2 + HF and other pollutant combinations, although Matsushima & Brewer (1972) had shown an additive effect on branch growth and leaf area of Sweet Orange (*Citrus sinensis* L.), and Mandl *et al.* (1975) had reported synergistic foliar injury to Maize (*Zea mays* L.) and Barley (*Hordeum vulgare* L.).

In their 1975 review, Reinert *et al.* stated: 'We now know that some plant species may respond differently to pollutant combinations than to single pollutants. However, our knowledge is fragmentary and many questions remain unanswered. . . Considering the knowledge base of 5–10 years ago, we are slowly moving toward expanding this base'. The balance of this chapter will attempt to demonstrate how successful this expansion has been to date.

PLANT RESPONSES TO POLLUTANT COMBINATIONS

Sulphur Dioxide and Ozone

Mixtures of SO_2 and O_3 (or ambient photochemical oxidants) were the first combinations of air pollutants to be studied with respect to their effects on plants, and have continued to be a focus for research. Following the demonstration by Menser & Heggestad (1966) that Tobacco leaves were injured more than additively by SO_2–O_3 mixtures (and that the injury threshold was reduced), Dochinger *et al.* (1970) showed a similar synergism on Eastern White Pine (*Pinus strobus* L.) to be the cause of the 'chlorotic dwarf' syndrome. Costonis (1973) subsequently showed that sequential exposures to the two gases one day prior to exposure to the mixture, predisposed plants to greater injury.

The first detailed study of this synergistic phenomenon was conducted on Tobacco by Macdowall & Cole (1971), who showed that it occurred below the threshold for SO_2 alone, but not below the O_3 threshold. Jacobson & Colavito (1976) confirmed this but found that an antagonist effect occurred in two bean cultivars. More than additive effects were reported for Quaking Aspen (*Populus tremuloides* Michx.), with injury occurring on sensitive clones at levels of either SO_2 or O_3 which were below the individual thresholds (Karnosky, 1976). Studies on Soybean (*Glycine max* (L.) Merr.) generally revealed antagonistic leaf injury responses (Hofstra & Ormrod, 1977; Beckerson & Hofstra, 1979*b*, 1980), but Heagle & Johnson (1979) showed that synergistic responses could occur where the concentration of SO_2 was at a low level which induced little injury by itself. In their experiments, this translated into SO_2 concentrations of less than 1.5 ppm for exposures lasting up to 3 hours—regardless of O_3 concentration, within the range 0.25 to 1.00 ppm O_3. At higher SO_2 concentrations, the responses were antagonistic—a trend that had already been observed in earlier studies (Tingey *et al.*, 1973).

While there are several reports that the symptoms observed in plants treated with the combination SO_2 and O_3 resemble those typical of O_3- rather than SO_2-injury, Tingey *et al.* (1973) found several exceptions. Hofstra & Ormrod (1977) reported that, in White Bean (*Phaseolus vulgaris* L. cultivar Sanilac), the mixture induced a chlorotic response which was totally unlike the necrotic lesions that are typical of the effect of O_3 or SO_2 alone.

In contrast to experiments with Soybean and White Bean, in which the dosages of SO_2 and O_3 used (Hofstra & Ormrod, 1977; Beckerson & Hofstra, 1979*b*, 1980) caused less than additive foliar injury, and a delay in its onset, comparable doses caused greater than additive injury to radish (*Raphanus sativus* L.) and Cucumber (*Cucumis sativus* L.). Synergism was also demonstrated in begonia (Gardner & Ormrod, 1976).

Recent studies have continued to demonstrate that effects of the combination of SO_2 and O_3 on foliar injury are not necessarily extended to effects of plant growth. Heagle *et al*. (1974) demonstrated trends of yield reduction and other growth characteristics of Soybean grown in $SO_2 + O_3$, without any apparent increase in foliar injury. But subsequently, Heagle & Johnson (1979) showed that, whereas dose combinations that resulted in greater-than-additive injury resulted in synergistic effects on plant fresh weight increase, combinations that led to antagonistic injury responses only induced antagonistic effects on weight-gain in one-half of the treatment combinations. Reinert & Weber (1980) have reported additive effects on Soybean growth.

In Begonia, on the other hand, dose-dependent synergistic or antagonistic effects on injury were usually mirrored in the effects on foliage weight (Gardner & Ormrod, 1976). Oshima (1978) employed a natural gradient of ozone doses to which were added 0.1 ppm SO_2 (6 hours per day) to determine the effects on yield of Kidney Beans (*Phaseolus vulgaris* L. cultivar California Small Red). He found a significant interaction that resulted in a 37% yield reduction when plants were exposed to a total O_3 dose of 51.4 ppm-hr over a 78-days' period. At a lower O_3 dose (28.2 ppm-hr), a 17% yield-reduction was not significant, while at higher O_3 doses, no interactions with SO_2 were observed.

With regard to physiological and biochemical changes that occur in plants subjected to combinations of SO_2 and O_3, several recent studies have been reported, each attempting to elucidate part of the mechanisms of plant response. Carlson (1979) showed that photosynthetic rates in two broad-leafed tree species were reduced synergistically. Studies of stomatal responses of Grape (Rosen *et al*., 1978) and Petunia (Elkiey & Ormrod, 1979*a*) revealed variable effects of combinations, or none. However, Beckerson & Hofstra (1979*a*) demonstrated a greater-than-additive effect on stomatal closure of the leaves of *Phaseolus vulgaris* L. cultivar Sanilac, which appeared to provide a partial explanation for the antagonistic effect of the combination on foliar injury. Soybean behaved somewhat similarly, but with Radish and Cucumber, both injury and leaf diffusive resistance were increased synergistically (Beckerson & Hofstra, 1979*b*).

In searching for other mechanistic explanations, Elkiey & Ormrod (1979*b*) were unable to demonstrate a clear relationship between injury and membrane permeability of Petunia leaf cells, but Beckerson & Hofstra (1980) subsequently showed that, in White Bean and Soybean (with antagonistic effects of SO_2 and O_3 on foliar injury), SO_2 reduced the increased membrane permeability caused by O_3 alone, while in Cucumber and Radish (with synergistic effects on injury), the combination caused a somewhat greater initial increase in permeability than did either pollutant alone.

At the biochemical level, Constantinidou & Kozlowski (1979), using high concentrations (2 ppm SO_2, 0.9 ppm O_3), found that the levels of non-structural carbohydrates, proteins, and lipids, in American Elm (*Ulmus americana* L.) seedlings showed a mixture of additive and less-than-additive responses. Beckerson & Hofstra (1979*c*) and Pratt (1980) have shown that the chlorophyll contents of leaves respectively of White Bean and Soybean are reduced synergistically. In addition, in Soybean, O_3 antagonizes S-accumulation.

With this fragmentary information, it is not yet possible to construct an overall model to account for all of the observed interactive effects of SO_2 and O_3, other than to implicate stomatal function and membrane permeability as probably being important components—since, as discussed in other chapters, they are implicated in the mechanisms of uptake and action of individual gaseous pollutants.

Sulphur Dioxide and Nitrogen Dioxide

The co-occurrence of SO_2 and NO_2 has long been associated with various industrial emissions and especially with those from coal-burning power-plants. As ambient concentrations of NO_2 rarely approach the injury threshold, concern over the presence of NO_2 in air stems largely from its potential interactions with other pollutants—particularly SO_2, as demonstrated by Tingey *et al.* (1971). They showed that SO_2 and NO_2 could act synergistically in inducing foliar injury in Soybean, Radish, Tobacco, various true beans, Oats (*Avena sativa* L.) and Tomato (*Lycopersicum esculentum* Mill.). The upper leaf surface injury usually resembled closely that caused by ozone.

Hill *et al.* (1974), however, found no evidence of other than additive effects on foliar injury to 87 species indigenous to the southwestern desert areas of the United States.

Several studies of plant growth under $SO_2 + NO_2$ stress have revealed interactions. Ashenden & Mansfield (1978), Ashenden (1979*a*), and Ashenden & Williams (1980), have reported that long-term exposures of the pasture grasses *Dactylis glomerata* L. (Orchard Grass or Cocksfoot), *Lolium multiflorum* Lam. (Italian Rye-grass), *Phleum pratense* L. (Timothy), and *Poa pratensis* L. (Common Meadow-grass or Kentucky Bluegrass), to weekly average concentrations of 0.068 ppm of SO_2 and/or NO_2, resulted in synergistic reductions in many growth parameters, e.g. total dry-weight, green leaf-weight, root weight, leaf area, number of tillers, and number of leaves. In a few cases, the reductions were additive or less than additive. Total growth of *D. glomerata* and *P. pratensis* was significantly reduced after 20-weeks' exposure to either SO_2 or NO_2 alone with synergistic and additive responses, respectively, resulting from the combination. NO_2 alone had no effect on *L. multiflorum* or *P. pratense*, and SO_2 alone had no effect of *L. multiflorum*. While these experiments provide no direct information about the thresholds of the growth responses, that of *L. multiflorum* at least was clearly lowered.

In contrast, Thompson *et al*. (1980) studied the effects of SO_2 + NO_2 combinations on ten species of plants native to the Mojave Desert and found considerable variability among the species, but no synergisms. Annuals were more severely affected than perennials, and tended to show antagonistic rather than additive effects. The study also revealed several subtle effects. For example, the combination stimulated lateral growth of *Encelia farinosa* Gray ex Torr. and *Erodium circutarium* (L) L'Her., caused increased the dry-weight in *Atriplex canescens* (Pursh) Nutt. and *Plantago insularis* Eastw., and resulted in increased flowering and survival of *Baileya pleniradiata* Harv. & Gray and *Phacelia crenulata* Torr. Species varied considerably in their sensitivities to the individual pollutants and combinations of them. Thus, *Chilopsis linearis* Cav. showed significant effects on growth only at relatively high levels of SO_2 (2.0 ppm), NO_2 (1.0 ppm), or both (2.0 ppm SO_2, 1.0 ppm NO_2), administered for 5 hours daily (25 h per week), while *Larrea divaricata* Cav. and *Ambrosia dumosa* (Gray) Payne were sensitive to mixtures as low as 0.22 ppm SO_2 + 0.11 ppm NO_2—but not to the individual gases at those concentrations.

Masaru *et al*. (1976) demonstrated a significant synergistic suppression of lily pollen tube growth by 0.24 ppm SO_2 and 0.12 ppm NO_2, whether administered for 30 or 60 minutes.

With regard to physiological effects, Ashenden (1979*b*) reported that, unexpectedly, the combination of 0.1 ppm each of SO_2 and NO_2 decreased the rate of transpiration of *Phaseolus vulgaris* L. leaves, although the individual gases caused stimulations. The rate of apparent photosynthesis of Alfalfa (*Medicago sativa* L. cultivar Ranger) was synergistically but reversibly reduced by combinations of SO_2 and NO_2 at concentrations of 0.25 ppm or less of each (White *et al*., 1974). As a consequence, the response thresholds for each gas were appreciably reduced when in combination. Bull & Mansfield (1974) reported an additive effect of the gases in reducing photosynthesis in seedling leaves of a Garden Pea (*Pisum sativum* L.).

However, it is at the biochemical level that advances of our understanding of the mechanisms behind these varied interactive responses to SO_2 and NO_2 are beginning to be made. SO_2 and NO_2 exercise synergistic effects on enzyme activities in leaves of a Garden Pea cultivar (*Pisum sativum* L. cultivar Feltham First) (Horsman & Wellburn, 1975). Ribulose-1, 5-diphosphate carboxylase activity was reduced and peroxidase activity increased, following exposures to mixtures of the gases within the ranges 0–2.0 ppm SO_2 and 0–0.1 ppm NO_2. Greater-than-additive increased activity of peroxidase also occurred following exposures to mixtures in the ranges of 0–0.2 ppm SO_2 and 0–1.0 ppm NO_2.

Recently, Wellburn *et al*. (1981) have focused attention on leaf enzymes concerned with nitrogen metabolism, and report that, while SO_2 alone had no effect on nitrite reductase (NiR) in the grasses *Dactylis glomerata* L. cultivar Aberystwyth 537, and the 'Helmshore' and 'Bell resistant' strains of *Lolium perenne* L. cultivar Aberystwyth 523, NO_2 significantly increased NiR activity after 9–13 days at 0.25 ppm. However, in combination, the presence of SO_2 completely prevented the NO_2-induction of NiR activity. This suppression of NiR induction also occurred in *Phleum pratense* L. cultivar Eskimo and *Poa pratensis* L. cultivar Monopoly. Thus, it appears that where additive effects of SO_2 and NO_2 occur, they do so because the normal induction of NiR by NO_2 is prevented by SO_2. As a consequence, the plants are unable to detoxify the NO_2 and are exposed to the harmful effects of both pollutants. However, this mechanism fails to explain greater-than-additive effects, and Wellburn *et al.* (1981) have hypothesized that these may be the result of both the failure to induce additional NiR activity and the combined effects of sulphite and nitrite in reducing the availability of reductant and consequently reducing the synthesis of adenosine triphosphate in leaf cells. It remains to be seen whether these mechanisms are also operating in the reported protection against SO_2 + NO_2-induced injury afforded by high CO_2 concentrations (Hou *et al.*, 1977).

Sulphur Dioxide and Hydrogen Fluoride

Combinations involving HF present an additional factor that is not found in other pollutant gas combinations, namely the fact that, when absorbed, the fluoride ion constitutes an accumulating, non-metabolizable or non-decomposable phytotoxicant. Both SO_2 and HF are emitted from a variety of industrial sources. In addition, emissions of HF are frequently accompanied by volatile silicofluorides; but no studies have been reported that were specifically concerned with the interactions of these other fluorine-containing compounds with SO_2 or other major gaseous pollutants.

Much of our knowledge of the effects of HF combinations comes from extensive studies conducted by various workers at the Boyce Thompson Institute for Plant Research, Ithaca, New York. Mandl *et al.* (1975) reported that a 7-days' exposure to a mixture of 0.15 ppm SO_2 and 0.60 ppb (parts per thousand million) HF resulted in similar amounts of foliar injury to Barley (*Hordeum vulgare* L. cultivar Dickenson) and Sweet Corn (*Zea mays* L. cultivar Marcross) to that produced by SO_2 alone. Kidney Bean (*Phaseolus vulgaris* L. cultivar Pinto) was not injured by any treatment. However, when the SO_2 concentration was reduced to 0.08 ppm and treatment extended to 27 days, the presence of 0.8 ppb HF resulted in greater-than-additive injury to Barley and Corn. The combination induced an unusual symptom on Corn, consisting of small elliptical bifacial lesions on the distal portion of older leaves. The more HF-tolerant Corn cultivar Surecross revealed reduced numbers of such lesions in the SO_2 + HF combination, and none at all when exposed to either pollutant alone.

McCune (in press) has summarized several other studies of SO_2 + HF combinations, but in no cases other than Corn and Barley were synergistic effects of injury observed. In *Gladiolus* cultivar White Friendship, HF prevented the appearance of SO_2-induced foliar lesions (Hitchcock *et al*., 1962). Susceptibility to SO_2-induced injury to Cotton (*Gossypium hirsutum* L.) was increased by pretreatment with HF, while decreased susceptibility occurred in Sunflower (*Helianthus annuus* L. cultivar Teddy Bear). Studies with White Pine (*Pinus strobus* L.) provided no evidence for non-additivity.

Effects on plant growth responses have proved to be as variable as those on foliar injury. McCune (in press) reports that concurrent exposures of Pinto Beans to 2.5 ppb HF and 0.4 ppm SO_2 (for 4 hours daily for 3 weeks) resulted in additive effects on pod weight, although HF alone caused a decrease while SO_2 alone caused an increase. Lower concentrations of longer duration had no effect on fresh- or dry-weights of Pinto Bean, Barley, or Sweet Corn.

Several studies have focused on the accumulation of fluoride by foliage, and McCune (in press) quotes the cases of the Alfalfa cultivars Iroquois and Saranac, in which foliar fluoride was significantly decreased by SO_2. There were smaller reductions of sulphur accumulation caused by HF. In a further cultivar, Mesa Sirsa, SO_2 again reduced foliar fluoride, but experiments with Perennial Rye grass (*Lolium perenne* L.), *Gladiolus* sp., and Sweet Corn, showed that the effect was dependent upon the concentrations of SO_2 and HF. Thus the effect only occurred at SO_2 concentrations of 0.5 ppm or greater in the presence of 12.5 ppb HF, but at 0.4 ppm SO_2 when HF was 5.5 ppb.

Elevated levels of atmospheric fluorides and SO_2 were shown to reduce cone development and increase seed abortion in Scots Pine (*Pinus sylvestris* L.) in the field, but the experimentation provided no evidence of interactive effects (Roques *et al*., 1980).

McCune (in press) presents limited evidence to support the view that varietal differences in the effects of SO_2 on F-accumulation in Alfalfa are explained by differences in SO_2-induced stomatal closure, but there appear to have been no investigations of SO_2 + HF combinations on other physiological processes, except that of Keller (1980). He found evidence for a synergistic reduction of CO_2 uptake of Norway Spruce (*Picea abies* (L.) Karst.) by 0.075 ppm SO_2 in the presence of increased tissue fluoride levels resulting from increased fluoride content of the rooting medium. SO_2 increased F-accumulation in roots and needles.

Thus our understanding of the interactive phenomena associated with SO_2 and HF combinations is extremely limited. The wide ranges of responses have yet to be fused into a comprehensive model, although McCune (in press) has made an initial attempt to describe the form that such a model should take. As it is applicable to all combinations, it will be discussed later in this chapter.

Ozone and Nitrogen Dioxide

Both ozone (O_3) and nitrogen dioxide (NO_2) are constituents of photochemical oxidant pollution, but few studies have been made of their combined effects on plants. An early report by Matsushima (1971), who found less-than-additive foliar injury in Tomato (*Lycopersicum esculentum* Mill) and Pepper (*Capsicum frutescens* L.), is of somewhat academic interest as the concentration of NO_2 employed was extremely high (15 ppm). Kress (1980) reported that the only significant interactions of O_3 and NO_2 on ten tree species occurred on *Liquidambar styraciflua* L. (Sweet Gum) and *Fraxinus americana* L. (White Ash). In both cases the effects were less-than-additive.

The course of photochemical oxidant generation frequently results in a build-up of NO_2 in ambient air (usually mid-morning) followed by an increase in O_3 (early and mid-afternoon). The effects of this sequence on the growth of Wheat (*Triticum aestivum* L. cultivar Sun), Bush Bean (*Phaseolus vulgaris* L. cultivar Pure Gold Wax), and Radish (*Raphanus sativus* L. cultivar Cherry Belle), have been studied in our laboratory (Runeckles *et al.*, 1978). Wheat and Radish were found to be somewhat similar in their gross responses, in that 0.1 ppm NO_2, administered daily from 0900 to 1200 h, was found to sensitize these species to the harmful effect of 0.1 ppm O_3 administered from 1200 to 1800 h. Although treatment with NO_2 alone was stimulatory, the combination resulted in less dry-matter accumulation in leaves and roots, than in O_3, with a particularly marked reduction in Radish hypocotyl growth. On the other hand, NO_2 pretreatment of Bush Beans led to a more complex response, as shown in Table I.

NO_2 alone resulted in a marked stimulation of top-growth, but O_3 alone caused significant inhibition. The sequence of NO_2 and then O_3 resulted in a simple additive response in terms of total top growth (i.e. the negative effects of O_3 were counteracted by the positive effects of NO_2), but this measure masked pronounced differences with respect to effects on individual organs. Thus pretreatment reduced the impact of O_3 on leaf growth and retention,

Table I   Effects of NO_2, O_3, and NO_2-O_3 sequential exposure, on the growth of *Phaseolus vulgaris* L. cultivar Pure Gold Wax. (Data are percentage changes from control after 40 days).

Treatment*	Total top dry weight	Green leaf dry weight	Senescence index†
NO_2	+27.8%	+ 3.6%	+15.6%
O_3	−69.1%	−87.5%	+75.6%
NO_2-O_3	−41.2%	−67.0%	+57.8%
Additive response	−41.3%	−83.9%	+91.2%

* NO_2: 0.1 ppm (0900 to 1200 h daily); O_3: 0.1 ppm (1200 to 1800 h daily); NO_2-O_3: sequential treatment involving both gases at the concentrations and times indicated. Controls in charcoal-filtered air.

† Ratio: senesced leaf weight/total leaf weight.

and resulted in less-than-additive effects on the weight of green leaf-tissue and the senescence index. These interactions undoubtedly reflect subtle effects of the individual gases and the sequence on plant development, the mechanisms for which have yet to be established. They also illustrate the complexities of interactive responses in which stimulations and inhibitions are acting concurrently or consecutively.

Ozone and Peroxyacetyl Nitrate

Peroxyacetyl nitrate (PAN), like O_3, is a product of photochemical reactions in the atmosphere. PAN frequently occurs simultaneously with O_3, although usually at only about one-tenth of the O_3 concentration (National Academy of Sciences, 1977a). In spite of the co-occurrence of these two air pollutants, few studies have been reported of their combined effects on vegetation. Kohut *et al.* (1976), and Kohut & Davis (1978), reported that O_3 and PAN behaved synergistically in the majority of experiments in which foliar injury of a hybrid Poplar (*Populus maximowiczii* × *trichocarpa*, clone 338) or of Pinto Bean (*Phaseolus vulgaris* L. cultivar Pinto III) was measured. However, in the case of the Pinto Bean, the synergistic response was only observed for the adaxial leaf surface. The pollutants acted antagonistically with regard to abaxial surface injury.

Similar studies by Kress (1972), involving sequential exposures of hybrid poplar, provided evidence of greater-than-additive foliar injury on 6-weeks-old cuttings, in early summer, but simple additive and less-than-additive responses were also observed at other times of the year, regardless of whether PAN preceded O_3 of *vice versa*. No evidence was provided to explain these variations in response.

Ozone and Hydrogen Fluoride

Mixtures of O_3 and HF have received some attention by workers at the Boyce Thompson Institute; their data have been summarized recently by McCune (in press). As in the case of SO_2–HF interactions, those of O_3 and HF are quite variable and illustrate species-differences and dose-dependence. O_3 caused decreased accumulation of F in Alfalfa (*Medicago sativa* L. cultivars Iroquois and Saranac) and Perennial Rye grass (*Lolium perenne* L.), but showed no effect on Tomato (*Lycopersicum esculentum* Mill. cultivar Bonny Best), Orchard Grass (*Dactylis glomerata* L.), Maize (*Zea mays* L. cultivars Marcross and Surecross), Timothy (*Phleum pratense* L.), Pinto Bean (*Phaseolus vulgaris* L. cultivar Pinto), mint (*Mentha* sp.), or *Coleus blumei* Benth., which had been exposed to concentrations of HF ranging from 0.9 to 19 ppb for as long as 30 days, and to concentration régimes of O_3

ranging from 0.2 ppm (8 hours daily for 12 days) to 0.05 ppm (4 hours daily for 9 days).

Leaf injury to the *Coleus* and mint showed a synergistic response, while in Tomato, the combination caused intercostal, bifacial necrosis—a symptomatology which was different from that caused by either pollutant alone. Few effects on growth were noted, although plant-mass of Tomato was reduced additively, and oxidant-induced depression of Alfalfa yield was reduced by concurrent exposure to HF, which by itself caused an increase in yield.

Again, the diversity of these findings prevents the advancement of a mechanistic explanation, although McCune (in press) suggests that they indicate in some cases, probable interactive effects on pollutant uptake, and in others, effects of one pollutant on susceptibility to the other.

Nitrogen Dioxide and Hydrogen Fluoride

The only studies of the effects of mixtures of NO_2 and HF which have been reported are those undertaken at the Boyce Thompson Institute, and summarized by McCune (in press), on *Phleum pratense* L., *Dactylis glomerata* L. and *Lolium perenne* L. The only response reported was uptake of fluoride, which was increased in *Phleum* and *Dactylis* after 8 days, when 2.5 ppb HF and 0.06 ppm NO_2 were administered for 4 hours per day. In the combination of 6.8 ppb HF and 0.6 ppm NO_2 (4 hours per day), accumulation of fluoride in *Lolium* was decreased after one day, but thereafter the effect disappeared. Thus, from these limited observations, it appears that NO_2 acts somewhat differently from SO_2, which was found either to decrease F-accumulation or to have no effect on it, although McCune (in press) suggests that these differences may merely reflect differences in the SO_2- or NO_2-dose.

Other Mixtures

There have been a few investigations of the effects of other air pollutant combinations on plants. Thus Capron & Mansfield (1976, 1977) studied the combination of nitric oxide (NO) and nitrogen dioxide (NO_2), as this combination frequently occurs in greenhouses in which CO_2-enrichment by combustion of propane or kerosene is practised. The effects of the two gases appeared to be independent and additive in causing reduced photosynthesis and growth of Tomato (*Lycopersicum esculentum* Mill. cultivar Moneymaker). The authors discuss the significance of the known oxidation of NO to NO_2 in air to their findings, and concede that the suggested additivitiy might be the result of complete conversion by the time the NO taken up had reached the reactive sites within the leaf.

The interaction of hydrogen sulphide (H_2S) and O_3 has been studied by

Coyne & Bingham (1978), who found in Kidney Beans (*Phaseolus vulgaris* L. cultivar GV50) that both diffusive conductance and apparent photosynthesis were reduced less than additively, i.e. the gases acted antagonistically.

Shinn *et al.* (1976) investigated the interactions of H_2S, carbon dioxide (CO_2), and methane (CH_4), on Lettuce (*Lactuca sativa* L. cultivar Climax) in connection with geothermal power development in southern California. Constant proportions of H_2S and CO_2 (1 : 15) were used in the gas mixture, and it was reported that H_2S concentrations of 0.55 and 1.1 ppm (in the presence of 8.25 and 16.5 ppm CO_2 above the atmospheric background concentration) stimulated photosynthesis, whereas 5.5 ppm H_2S (and a 75 ppm CO_2 excess) caused a reduction. Thus there is the possibility that CO_2 is countering the effect of low concentrations of H_2S, as suggested by Hou *et al.* (1977) to be the case with SO_2–NO_2 mixtures, albeit at considerably higher enrichments with CO_2. Recently Carlson (1982) has shown that 0.25 ppm SO_2 was more inhibitory to the growth of several C3 species at normal atmospheric CO_2 levels than at higher CO_2 levels, whereas elevated CO_2 (600 or 1,200 ppm) increased SO_2-induced inhibition of C4 species.

Ternary mixtures of air pollutants have received virtually no attention, which is perhaps not surprising in view of the diversity of responses which abound in the studies of binary mixtures. However, McCune (in press) reports that in Pinto Beans (*Phaseolus vulgaris* L. cultivar Pinto) subjected to SO_2 and HF combinations in open-top chambers into which intrusions of ambient oxidants could occur, the incidence of oxidant-induced injury was unaffected by HF—but was increased by SO_2, with the combination of HF and SO_2 apparently acting in an additive manner, to cause an intermediate response.

No discussion of the effects of interactions of air pollutants on plants can conclude without mention of the few studies that have been undertaken on the effects of combinations of gaseous and particulate pollutants. Other chapters are devoted to the effects of individual gaseous pollutants, and to the effects of heavy-metals deposited from the air. However, Krause and Kaiser (1977) found that, although 0.08 ppm SO_2 increased the foliar injury caused by CdO, PbO, CuO, and MnO, dusts to Lettuce (*Lactuca sativa* L.), Radish (*Raphanus sativus* L.), and Foxtail Millet (*Setaria italica* Beauv.), SO_2 treatment had no effect on the uptake and translocation of the metals, or on their inhibition of growth.

On the other hand, Czuba & Ormrod (1974), and Ormrod (1977), have shown that cadmium, zinc, and nickel (all supplied *via* the root system), may, at low concentrations, increase O_3-induced injury in several species, but that at high concentrations (of Cd), O_3-induced injury to Garden Pea (*Pisum sativum* L. cultivar Laxton's Progress) was reduced. Lamoureaux & Chaney (1978) observed different interactions between Cd and SO_2, in that at low concentrations of Cd (taken up by excised leaves), 2 ppm SO_2 (30 minutes' exposure) resulted in no change in net photosynthesis in Silver Maple (*Acer saccharinum* L.) from controls, whereas at high Cd concentrations, an 83% reduction occurred. However, these isolated observations permit no generalizations to be made.

EPILOGUE

The foregoing review illustrates clearly the currently fragmented information that is available about the effects of mixtures or sequences of air pollutants on plants. As pointed out above, the permutations and combinations of species, pollutants, dosages, and other variables, might be expected to produce such a situation of uncertainty. In order to avoid endless repetitions of the type of experimentation which has led to this situation that would undoubtedly add to our knowledge but probably do little to increase our understanding of such interactions clearly an attempt must be made to develop a *modus experientiae* aimed at the development of models that have general applicability to interactive effects.

One such attempt has recently been made by McCune (in press), who stresses the importance of viewing the subject in terms of 'response surfaces', over which less-than-additive, additive, or greater-than-additive, effects may occur—depending upon the particular pollutant concentrations and responses under study. Viewed in this way, the often apparently conflicting results obtained with different pollutant concentrations and different species, begin to fall into patterns. Such patterns may not be easy to discern—because of the simple experimental designs used in much of the experimentation, which involve too few levels of pollutant stress, and also because of the emphasis which apparently is all too frequently placed on the need to demonstrate greater-than-additive effects.

McCune's surfaces for SO_2 and HF interactions are based on the extensive studies conducted at the Boyce Thompson Institute, but are applicable to other interactions. Thus the response surface which depicts the effect of SO_2 dose on the apparent 'effectiveness' of HF on several species, suggests that synergistic effects will occur at low SO_2 doses, and antagonistic effects at high SO_2 doses. Evidence from studies of SO_2–O_3 interactions (Tingey *et al.*, 1973; Heagle & Johnson, 1979), of SO_2–NO_3 interactions (Tingey *et al.*, 1971; Ashenden & Mansfield, 1978; Ashenden, 1979*a*; Ashenden & Williams, 1980), and of O_3–heavy-metal interactions (Ormrod, 1977), suggest that these interactions fit similar response surfaces. Indeed, it may be that all such interactions are interrelated *via* the specifics of the chemistries of the different pollutants. Thus, following McCune's suggestion, the scheme depicted in Fig. 1 suggests that the major pollutants fall along a spectrum of transience–persistence. Ozone and HF fall at either end of such a spectrum, with NO_2 and SO_2 in between.

It should be noted that this spectrum also covers the range from area to

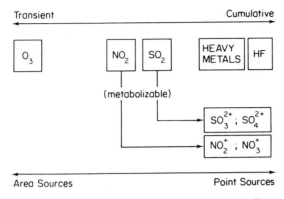

Fig. 1 Interrelationships between major air pollutants which may influence the effects of their combinations on plants (after McCune, in press).

point sources, which may have additional importance with regard to the type of concentration profiles of interacting pollutants that influence response.

Isolated observations, or those covering limited ranges of pollutant dose, will obviously be difficult to interpret in terms of surfaces—because of the lack of data available to define such surfaces adequately for individual species. Furthermore, it is to be expected that different species will reveal different response-surface characteristics—especially where major differences in anatomy, morphology, physiology, or biochemistry (e.g. C3 *versus* C4 plants) occur. But it is important to recognize the value of placing studies of the interactions of pollutants in such a perspective, in order to provide more structure to the investigations of the future—in the expectation that their outcomes will be relevant to the problems at hand, and not merely prove to be interesting academic exercises.

It should also be stressed that renewed attention needs to be focused upon experimentation that is designed to elucidate the mechanisms involved in pollutant interactions, and not simply to collect empirical data on effects. Understanding of the mechanisms involved is, in the final analysis, the cornerstone on which general-response models can be built. Without such understanding, extrapolation to other species and situations is at best hazardous, at worst indefensible.

Finally, it will have been noted that there are some major gaps in our present knowledge of pollutant interactions. For example, there appear to have been no definitive studies of pollutant interactions on cryptogamic species, although these are extremely important components of many ecosystems. Furthermore, ecosystem responses to combinations of pollutants, involving higher plants and their interactions with the animal kingdom, are virtually unknown. These deficiencies were noted by Reinert *et al.* (1975), particularly in relation to the need to develop acceptable standards or objectives of ambient air quality based upon the reality of pollutant combinations and an understanding of their effects. They also remarked: 'Considering the knowledge base of 5–10 years ago, we are slowly moving toward expanding this base'. Further progress has been made over the last 8 years, and it is heartening to see that not only is our base of knowledge continuing to expand, but a beginning has been made in integrating this knowledge into an understanding of effects of pollutant combinations.

SUMMARY

Different air pollutants frequently occur in mixtures or sequences as a result of which their effects on vegetation may differ from those of the individual pollutants. These interactive effects may be antagonistic, additive, or greater-than-additive (synergistic).

At the present time, there is no clear understanding of the likelihood that a particular type of interaction may occur when a given species is subjected to combinations of pollutants. Few generalizations are possible, because the evidence for specific types of interaction is frequently conflicting, even within the effects reported for a single species exposed to a given combination of pollutants. Nevertheless, there is sufficient evidence for the occurrence of synergistic effects of mixtures such as $SO_2 + NO_2$, $SO_2 + O_3$, and $SO_2 + HF$ to indicate that interactive effects may seriously affect the growth of many crop species. In such instances the effects are ultimately revealed in terms of reductions in yield, but numerous studies have shown that interactions may affect various individual physiological and biochemical processes and activities such as stomatal function, photosynthesis, gas exchange, water relations, and the activities of individual enzymes and metabolic pathways.

Much remains to be learned as to the scale and importance of such interactions in terms of effects on vegetation under ambient, field conditions, in order to improve the criteria upon which air quality standards are based.

REFERENCES

Ashenden, T. W. (1979a). The effects of long-term exposures to SO_2 and NO_2 pollution on the growth of *Dactylis glomerata* L. and *Poa pratensis* L. *Environ. Pollut.*, **18**, pp. 249–58.

Ashenden, T. W. (1979b), Effects of SO_2 and NO_2 pollution on transpiration in *Phaseolus vulgaris* L. *Environ. Pollut.*, **18**, pp. 45–50.

Ashenden, T. W. & Mansfield, T. A. (1978). Extreme pollution sensitivity of grasses when SO_2 and NO_2 are present in the atmosphere together. *Nature* (London), **273**, pp. 142–3.

Ashenden, T. W. & Williams, I. A. D. (1980). Growth reductions in *Lolium multiflorum* Lam. and *Phleum pratense* L. as a result of SO_2 and NO_2 pollution. *Environ. Pollut.* (Ser. A), **21**, pp. 131–9.

Beckerson, D. W. & Hofstra, G. (1979*a*) Stomatal responses of White Bean to O₃ and SO₂ singly or in combination. *Atmos. Environ.,* **13**, pp. 533–5.

Beckerson, D. W. & Hofstra, G. (1979*b*).Response of leaf diffusive resistance of Radish, Cucumber, and Soybean, to O₃ and SO₂ singly or in combination. *Atmos. Environ.,* **13**, pp. 1263–8.

Beckerson, D. W. & Hofstra, G. (1979*c*). Effects of sulphur dioxide and ozone singly or in combination on leaf chlorophyll, RNA, and protein, in White Bean. *Can. J. Bot.* **57**, pp. 1940–5.

Beckerson, D. W. & Hofstra, G. (1980). Effects of sulphur dioxide and ozone singly or in combination on membrane permeability. *Can. J. Bot.,* **58**, pp. 451–7.

Bull, J. N. & Mansfield, T. A. (1974). Photosynthesis in leaves exposed to SO₂ and NO₂. *Nature* (London), **250**, pp. 443–4.

Capron, T. M. & Mansfield, T. A. (1976). Inhibition of net photosynthesis in Tomato in air polluted with NO and NO₂. *J. Exp. Bot.,* **27**, pp. 1181–6.

Capron, T. M. & Mansfield, T. A. (1977). Inhibition of growth of Tomato by air polluted with nitrogen oxides. *J. Exp. Bot.,* **28**, pp. 112–6.

Carlson, R. W. (1979). Reduction in the photosynthetic rate of *Acer, Quercus,* and *Fraxinus,* species caused by sulphur dioxide and ozone. *Environ. Pollut.,* **18**, pp. 159–70.

Carlson, R. W. (1982). The influence of elevated atmospheric CO₂ and SO₂ on the growth and photosynthesis of early successional species. Pp. 489–91 in *Effects of Gaseous Air Pollution in Agriculture and Horticulutre* (Ed. M. H. Unsworth & D. P. Ormrod). (Proc. 32nd School in Agricultural Sciences, University of Nottingham School of Agriculture.) Butterworths, London, England, UK: xiv + 532 pp., illustr.

Constantinidou, H. A. & Kozlowski, T. T. (1979). Effects of sulphur dioxide and ozone on *Ulmus americana* seedlings, II: Carbohydrates, proteins, and lipids. *Can. J. Bot.,* **57**, pp. 176–84.

Costonis, A. C. (1973). Injury of Eastern White Pine by sulphur dioxide and ozone alone and in mixtures. *Eur. J. For. Path.,* **3**, pp. 50–5.

Coyne, P. I. & Bingham, G. E. (1978). Photosynthesis and stomatal light responses in Snap Beans exposed to hydrogen sulfide and ozone. *J. Air Pollut. Control Assn.,* **28**, pp. 1119–23.

Czuba, M. & Ormrod, D. P. (1974). Effects of cadmium and zinc on ozone-induced phytotoxicity in Cress and Lettuce. *Can. J. Bot.,* **52**, pp. 645–8.

Dochinger, L. S., Bender, F. W., Fox, F. O. & Heck, W. W. (1970). Chlorotic dwarf of Eastern White Pine caused by ozone and sulphur dioxide interaction. *Nature* (London), **225**, p. 476.

Elkiey, T. & Ormrod, D. P. (1979*a*). Leaf diffusion resistance responses of three Petunia cultivars to ozone and/or sulfur dioxide. *J. Air Pollut. Control Assn,* **29**, pp. 622–5.

Elkiey, T. & Ormrod, D. P. (1979*b*). Ozone and/or sulphur dioxide effects on tissue permeability of Petunia leaves. *Atmos. Environ.,* **13**, pp. 1165–8.

Gardner, J. O. & Ormrod, D. P. (1976). Response of the Rieger begonia to ozone and sulphur dioxide. *Sci. Hortic.,* **5**, pp. 171–81.

Guderian, R. (1977). *Air Pollution.* Springer-Verlag, Berlin–Heidelberg–New York: vi + 127 pp., illustr.

Heagle, A. S. & Johnson, J. W. (1979). Variable responses of Soybeans to mixtures of ozone and sulfur dioxide. *J. Air Pollut. Control Assn,* **29**, pp. 729–32.

Heagle, A. S., Body, B. E. & Nealy, G. E. (1974). Injury and yield responses of Soybean to chronic doses of ozone and sulfur dioxide in the field. *Phytopathology,* **64**, pp. 132–6.

Hill, A. C., Hill, S., Lamb, C. & Barrett, T. W. (1974). Sensitivity of native desert vegetation to SO_2 and to SO_2 and NO_2 combined. *J. Air Pollut. Control Assn*, **24**, pp. 153–7.

Hitchcock, A. E., Zimmerman, P. W. & Coe, R. R. (1962). Results of ten years' work (1951–1960) on the effects of fluorides on Gladiolus. *Contrib. Boyce Thompson Inst. Plant Res.*, **21**, pp. 303–44.

Hofstra, G. & Ormrod, D. P. (1977). Ozone and sulfur dioxide interaction in White Bean and Soybean. *Can. J. Plant Sci.*, **57**, pp. 1193–8.

Horsman, D. C. & Wellburn, A. R. (1975). Synergistic effect of SO_2 and NO_2 polluted air upon enzyme activity in pea seedlings. *Environ. Pollut.*, **8**, pp. 123–33, fig.

Hou, L-Y., Hill, A. C. & Soleimani, A. (1977). Influence of CO_2 on the effects of SO_2 and NO_2 on Alfalfa. *Environ. Pollut.*, **12**, pp. 7–16.

Jacobson, J. S. & Colavito, L. J. (1976). The combined effect of sulfur dioxide and ozone on bean and tobacco plants. *Environ. Exper. Bot.*, **16**, pp. 277–85.

Karnosky, D. F. (1976). Threshold levels for foliar injury to *Populus tremuloides* by sulfur dioxide and ozone. *Can J. For. Res.*, **6**, pp. 166–9.

Keller, T. (1980). The simultaneous effect of soil-borne NaF and air pollutant SO_2 on CO_2-uptake and pollutant accumulation. *Oecologia*, **44**, pp. 283–5.

Kohut, R. J. & Davis, D. D. (1978). Response of Pinto Bean to simultaneous exposure to ozone and peroxyacetylnitrate. *Phytopathology*, **68**, pp. 567–9.

Kohut, R. J., Davis, D. D. & Merill, W. (1976). Response of hybrid poplar to simultaneous exposure to ozone and PAN. *Plant Dis. Rep.*, **60**, pp. 777–80.

Krause, G. H. M. & Kaiser, H. (1977). Plant response to heavy metals and sulphur dioxide. *Environ. Pollut.*, **12**, pp. 63–71.

Kress, L. W. (1972). *Response of Hybrid Poplar to Sequential Exposures of Ozone and PAN.* M.S. Thesis, Pennsylvania State University, University Park, Pennsylvania, USA; iv + 39 pp. (mimeogr.).

Kress, L. W. (1980). Effect of O_3 and O_3 + NO_2 on growth of tree seedlings. P.239 in *Proceedings of Symposium of Effects of Air Pollutants on Mediterranean and Temperate Forest Ecosystems.* Rept PSW-43, Pacific Southwest Forest and Range Experiment Station, Berkeley, California, USA: iii + 256 pp., illustr.

Lamoureaux, R. J. & Chaney, W. R. (1978). Photosynthesis and transpiration of excised Silver Maple leaves exposed to cadmium and sulphur dioxide. *Environ. Pollut.*, **17**, pp. 259–68.

McCune, D. C. (in press). Terrestrial vegetation air pollutant interactions: Gaseous pollutants—Hydrogen fluoride and sulphur dioxide. In *Air Pollutants and Their Effects on the Terrestrial Ecosystems* (Ed. S. V. Krupa & A. H. Legge). Proc. Internat. Conference on Current Status and Future Needs of Research on Effects and Technology, Banff, Alberta.) John Wiley & Sons, Chichester, England, UK.

Macdowall, F. D. H. & Cole, A. F. W. (1971). Threshold and synergistic damage to Tobacco by ozone and sulfur dioxide. *Atmos. Environ.*, **5**, pp. 553–9.

Mandl, R. H., Weinstein, L. H. & Keveny, M. (1975). Effects of hydrogen fluoride and sulphur dioxide alone and in combination on several species of plants. *Environ. Pollut.*, **9**, pp. 133–43.

Masaru, N., Syozo, F. & Saubro, K. (1976). Effects of exposure to various injurious gases on germination of Lily pollen. *Environ. Pollut.*, **11**, pp. 181–7.

Matsushima, J. (1971). On the composite harm to plants of sulphurous acid gas and oxidant. *Sangyo Kogai*, **7**, pp. 218–24.

Matsushima, J. & Brewer, R. F. (1972). Influence of sulfur dioxide and hydrogen fluoride as a mix or reciprocal exposure on *Citrus* growth and development. *J. Air Pollut. Control Assn*, **22**, pp. 710–3.

Menser, H. A. & Heggestad, H. E. (1966). Ozone and sulfur dioxide synergism: injury to Tobacco plants. *Science*, **153**, pp. 424–5.

Middleton, J. T., Darley, E. F. & Brewer, R. F. (1958). Damage to vegetation from polluted atmospheres. *J. Air Pollut. Control Assn*, **8**, pp. 9–15.

National Academy of Sciences (1977*a*). *Ozone and Other Photochemical Oxidants.* (Committee on Medical and Biologic Effects of Environmental Pollutants.) National Academy of Sciences, Washington, DC, USA: vii + 719 pp., illustr.

National Academy of Sciences (1977*b*). *Nitrogen Oxides.* (Committee on Medical and Biologic Effects of Environmental Pollutants.) National Academy of Sciences, Washington, DC, USA: vii + 333 pp., illustr.

Ormrod, D. P. (1977). Cadmium and nickel effects on growth and ozone sensitivity of Pea. *Water, Air, Soil Pollut.*, **8**, pp. 263–70.

Ormrod, D. P. (1978). *Pollution in Horticulture.* Elsevier, Amsterdam–Oxford–New York: xi + 260 pp., illustr.

Ormrod, D. P. (1982). Air pollutant interactions in mixtures. Pp. 207–31 in *Effects of Gaseous Air Pollution in Agriculture and Horticulture* (Ed. M. H. Unsworth & D. P. Ormrod). (Proc. 32nd School in Agricultural Sciences, University of Nottingham School of Agriculture.) Butterworths, London, England, UK: xiv + 532 pp., illustr.

Oshima, R. J. (1978). *The Impact of Sulfur Dioxide on Vegetation: a Sulfur Diox-ide–Ozone Response Model.* (Final Rept to Calif. Air Resources Board. Agreement No. A6-162-30.) Statewide Air Pollution Research Center, University of California, Riverside, California, USA: 91 pp.

Pratt, G. C. (1980). *Interactive Effects of Ozone and Sulfur Dioxide on Soybeans.* M.S. thesis, University of Minnesota, St Paul, Minnesota, USA: iii + 91 pp., illustr. (mimeogr.).

Reinert, R. A. & Weber, D. E. (1980). Ozone and sulphur dioxide-induced changes in Soybean growth. *Phytopathology*, **70**, pp. 914–6.

Reinert, R. A., Heagle, A. S. & Heck, W. W. (1975). Plant responses to pollutant combinations. Pp. 159–77 in *Responses of Plants to Air Pollution* (Ed. J. B. Mudd & T. T. Kozlowski). Academic Press, New York–San Francisco–London: xii + 383 pp., illustr.

Roques, A., Kerjean, M. & Auclair, D. (1980). Effets de la pollution atmosphérique par le fluor et le dioxyde de soufre sur l'appareil reproducteur femelle de *Pinus sylvestris* en Forêt de Roumare (Seine-Maritime, France). *Environ. Pollut.* (Ser. A), **21**, pp. 191–201.

Rosen, P. M. Musselman, R. C. & Kender, W. J. (1978). Relationship of stomatal resistance to sulfur dioxide and ozone injury in grapevines. *Sci. Hortic.*, **8**, pp. 137–42.

Runeckles, V. C., Palmer, K. & Giles, K. (1978). Effects of sequential exposures to NO_2 and O_3 on plants. *3rd Int. Congr. Plant Pathology, Munich, Abstracts*, p. 343.

Shinn, J. H., Clegg, B. R., Stuart, M. L. & Thompson, S. E. (1976). Exposure of field-grown Lettuce to geothermal air pollution: Photosynthetic and stomatal responses. *J. Environ. Sci. Health*, **A111**, pp. 603–12.

Thomas, M. D., Henricks, R. H. & Hill, G. R. (1952). Effect of air pollution on plants. Pp. 41–65, in *Air Pollution* (Ed. L. C. McCabe), McGraw-Hill, New York–St. Louis–San Francisco–Dusseldorf–London–Mexico–Panama–Sydney–Toronto: xv + 422 pp., illustr.

Thompson, C. R., Kats, G. & Lennox, R. W. (1980). Effects of SO_2 and/or NO_2 on native plants of the Mojave Desert and Eastern Mojave–Colorado Desert. *J. Air Pollut. Control Assn*, **30**, pp. 1304–9.

Tingey, D. T. & Reinert, R. A. (1975). The effect of ozone and sulphur dioxide singly and in combination on plant growth. *Environ. Pollut.,* **9**, pp. 117–25.

Tingey, D. T., Reinert, R. A., Dunning, J. A. & Heck, W. W. (1971). Vegetation injury from the interaction of nitrogen dioxide and sulfur dioxide. *Phytopathology,* **61**, pp. 1506–11.

Tingey, D. T., Reinert, R. A., Dunning, J. A. & Heck, W. W. (1973). Foliar injury responses of eleven plant species to ozone–sulfur dioxide mixtures. *Atmos. Environ.,* **7**, pp. 201–8.

Valley, S. L. (1965). (Ed.) *Handbook of Geophysics and Space Environment.* McGraw-Hill, New York – St Louis – San Francisco – Dusseldorf – London – Mexico – Panama – Sydney – Toronto: xvi + 696 pp., illustr.

Wellburn, A. R., Higginson, C., Robinson, D. & Walmsley, C. (1981). Biochemical explanations of more than additive levels of SO_2 + NO_2 upon plants. *New Phytol.,* **88**, pp. 223–37.

White, K. L., Hill, A. C. & Bennett, J. H. (1974). Synergistic inhibition of apparent photosynthesis rate of Alfalfa by combinations of sulfur dioxide and nitrogen dioxide. *Environ. Sci. Technol.,* **8**, pp. 574–6.

Air Pollution and Plant Life
Edited by M. Treshow
© 1984 John Wiley & Sons Ltd.

CHAPTER 12

Responses of Lichens to Atmospheric Pollution

FRANKLIN K. ANDERSON

Senior Environmental Scientist, Ford, Bacon and Davis, Inc., Salt Lake City, Utah 84108

&

MICHAEL TRESHOW

Department of Biology, University of Utah, Salt Lake City, Utah 84112, USA

BACKGROUND

Among the many organisms, communities, and agricultural situations, that have been studied in connection with air pollution, are lichens—modest, plant-like organisms that can colonize difficult situations where other plants cannot exist, such as rock surfaces and the bark of tree trunks and branches. Lichens, although looking like individual plants, are in reality organisms composed of two separate and unrelated entities that live together for their mutual benefit—an alga and a fungus. These composite organisms, which are not rooted in the manner of higher plants, receive their nutrients and moisture mainly from the atmosphere.

Lichens are at once among the most ubiquitous, yet enigmatic, of living organisms. Although they embrace one of the largest groups of described fungi, some 15,000 to 20,000 species, they are not very well known. Few can be identified to species with any confidence, except by someone who has a determined persistence in their study. Lichens will not grow well in captivity or under artificial conditions, and this makes them poor laboratory subjects. They are not prestigious organisms and do not command the attention of funding agencies in the manner that economically important plants do. Lichens are studied intensively, to be sure, but mainly by a relatively few, highly competent and dedicated specialists. Among biologists at large they are scarcely noticed at all, apart from the fairly recent attention they have attracted with respect to air pollution.

Lichens often grow in the same habitats as mosses and liverworts, and sometimes resemble them in growth forms and ecological responses. Indeed, the methodology of their herbarium care and the techniques of studying their gross morphology, are practically identical with those of mosses and liver-worts. It has been pointed out (Gilbert, 1969) that the bryophytes and lichens are even remarkably similar in their responses to sulphur dioxide—the out-standing atmospheric pollutant associated with their mutual decline. Appar-ently these modest organisms which, as we have already noted, can colonize difficult habitats that other plants cannot, do so by means of adaptations which simultaneously predispose them to damage by air pollution.

Except for their intolerance of city and industrial environments, lichens are renowned for their ability to withstand harsh living conditions. H. E. Jacques (1958), in a popular handbook, typifies this opinion in his statement that 'lichens would likely rate as the world's sturdiest plants . . . For sheer hardi-ness and resistance to unfavourable conditions they have no equal'.

They grow farther north, farther south, and higher on mountains, than almost any other plant group (Ahmadjian, 1974). In his excellent review of the lichen symbiosis, Ahmadjian (1967) reports experiments in which lichens were able to survive and measurably respire after being frozen for 18 hours at minus 183°C in liquid oxygen. Becquerel (1951), who speculated on the possibility of lichens surviving for thousands of years under Antarctic ice, experimented with even colder temperatures near to absolute zero (−273°C). The lichen *Xanthoria parietina* not only survived this low temperature, but in other experiments retained its viability after being held for six years in a vacuum (Becquerel, 1948). Similar evidence of remarkable durability and resistance to damage could be cited for elevated temperatures, desiccation, mineral and nutrient stress, light intensity (both high and low extremes), hostile habitats, and several other adverse living conditions—except one: the altered, polluted, and unnatural, environment of city and industrial areas.

Much notice has been taken of the absence or decline of lichens in urban areas, and the general consensus among research workers is that lichens are unable to withstand polluted air—particularly air containing sulphur dioxide. This single characteristic, more than any other, has brought a prominent profile to lichen research during recent years as this interest in lichens has coincided with the recent upsurge in popular interest and concern for environmental matters in general.

The correlation of lichen decline with increased pollution is not, however, a newly recognized phenomenon. Annie Lorrain Smith (b.1854, d.1937), in her classic treatise on lichens (1921), which is still unsurpassed as a basic source-book in general lichenology, opens her work (p. xxiii, para. 2) with a state-ment of the fact as a simple article of faith:

'Lichens abound everywhere, from the sea-shore to the tops of high mountains, where indeed the covering of perpetual snow is the only barrier to their advance; but owing to their slow growth and long duration, they are more seriously affected than are the higher plants by chemical or other atmospheric impurities, and they are killed out by the smoke of large towns: only a few species are able to persist in somewhat depauperate form in or near the great centres of population or of industry.'

This scarcity of lichens in polluted areas is a well-documented phenomenon. For example, more than a century ago the lichens of Paris were singled out for comment in this context in a pioneering paper (1866) by Finnish lichenologist W. Nylander (b.1822, d.1899), who made a list of the lichens growing on the trunks of chestnut trees in the Luxenbourg Gardens in Paris. The list included about 35 species, many of which Nylander observed to be poorly developed or sterile. He then wrote:

'Les lichens donnet à leur manière la mesure de salubrité de l'air et constituent une sorte d'hygromètre très sensible'. [Lichens give, in their own fashion, a measurement of the purity of the air and constitute a kind of very sensitive instrument for measuring this quality.]

Three decades later L'Abbe Hue (in 1898, *see* A. L. Smith, 1921), could find no lichens at all in the same area.

Another early observation was made by A. J. Grindon, in his *Manchester Flora* published in 1859 (in Gilbert, 1973*b*). Grindon noted that the quantity of lichens was 'much lessened of late years through the cutting down of old woods and the influx of factory smoke . . .'.

In 1879, J. A. Johnson was unable to find either foliose or fruticose lichens in Gibside Woods 8 km west of Newcastle (Gilbert, 1970). By 1885, J. Crombie observed a diminution of lichens near London, as well as their extinction elsewhere in England as a result of extensive urbanization (Hawksworh & Rose, 1970).

At the turn of the century, J. Arnold described the declining condition of lichens in Munich (cf. LeBlanc & Rao, 1973*a*). Numerous examples exist of the gradual disappearance of lichen vegetation in cities around the world. LeBlanc & Rao (1973*a*) summarize many reports from such countries as Sweden, Germany, Belgium, Finland, Hungary, France, Poland, Yugoslavia, Austria, England, Ireland, Czechoslovakia, New Zealand, Canada, Wales, and the United States.

Similar remarks, based on contemporary observations, have described the demise of lichens in many other urban areas over the ensuing decades (in Hawksworth, 1971). The interactions of air pollutants and lichens have been the subject of excellent reviews in recent years (e.g. Ferry *et al.*, 1973; LeBlanc & Rao, 1975). One of the clearest is that by Hawksworth & Rose (1976).

No detailed accounts of the lichen flora of any town or city, with respect to air pollution, antedates these early observations by Nylander, Grindon, and others in the mid-1800's (Laundon, 1973; LeBlanc & Rao, 1973*a*), although air pollution itself was both objected to and condemned much earlier —indeed even as early as 1273, when a prohibition against the burning of coal was enacted in London (cf. Lodge, 1969). Since 1866, many studies of the distribution, abundance, and condition, of lichen species relative to the degree of air pollution, have been conducted. With a few exceptions, however (Sernander, 1926; Vareschi, 1936, 1953; and others), only those done since the mid-1950s have been adequately detailed, comprehensive, and quantitative (Hawksworth *in* Ferry *et al.*, 1973).

Nearly all such studies have borne out Nylander's original intuition (1866), although differing viewpoints have been expressed concerning the actual agent of lichen decline in polluted areas. Thus, the reputation that lichens have as bioindicators of air pollution is not new; what is new is the present-day concern for the environment itself, and this has led to a new evaluation of the role of lichens in the assessment of environmental degradation.

LICHEN BIOLOGY

Composite Organisms

Lichens, although they appear to be independent plants, are associations of fungi and algae that are living together in a symbiotic relationship. Lichenologists today mostly regard the union as one of the most successful examples of mutualism to be found in Nature (Ahmadjian, 1974). Mutualism is a condition in which both partners of a symbiotic union derive benefit from the combined association. The combined symbionts take on a new growth-form which is unlike that of either the mycobiont (the fungus component) or the phycobiont (the alga component), were they to grow independently.

About 32 genera of algae participate in the formation of lichen thalli, 21 genera of green algae, and 11 genera of blue-green algae (Hale, 1967). The fungi of lichens are nearly all Ascomycetes; a handful only are known to be Basidiomycetes. Although the algae sometimes provide useful taxonomic characters for determining lichens, the fungus is the dominant member of the association, forms the 'envelope', and is the only one that produces any reproductive units sexually. Thus, the fungus accounts for nearly all the important morphological and reproductive features of the lichen thallus, and accordingly the systematics of lichens follows the fungus.

Growth-forms

While it is true that the relationships among lichens depend on the fungus alone, and the lichen species are today being treated taxonomically as fungi, the ecology of lichens is governed to a great extent by their growth-forms and the physiological response of the combined organisms to the environment. Thus, although the older lichen classification schemes that relied on growth-forms may be obsolete, the old artificial growth-form categories remain useful and even indispensable to the study of lichens.

The three main growth-form categories for lichens are termed 'fruticose', 'foliose', and 'crustose' or 'crustaceous'. There are also many intergrading forms that make for more of a continuum of growth-forms than clear categories. Indeed, to those three general growth-form terms could be added others: squamulose, gelatinous, and filamentous, for example. The true relationships among lichens cut across the growth-form categories in sometimes bewildering ways, so that phylogenetic ties are scarcely evident from the superficial form and structure of the thallus.

Lichen Structure

The growth-form terms relate to the distribution of tissues within the lichen thallus:

A. *Fruticose* lichens are usually shrubby in appearance, string-like, strap-shaped, or otherwise upright or pendent forms, that have no clear 'upper' and 'lower' surfaces. They have, rather, an 'interior' and an 'exterior'. The external layer of a lichen thallus, or surface, is termed the cortex. Beneath it is a layer of fungus-enmeshed algal cells called the algal layer. Metabolic activity in the lichen is greatest in the algal layer, and it is this area that suffers the most damage during the first stages of exposure to sulphur dioxide fumigation (Skye, 1968). Below the algal layer is a region of cottony, loosely woven fungal hyphae free from algal cells and called the medulla. The medulla appears to function as a storage organ (D. C. Smith, 1960), and it is here that unique lichen substances called lichen acids are produced and deposited, the deposition apparently being on the external surfaces of the fungal hyphae. Sugars produced during photosynthesis in the algal layer are rapidly transported to the medulla, where they are stored as mannitol (Saunders, 1970). Thus, considering these tissue regions in a lichen thallus, a cross-section of a fruticose lichen would be more or less circular, with the central part either hollow or filled with the medulla tissue. The algal layer would lie around the circumference of the section, just under a thin cortex.

B. *Foliose* lichens also possess the same internal tissues, but have them arranged in a flat, leaf-like fashion with a definite upper surface (the side containing the algal layer) and a lower surface (the side containing the fungal layer). The lower surface of foliose lichens has various structures that are not encountered in fruticose lichens; for example, there is a lower cortex, usually coloured very differently from the mineral-grey to pale-green upper surfaces that most lichens have: white, brown, or black, are the usual colours of lower cortex tissues. There may be a dense to sparse distribution of hairlike processes on the lower cortex, called, variously, tomentum or rhizines. Although root-like in appearance, and even sometimes called rootlets, rhizines and tomentum do not function as organs of absorption but only as organs of attachment. When hairs occur around the margins of the thallus lobes, or around the spore-producing fruiting structures called apothecia, they are called cilia or fibrils. Bare patches, holes, pores, and depressions, can be found on either surface—usually the lower—and these are called cyphellae, pseudocyphellae, or pores, depending on their structure and size. The lower surface is also sometimes provided with anastomosing series of ridges and folds called veins. Lichens have no waxy cuticle like that of higher plants, to isolate their internal tissues from the outside air. Nor do they have stomata that can be closed during the night or in times of stress. The interior of a lichen thallus is thus largely exposed to the environment.

C. *Crustose* lichens are so tightly appressed to their growing surface, called the substrate, that no lower cortex or surface can be detected at all. Thus, although they resemble foliose forms superficially from above, they cannot be lifted free from the surface upon which they are growing, without damaging them. Many of these crustose forms appear almost to have been painted by some imaginative artist onto the surfaces on which they grow in varied rich shades of yellow, orange, green, or black.

D. There are some foliose lichens that have no visible lower cortex tissues but have only exposed medulla on the lower surface. In this they resemble crustose lichens, but as they are not permanently attached they are not, of course, crustose lichens. Among the lichens having this feature are even some that are traditionally called fruticose; this is because of the nature of their fruiting structures, which are elevated on a goblet-shaped pedestal (the podetium) arising from either a crustose primary thallus or a thallus that is termed *squamulose* because it consists of squamules. These are typically small, scale-like or even minutely leaf-like, flecks of foliose thallus that lack a lower cortex. When the squamules collectively form the primary thallus of a lichen, the growth-form is termed *squamulose*—a category ranking with foliose, fruticose, and crustose. When the squamules are distributed along the axis or branches of an erect podetium or fruticose thallus, they are sometimes called phyllocladia. The main groups of lichens exhibiting squamules, podetia, and phyllocladia, are several genera centred around the great genus *Cladonia,* which are therefore often called the cladoniform lichens.

Fruticose, foliose, crustose, and squamulose, lichens all partake of one common feature: their tissues are arranged in layers, with the phycobionts in a definite layer near the upper (or 'outer') cortex. These growth-forms were formerly grouped together as a category called the stratose lichens. Phycobionts among these may be either green algae or blue-greens, but the greens predominate.

There is another type of lichen thallus that lacks the stratified tissue layers but instead consists of a uniformly intertangled mass of both algal and fungal filaments. These are the gelatinous lichens, and their algal symbionts are all blue-green algae that exercise a greater influence upon the growth-form of the thallus than occurs in the stratified lichens. Gelatinous lichens, however, are relatively few in number of species compared with the other forms. They are nevertheless important, because many can fix atmospheric nitrogen, which eventually becomes available to other plants.

Vegetative Reproduction

Many lichen species among the principal growth-form categories—the foliose, fruticose, and crustose, lichens—possess a number of unique morphological features that are probably functional mainly in vegetative propagation. The most prominent and important among these are structures called isidia and soredia. Isidia are wart-like, knobby, or elongate, projections that grow from the upper and lower cortex and from margins of the lichen thallus. Their presence, size, shape, location, distribution, and density, are all important taxonomically. In vegetative reproduction they break off and become scattered, carrying with them both algal and fungal portions of the thallus, and so permitting the establishment of new colonies.

Soredia are cottony bits of tangled hyphae enclosing a few algal cells. Resembling little balls of fluff, they erupt through pores, or through larger areas called soralia—to be blown about by the wind, or distributed by trickling rain or even carried about by ants (Ainsworth, 1971). Eventually they, too, can grow to become new lichen thalli. The location, size, and sometimes the colour, of the soralia (as well as their mere presence or absence) are important taxonomic characters.

It has been suggested by some workers (Laundon, 1967; Saunders, 1970) that, in the presence of sulphur dioxide or other air pollutants, it becomes increasingly difficult for lichens to colonize new surfaces. It is noteworthy that the algal cells of soredia are very much more exposed to the atmosphere than those of an intact lichen thallus. Sulphur dioxide attacks the chloroplasts of lichen algal cells and interrupts photosynthetic processes (Rao & LeBlanc, 1966; Showman, 1972; Nash, 1973). Such an effect could account for the failure of new lichens that propagate by means of soredia, to become established in city and industrial areas.

Sexual Reproduction

Sexual reproduction in lichens is little known. Ascospores are produced, which germinate in culture experiments, and the cultured colonies exhibit the type of variability expected from genetic processes. The process of lichenization, or the invasion of free-living algal cells by lichen fungal mycelia, has been observed in culture too, but the extent to which this occurs in Nature is not known. The mycobionts of several hundred lichens have been grown in culture from spores, but the typical growth-form of the lichenized species is never produced by the isolated fungal component.

Ascocarps—the spore-producing 'fruiting' structures—are frequently observed in cultured mycobionts, but the ability to fruit is gradually lost as the culture ages (Ahmadjian, 1967). Isolated algal cells grown in culture are capable of zoospore production, a reproductive process found among the Green Algae, but this never happens in an intact lichen thallus. The fungal component, however, fruits abundantly in combined lichen thalli—but seems almost to require the presence of the symbiotic Alga to continue this process.

LICHEN COMMUNITIES

Habitats

In addition to growth-form categories, lichens may conveniently be categorized into habitat groups. Some species have the capability of occupying several different substrates, but most are restricted to one basic type. Thus, all soil-inhabiting lichens are termed 'terricolous'; those that grow on bare rock surfaces are 'saxicolous' lichens (sometimes the term 'epipetric' is used); and those that are found occupying the bark of trees are called 'corticolous' (and/or 'epiphytic') lichens.

These three categories are the most widely used by lichenologists, but others exist: epiphyllous lichens, for example, grow on the leaves of plants; lignicolous lichens grow on decorticated (bare, no bark) wood such as rotting logs, shingles, fence-posts, boards, and other wood surfaces, Some lichens, termed omnicolous, can occupy several habitats indiscriminately, ranging from those mentioned above to brick, asphalt, asbestos roofing tiles, cement, paint, sea shells, bone, and even glass. A number of epizooic lichens have been discovered that live on the backs of male Galapagos tortoises; others even occupy the wing-covers of certain large flightless weevils (beetles) of tropical areas. A very few lichens are completely aquatic; others exist only in the driest of deserts, where they must obtain all their moisture from the air. Of course, just as for all plants and animals, there are lichens restricted to the tropics, to tundra, to temperate regions, to particular types of rock substrates (such as calcareous rock, e.g. limestone), or to other restricted habitats, but in general lichens are widely distributed and depend less upon the substrate than do rooted plants.

All lichens respond strongly to the microclimatic conditions that occur in their habitats, even more than to the larger continental or seasonal patterns in the climate. The vertical distributions of lichens on tree trunks, for example, are very pronounced because of the great variation that can exist in such environmental factors as light intensity, moisture, relative humidity, and others—even though the long-term weather pattern is the same for the whole tree. This sensitivity to microclimatic variation may relate to the problem of lichens *versus* air pollution so that, besides being fascinating subjects for study in their own right, there is urgent need to investigate the ecology of lichens in healthy and polluted environments.

The phytosociology of lichens is probably one of the most thoroughly studied aspects of lichen biology. The growth-forms, substrate preferences, and microclimatological responses, of lichens combine to produce a kaleidoscope of distinctive communities, many of which in temperate regions have been surveyed, described, and formally named. Many of these communities are stable and persistent for as long as they remain undisturbed.

Succession

As with higher plants, lichen communities develop by degrees, through various stages of a successional sequence. For example, considering only corticolous lichens, the twigs, branches, and trunks, of trees are often inhabited by a rich lichen flora. As the tree grows, the new regions of the trunk and branches are gradually occupied by a succession of lichens that begin to appear on about the third internodes from the growing-tips (Degelius, 1964; Hale, 1967). Foliose lichens appear first, occupying the youngest internodes. On internodes that are from one to four years older still, crustose lichens begin to appear. Some ten years later, fruticose lichens finally begin to take over.

Eventually, each internode ages, increases in diameter, yields its terminal or peripheral position on the tree to younger, newer internodes, and alters in its physical and environmental characteristics. The sum of these changes represents a sequence of changing microhabitats each of which is commonly occupied by different lichen associations. At length the lower parts of the tree become shaded by the crown, the lower branches die, and the process of change becomes more or less arrested. The lichen community of the lower branches likewise becomes stable and may be considered to be a kind of climax stage.

From base to crown, a growing tree presents a complex gradient of interlocking microhabitats and corresponding 'seral' stages of the final lichen community. The result is a vertical distribution-pattern of lichens, in which the community structure of those near the ground is very different from those higher up. In addition to this vertical pattern, the mosaic of lichen communities in an extensively forested region is correlated with the density and species composition of the trees themselves (Hale, 1967).

In such an area, the total biomass of lichens can be large. One study reported 350 to 450 pounds per acre (393–505 kg per ha) (fresh weight) of the large, foliose lichen *Lobaria oregana* growing in the crowns of large Douglas (*Pseudotsuga menziesii*) trees (Denison 1973). This lichen contributed between 1.8 and 10 pounds (0.8 and 4.5 kg) of nitrogen per year per acre (0.408 ha) to the forest ecosystem, because it is one of the lichen species whose phycobiont is the blue-green *Nostoc*, which is capable of fixing atmospheric nitrogen. This is an impressive example of one of the roles of lichens in Nature—one that has only recently been attracting the attention of environmentalists and ecologists. The importance of determining the pristine composition of forest lichen communities, and of discovering their responses to the effects of atmospheric pollution, is obvious.

Lichens as Bioindicators of Air Pollution

The case of lichens with respect to air pollution is somewhat different from that of higher plants. Although they have their ecological roles, lichens are scarcely important economically (Llano, 1944, 1948, 1951; LeBlanc & DeSloover, 1970; Richardson, 1975), at least not in the way forests of timber are, or the thousands of acres of crop and forage plants that have sometimes been damaged in air pollution incidents. The decline of lichens in polluted regions is, to the average citizen, if he notices it at all, more an aesthetic loss than a real one.

To the biologist or environmentalist the decline of lichens is of course a tragedy. But even the specialist often studies the stunted and afflicted lichen populations of polluted areas, with an eye more to determining the extent and intensity of the pollution itself than investigating the biological responses of the struggling survivors. There is a feeling among research workers (Hawksworth, 1971; LeBlanc *et al.*, 1972*a*, 1972*b*; LeBlanc & Rao, 1973*a*; Nash, 1974; and others) that lichens, especially epiphytic lichens, are particularly valuable as bioindicators of air pollution, and that the use of lichens can provide quick, easy, and inexpensive, access to information about the extent or urban and industrial atmospheric effects.

SULPHUR DIOXIDE EFFECTS ON LICHENS

Smelters and Sulphur Oxides

Sulphur dioxide provides the greatest threat to lichens, and the largest sources of SO_2 are the most crucial. While the coal-fired hearths of the city have made their contribution to decreasing lichen populations in urban areas, the smelters and the power plants, with their huge furnaces, have emitted large amounts of SO_2, which have long been taking their toll. In the early days of smelters, the impact of SO_2 on the higher plants was so overwhelming that little attention was paid to the demise of the lichen flora. It was not until 1958 that detailed studies of lichens were finally undertaken in the vicinity of specific industrial sources (Skye, 1958; LeBlanc & Rao, 1966; Schönbeck, 1969).

Lichens have no waxy cuticle or other outer layer to protect them from environmental hazards or reduce water-loss, nor have they any openings through which gases pass. Their entire surface is exposed 24 hours a day to the diffusion of any chemicals in the air or on or in the substrate. Such chemicals as gain entrance to the lichen thallus may then accumulate in the tissues, and any to which the tissues are not adapted can prove harmful when the concentrations reach excessive levels. Moreover, lichens have no deciduous parts which, in the manner of the leaves of trees, can be dropped when killed by toxic substances. The foliose and fruticose lichens have the most distinct algal and fungal layers. In the algal layer, metabolic activity is the greatest, and it is this same algal layer that is most sensitive to air pollution damage. Sulphur dioxide, for instance, attacks the chloroplasts of the algal cells, interrupting their photosynthetic process.

Lichen Sensitivity

Much of the scientific literature dealing with the effects of air pollution on lichens suggests that lichens may be more sensitive than higher plants (LeBlanc, 1969; Hawksworth, 1971; LeBlanc *et al.*, 1974). Evidences are cited to the effect that extremely low ambient air levels of sulphur dioxide are sufficient to cause a dramatic decline among lichens.

One quantitative scale for the estimation of sulphur dioxide air pollution in England and Wales, using epiphytic lichens (Hawksworth & Rose, 1970), indicated that all epiphytic lichens were absent (zone 0) at mean annual ambient air levels of 170 μg SO_2/m^3 (0.06 ppm). A series of severely lichen-depleted 'struggle zones' (zones 1 to 5) existed when ambient levels exceeded about 60 μg SO_2/m^3 (0.02 ppm). According to this scale, effects are measurable in terms of lichen abundance and vigour at concentrations as low as 30 μg SO_2/m^3 (0.01 ppm).

It is difficult to assess the true impact of sulphur dioxide from such scales, because they do not give any information about the number, intensity, or duration, of the high-level peaks of SO_2 exposure. Consequently, a great controversy still continues over the question of whether sensitive lichens are responding to long-term, low levels of sulphur dioxide or to relatively infrequent high concentrations that occur during 'fumigations' caused by air stagnation or other atmospheric effects. Whichever is true, lichens are usually characterized as 'sensitive' because they really do disappear from polluted areas. Although this is particularly true for epiphytic lichens, it is difficult to assess, as yet, the impact of sulphur dioxide from the studies so far reported (Nash, 1973).

Part of the uncertainty about lichen sensitivity results from the lack of reliable SO_2 measurements, which are often reported only as annual means. However, it is not yet known whether lichens are sensitive to long-term low-level SO_2 exposures, or only to the relatively infrequent peak values that sometimes occur. Lichen populations in cities and around industrial complexes have been shown to be limited or non-existent, and normal species-abundance and diversity appear only as one travels to unpolluted regions away from these areas (DeSloover & LeBlanc, 1968; Hawksworth & Rose, 1970; and others). This work has been extensively reviewed (Hawksworth, 1971; Ferry *et al.*, 1973; Gilbert, 1973*a*, 1973*b*; Nash, 1976; and others). Although some workers have suggested high temperatures, urban drought (J. Rydzak in LeBlanc & Rao, 1973*a*), or complex 'city effects' (Brodo, 1966, 1968), as the cause of lichen stress, most now attribute lichen damage to air pollution—particularly SO_2. However, fluorides and heavy metals have been implicated in a few studies (e.g. Nash, 1975).

In North America, some of the most detailed studies of lichen distributions near a pollution source have been made by LeBlanc and his collegues around Wawa (Rao & LeBlanc, 1967) and Sudbury (LeBlanc & Rao, 1966; LeBlanc *et al.*, 1972*a*) in Ontario, Canada. Similar studies have also been made in Europe (Skye, 1968; Hawksworth, 1973; Laundon, 1973; Johnson & Søchting, 1976).

SO₂ Dose versus Lichen Response

In reality, the main result of these studies was merely to show that lichen distributions around these sources were correlated with the SO_2 concentrations in the air (Nash, 1976). These concentrations were reported in the form of long-term averages, and revealed nothing concerning the threshold of injury for lichens. In some of the studies cited above, not only was the lichen flora decimated near the pollution source, but the forest trees were also killed; yet no one suggested that the trees were exhibiting an unusual sensitivity to the same low levels of sulphur dioxide as was suggested for the lichens of the same region.

Naturally, when lichen decline is attributed to such low annual averages as 0.02 ppm, for example, there is an understandable inclination to interpret the relationship directly. An annual average of 0.02 ppm can be attained in many ways near a pollution source. For example, a low SO_2 concentration of 0.02 ppm for 24 hours a day, 365 days a year, will yield that average. So will a concentration of 0.5 ppm for 8 hours a day on only 45 days in a year. This latter set of conditions would be considered a fairly highly polluted atmosphere.

Few experiments have been conducted with lichens in controlled atmospheres containing known levels of sulphur dioxide at realistic concentrations, so the question of dose–response in lichens remains as yet largely unanswered. When higher plants are investigated in environmental chambers with respect to specific pollutants, the experimental doses (time and concentrations) used and the responses observed are commonly reported in the scientific literature. These studies usually correspond to the high-level, relatively short-duration peaks that occur near urban or industrial areas. The distinction between this type of report and that given for lichens is usually overlooked, and many people promptly put lichens at the top of the sensitivity list because their decline is reported to result from such low annual average levels of sulphur dioxide. Gilbert (1973*b*) says that such scales still need to be evaluated and adjusted because 'it is still a matter of opinion whether distributional data (for lichens) reflect mean or peak concentrations experienced at a site'.

It should be emphasized that most of the work suggesting extreme sensitivity of lichens to SO_2 has been conducted under one or more of the following qualifying conditions, some of which have no application to either the environmental conditions or the lichen flora of areas other than that in which the study was conducted.

(1) Studied areas exhibiting lichen decline have usually been in regions of high humidity and ample rainfall. These conditions permit luxuriant lichen growth and also increase the likelihood of SO_2 damage (Hawksworth, 1973; Sundström & Hällgren, 1973; and others).

(2) Such studies have dealt almost exclusively with epiphytic lichens growing above the ground on tree trunks and branches (Harris, 1972; LeBlanc *et al.*, 1972*a*). This type of lichen flora is largely absent from arid regions and even from wetter mountainous regions throughout the US West and Southwest. The same is true for many other regions in the world.

(3) The documented cases of lichen decline have often occurred in regions sustaining simultaneous extensive habitat destruction, usually associated with city growth or smelter fumigations that have killed or damaged not only lichens but other plants as well (Laundon, 1973). Habitat destruction includes the removal of host trees by logging, damage by fumigation, fire, urban sprawl, highway building, agricultural practices, and so on. Many of these are difficult to separate from the effects of air pollution. For example, in England Gilbert (1973*b*) reported the impact of SO_2 on lichens to be greatest-to-least in the following sequence: bombed sites, woods, asbestos roofs, reservoirs, heaths and commons, sewage farms,

private gardens, parks, old brick walls, and finally churchyards and cemeteries. This is precisely the sequence of greatest-to-least habitat disturbance (remembering that English woods are heavily cut and the heaths overgrazed).

(4) Most cases of lichen decline in polluted areas have been reported without adequate monitoring of ambient SO_2 levels. For example, (a) annual or monthly mean values are reported instead of peak values and durations; (b) zero values from monitoring stations are often included in the computation of the averages; (c) monitoring instruments with inherent limitations have often been used, such as lead candles, sulphation plates, and outdated continuously monitoring instruments that do not respond to levels below about 0.1 ppm SO_2.

(5) In many cases reporting fumigation studies, the chambers used were merely boxes, Petri dishes, flasks, or desiccator jars, without conditioning or monitoring equipment attached. Many studies termed 'fumigations' have been conducted with lichen pieces submerged in aqueous sulphite solutions. These have little relevance to ambient air conditions (Puckett *et al.*, 1973). In one of the best-controlled fumigation studies with lichens and SO_2, Nash (1973) utilized continuous-flow chambers and concluded that lichens are 'no more sensitive than higher plants'.

(6) Lichen sensitivity is also lacking in accurate documentation because it is not even known how many days per year per given habitat the lichens are (a) moist and active metabolically, and (b) simultaneously subjected to SO_2 fumigations.

The best guide for determining the SO_2 concentrations that damage lichens would be the results from quantitative studies done with controlled atmospheres containing known, realistic quantities of SO_2. In Nash's study (1973) mentioned above, he concluded that 'the limit of lichen susceptibility to short-term SO_2 fumigations appears to be about 0.5 ppm'. Nash's fumigation durations were 12 hours, and his SO_2 levels ranged from 0.5 to 4 ppm. 'If this is generally true,' he continues, 'then lichens appear to be no more sensitive to direct SO_2 fumigations than do higher plants.' A major obstacle to performing long-term, low-level SO_2 fumigations is the reluctance of lichens to thrive in captivity—even in clean air. We do not know much about the culture of intact lichens.

The SO_2 concentrations that are required to injure lichens have still not been clearly established. According to Gilbert (1973*b*), it is still 'a matter of opinion whether distributional data reflect mean or peak concentrations experienced at a site'. It appears that the nature of the substrate and its buffering capacity, the time and season of SO_2 fumigation, the height of the lichen above the ground, and other factors, as well as the inherent tolerance of the lichen species, may influence the sensitivity. The impact is sometimes thought to involve mostly young stages—including germination of spores—preventing the establishment of colonies.

In some lichen species that grow outwards from the point of initial colonization, when a colony reaches a certain size, the central section often begins to erode, and young colonies of the same or a second species begin to colonize this centre. This is a form of lichen succession on rock known as 'recycling'. If recycling is present in unpolluted control areas, but absent from polluted sites, it may be an indication that the young colonies are the most sensitive. Reappearance of recycling in a recently cleaned-up polluted area would be evidence for the effectiveness of the pollution controls.

One study exemplifying the sensitivity of the reproductive phase showed that undisturbed old limestone grave-markers in a London churchyard still had large, healthy lichen thalli, while newer gravestones had no lichens (Laundon, 1967). This suggested that the reproductive units were more sensitive than the mature thalli. Pollutant-tolerant species occurred on over 80% of the stones. A more sensitive species, *Caloplaca heppiana,* occurred on 90% of the stones erected before 1751, whereas none at all occurred on stones dated after 1901. Hale (1967) also reported that reproductive stages were most sensitive to stress and especially to changes in acidity.

Some Suggested Mechanisms for Lichen Sensitivity

The importance of the substrate is also demonstrated by relative sensitivity of lichens on different substrates. For example, Gilbert (1965) reported that species numbers first showed an increase on asbestos $1\frac{1}{2}$ miles (2.4 km) from Newcastle, on sandstone $3\frac{1}{3}$ miles (5.3 km) from the city, and on Ash trees (*Fraxinus excelsior*) $4\frac{1}{2}$ miles (7.2 km) away. This was consistent with the marked decrease of the pH (i.e. increase in acidity) of bark to pH 3.1 as the city was approached. No lichens grew on the bark of Ash trees having pH values below 4.0 (Gilbert, 1970). The substrates that are lowest in buffering capacities are the first to lose their epiphytic flora. Apparently, calcareous substrates such as limestone rocks, mortar in walls, and asbestos, provide a special niche affording built-in chemical protection from acidity originating with SO_2.

Robitaille et al. (1977) consider acid rain to be a major factor in acidifying tree bark and the lichens that grow on it, thus restricting growth of epiphytic lichens. In their study, there was a correlation between the concentration of atmospheric SO_2 and the acidity of tree bark. The pH of bark in the control area (20.4 km south-west of the smelter) was 4.03, whereas 6.5 km from the smelter the pH was 3.66.

Beetham & Hedger (1978) concluded that lichen damage was not directly caused by a reduction in pH value or an increase in sulphate content, but that dry deposition of sulphur dioxide was probably the most important factor.

Lichen flora is least affected by SO_2 where the substrate has a basic reaction. Lichens persist longest on the bark of ash (*Fraxinus* spp.) and elm (*Ulmus* spp.) trees that are most alkaline. Brodo (1966) has observed in New York that the most city-tolerant lichens were species growing on concrete and mortar. Lichens growing on alkaline substrates can tolerate airborne concentrations of SO_2 many times higher than those growing on more acid substrates.

METHODS OF INVESTIGATION

Methods used in the above and similar studies were much the same as those developed to study urban pollution. First there was the procedure of determining the species of lichens that were present, and their relative numbers. This permitted mapping their distribution. A second method was rather similar but treated the total lichen community so that the percentage of cover and other ecological parameters were considered as well as the species richness. Thirdly, lichens could be taken from unpolluted areas and transplanted or placed in the study area. Disks of tree bark bearing epiphytic lichens were cut or punched out and transferred to posts or other structures located in polluted areas. Their responses were studied by periodically photographing them. A fourth principle approach was to bring lichens into the laboratory and expose them to different concentrations of SO_2, either in a gaseous atmosphere or in a liquid medium.

Distribution Studies

The lichen flora especially of epiphytic forms, in the larger countries around the world, has been mapped to indicate its distribution. Follow-up work could then show any changes that might have occurred. Most of these reports suggested that air pollutants were associated with the disappearance of lichens, but the relation has only been clearly established since the mid-1950s.

The methods used to map lichen distribution are much the same throughout the world. Generally, the numbers of lichen species (the diversity) present on a particular kind of substrate, and/or the frequency with which a species occurs in the community, and/or the percentage of area it occupies (the cover), and/or its average number of individuals per unit area (the density), is/are determined. The more thorough studies incorporate all of these values. Current studies also include photographs of lichen communities taken periodically from fixed points.

In order to be valid, comparisions must be made only between the same species growing on the same substrate and with the same orientation to the sun. DeSloover & LeBlanc (1968) provide an excellent discussion of methodology. On the basis of these types of data, the area under study is delineated into zones based on pollution data. Isopleths are then drawn, denoting the various zones. These zones generally have an elliptic shape with the long axis extending in the direction of the prevailing winds. This approach was developed in 1926 by Sernander, who first described pollution damage to lichens in terms of a lichen desert, a struggle zone, and a normal zone. He coined the term lichen desert to describe areas almost devoid of lichens. Some work has gone beyond this (e.g. Rose, 1973), to show not only a correlation between reduction in lichen populations and increased SO_2 pollution, but also to indicate the average SO_2 concentration in the air from the lichen data.

Laundon (1967) divided London into zones called central, inner urban, suburban, and green-belt, rings. Only one species, *Lecanora dispersa,* grew in the innermost, central zone, which was regarded as a lichen desert. Some 32 species or 52% of the lichen flora were found within the inner urban area, and 92% within the suburban ring. Skye (1968) divided the Stockholm region similarly, and also showed an inverse correlation between SO_2 concentration and the numbers of lichen species present.

The impoverishment of lichens in cities has also been blamed on the dryness of the air and the higher temperatures found in cities. Rydzak (1969), the principal proponent of this thesis, provided some cogent arguments; but their validity has been seriously questioned as reviewed in modern studies by Coppins (1973). Perhaps the most significant argument againt this drought-effect hypothesis is that lichens have suffered equally both in naturally dry and extremely humid parts of the world, provided the cities were industrialized. Also within city limits, industrial areas had fewer lichens than residential ones, despite similar levels of humidity. Barkman (1958) concluded that, while the aridity was a contributing and perhaps interacting factor, toxic gases were more critical, especially beyond the towns.

Not every species of lichen is harmed in cities. Resistant species move in to fill voids left by intolerant ones. *Lecanora conizaeoides* is the most common of the resistant species. Once relatively infrequent, it is now widespread throughout Europe (Gilbert, 1973*b*). Other species may be still more resistant, depending on the substrate—*Lecanora dispersa* on basic substrates, for example.

Barkman (1969) described how the Netherlands lichen flora had lost 27% of its epiphytic lichens and 15% of its epiphytic bryophytes. Some epiphytic species that had disappeared from the Netherlands over the past century included *Caloplaca cerina, Lobaria pulmonaria, Parmelia vittae, Ramalina pollinaria, Pyrenola nitida,* and several species of *Usnea.* Some reductions in species richness characterize conditions in all European urban areas where the lichen flora has been mapped.

The real question was, which toxic gases were most responsible for the lichen demise? Consistently, the effects on lichen coincided most closely with denser urbanization and the increased use of coal to heat the houses. Industries in cities, burning even more coal, also contributed. Thus the coal smoke was implicated and, most critically, the SO_2 in it. Numerous mapping studies in urban and industrial areas have repeatedly demonstrated that lichen distributions correlate with SO_2 concentrations in the air.

Rather than plot or map the distribution of all the lichen species in an area, it has been practicable to study only a representative number of the more sensitive indicator species. Skye first suggested this method in 1958. He used the species *Evernia prunastri*, *Ramalina farinacea*, *Xanthoria parietina*, *Parmelia acetabulum*, and *Anaptychia ciliaris*.

Index of Atmospheric Purity

DeSloover & LeBlanc (1968), and LeBlanc & Rao (1975), expanded their lichen distribution studies to involve a detailed mapping using an 'index of atmospheric purity' (IAP). The IAP can be summarized in the following formula:

$$\text{IAP} = \sum_{n}^{1} \frac{Qf}{10}$$

where Q is the Ecological Index, which is determined by adding the numbers of companion species present at all investigated sites; f = frequency-coverage (or abundance) value of each species, expressed in a number on a numerical scale chosen by the investigator; n = the total number of species at a station

The sum of Qf is divided by 10 to reduce it to a manageable figure. The indexes from all the stations are then plotted on a map, and isopleths are drawn between similar points to provide isotoxic zones.

This method was also applied to epiphytic lichens and mosses in studies in Montreal, Canada to elucidate the relationship between industrialization and lichen distribution (LeBlanc & DeSloover, 1970).

Ecological Investigations

Lichens have very limited ability to respond to abrupt environmental change. Their point of greatest abundance and diversity is reached in moist areas that have remained stable over a long period. Urbanization and industrialization, with their impacts on habitats, substrates, and air quality, have seriously impaired, disrupted, and impoverished, lichen communities. But as with the individual species, not all lichen communities are equally sensitive to air pollution. Composition of both the community and the substrate is important. Epiphytic communities are the first to disappear; those on limestone are commonly the last remaining (Gilbert, 1977).

Very few lichen species occur in the centres of old towns, where the trees and old stone walls may remain completely uncolonized (Gilbert, 1973b). In

Britain, only *Lecanora dispersa, Candelariella aurella,* and *Lecania erysibe,* plus a few species on calcareous substrates, are common in old towns. Within the stressed urban and industrial environments, tolerant forms tend to persist whereas sensitive ones do not.

Anderson & Treshow (1978) found that a single exposure to SO_2 may perturb a lichen community. They fumigated *in situ* lichens growing on Lodgepole Pine (*Pinus contorta*) bark, using SO_2 at concentrations of 0.5, 2.0, and 5.0, ppm for 2 and 6 hours. Lichens of the pre-treatment community occupied 69.2% of the substrate and non-lichens 1.6%. The remaining surface was unoccupied. During the next 3 years, the change in percentage lichen cover was negatively correlated to the SO_2 time-concentration dose. The linear regression of change in lichen cover *versus* SO_2 dose was negatively correlated (-0.87), although there was no visible injury to the lichens.

In one early study around a smelter, Rao & LeBlanc (1967) inventoried epiphytic vegetation at 64 sites at increasing distances from an iron sintering plant at Wawa, Ontario. Seventy-three lichen and thirty-nine bryophyte species were recorded. Lichen richness was noted against variations in pH and in sulphate content of surface water, soil, and vegetation. A marked reduction in total epiphytic vegetation was found to be correlated with a rise in the concentration of SO_2 in the air, and of sulphate in the water and soil.

Transplant Studies

Transplant studies have been especially useful in indicating the extent of impact of emissions from specific, suspected air pollution sources. The methods are roughly analogous to the bioassay methods in which higher plants are used as indicators and monitors of the presence and effects of air pollution (Feder & Manning, 1979). Changes that might develop over time in the appearance, growth, or physiology, of the lichens can be recorded. The extent of change, and the time that it takes, can provide a general measure of the amount of pollution. Lichens growing on any substrate may be used, but foliose-corticolous forms have been utilized most often. Areas of bark a few centimetres in diameter are punched out from trees growing in unpolluted areas and attached to trees or posts in the area to be studied.

This approach was first applied by Brodo in New York (1961, 1966). The closer the transplants were set to New York City, the sooner they died. Nearest the city, the margins of the thallus yellowed, bleached, and died within 3 to 4 months. Lichens transplanted to trees at the east end of Long Island thrived.

LeBlanc & Rao (1973*b*) transplanted bark disks supporting lichens and bryophytes to trees growing near the Sudbury nickel smelter. Injury symptoms were clearly correlated with SO_2 concentrations at various distances from the source. The lichens progressively changed in colour from grey-green to whitish as the heavily polluted area was approached. Plasmolysis and death were apparent, and within a year most of the epiphytes at all near the smelter were dead. Reduced thallus thickness was one of the earlier symptoms, together with chlorosis from the chloroplast breakdown. Even before this, however, lichen reproductive structures may be impaired or fail to develop. Transplanting was not the cause of these symptoms, as controls remained healthy following a similar transfer to trees in an area unaffected by the smelter emissions.

Studies in Germany (Schönbeck, 1969) yielded similar results. Bark disks bearing the lichen *Hypogymnia physodes* were nailed onto wooden frames at numerous sites in the Ruhr Valley. The lichens shrivelled and died within a few weeks at the most polluted sites, whereas lichens at sites increasingly distant from the industrial centres remained normal for several months.

Laboratory Studies

Studying the response of lichens to SO_2 and other pollutants under controlled laboratory conditions also has provided answers to some key questions. Sulphur dioxide acts chemically on plant cells in much the same general way, regardless of the plant group or species; so it is with the green, chloroplast-containing cells of the algal component of lichens.

Some of the earliest laboratory studies with lichens were conducted by Pearsen & Skye (1965). Exposing lichens to high concentrations of SO_2 in flasks kept in growth chambers, these authors found, as expected, that the photosynthetic rate decreased gradually as either the concentration of the gas or the duration of exposure was increased. However, the equivalent SO_2 concentrations of 100 to 100,000 ppm used in these experiments were far too high to be of interpretative value. Respiration increased up to a point; but at the high concentrations, both photosynthesis and respiration dropped rapidly.

Gas exchange, or oxygen uptake, is a common method of determining physiological disfunction, and it has also been utilized in lichen studies (Turk *et al.*, 1974). There was a marked decrease in assimilation following exposure to 0.2 ppm sulphur dioxide for 14 hours at 10°C.

Photosynthesis, i.e. the rate of carbon fixation by the algal lichen component, is also affected by SO_2. The sensitivity varies with the species, but is most pronounced with increasing acidity and SO_2 concentration (Richardson, 1975).

Effects of SO_2 on chlorophyll were studied by Rao & LeBlanc (1966), who exposed lichen thalli to 5 ppm SO_2 gas for 24 hours. Microscopic examination of the algal cells revealed bleaching of the chloroplasts, permanent plasmolysis, and brown spots on the chloroplasts. The injury was most conspicuous at the higher atmospheric humidities of e.g. 92%. These authors believed that the injury resulted from the presence of sulphurous acid (H_2SO_3) and sulphate ion (SO_4^{2-}) in the thallus. The presence of magnesium ions in the extracts, and the brown spots on the chloroplasts, suggested that magnesium had been removed from the chlorophyll, converting it to phaeophytin-*a*. Even a slight degradation of the small amount of chlorophyll present is likely to impair the metabolism of the cell and disrupt the association between the lichen and fungal components of the lichen. This work was valuable in interpreting mechanisms of action, but the SO_2 concentrations used were too high to relate to actual situations.

The response of isolated lichen algal cells as well as whole lichens was studied by Showman (1972), who exposed lichens to 2–4 ppm SO_2 in dry air and to 6 ppm at 50% relative humidity. The cultured algal cells were physiologically injured, but not the whole lichens. Respiration was also decreased.

THRESHOLD CONCENTRATIONS

The critical question to be asked is, at what concentrations of SO_2 or other pollutants are lichens injured? There is no simple answer. As with higher plants, environmental factors modify sensitivity tremendously. Moisture, for instance, is especially critical as, when they are dry, lichens are extraordinarily tolerant. Secondly, the species must be known, as individual species range in tolerance from highly sensitive to resistant; here again, it is similar to the situation with higher plants.

Thus we see a range in sensitivity from the very tolerant crustose *Lecanora conizaeoides* and the relatively resistant foliose *Parmelia saxatilis* to several species of the highly sensitive lichen genera *Usnea, Ramalina,* and *Evernia.*

Most of the work that has been directed towards determining injury thresholds has utilized long-term average SO_2 concentrations; these are the average concentrations over a growing season or a year. Unfortunately, the peak concentrations, occurring over a period of commonly only a few minutes or hours, may be equally or more important. These are rarely recorded.

LeBlanc & Rao (1973*b*) averaged SO_2 concentrations over a 6-months' period and compared the effects obtained with those on transplanted lichens. They found that average concentrations above 0.154 ppm could cause acute injury if only periods when SO_2 was actually present were averaged. Concentrations above 0.079 ppm caused chronic injury. When the total 6-months' period was considered, concentrations above 0.006 ppm could be injurious.

A long-term average value can be achieved in an almost infinite number of ways. Moreover, it is misleading to use long-term average SO_2 values as direct correlates with lichen damage if there were also periods of high-level SO_2 concentrations during the fumigations. LeBlanc (1969) studied the epiphytic flora found on the bark of *Populus balsamifera* at distances ranging from 5 to 40 miles (8 to 64 km) from a nickel smelter at Sudbury, Ontario. Four pollution zones were described based on average SO_2 concentrations over a 12-years' period. Zone I, averaging 0.03 ppm, had only 2 species of lichens; Zone II, with 0.02–0.03 ppm, had 3 species; Zone III, with 0.01–0.02 ppm, had 12 species present; and Zone IV, had 28 species. However, peak intensities of 8.1 ppm SO_2 were recorded in Zone I, and occasional peaks up to 1.3 ppm were observed even in Zone IV. Also, there were 257 hours during which SO_2 readings exceeded 1.0 ppm, mostly occurring in Zones I and II.

The above is in marked contrast to the tolerance of such species as *Lecanora dispersa*, which thrives in central London where mean annual SO_2 concentrations of 0.14 ppm prevail (Laundon, 1967).

The importance of atmospheric humidity, as well as of genetic differences, to a threshold of injury, was demonstrated in the field by Marsh & Nash (1979), who considered 159 lichen species, of which 58% were crustose forms, 38% were foliose, and the remaining 4% fruticose. The 8 most common species were exposed to 0.5 ppm SO_2 for 8 hours. All exhibited significant reductions in respiration, *Caloplaca trachyphylla, Dermatocarpon miniature,* and *Lecanora muralis*, being the most sensitive. Moistened lichens exhibited no reduction in respiration following 0.1 ppm exposure for 8 hours.

Fenton (1964) reported that even the most tolerant species, *Lacanora conizaeoides,* could not survive in Belfast, Ireland, when average SO_2 concentrations exceeded 0.03 ppm over a year. No peak SO_2 values or their durations were reported.

Rao & LeBlanc (1966) found that a 1-hour exposure to 5 ppm SO_2 at a high atmospheric humidity caused significant morphological and physiological changes in *Xanthoria parietina*.

Gilbert (1965) reported that no epiphytic lichens could survive in Newcastle if the SO_2 concentration was more than 0.10–0.15 ppm.

LeBlanc & Rao (1975) concluded that mean annual concentrations as low as 30 $\mu g/m^3$ (0.010 ppm) may injure sensitive species of such genera as *Usnea, Lobaria, Ramalina,* and *Cladonia*. It may be, though, that short bursts of high SO_2 concentrations are really most critical.

FACTORS UNDERLYING SENSITIVITY

Generally speaking, the more of the thallus of a lichen species that is exposed to the air, the more sensitive will that species be. Thus, the fruticose forms tend to be highly sensitive, the foliose to be next, while the crustose, with their thallus closely appressed to the substrate, are commonly the most tolerant.

Unlike higher plants, lichens absorb over their entire surface and can accumulate substances from water or air rather rapidly when moist (Tuominen & Jaakkola, 1973). This circumstance may contribute markedly to their sensitivity. Dry lichens, on the other hand, are relatively resistant to SO_2. Furthermore, the absence of stomata, which in higher plants are commonly closed during the night, renders lichens subject to SO_2 absorption through every hour of the day and night.

Another weak link in the stability of lichens may lie in the relationship of the two biotic components. Should the lichen partners be affected unevenly, the symbiosis may well be disrupted. D. C. Smith (1960) suggests moreover that the low metabolic rate of lichens limits their growth rate, making them responsive to any disruption. Their potential for accumulating toxic substances from much diluted solutions could also be important, as could their almost complete dependence on rainfall and ambient air for nutrients.

The efflux of potassium from lichens exposed to SO_2 has been shown to be another sensitive response (Thomassini *et al.*, 1977). This could alter the membrane permeability and interfere with electron and ion transport.

FLUORIDES

Pollution by fluoride is not as common as that by SO_2, but its impact on lichens can be equally devastating. The same decline in lichen richness and abundance that is found around SO_2 sources has been observed also near sources of fluoride, most noticeably near aluminium reduction plants (Gilbert, 1971) but also near phosphate plants (Takala *et al.*, 1978). Nash (1971) has also shown responses near a titanium plant. Doubtless there have been effects near ceramics factories and other fluoride sources, but these seem to have gone unreported.

In all cases, a decline in lichen communities has been related to fluoride accumulations. Transplants of healthy lichens into fluoride-affected areas have resulted in their demise. As with higher plants, the critical threshold for effects can be variable; but in general it seems to be in the accumulation-range of 20 to 80 ppm in the thallus depending on the species and environment (Gilbert, 1971; Nash, 1971).

Essentially the same methods of study described for SO_2 have been applied in fluoride situations: field observations, photography applied to distribution, mapping studies, and transplant experiments. But there have been few laboratory fumigation studies. In one such study, Nash (1971) exposed lichens to the relatively high concentration of 3.9 ppb (parts per billion) (4 $\mu g/m^3$) for 9 days. Thalli became chlorotic and bleached during the exposure period. Similarly, Comeau & LeBlanc (1972) exposed *Hypogymia physodes* to 13–130 ppb HF (11–106 $\mu g/m^3$). Exposure durations exceeding 36 hours caused chlorotic spots and curling the margins of the thallus.

As the lichen species found at the various areas that have been studied generally differ, it is not possible to rank the order of sensitivity as has been done for higher plants. However, Gilbert (1973*a*) has summarized the data available from different locations. These indicated several of the more sensitive species, but as neither the air nor lichen tissue fluoride concentrations were available, let alone both, no valid comparisons could be made. Further-more, it is most significant that individuals of the same species, even when growing side-by-side, can show marked differences—probably genetic—in sensitivity. Thus one individual could become bleached and moribund while others were completely unaffected (Gilbert, 1973*b*).

One of the most detailed ecological studies of the effects of fluorides on lichens was conducted by Donald Perkins and his colleagues at the Institute of Terrestrial Ecology in Bangor, Wales (Perkins *et al.*, 1980). They photo-graphed and mapped saxicolous lichen communities on stone walls, and cor-ticolous forms mostly on Hawthorn (*Crataegus monogyna*) and Blackthorn or Sloe (*Prunus spinosa*) in hedgerows, over an 8-years' period at distances up to 12 km from an aluminium plant near Holyhead in North Wales. Atmospheric and lichen fluoride concentrations were also determined. Before operations began, corticolous and saxicolous species of *Ramalina* contained an average of 9 and 16 ppm fluoride, respectively. After aluminium production began, in 1969, fluoride concentrations increased each year, reaching 241 ppm in *R. siliquosa* by 1977 within 2 km of the source and up to 27 ppm at 8 to 12 km distant. The greatest accumulation was at the nearest, most exposed sites. Accumulation was attributed to surface contamination, active uptake, ion exchange, and chelation.

In the same studies of Perkins *et al.* (1980), it was observed that the cor-ticolous species declined the most rapidly, suffering the greatest reduction in abundance. Their representation declined at sites where the thalli accumu-lated over 50 ppm fluoride. By 1975, 6 years after aluminium production began, five sites were entirely devoid of lichens; by 1977 the lichenless sites had increased to 10.

OZONE

Ozone has been recognized as a serious air pollutant affecting higher plants since the 1960s (e.g. Harward & Treshow, 1975; Treshow, 1980), but only since 1977 has its impact on lichens been recognized (Rosentreter & Ahmad-jian, 1977). However, these last authors found no decrease in chlorophyll concentration in *Cladonia arbuscula* or *C. stellaris* when exposed to 0.1 to 0.8 ppm ozone for a week. There still have been few studies of the effect of ozone on lichens, and none that we know of reporting the possible SO_2–O_3 interac-tions which are well known for higher plants (*see* preceding chapter).

Nash & Sigal (1979) chose *Parmelia sulcata* for study as it was absent from areas of high oxidants, and *Hypogymnia enteromorpha* as it showed marked signs of deterioration in the same areas. Specimens of the *Parmelia*, fumi-gated with ozone at 0.5 ppm, exhibited a significant decline after 3 hours of such exposure, whereas the *Hypogymnia* was inhibited only at 0.8 ppm. While there were no definite visible symptoms in either case, algal cells examined microscopically became yellow-green and amber, respectively.

One of the few field studies in which lichens were shown to be sensitive to oxidants was conducted by Nash & Sigal (1980). The Lichen populations in the San Bernardino mountains next to Los Angeles were compared with those in the Cuyama Rancho State Park (some 80–100 miles to the south and serving as a control region). An interesting comparison of the flora was made with that found in a study by H. E. Hasse (1913). Of the 91 lichen species reported in the earlier study, only 34 were found in the later investigation which was conducted over 3 summers. Lichen species on conifers and oaks (*Quercus* spp.) totalled 16 species in 1913 of which 8 still occurred on such trees; in the Cuyamas, 15 or 16 species remained. Significantly, the only nitrogen-fixing lichen present in 1913, *Collema nigrescens,* was completely absent in the late 1970s in the San Bernardino mountains. Other species had declined in abundance, but a few had apparently not been affected.

TRACE ELEMENTS

The vast quantities of SO_2 and other gaseous effluents that are emitted by smelters, have tended to draw attention away from particulate emissions, of which trace elements, including metals, are a major component (*see also* Chapter 13). Where attention has been focused on such matters, it has involved primarily the higher plants.

Studies of heavy-metal toxicity can best be determined after a facility has stopped operating. Since the particulates tend to be dispersed over only a few miles' (1 mile = 1.6 km) distance, restricted plant growth would be expected close to the site. But in such situations, restoration of new plant cover may be seriously impeded for many years after any SO_2 emissions have ceased.

Numerous studies have demonstrated the ability of lichens to accumulate trace elements, including heavy metals. These were reviewed by James (1973), by Brown (1976), and by Nash & Sigal (1980).

Lawrey & Hale (1979) found that lead, accumulating in lichens along highways at concentrations exceeding about 1,000 ppm, were toxic to the lichens. In a more detailed study, Nash (1972) demonstrated that species richness was reduced from 77 to 7 as a zinc smelter in eastern Pennsylvania was approached. Sulphur dioxide here had been controlled, but metal concentrations included 20–200 ppm zinc and 1–30 ppm cadmium. Physiological studies showed that cadmium was toxic to the lichens when 300–500 ppm accumulated in the thallus. Laboratory studies showed a reduced net assimilation rate in *Cladonia* and *Lasallia* at high cadmium or zinc levels.

Brown (1976) discussed the substantial absorption of lead, copper, zinc, nickel, and cobalt, by *Cladonia rangiformis,* to illustrate cation selectivity and competition that can be achieved by a passive uptake mechanism. Many lichens characteristically absorb inorganic cations far in excess of their requirements. James (1973) summarized previous work that showed lichens to accumulate zinc up to 10,000 ppm, chromium to 1,000 ppm, and nickel in

Parmelia centrifuga up to 600 ppm—all dependent on the rock substrate. Lichens may have a particular affinity for iron. Over 90,000 ppm was reported in *Peltigera rufescens* near steel smelters. The ability of lichens to accumulate such high concentrations make them especially valuable for fallout studies around smelters and other sources.

Considering such high accumulations, the tolerance would seem remarkable. O. L. Lange & I. Zeiger (in James, 1973) suggest that this may be due to (1) inherent cytoplasmic tolerance, (2) cytoplasmic immobilization and inactivation involving detoxification of cations by means of chemical combination, and/or (3) the deposition of cations in regions outside the cytoplasmic membrane (e.g. the occurrence of lead at sites in the hyphal walls).

Summary

The scarcity of lichens in polluted areas is a well-documented phenomenon that has led to a general conclusion that this group of organisms is among the most highly sensitive to air pollution. They have no waxy layer to protect them from environmental stress or water-loss, and their entire surface is continually exposed to any chemicals in the atmosphere. Other reasons for lichen sensitivity may lie in the relationship between the component organisms, their low metabolic rate, or the efflux of potassium from lichens exposed to SO_2 that could alter membrane permeability. The actual sensitivity of lichens, however, is difficult to assess, as reliable, short-term pollutant measurements are rarely available.

It appears that while epiphytic lichens in regions of high humidity and ample rainfall are highly sensitive, lichens as a group, including crustose species, may be no more sensitive than higher plants. Most quantitative studies indicate that the most sensitive species are impaired at short-term SO_2 concentrations of the order of 1300 $\mu g/m^3$. Studies with fluoride and ozone also indicate sensitivity thresholds comparable with those of flowering plants.

References

Ahmadjian, Vernon (1967). *The Lichen Symbiosis.* Blaisdell, Waltham, Massachusetts, USA: viii + 152 pp., illustr.

Ahmadjian, Vernon (1974). Lichen. *Encyclopaedia Britannica,* 15th edn, Vol. 10, pp. 882–8.

Ainsworth, G. C. (1971). *Ainsworth & Bisby's Dictionary of the Fungi,* 6th edn. Commonwealth Mycological Institute, Kew, Surrey, England, UK: 663 pp.

Anderson, F. K. & Treshow, M. (1978). Prediction of air pollution impact on lichen communities. Abstr. *3rd Internat. Congress of Plant Pathology,* Munich, 16–23 August 1978, 352 pp.

Barkman, J. J. (1958). *Phytosociology and Ecology of Cryptogamic Epiphytes,* Van Gorcum, Assen, The Netherlands: 658 pp.

Barkman, J. J. (1969). The influence of air pollution on bryophytes and lichens. Pp. 197–209 in *Air Pollution: Proceedings of the First European Congress on Influence of Air Pollution on Plants and Animals*. (Wageningen, The Netherlands). 1968. 419 pp., illustr.

Becquerel, P. (1948). Reviviscence du *Xanthoria parietina* desséché avec sa faune, six ans dans le vide et deux semaines a – 189°C: ses consequences biologiques. *Compt. Rend. Acad. Sci.*, **226**, pp. 1413–5.

Becquerel, P. (1951). La suspension de la vie des algues, lichens, mousses aux confins du zéro absolu et rôle de la synthèse réversible pour leur survie dégelé expliquant l'existence de la flore polaire et des hautes altitudes. *Compt. Rend. Acd. Sci.*, **232**, pp. 22–5.

Beetham, P. A. & Hedger, J. N. (1978). The effects of low concentrations of sulphur dioxide, *in vivo*, on epiphytic lichens Abst. *3rd Internat. Congress of Plant Pathology*, 16–23 August 1978. 352 pp.

Brodo, Irwin M. (1961). Transplant experiments with corticolous lichens using a new technique. *Ecology,* **42**(4), pp. 838–41.

Brodo, Irwin M. (1966). Lichen growth and cities: A study on Long Island, New York. *The Bryologist,* **69**(4), pp. 427–49.

Brodo, Irwin M. (1968). *The Lichens of Long Island, New York: A Vegetational and Floristic Analysis*. PhD thesis, Michigan State University, East Lansing, Michigan, USA.

Brown, D. H. (1976). Mineral uptake by lichens. Pp. 419–39 in *Lichenology Progress and Problems* (Ed. D. H. Brown, D. L. Hawksworth & R. H. Bailey). Academic Press, London, England, UK: xii + 551 pp.

Comeau, G. & LeBlanc, F. (1972). Influence de l'ozone et de l'anhydride sulfureux sur la régénération des feuilles de *Funaria hygrometrica* Hedw. *Natur. Canad.,* **98**(3), pp. 347–58.

Coppins, B. J. (1973). The drought hypothesis. Pp. 124–42 in *Air Pollution and Lichens* (Ed. B. W. Ferry, M. S. Baddeley & D. L. Hawksworth). University of Toronto Press, Toronto, Ontario, Canada: viii + 389 pp., illustr.

Degelius, G. (1964). Biological studies of epiphytic vegetation on twigs of *Fraxinus excelsior*. *Acta Horti Gotoburg.,* **27**, pp. 11–55.

Denison, W. C. (1973*b*). Life in tall trees. *Scientific American,* **228**(6), pp. 74–80.

DeSloover, Jacques & LeBlanc, Fabius (1968). Mapping of atmospheric pollution on the basis of lichen sensitivity. Pp. 42–56 in *Proc. Symp. Recent Adv. for Tropical Ecology* (Eds Misra, R. & Gopal, G.).

Feder, W. A. & Manning, W. J. (1979). Living plants as indicators and monitors. Ch. 9 (29 pp.) in *Handbook of Methodology for the Assessment of Air Pollution Effects on Vegetation*. Air Poll. Control Assoc. Info. Report No. 3, Pittsburg, Pennsylvania, USA: 380 pp.

Fenton, A. F. (1960). Lichens as indicators of atmospheric pollution. *Irish Natur. J.,* **13**, pp. 153–9.

Fenton, A. F. (1964). Atmospheric pollution of Belfast and its relationship to the lichen flora. *Irish Natur. J.,* **14**, pp. 237–245.

Ferry, B. W., Baddeley, M. S. & Hawksworth, D. L. (Ed.) (1973). *Air Pollution and Lichens* Athlone Press, London, England, UK: viii + 389 pp., illustr.

Gilbert, O. L. (1965). Lichens as indicators of air pollution in the Tyne Valley. Pp. 35–47 in *Ecology and the Industrial Society: 5th Symposium, British Ecological Society* (Ed. G. T. Goodman). Blackwell Scientific Publications, Oxford, England, UK.

Gilbert, O. L. (1969). The effect of SO_2 on lichens and bryophytes around Newcastle upon Tyne. Pp. 233–35 in *Air Pollution: Proceedings of the First European Congress on the Influence of Air Pollution on Plants and Animals*. (Wageningen, The Netherlands, 1968). 419 pp., illustr.

Gilbert, O. L. (1970). Further studies on the effect of sulphur dioxide on lichens and bryophytes. *New Phytol.*, **69**, pp. 605–27.

Gilbert, O. L. (1971). The effect of airborne fluorides on lichens. *The Lichenologist*, **5**, pp. 26–32.

Gilbert, O. L. (1973*a*). The effect of airborne fluorides. Pp. 176–91 in *Air Pollution and Lichens* (Ed. B. W. Ferry, M. S. Baddeley & D. L. Hawksworth). University of Toronto Press, Toronto Ontario, Canada. viii + 389 pp., illustr.

Gilbert, O. L. (1973*b*). Lichens and air pollution. Ch. 13 (pp. 443–72) in *The Lichens* (Ed. V. Ahmadjian & M. E. Hale). Academic Press, New York & London: xiv + 679 pp., illustr.

Gilbert, O. L. (1977). Lichen conservation in Britain. Ch. 11 (pp. 415–36) in *Ecology* (Ed. M. R. D. Seaward). Academic Press, London, England, UK: 550 pp., illustr.

Hale, Mason E., jr (1967). *The Biology of Lichens*. Edward Arnold, London, England, UK: vii + 176 pp., illustr.

Harris, G. P. (1972). The ecology of corticolous lichens. *J. Ecol.*, **60**, pp. 19–40.

Harward, M. & Treshow M. (1975). Impact of ozone on the growth and reproduction of understorey plants in the aspen zone of western USA. *Environmental Conservation*, **2**(1), pp. 17–23, 3 figs.

Hasse, H. E. (1913). The lichen flora of Southern California. *Contrib. US Nat. Herb.*, **17**, pp. 1–132.

Hawksworth, D. L. (1971). Lichens as litmus for air pollution: A historical review. *Internat. J. Environm. Stud.*, **1**, pp. 281–96.

Hawksworth, D. L. (1973). Mapping studies. Pp. 38–76 in *Air Pollution and Lichens* (Ed. B. W. Ferry, M. S. Baddeley & D. L. Hawksworth). Athlone Press, London, England, UK: viii + 389 pp., illustr.

Hawksworth, D. L. & Rose, F. (1970). Qualitative scale for estimating sulphur dioxide air pollution in England and Wales using epiphytic lichens. *Nature* (London), **227**, 145–8.

Hawksworth, D. L. & Rose, F. (1976). *Lichens as Pollution Monitors*. Edward Arnold, London, England, UK: 60 pp., illustr.

Jacques, H. E. (1958). *How to Know the Economic Plants*. William C. Brown, Dubuque, Iowa, USA: 174 pp.

James, P. W. (1973). The effect of air pollutants other than hydrogen fluoride and sulphur dioxide on lichens. Pp. 143–75 in *Air Pollution and Lichens* (Ed. B. W. Ferry, M. S. Baddeley & D. L. Hawksworth). University of Toronto Press, Toronto, Ontario, Canada: viii + 389 pp., illustr.

Johnson, I. & Søchting, U. (1976). Distribution of cryptogamic epiphytes in a Danish city in relation to air pollution and bark properties. *The Bryologist*, **79**, pp. 86–92.

Laundon, J.R. (1967). A study of the lichen flora of London. *The Lichenologist*, **3**, pp. 227–327.

Laundon, J. R. (1973). Urban lichen studies. Pp. 109–123, in *Air Pollution and Lichens* (Ed. B. W. Ferry, M. S. Baddeley & D. L. Hawksworth). University of Toronto Press, Toronto, Ontario, Canada: viii + 389 pp., illustr.

Lawrey, J. D. & Hale, M. E., jr (1979). Lichen growth responses to stress induced by automobile exhaust pollution. *Science*, **204**, pp. 423–4.

LeBlanc, Fabius (1961). Influence de l'atmosphère polluée des grandes agglomérations urbaines sur les épiphytes corticoles. *Rev. Canad. Biol.*, **20**(4), pp. 823–7.

LeBlanc, Fabius (1969). Epiphytes and air pollution. Pp. 211–21 in *Air Pollution: Proceedings of the First European Congress on the Influence of Air Pollution on Plants and Animals.* (Wageningen, The Netherlands, 1968): 419 pp., illustr.

LeBlanc, Fabius & DeSloover, Jacques (1970). Relation between industrialization and the distribution and growth of epiphytic lichens and mosses in Montreal. *Can. J. Bot.,* **48**(7), pp. 1485–96.

LeBlanc, Fabius & Rao, D. N. (1966). Réaction de quelques lichens et mousses épiphytiques a l'anhydride sulfureux dans la région de Sudbury, Ontario. *The Bryologist,* **69**(3), pp. 338–46.

LeBlanc, Fabius & Rao, D. N. (1973*a*). Evaluation of the pollution and drought hypotheses in relation to lichens and bryophytes in urban environments. *The Bryologist,* **76**(1), pp. 1–19.

LeBlanc, Fabius & Rao, D. N. (1973*b*). Effects of sulphur dioxide on lichen and moss transplants. *Ecology,* **54**(3), pp. 612–7.

LeBlanc, Fabius & Rao, D. N. (1975). Effects of air pollutants on lichens and bryophytes. Pp. 231–72 in *Responses of Plants to Air Pollutants* (Ed. B. H. Mudd & T. T. Kozlowski). Academic Press, New York, NY, USA: 383 pp.

LeBlanc, Fabius, Rao, D. N. & Comeau, Gilbert (1972*a*). The epiphytic vegetation of *Populus balsamifera* and its significance as an air pollution indicator in Sudbury, Ontario. *Can. J. of Bot.,* **50**(3), pp. 519–28.

LeBlanc, Fabius, Rao, D. N. & Comeau, Gilbert (1972*b*). Indices of atmospheric purity and fluoride pollution pattern in Arvida, Quebec. *Can. J. of Bot.,* **50**, pp. 991–8.

LeBlanc, Fabius, Robitaille, Gilles & Rao, D. N. (1974). Biological response of lichens and bryophytes to environmental pollution in the Murdochville Copper Mine Area, Quebec. *J. Hattori Bot. Lab.,* **38**, pp. 405–33.

Llano, G. A. (1944). Lichens: Their biological and economic significance. *Bot. Rev.,* **10**, pp. 1–65.

Llano, G. A. (1948). Economic uses of lichens. *Economic Botany,* **2**, pp. 15–45.

Llano, G. A. (1951). Economic uses of lichens. *Smithsonian Report for 1950,* pp. 385–422.

Lodge, J. P., Jr (1969). *The Smoake of London: Two Prophecies.* Maxwell Reprint Co., Elmswood, New York, NY, USA: xvi + 56 pp., illustr.

Marsh, J. E. & Nash, T. H., III (1979). Lichens in relation to the Four Corners power-plant in New Mexico. *The Bryologist,* **82**, pp. 20–8.

Nash, T. H., III (1971). Lichen sensitivity to hydrogen fluoride. *Bull. Torrey Bot. Club.,* **98**, pp. 103–6.

Nash, T. H., III (1972). Simplification of the Blue Mountain lichen communities near a zinc factory. *The Bryologist,* **75**, pp. 315–24.

Nash, Thomas H., III (1973). Sensitivity of lichens to sulfur dioxide. *The Bryologist,* **76**(3), pp. 333–9.

Nash, Thomas H., III (1974). Lichens of the Page environs as potential indicators of air pollution. *J. Ariz. Acad. Sci.,* **9**, pp. 97–101.

Nash, Thomas, H., III (1975). Influence of effluents from a zinc factory on lichens. *Ecol. Monogr.,* **45**, pp. 183–98.

Nash, Thomas, H., III (1976). Sensitivity of lichens to nitrogen dioxide fumigations. *The Bryologist,* **79**(1), pp. 103–6.

Nash, T. H., III & Sigal, L. L. (1979). Gross photosynthetic response of lichens to short-term ozone fumigations. *The Bryologist,* **82**, pp. 280–5.

Nash, T. H., III & Sigal, L. L. (1980). Sensitivity of lichens to air pollution with an emphasis on oxidant air pollutants. Pp. 117–24. in *Proc. Symp. on Effects of Air Pollutants on Mediterranean and Temperate Forest Ecosystem.* (Riverside, California, USA.) 256 pp.

Nylander, W. (1866). Les lichens du jardin du Luxembourg. *Bull. Soc. Bot. R.,* **13**, pp. 364–72.

Pearsen, Lorentz & Skye, Erik (1965). Air Pollution affects pattern of photosynthesis in *Parmelia sulcata,* a corticolous lichen. *Science,* **148**, 1600–2.

Perkins, D. F., Millar, R. O. & Neep, P. E. (1980). Accumulation of airborne fluoride by lichens in the vicinity of an aluminium reduction plant. *Environ. Pollut.,* **21**, pp. 155–68, illustr.

Puckett, K. J., Nieboer, E., Flora, W. P. & Richardson, D. H. S. (1973). Sulphur dioxide: Its effects of photosynthetic ^{14}C fixation in lichens and suggested mechanisms of phytotoxicity. *New Phytol.,* **72**, pp. 141–54.

Rao, D. N. & LeBlanc Fabius (1966). Effects of sulfur dioxide on the lichen Algae, with special reference to chlorophyll. *The Bryologist,* **69**(1), pp. 69–75.

Rao, D. N. & Leblanc, Fabius (1967). Influence of an iron-sintering plant on corticolous epiphytes in Wawa, Ontario. *The Bryologist,* **70**(2), pp. 141–57.

Richardson, D. H. S. (1975). *The Vanishing Lichens.* David & Charles, Newton Abbot, England, UK: 231 pp.

Robitaille, Gilles, LeBlanc, Fabius & Rao, D. N. (1977), Acid rain, a factor contributing to the paucity of epiphytic cryptogams in the vicinity of a copper smelter. *Rev. Bryol. Lichenol.,* **43**, pp. 53–66.

Rose, F. (1973). Detailed mapping in southeast England. Pp. 77–88 in *Air Pollution and Lichens* (Ed. B. W. Ferry, M. S. Baddeley & D. L.Hawksworth). Athlone Press, London, England, UK: viii + 389 pp., illustr.

Rosentreter, R. & Ahmadjian, V. (1977). Effect of ozone on the lichen *Cladonia arbuscula* and the *Trebouxia* phycobiont of *Cladonia stellaris. The Bryologist,* **80**, pp. 600–5.

Rydzak, J. (1959). Influence of small towns on the lichen vegetation, Part VII: Discussion and general conclusions. *Ann. Univ. Mariae Curie-Sklodowska (Lublin),* Sect. C, **13**, pp. 275–322.

Rydzak, J. (1969). Lichens as indicators of the ecological conditions of the habitat. *Ann. Univ. Mariae Curie-Sklodowska (Lublin),* Sect. C, **22**, pp. 131–64.

Saunders, Peter J. W. (1970). Air pollution in relation to lichens and Fungi. *Lichenologist,* **4**, pp. 337–49.

Schönbeck, Helfried (1969). Eine Methode zur Erfassung der biologishen Wirkung von Luftverunreinigungen durch transplantierte Flechten. *Staub-Reinhaltung de Luft,* **29**(1), pp. 14–8.

Sernander, R. (1962). *Stockholm Natur.* Uppsala: Almqvist & Wiksells.

Showman, Ray E. (1972). Residual effects of sulfur dioxide on the net photosynthetic and respiratory rates of lichen thalli and cultured lichen symbionts. *The Bryologist,* **75**(3), pp. 335–41.

Skye, E. (1958). Luftföroreningars inverkan på busk-och bladlavfloran Kring skifferoljeverket i Närkes Kvarnøtorp. *Svensk. Bot. Tidskr.,* **52**, pp. 133–90.

Skye, E. (1965). Botanical indications of air pollution. The plant cover of Sweden. *Acta Phytogeographica Suecica,* **50**, pp. 285–7 + bibliography.

Skye, E. (1968). Lichens and air pollution, a study of cryptogamic epiphytes and environment in the Stockholm region. *Acta Phytogeographica Suecica,* **52**, 123 pp.

Smith, Annie Lorrain (1921). *Lichens.* Cambridge University Press, London, England, UK: xxiii + 464 pp., illustr.

Smith, D. C. (1960). Studies in the physiology of lichens. *Ann.Botany,* **24**, pp. 52–62 and 172–99.

Smith, D. C. (1962). The biology of lichen thalli. *Biol. Rev., 37*, pp. 537–70.

Sundström, Karl-Ragnar & Hällgren, Jan-Erik (1973). Using lichens as physiological indicators of sulfurous pollutants. *Ambio, 2*, pp. 13–20.

Takala, K., Kauranen, P. & Olkkonen, H. (1978). Fluorine content of two lichen species in the vicinity of a fertilizer factory. *Ann. Bot. Fennici, 15*, pp. 158–66.

Thomassini, F. D., Lavoie, P., Puckett, K. J., Nieboer, E. & Richardson, D. H. S. (1977). The effect of time of exposure to sulphur dioxide on potassium ions from and photosynthesis in the lichen *Cladonia rangiferina* (L). Harm. *New Phytol., 79*, pp. 147–55.

Treshow, Michael (1980). Pollution effects on plant distribution. *Environmental Conservation, 7*(4), pp. 279–86, 5 figs.

Tuominen, Y. & Jaakkola, T. (1973). Absorption and accumulation of mineral elements. Pp. 185–224, in *The Lichens* (Eds V. Ahmadjian & M. E. Hale). Academic Press, New York & London: xiv + 697 pp., illustr.

Turk, R., Wirth, V. & Large, O. L. (1974). Carbon dioxide exchange measurements for determination of sulphur dioxide resistance of lichens. *Oecologia, 15*, pp. 33–64.

Vareschi, V. (1936). Die epiphytenvegetation von Zürich (Epixylenstudien II). *Ber. Schweiz. bot. ges., 46*, pp. 446, 488.

Vareschi, V. (1953). La influencia de los Bosques y Zagrebu, *Glasn. biol. Seke hru. prirodsl* Društ., ser IIB, *7*, pp. 99–100.

Air Pollution and Plant Life
Edited by M. Treshow
© 1984 John Wiley & Sons Ltd.

CHAPTER 13

Impact of Trace Element Pollution on Plants

DOUGLAS P. ORMROD

Department of Horticultural Science, University of Guelph, Guelph, Ontario N1G 2W1, Canada

TRACE ELEMENTS AS ATMOSPHERIC POLLUTANTS

The term 'trace element' is used in this chapter rather than 'heavy metal'—a term widely used but meaning different things to different scientists. 'Trace elements' are more comprehensive, including as they do both essential and non-essential, metallic and non-metallic, elements, although relatively few of the trace elements are of direct interest as pollutants. The degree of ambiguity and disagreement in element classification is such that a nomenclature system has been proposed which replaces the term heavy metals by a classification of metal ions as oxygen-seeking, nitrogen/sulphur-seeking and borderline, grouping the trace elements according to their binding preferences (Nieboer & Richardson, 1980). Such a classification is related to atomic properties and to the solution chemistry of metal ions.

The information in this chapter is concerned with airborne trace elements and their effects on plants. A few references from the vast literature on soil and water trace element pollution will be used only to illustrate some plant responses to atmospheric pollution that may occur, but for which the best current data are available only for soil or water pollution.

Most of the trace elements exist in the atmosphere in particulate form only, and are associated, mainly as metal oxides, with this fraction (Purves, 1977). A few are present in both gaseous and particulate forms (White & Turner, 1970). Airborne particles, whether solid, liquid, or solid and liquid, are referred to as aerosols. They range in size from *ca.* 0.0005 to 500 μm in diameter (Smith, 1976*b*), the smallest size being limited by coalescence caused by Brownian movement and the largest by gravitational fallout. Small aerosols are usually electrically charged and frequently attach themselves to other aerosols.

Aerosols in the 0.1 to 1 μm range frequently represent gases that have condensed to form non-volatile products. Aerosols are hygroscopic; their size and density vary widely with changing humidity, so their terminal velocity also varies, affecting their ability to impact. Aerosols are usually not formed of a single element or compound, but are, rather, a heterogeneous mixture of components (Lee *et al.*, 1972). It is therefore difficult to make accurate measurements of trace elements in aerosols (Zoller *et al.*, 1973).

The composition of aerosols is extremely varied; inorganic or organic, if the latter, then either viable or non-viable (Smith, 1976*b*). Inorganic aerosols usually contain numerous elements, the most common being Si, Ca, Al, and Fe. The trace element concentration in aerosols is related somewhat to particle size, with a concentration distribution among discrete size fractions. In one study, Fe had the highest concentration (3–5%) in aerosols of less than 3 μm diameter, followed closely by Pb (1–2%) (Lee *et al.*, 1972). For aerosols of which the components are less than or equal to 0.5 μm in diameter, Pb was highest (2–4%) and Fe markedly lower (less than 1%). Lead and vanadium were associated with submicrometre-sized particles, while particles with high Fe had the greatest median diameter; other trace elements were in between. Particle size, along with concentration and composition of the constituents, determines the extent of particle–particle and particle–gas interactions, as well as the formation of atmospheric precipitation (Lee & Goranson, 1972). Physicochemical conversions of pollutants occur within the atmosphere (Saunders, 1973). For example, SO_2 interacts with aerosols to produce ammonium sulphate, various metallic sulphates, and sulphuric acid.

Aerosols are removed from the atmosphere by wet deposition, dry deposition or sedimentation, diffusion, and impaction on solid objects. Wet deposition dominates in the removal of aerosols, while dry deposition is most important in the removal of gases from the atmosphere (Saunders, 1973). In

Table I   Zn, Pb, and Cd, in Oak and Elm leaves collected from side of tree exposed to smelter and from side of tree sheltered from smelter (ppm dry-weight basis).*

Species of	Zn	Pb	Cd
		Exposed	
Oak (*Quercus robur*)	6,200	6,800	52.5
Elm (*Ulmus glabra*)	1,600	875	10.0
		Sheltered	
Oak (*Quercus robur*)	7,000	7,000	50
Elm (*Ulmus glabra*)	2,150	1,170	12.5

* After Little & Martin (1972).

wet deposition, aerosols are removed from the atmosphere by rainfall. Aerosols greater than 0.1 μm in diameter are necessary for raindrop formation, which starts by condensation of such aerosols. Vegetation forms a layer of rough surfaces which intercept the aerosols and also cause turbulence. This turbulence results in changes in wind velocity that affect aerosol deposition.

Understanding impaction and, to a lesser extent, sedimentation on vegetation involves understanding the ways in which plant configurations affect turbulence and wind velocity profiles. Wind speed, and thus particle concentration, decreases rapidly from the edge of the vegetation canopy. Concentrations of trace elements are usually higher on the sheltered than on the exposed sides of trees, owing to greater deposition of aerosols from slower-moving air (Table I) (Little & Martin, 1972). Dochinger (1980) found that tree plantings were effective in intercepting suspended particles and reducing the dust content of the air. The relative contributions of aerodynamic effects on wind-speeds, and the apparent cleansing efficiency of trees, has not been determined.

SOURCES AND ROLES OF AIRBORNE TRACE ELEMENTS

The trace elements include a large number of elements found in small amounts in the environment. Only a few of these can cause plant injury under certain environmental conditions, thus earning a designation as trace element pollutants. The number of such elements that are known to be injurious as atmospheric pollutants is even smaller. Most information on trace element toxicity to plants stems from studies involving metal ore deposits, mining, smelting, and other industrial operations, or pesticide use (Smith, 1973). Emissions from coal-fired power plants may also contribute to the trace element content of vegetation in the vicinity (Wangen & Turner, 1980). Trace element contamination of plants on roadsides and in urban areas has also been studied extensively.

Aerosols containing trace elements arise predominantly from industrial activities, coal and petroleum combustion, chemical reactions between trace gases, and sea spray and mineral dusts raised from the ocean and land by winds (Lee *et al.*, 1972; Pierson *et al.*, 1973). Human activity is the primary source of such elements as Pb, Zn, Cu, Fe, Mn, and Ni, in atmospheric pollution (Lazrus *et al.*, 1970). Analysis of industrial smokes reveals that they are significant potential sources of atmospheric contamination by As, At, Cu, F, Pb, Mn, and Ni (Bowen, 1977). (At is astatine, a radioactive element of short life that can only be made artificially. It is the heaviest element of the halogen group and was formerly known as Alabamine and Anglo-Helvetium.)

Aerosols may also originate from industrial activity indirectly through the wind-blowing of contaminated soils and wastes from former industrial sites

(Goodman & Roberts, 1971). Aerosols may also descend from sources beyond the atmosphere and arise from the eruption of volcanoes. While many of the trace element aerosols in the atmosphere may arise from human activity, environmental pollution is a minor factor in the soil distribution of trace elements compared with the role of parent material, soil profiles, soil management, and climate (Lagerwerff, 1967). Trace element contamination of soil from industrial sources of atmospheric pollution can lead to accumulation of contaminant elements in the surface horizon, because many trace elements tend to remain near the surface after deposition (Purves, 1977).

Aerosol particles larger than 10 μm in diameter may result from natural and mechanical processes such as wind erosion, grinding, or spraying (Pierson *et al.*, 1973; Smith, 1976*b*). Soil particles, process dust, industrial combustion particles, and marine salt particles are generally between 1 and 10 μm in diameter. Usually, industrial processes involving combustion will yield small particles (less than 1 μm diameter) formed by condensation.

Aerosol concentrations vary with the activity and distance of the source, the height of sampling, the meteorological conditions prevailing, the distance from 'sinks', and the size of the particles involved (White & Turner, 1970). Larger particles return rapidly to Earth by sedimentation, resulting in less widespread dispersion than of smaller particles, which may be raised by updrafts and turbulence to higher altitudes, so that their dispersal is more widespread (Pierson *et al.*, 1973). Such widespread dispersal is illustrated by a study in which substantial concentrations of trace elements in bulk rainfall in a rural mountainous area were a result of the continental downwind position of the area (Schlesinger *et al.*, 1974). Trace elements of artificial origin have low deposition velocities, corresponding to their small diameters. Deposition velocities for Zn, Pb, and Cd aerosols on grass have been estimated from uptake rates and atmospheric concentrations (Schwela, 1979). Dispersion and deposition patterns are influenced by stack height, wind velocities, meteorological stability, rainfall and terrain (Evans *et al.*, 1980). Where electrostatic precipitators and other ash collectors are installed, the degree of contamination by trace elements will depend on the effectiveness of ash disposal practices. Prevailing wind conditions strongly affect the distribution and deposition of airborne trace metals on vegetation (Little & Martin, 1972).

The eventual removal of trace elements by precipitation in rainfall or gravitational fallout to vegetation, soil, and water surfaces, tends to result in restoration of an unpolluted state by natural processes (Purves, 1977). The atmosphere provides the route through which trace elements, discharged in smoke and fumes, contaminate vegetation and other terrestrial surfaces.

Specific comments on some of the trace elements of concern in vegetation studies follow. For more detailed descriptions of sources and roles of the major trace elements, the reader is urged to consult reviews such as that on lead, mercury, and cadmium (Lagerwerff, 1972), and monographs on individual elements such as those prepared by the National Academy of Sciences (1972–77).

Aluminium: Knowledge of Al toxicity is based on studies making use of excesses in nutrient solutions. There have been no recorded studies of pertinent aerosol effects, even though some particulates are known to have substantial Al content. Aluminium toxicity is manifested by restricted and abnormal root growth, leaf margin chlorosis, and defoliation (Ormrod, 1978).

Arsenic: Vegetation in the vicinity of secondary Pb smelters may be contaminated with As, because small amounts of As are alloyed with Pb in the production of battery plates and grids (Temple *et al.*, 1977). This source of As did not result in above normal As concentration in the edible parts of vegetables grown near a smelter, even though other vegetation was contaminated. Other injurious effects have been identified in soil studies which have indicated that elevated As is associated with needle abscission and death of fine roots in certain conifers, stunting of plant growth, and sparse mycorrhizal development (Ormrod, 1978).

Cadmium: Cadmium is a medical as well as vegetational problem because it causes cardiovascular and hypertension disorders in man (Fassett, 1975). Atmospheric Cd sources are primarily metal-processing, or impurities in Zn-base oil and rubber additives that are released when diesel and heating oils are burned or vehicle tyres wear (Lagerwerff, 1971; Haghiri, 1973). All industrial processes involving 'technical' Zn involve Cd, as both occur in the raw materials. Roadside vegetation concentrations of Cd decrease away from traffic, owing to fallout of the larger particles nearer the roadway (Lagerwerff & Specht, 1970). Phosphate fertilizers and pesticides may also contribute significantly, making Cd a major environmental contaminant (M. B. Bazzaz & Govindjee, 1974). Cadmium-containing aerosol particles settle with dust and precipitation. Their distribution is strongly affected by prevailing wind conditions (Little & Martin, 1972). Deposited on vegetation or soil, the Cd enters the food chain by uptake through the roots or by foliar absorption or adherence (Lagerwerff, 1967).

Cadmium toxicity symptoms have been determined in mineral nutrition experiments and include reduced plant growth and chlorosis (Ormrod, 1978). Photosynthesis and transpiration rates are reduced (F. A. Bazzaz *et al.*, 1974*a*, 1974*b*), though spray applications of Cd to shoots of Pin Oak (*Quercus palustris*) caused no overt foliar or root symptoms even at a 100 ppm rate, which resulted in approximately 200 ppm Cd in the foliage and 50 ppm in twigs (Russo & Brennan, 1979). Foliar injury by Cd apparently depends on root injury, suggesting that Cd interferes with the uptake of some essential elements.

Chromium: There has been little study of this element in the present context, though toxicity levels in nutrient solution have been established and some injurious effects noted (Ormrod, 1978).

Copper: Even though Cu is an important air contaminant, most studies of toxicity effects have been in nutrient or soil culture. An essential element for plants, Cu at toxic levels causes growth retardation. High levels of Cu have been found in vegetation growing near to Cu or Ni—Cu smelters (Hutchinson & Whitby, 1974). Species differences have been demonstrated in threshold concentrations for injury (Beavington, 1975a). Copper levels in vegetation may also be elevated in urban areas (Purves, 1967; Burton & John, 1977).

Lead: There appears to have been more research conducted on sources of atmospheric Pb than on those of all other trace elements put together. Of the many trace elements released by human activity, Pb is considered the one most likely to approach human health hazard limits (Lazrus *et al.*, 1970). The major sources of Pb contamination are internal-combustion engines burning leaded gasoline, metal smelting plants, lead arsenate pesticides, phosphate fertilizers, Pb-based paints, and spent Pb shot from hunting activities (Goldsmith *et al.*, 1976). Fallout of particles of leaded paint may constitute a contamination hazard in greenhouses (Lagerwerff *et al.*, 1973).

The isotopic composition of Pb aerosols is similar to that of Pb additives isolated from gasoline, which are the largest contributors to atmospheric Pb pollution (Chow & Earl, 1970). About 75% of the metallic Pb in gasoline is emitted in automobile exhaust in the form of Pb aerosols (Smith, 1971). The combustion by motor vehicles of Pb alkyl derivatives in gasoline has resulted in extensive contamination near roadways (Cannon & Bowles, 1962; Chow, 1970; Motto *et al.*, 1970; Page & Ganje, 1970; Tyler, 1972). The advent of Pb-free gasoline should ultimately alleviate the Pb pollution problem. The increased Pb burden of plants, largely as surface deposition in the vicinity of roadways, may be 5–20, 50–200, and even 100–200 times baseline Pb levels for unwashed crops, grass, and trees, respectively (Smith, 1976a).

Numerous studies have been conducted to document the nature and distribution of Pb pollution arising from burning leaded gasoline. The composition of Pb aerosols from exhaust gas changes from $PbBr_2$, PbBrCl, $Pb(OH)Br$, $(PbO)_2PbBr_2$, and $(PbO)_2PbBrCl$ initially, to carbonates, oxycarbonates, and oxides, apparently by a simple acid–base reaction, with CO_3^{2-} and OH^- displacing the Br- and, in some cases, the Cl-ion from the Pb halides initially in the exhaust (Haar & Bayard, 1971). Reactions of the insoluble oxides with atmospheric SO_2 may maintain the Pb as the relatively insoluble sulphate (Lazrus *et al.*, 1970), but oxidation cannot be ruled out completely. The particles of inorganic Pb salts commonly range from 1 to 5 μm in diameter (Smith, 1971, 1976a). The Pb content of roadside atmospheres is elevated above non-roadside ones by these aerosols. The amount of Pb in the air is related to traffic volume, proximity to the highway, engine acceleration (*versus* constant speed), and wind direction (Daines *et al.*, 1970). The gravitational sedimentation and impaction on vegetation of coarse fractions is responsible for the high Pb contamination of vegetation and soils within about 35 m of roadways (Chow, 1970; Smith, 1971).

Areas of greater traffic density have vegetation with higher Pb levels (Table II) than others (Goldsmith *et al.*, 1976; Smith, 1976a; Wheeler & Rolfe, 1979), and the downwind side of the highway has the highest air and vegetation concentration of lead particulates (Schuck & Locke, 1970). The Pb content of vegetation decreases rapidly as distance from the highway increases (Smith, 1976a), the decreases in Pb levels in the vegetation being approximately an exponential function of distance (Ward *et al.*, 1975). Evaluation of this pattern has been extended to the development of a double exponential function for Pb distribution (Wheeler & Rolfe, 1979). The first exponent is associated with the larger particles that settle out rapidly (Chow & Earl, 1970) within about 5 m of the highway, and the second with smaller particles that settle out more slowly, within 100 m of the highway. The Pb contained in these smaller particles may be more soluble and thus have a greater effect on plants. In addition, there is a third fraction that does not decrease nearly as rapidly. The smallest Pb aerosols travel long distances in

Table II   The distribution of Pb in vegetation along highways of central Illinois (ppm Pb).*

Distance from highway (m)	Vehicles/day			
	550	1,500	2,300	8,100
0.3	20	29	32	196
1	14	19	22	76
5	12	17	19	46
10	11	15	16	30
15	11	12	14	26
20	10	11	13	21
25	9	10	12	17
30	9	9	11	16
40	8	10	11	13
50	8	9	9	10
100	8	9	9	10
200	8	9	9	9

* After Wheeler & Rolfe (1979).

global air currents (Chow, 1970). This persistent fraction is considered to be a significant portion of the exhaust-emitted Pb (Ward *et al.*, 1975).

Tree sampling analysis (leaves, bark, and trunk cores) indicated that the direction of the prevailing wind influences the distribution of emitted Pb. Also, lead levels are higher on the sides of trees facing the traffic (Smith, 1971; Ward *et al.*, 1975). Levels of Pb in leaves of Perennial Rye-grass (*Lolium perenne*) and Radish (*Raphanus sativus*) were related to atmospheric Pb concentrations in the vicinity of highways, but Pb in Radish roots was unrelated to Pb in the air (Table III) (Dedolph *et al.*, 1970). Grasses growing adjacent to highways have greater Pb concentrations than those living away from roadways (Chow, 1970), as do leafy vegetables (Preer *et al.*, 1980) and trees (Smith, 1971). Forage grasses may contain enough Pb to pose a threat to the health of grazing animals (Graham & Kalman, 1974).

The Pb concentration of urban vegetation may be high and even exceed that of vegetation on primary highways (Smith, 1973). Lead contamination may also be high in industrial areas. Lead contents of vegetables and fruit were 10 to 100 times normal in the vicinity of a Pb-smelting plant (Auermann *et al.*, 1976).

Much of the Pb contamination of vegetation is present as removable surface dust (Motto *et al.*, 1970). The abundance, distribution, size, and chemistry of Pb-containing particulates on leaves has been studied (Koslow *et al.*, 1977). Lead particles were infrequently encountered among the many particles associated with leaf surfaces. They were generally in either microcrystalline or crystalline microaggregate arrangements with their maximum dimension 1.0 to 25 μm. The presence of sulphates and phosphates rather than halides, and the size distribution, suggest that alterations in Pb particle chemistry and aggregation may occur in the atmosphere or on the leaf.

Table III   Lead concentrations in radish leaves and roots grown under different atmospheric lead environments.*

Distance from road (m)	Tissue of	Pb concentration		
		Soil (μg/g dry wt)	Atmospheric (μg/m^3)	Tissue (μg/g dry wt)
12	Leaves	82	2.32	16.4 a[†]
	Roots			0.7 d
36	Leaves	45	1.71	10.8 b
	Roots			0.8 d
155	Leaves	27	1.07	4.8 c
	Roots			0.8 d

[†] Means followed by the same letter are not significantly different.
* After Dedolph *et al.* (1970).

Mercury: Injury from this element appears as necrotic spots in the interveinal areas on the underside of leaves (Waldron & Terry, 1975). Mercury is released into air by diffusion from soil, transpiration and decay of vegetation, and heating processes (Kothny, 1973). Most Hg gas is absorbed into aerosols, which are ultimately removed from the air by dry and wet deposition. Most atmospheric Hg in urban areas is contributed by processes, involving heat, which vapourize Hg contained in fuel and raw materials. Sources include steel plants, ore smelters, power plants, ceramic and cement manufacturing, and space heating. Natural sources include vegetation, soil, natural steam wells, and forest fires. Agriculture contributes through Hg-containing pesticide sprays and slow release of residual Hg. Mercury compounds in the soil are reduced to elemental Hg which is liberated as vapour (Waldron & Terry, 1975). Mercurial pesticides are also reduced in the soil, particularly if there is a high content of organic matter. Release of Hg vapour from soil into a closed greenhouse could have direct toxic effects on plants.

Nickel: The use of Ni-containing gasoline, and abrasion of Ni-containing automobile parts, may explain the detectable gradient of Ni at sites near dense traffic. Tissue Ni concentrations and toxicity symptoms of several species have been described (Table IV), as well as the effects of environmental factors on toxicity (Hunter & Vergnano, 1952).

Selenium: Selenium is an essential element for animals but not plants. The primary sources of Se in the general environment are volcanic eruptions and metallic sulphides associated with burning (Lakin, 1973). Secondary sources

Table IV   Distribution of nickel in moderately and severely affected Oat plants at the flowering stage.*

Plant Part	Ni conc. (ppm)	
	Moderately affected	Severely affected
Old leaf-blades	57	199
Fully expanded leaf-blades	54	162
Young leaf-blades	87	159
Old leaf-sheaths	21	83
Fully expanded leaf-sheaths	16	81
Young leaf-sheaths	28	101
Lower third of stems	18	36
Middle third of stems	22	68
Upper third of stems	74	168
Flowers	89	151
Peduncles	32	108

* After Hunter & Vergnano (1952).

include volatilization from biological sinks in which it has accumulated. The burning of coal and petroleum gives rise to aerosols containing Se and SeO$_2$.

Vanadium: The major natural sources of V are marine aerosols produced by bubbles bursting at sea-surface and dust from wind erosion of rocks and soil (Zoller *et al.*, 1973). Vanadium in the general environment also originates from man's activities, fuel combustion being the most likely source; as a result there may be significant concentrations of V in plant tissues.

Trace element mixtures: In many studies the trace elements have been considered as groups rather than individually. Those arising from industrial activity and from combustion of fossil fuels include substantial quantities of Zn, Cu, Cd, Cr, Ni and V (Tyler, 1972). Contamination of vegetation by Pb, Cd, Cu, and Ni, in industrial regions is mainly by aerial deposition, with motor vehicles the primary source (Burton & John, 1977). Pb, Cd, Ni, and Zn, are found in gasoline, motor oil, and car tyres, with a consequence that concentrations of these elements are elevated in vegetation samples from roadsides and decrease with distance from traffic (Lagerwerff *et al.*, 1973). Concentration depends strongly on the plant part considered, garden fruits being lower in trace elements than leafy or root vegetables (Pierson *et al.*, 1973). Species differences among leafy vegetables have also been observed. Contamination could be high enough to affect N mineralization in the soil, and could also have adverse effects on grazing animals (Table V).

Pb, Cd, Ni, and Zn, are easily volatilized at the temperature of common industrial processes; consequently they will usually be present in the atmosphere, especially near urban and industrial areas (Lagerwerff & Specht, 1970). Marked enhancement of Cu, B, and Pb in soils of urban areas, whether small towns or large urban–industrial areas, suggests the general enhancement of trace element contents of garden plants in built-up areas (Purves, 1967). The atmosphere also contains Cd, Pb, and Zn as radioisotopes from nuclear detonations. They can all reach the plant cover in precipitation as well as by dry deposition of aerosols. Unwashed leaves of White Clover (*Trifolium repens*) and Paspalum (*Paspalum dilatatum*), grown in urban areas, had high Zn and Cu, which correlated with proximity to a steel works (Beavington, 1975*a*). Plants in urban parkland contain higher levels of Pb and Zn than those in rural areas (Purves, 1977).

Analyses of rain water may reveal the nature of trace element pollution. Industrial regions may receive regular aerial additions of V, Mn, Cr, Ni, Co, and Mo that must be taken into account in the evaluation of trace element balance in the area (Hallsworth & Adams, 1973). Analyses of city trees may reveal patterns of trace element pollution (Table VI). In one urban area, tree analyses indicated that Li, Cr, Ni, Fe, Pb, Na, and Zn were all present in above normal amounts (Smith, 1973). Woody plants in a city may have ele-

Table V   Trace element content of hay, grass, and soft tissues of horses, from trace element-affected and control sites.*

	Zn	Pb	Cd	Cu	Ni
Affected hay (ppm dry weight)	166	86	7.6	5.7	5.5
Affected grass (ppm dry weight)	360	105	9.9	14.3	13.8
Control grass (ppm dry weight)	58	18	1	6	5
Kidney (ppm wet weight)					
Affected horse	78	40	330		
Control horse	25	3.1	35		
Liver (ppm wet weight)					
Affected horse	143	12	7.5		
Control horse	34	9	1.6		
Lung (ppm wet weight)					
Affected horse	25	1.6	0.3		
Control horse	10	1.3	< 0.1		

* After Goodman & Roberts (1971).

vated Pb and Hg contents (Smith, 1972). Others have found Zn, Cu, Pb, and Cd, accumulation to high levels in urban soils, with concentrations above thresholds causing inhibition of growth (Beavington, 1973). Concentrations were sufficient for domestic vegetable gardens in the area to be regarded as possible health hazards. Aerial contamination accounted for more than 40% of the Pb, Cd, and Zn contents of Radish tops grown near a busy highway, and only Zn was translocated to the roots to a significant extent (Lagerwerff, 1967).

Metal oxides escaping from Zn smelters have contaminated vegetation with Zn, Cd, Cu, and Pb, with concentrations of Zn and Cd far above their thresholds for injury (Buchauer, 1973). High levels of Cu, Zn, Pb, Cd, Ni, and Fe have been reported in leafy vegetables grown in domestic gardens in close proximity to a Cu smelter (Beavington, 1975b). Analyses of vegetation in the vicinity of a Ni–Cu smelter revealed excessive levels of Ni, Cu, and Al in the foliage (Hutchinson & Whitby, 1974). Dustfall–rainfall collections and analyses corroborated the source of these contaminants as aerial.

A diverse trace element mixture resulting from the combustion of coal is discharged into the atmosphere as vapour or small aerosols (Klein *et al.*,

Table VI   Average and maximum trace element contamination of washed leaves of three deciduous tree species in New Haven, Connecticut, USA.*

	Leaf contamination (µg/g dry wt)	
	Average	Maximum
Present in 'normal amounts'		
Cadmium	1.5	2.2
Copper	8.7	18
Manganese	469	1,311
Present in 'slightly above normal amounts'		
Aluminium	503	783
Chromium	2.8	7.4
Nickel	10	19
Present in 'above normal amounts'		
Iron	404	791
Lead	110	275
Sodium	373	927
Zinc	130	265

* After Smith (1976*b*).

1975). The mass of aerosols is drastically reduced by precipitators, but As, Br, Cd, Cl, Hg, Pb, Sb, Se, and Zn are substantially enriched in the surrounding atmosphere. Of these, Br, Cl, and Hg are probably discharged as gases.

SORPTION OF AIRBORNE TRACE ELEMENTS BY VEGETATION

Various relationships do exist among trace elements, and sorption and accumulation patterns often exhibit similarities. Much information may be obtained by studying trace elements along with major elements (Tyler, 1972). Plant surfaces become contaminated with aerosols containing numerous trace elements (Smith *et al.*, 1978), and leaf pathogen growth can be affected (Table VII). These aerosols adhere to the plant, and some of the trace elements may be absorbed (Haghiri, 1973). Lead remains largely as a superficial deposit or topical aerosol coating on plant surfaces (Schuck & Locke, 1970; Zimdahl, 1976), while Zn and Cd penetrate at least partially into the leaf (Little & Martin, 1972). Foliar uptake is generally limited, with greater uptake by roots from soil than by leaves exposed to simulated wet deposition (Haar *et al.*, 1969). Uptake from soil may be further enhanced by the addition of chelates (Marten & Hammond, 1966).

Foliar uptake of Pb has had particularly thorough study (Zimdahl, 1976). If Pb salts are impacted on the plant surface in an insoluble form, they probably remain on the outside of the epidermis (Smith, 1971). If they are deposited in

Table VII   Linear growth (mm) following 8 days' incubation at 24°C of the parasitic leaf surface fungus *Gnomonia platani* on potato dextrose agar plates amended with trace element nitrate salts.*

Dose†	Trace element					
	Al	Fe	Mn	Ni	Pb	Zn
0	44.3 ± 0.7	38.7 ± 7.0	41.5 ± 0.6	35.3 ± 0.7	33.0 ± 0.7	44.8 ± 0.5
1	43.5 ± 0.9	34.0 ± 0.1	39.0 ± 0.8	39.5 ± 0.8	39.3 ± 0.5	43.3 ± 0.6
2	0	35.7 ± 0.6	39.0 ± 0.5	36.5 ± 0.2	38.7 ± 0.6	40.8 ± 0.9
3	0	0	36.0 ± 0.4	36.5 ± 0.2	38.2 ± 0.3	41.5 ± 0.7
4	0	0	12.7 ± 0.8	30.8 ± 0.5	0	8.5 ± 0.8
5	0	0	0	30.7 ± 0.9	0	0

* After Smith *et al.* (1978).
† 0 = control; 1 = 1/10x; 2 = 1/5x; 3 = 1x; 4 = 5x; 5 = 10x leaf burden.

a soluble form (e.g. PbClBr), or are rendered soluble after impact, they are probably taken into the plant through stomata and other openings. Only extremely small amounts of Pb can penetrate the cuticle, even after extended exposure (Arvik & Zimdahl, 1974). Removal of waxes permits increased penetration of some cuticles, and penetration seems more related to the species of plant than to the thickness of its cuticle. Environmental variables and plant age, as well as species, are important determinants of the amount of foliar uptake. Species differences may be due to cuticle morphology, wax stratification, number and distribution of ectodesmata, ion-binding sites, and number or density of trichomes (Krause & Kaiser, 1977).

The fruiting and flowering parts of plants contain the smallest concentrations of Pb of any parts, and are the least affected by changes in the amounts of Pb supplied (Motto *et al.*, 1970). Lead in roots and new shoot growth is also little affected by Pb aerosol application (Arvik & Zimdahl, 1974). In one study, the edible portion of Carrots (*Daucus carota*), Maize = Corn (*Zea mays*), Potatoes (*Solanum tuberosum*), and Tomatoes (*Lycopersicum esculentum*), contained the least Pb of all parts, but the leaves of Lettuce (*Lactuca sativa*) contained the most (Motto *et al.*, 1970). In another study, Pb was found only in the outer leaves of Cabbage (*Brassica oleracea* var. *capitata*) but there was no detectable Pb in the interior of the cabbage head (Schuck & Locke, 1970).

The ability of washing procedures to remove trace elements from leaves extraction procedures were used on leaves contaminated with aerial fallout of Zn, Pb, and Cd (Little, 1973). Large proportions were removable, but there was considerable variation among species. Leaf cuticle structure, foliar absorption, and uptake from soil, all have a role. Generally, at least 50% of the Pb can be removed by simply washing with water (Schuck & Locke, 1970), although 80% of the Pb was removed from tomato fruit surfaces. A chloro-form rinse or mild acid washes removed all the Pb that had been added to plants as PbBrCl aerosols, indicating that even this form of Pb remained as a coating on leaves (Arvik & Zimdahl, 1974).

In another study, leaves and twigs of Bramble (*Rubus fruticosus* ssp.) and Rhododendron (*Rhododendron ponticum*) were collected beside a major road and washed in a variety of chemicals, of which the most efficient proved to be 2NH$_4$-EDTA (Flanagan *et al.*, 1980). Large amounts of Pb and Zn had accumulated on the leaf surfaces, and Bramble leaves, which are pubescent, collected much more than Rhododendron leaves, which are smooth and waxy (Table VIII). Unwashed tree foliage proved to have from 2 to 6 times as much Zn and Cd as leaves that had been thoroughly washed prior to analysis (Buchauer, 1973).

The mechanism of absorption of trace elements, and their subsequent translocation, will determine their ultimate distribution within the plant. Zn, Cu, Mn, Fe, and Mo are intermediate between extremes in their capabilities for absorption and mobility, with decreasing mobility in the order given (Bukavoc & Wittwer, 1957). Trace element aerosols may enter leaves, to some extent, directly through the stomata (Buchauer, 1973). Many aerosol particles of industrial origin are less than 1 μm in diameter, and stomatal openings range from 5 to 30 μm, so that entry through them is possible.

The mechanisms involved in bringing trace elements into solution on leaf surfaces are not well known. Some of the trace elements, in the form of relatively insoluble oxides, may be absorbed through the leaf surface in solution (Haghiri, 1973; Krause & Kaiser, 1977). Uptake is independent of con-current SO$_2$ exposure, though CO$_2$-water films, or excretion or leaching of organic cell material, as well as guttation, could be involved (Krause & Kaiser, 1977).

Table VIII   Lead and zinc values of unwashed leaves and twigs (μg per g dry weight).*

	Bramble		Rhododendron	
	5 m	50 m from road	5 m	150 m from road
		Leaves		
Pb	102.3 ± 13.7†	15.9 ± 10.9	18.8 ± 8.7	25.5 ± 4.8
Zn	81.3 ± 11.8	44.9 ± 5.9	38.8 ± 16.5	17.4 ± 6.5
		Twigs		
Pb	26.4 ± 9.8	17.7 ± 9.8	32.1 ± 7.5	8.0 ± 2.3
Zn	84.7 ± 18.8	81.3 ± 10.4	47.8 ± 5.5	24.3 ± 2.9

* After Flanagan *et al.* (1980).
† Mean values ± standard deviation.

Under natural conditions, deposits of trace element aerosols on the surfaces of leaves may be removed by rain, wind, sloughing of cuticle wax, or absorption by the leaf and subsequent translocation (Little, 1973). The removal by rain washing of trace elements adhering to leaves, will depend on the nature of the leaf surfaces and characteristics of the trace elements (Haghiri, 1973). The aerosols may also be splashed onto other leaves or plants, or onto the soil, either in the dissolved state or as particulates.

INTERACTIONS OF TRACE ELEMENTS

In a complex system of interacting trace elements, each single element has to be considered alone and in combination if we are to estimate its contribution to vegetation injury (Krause & Kaiser, 1977). Trace elements in a mixture may interact to give a different plant response when compared with the single elements, but there has been little research on this aspect of trace element atmospheric pollution. A Ni–Cu interaction has, however, been identified in the appearance of necrotic injury symptoms (Hunter & Vergnano, 1952). A combination of Cd and Cu in plant tissue causes growth retardation, but it is not known whether combined effects are different from single effects (Krause & Kaiser, 1977).

Microorganisms on leaves are exposed to mixtures of trace elements from the atmosphere, and they may be particularly vulnerable to some of these elements (Smith *et al.*, 1978). Trace element relationships with foliar pathogens, or with microbes that may influence foliar pathogens, are important in affecting the occurrence of plant disease. Growth inhibition by (or as a result of the absence of) trace elements in Nature may alter competitive abilities or population structures of plant surface microbes (Smith, 1977). Six trace elements were evaluated in dose–response tests on fungi isolated from natural leaves (Smith *et al.*, 1978). Of Al, Fe, Mn, and Ni, only Al and Fe were fungitoxic—but only at a high dosage, so that direct inhibition of leaf-surface fungal growth by trace element foliar contamination appears unlikely.

Other evidence for interactions of trace elements in affecting microorganisms comes from studies of decomposition of forest litter in the vicinity of metal-processing industries emitting Cu, Zn, Cd, Ni, or Pb (Rühling & Tyler, 1973; Coughtrey *et al.*, 1979; Freedman & Hutchinson, 1980). The general decomposition rates of surface litter in certain size ranges at acidic forest soil sites are depressed by moderate concentrations of trace elements if there is an adequate water supply.

INTERACTION OF TRACE ELEMENTS WITH OTHER ATMOSPHERIC POLLUTANTS

The presence of elevated concentrations of trace elements in plant tissue may alter the plant's response to gaseous pollutants. Ozone toxicity to Garden or

Table IX   Ozone-induced leaf injury on pea plants (*Pisum sativum* cultivars) grown
in nutrient solution containing NiSO$_4$ or CdSO$_4$.*

Pea cultivar	NiSO$_4$			CdSO$_4$			
	0	10	100 mol	0	1	10	100 mol
Laxton's Progress	2.3† a‡	2.3 ab	2.8 abc	2.3 b	2.6 bc	2.5 bc	0.6 a§
Dark skin Perfection	2.6 ab	3.2 bc	3.6 c	3.0 c	3.6 cd	3.8 d	0.3 a

* After Ormrod (1977).
† Injury 4 days after 50 pphm O$_3$ for 6 h using 0 to 5 scale with 0 = no injury.
‡ Means followed by the same letter, within NiSo$_4$ or CdSO$_4$, are not significantly different
  according to Duncan's multiple range test at P = 0.05.
§ Growth of plants was retarded by this concentration of CdSO$_4$.

Field Pea (*Pisum sativum*) plants was enhanced by levels of tissue Cd or Ni that
did not themselves retard growth (Table IX) (Ormrod, 1977). When growth
was retarded by the trace elements, O$_3$-induced injury was less than in control
plants. Ozone responses differed among SO$_2$, Cl$_2$, and NO$_3$ salts of Cd or Ni,
and there were cultivar differences in both O$_3$ and trace element responses.
Similarly, O$_3$ sensitivity of Garden Cress (*Lepidium sativum*) and Lettuce was
enhanced by foliar sprays or the addition of Cd or Zn salts to the rooting
medium (Table X) (Czuba & Ormrod, 1974). Foliar application produced
less effect on O$_3$ phytoxicity than did soil application. The interaction of trace
elements and O$_3$ has been corroborated by other studies in which O$_3$ sensitivity of Pinto Bean (*Phaseolus vulgaris* cultivar) plants increased as available
soil Zn increased (Table XI) (McIlveen et al., 1975).

Interactions of trace elements with SO$_2$ have also been demonstrated, but it
is not well known how low levels of SO$_2$ affect metal uptake and alter plant
response (Krause & Kaiser, 1977). Some SO$_2$ in solution on the leaf surfaces

Table X   Total chlorophyll in Garden Cress or Lettuce leaves treated with Cd
(100 ppm) or Zn (100 ppm) and exposed to O$_3$ (30–35 pphm for 6 h)
(mg/g fresh-weight).*

	Garden Cress	Lettuce
No O$_3$, no Cd or Zn	3.34 cd†	1.35 c
O$_3$, no Cd or Zn	2.69 b	1.32 c
O$_3$, Cd leaf spray	2.88 bc	1.13 bc
O$_3$, Zn leaf spray	3.49 d	0.92 b
O$_3$, Cd root application	2.15 a	0.23 a
O$_3$, Zn root application	2.88 bc	1.19 bc

* After Czuba & Ormrod (1974).
† Mean separation within columns at P = 0.05.

Table XI   Influence of various concentrations of soil Zn on growth and ozone sensitivity of Pinto Bean (*Phaseolus vulgaris* cultivar).*

Zn conc. (μg/g)		Shoot dry wt (g)	Ozone injury (%)
Added	Available in soil		
0.0	0.60	0.74 a†	21 a
18.0	0.66	0.69 a	23 ab
45.0	1.55	0.62 a	28 ab
135.0	4.18	0.70 a	32 b

* After McIlveen *et al.* (1975).
† Means followed by the same letter in the same column are not significantly different.

is irreversibly converted to sulphate in the presence of alkaline trace element contaminants (Rühling & Tyler, 1971), and sulphates of trace elements may be formed (Lazrus *et al.*, 1970). The combined effects of SO_2 and several trace elements have been evaluated by dusting with a mixture of metal oxides, with composition based on analysis of dust samples in an industrial area, and then exposing plants to SO_2 (Krause & Kaiser, 1977). The presence of SO_2 did not influence uptake or translocation of trace elements, but foliar injury caused by trace elements was increased by SO_2 (Table XII). The mode of

Table XII   Effect of a dust containing Cd-Pb-Cu-Mn upon yield (dry weight) and foliar/flower injury with and without SO_2 exposure.*

Species	Treatment	Relative yield %	Relative injury %
Lattuca sativa leaf	Control	100.0 a‡	0.0 a
	Control + SO_2	101.0 a	0.4 a
	Dust†	94.4 b	20.2 b
	Dust + SO_2	92.6 b	31.9 c
Setaria italica leaf	Control	100.0 a	0.0 a
	Control + SO_2	111.5 b	0.5 a
	Dust†	63.8 c	32.9 b
	Dust + SO_2	69.3 c	41.1 c
Tagetes sp. flowers	Control		0.0 a
	Control + SO_2		0.0 a
	Dust†		18.4 b
	Dust + SO_2		31.0 c

* After Krause and Kaiser (1977).
† Total application of 5.2 mg Cd, 488 mg Pb, 40.8 mg Cu, 72.8 mg Mn/m², as a dust mixture.
‡ Mean separation within species at $P = 0.05$.

Table XIII   Transpiration and net photosynthesis rates of excised Silver Maple (*Acer saccharinum*) leaves exposed to Cd for 45 h followed by exposure to SO_2 for 30 min.*

Cd (ppm)	Transpiration mg H_2O per dm^2 per h			Net photosynthesis mg CO_2 per dm^2 per h			SO_2 (ppm)
	0	1	2	0	1	2	
0	91.3 c†	66.7 d	59.7 d	2.12 b	1.10 cd	0.91 de	
5	109.6 a	107.8 ab	101.2 abc	2.90 a	1.99 b	2.01 b	
10	87.6 c	69.2 d	63.3 d	2.01 b	1.06 cd	1.08 cd	
20	71.2 d	63.1 d	67.0 d	1.34 c	0.58 ef	0.36 f	

* After Lamoreaux & Chaney (1978).
† Values followed by the same letter, within transpiration or net photosynthesis, are not significantly different at the 0.05 level using Duncan's multiple range test.

Table XIV   The effect of Cu in the rooting medium on Barley (*Hordeum vulgare*) exposed to SO_2 (1.00 ± 0.10 ppm) for 13 h.*

Cu ppm	Dry wt/ plant g	Area of leaf number 3 cm^2	Cu content ppm Tops	Roots	Percentage injury 1 exposure	2 exposures
2	0.36	139.7	60	163	2.7	9.0
10	0.25	138.5	61	249	1.7	9.3
50	0.20	111.0	59	1,040	0	3.1
100	0.15	64.7	171	2,272	0	1.5

* After Toivonen & Hofstra (1979).

action may be that SO_2-induced changes in membrane permeability result in the release of bound cations which could enter the cytoplasm to cause injury without increasing the rate of trace element uptake.

Cadmium interacts with SO_2 in affecting photosynthesis and transpiration of Silver Maple (*Acer saccharinum*) leaves (Lamoreaux & Chaney, 1978). Reductions caused by Cd were greater in the presence of SO_2, and diffusive resistances of leaves were increased by both Cd and SO_2 (Table XIII). In other research, less SO_2 injury to Barley (*Hordeum vulgare*) occurred when Cu was added to the rooting medium (Table XIV) (Toivonen & Hofstra, 1979). Stomatal diffusive resistance was increased by the Cu, thereby decreasing SO_2 uptake. Accordingly, plants growing in mining and smelting areas may have less sensitivity to SO_2 because of elevated levels of Cu in their tissues.

MODE OF ACTION OF TRACE ELEMENTS

There has been considerable interest in the metabolic effects of trace elements and the nature of tolerance to trace elements (Peterson, 1969;

Antonovics *et al.*, 1971; Vallee & Ulmer, 1972; Wainwright & Woodhouse, 1975). Such studies are subject to certain inherent problems. Thus it is difficult to determine the amounts of trace elements in biologically active forms in cell cytoplasm (Buchauer, 1973), while surface contamination may exist even after thorough washing procedures have been carried out. Also, aerosols impacting on foliage and entering plants directly may be less biologically active in terms of producing toxicity symptoms than are trace elements entering plants *via* their roots. The uptake of Cd *via* the root system was more than 16 times as efficient as foliar absorption in causing injury, even though foliar-applied Cd was readily translocated (Haghiri, 1973). Lead may be an exception to this general rule, as large amounts of Pb can be taken up by plants' roots but immobilized by dictyosome vesicles and deposited in the cell wall (Malone *et al.*, 1974; Zimdahl, 1976). This mode of action of Pb tolerance is intriguing and apparently unique. The dictyosomes of the cytoplasm have a much higher affinity for Pb than do mitochondria. The Pb moves through the cytoplasm until it encounters a sink—a dictyosome vesicle. Encased Pb deposits surrounded by membranes migrate towards the outside of the cell, where the membranes surrounding the deposit fuse with the plasmalemma and then the cell wall. The result is the concentration of Pb deposits in the cell wall outside the plasmalemma.

Toxicity varies greatly among trace elements, and toxicity sequences have been determined for some species. The sequence for Barley plants is Hg > Pb > Cu > Cd > Ni > Zn (Oberländer & Roth, 1978). The toxicity of some trace elements may be so great that plant growth is retarded before large quantities of the element can be translocated (Haghiri, 1973).

Low concentrations of trace elements have profound effects on plant processes, including photosynthesis and respiration. For example, Pb concentrations as low as 1 ppm inhibit these processes (Zimdahl, 1976). The site of action may be at both the physiological and biochemical levels. Relatively low concentrations of Pb, Cd, Ni and Tl inhibit photosynthesis and transpiration due to interference with stomatal function (Fig. 1) (F. A. Bazzaz *et al.*, 1974a, 1974b). Thallium was most effective, probably because of its similarity to K. These and other trace elements also interfere with mitochondrial respiration and photosynthesis by isolated chloroplasts.

Inhibition of photosynthetic electron transport by Pb salts in isolated chloroplasts has been demonstrated, the primary site of inhibition being on the oxidizing site of Photosystem II (Miles *et al.*, 1972). Cadmium is a particularly potent inhibitor of photosynthesis in chloroplasts; low concentrations strongly inhibit Photosystem II activities, in addition to changing the concentration and composition of pigments (M. B. Bazzaz & Govindjee, 1974). Cadmium effects on respiration are likely to be early in electron transport, and sulphydryl groups are likely to be involved in Cd–membrane interactions in mitochondria (Miller *et al.*, 1973). Mitochondria take up Cd, Zn, Co, Ni, and Mn about equally, while their sorption of Pb is about 10

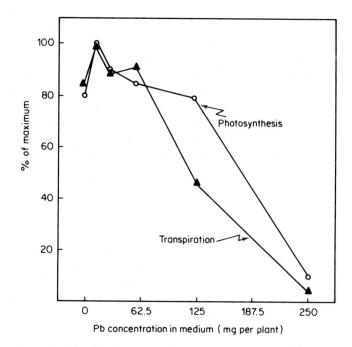

Fig. 1 Relationship between apparent photosynthesis and transpiration of Soybean and Pb concentration in the rooting medium. Rates are expressed as percentage of maximum. Reproduced from F. A. Bazzaz *et al.*, *Journal of Environmental Quality*, **3** (1974), pp. 156–158, by permission of the American Society of Agronomy.

Table XV   Effect of Cd on N-fixation of Red Alder (*Alnus rubra*) plants, previously nodulated and grown for 11 weeks with the indicated concentrations of $CdCl_2$ in N-free nutrient solution in a summer experiment.*

Conc. $Cd Cl_2$	Dry weight leaves	Nitrogenase activity	N fixed†	No. of nodules	Cd conc. Leaves	Roots	Nodules
μm	g	μmoles	mg per plant	per plant	μg per g of dry-wt		
136	0.21	3.7	2.9	69	74.1	5,584.1	1,279.0
54.5	0.41	5.5	11.2	115	66.2	2,767.3	810.1
27.3	1.11	8.0	60.8	86	55.0	2,331.3	435.5
10.9	2.80	17.9	132.7	120	40.1	1,307.4	199.1
0.0	4.50	36.7	227.8	259	0.0	0.0	0.0

* After Wickliff *et al.* (1980).
† Formed per g of dry nodule per hour.

times as high, though precipitate formation may be the cause of this higher value (Smith *et al.*, 1978). Inhibition of oxidation, depending on ion and concentration, was observed for succinate, NADH, malate, and pyruvate. Substantial swelling of mitochondria was noted following Cd and Pb additions, suggesting a non-specific increase in mitochondrial membrane permeability due to the cations. Sites of action may be common to several cations. Cadmium can inhibit nitrogenase activity and nodulation in Red Alder (*Alnus rubra*) (Table XV) (Wickliff *et al.*, 1980). Growth and nitrate reductase activity are also depressed. Inhibition of N-fixation by the addition of 1 ppm of trace element in nutrient solution has been demonstrated for Cu and Zn with Yellow Sweet-clover (*Melilotus officinalis*), for Cd, Cu, and Zn, with Alfalfa (*Medicago sativa*), and for Cd and Zn with Soybean (*Glycine max*) (Sheridan, 1979).

GENETICS OF TRACE ELEMENT SENSITIVITY

Rapid evolution of insensitive populations of certain species has been noted at trace element polluted sites. *Agrostis tenuis* (Common Bentgrass) and *Dactylis glomerata* (Orchard Grass) produce a low frequency of tolerant survivors of Cu-contaminated soil (Gartside & McNeilly, 1974). *Agrostis stolonifera* (Fiorin) populations have an increased frequency of Cu-insensitive individuals as the age of the population increases (Wu *et al.*, 1975). The insensitive individuals exist at very low frequency in populations growing in unpolluted areas, and are selected at both seedling and adult stages in the presence of Cu. Some individuals of *Agrostis tenuis* populations are insensitive to Cu and Zn, but the insensitivities are independent (Walley *et al.*, 1974) in that the metal insensitivity is considered to be a continuously varying

Table XVI   Copper tolerance of individual survivors from Cu-screening experiments compared with various other materials.*

Source	Mean tolerance†
Normal population unselected (seed grown on normal soil)	5.6
Normal population selected (survivors from various copper waste–soils mixtures)	58.2
Pasture population (individuals from a pasture)	1.2
Mine population (individuals from a copper mine)	75.5

* After Walley *et al.* (1974).
† Based on the ability of tillers to root in a solution containing copper.

character with a threshold, and high selection pressure picks out rare gene combinations for insensitivity, resulting in a rapid and marked change in the genetic constitution of the population (Table XVI).

The above wide range of genetic combination is in accordance with the cross breeding nature of *Agrostis*. The insensitivity has been found to be due to complexing of trace elements by substances in cell walls (Walley *et al.*, 1974). Accumulation of As to high levels by plants growing on mine and smelter waste has also been noted (Porter & Peterson, 1977). *Agrostis tenuis* plants are tolerant of arsenate, but not of arsenite in solution, perhaps as an adaptation to the predominance of arsenate in waste soils.

DETECTION OF TRACE ELEMENT POLLUTION

Plant tissue analysis is an alternative to mechanical air sampling and analysis methods. Data are available on the relationship of tissue concentrations of many trace elements to phytotoxicity (Table XVII)—*see* Cottenie *et al.* (1976), whose approaches include leaf content *versus* decreases in yield for Zn, Cu, Cd, Cr, Co and F (Table XVIII); threshold concentrations for visible injury on 13 species for Zn, Pb, Cu, Cd, Ni, Co and Mn; concentrations of many trace elements in grasses growing near metal smelting industries and near a metallurgical plant; and trace elements in vegetables of different origin. Plant tissue analysis for total foliar trace element content may be inadequate to gauge the severity of aerosol pollution where root sorption of trace elements could occur as well as foliar uptake (Buchauer, 1973; Little, 1973).

Table XVII   Trace element concentrations in dry matter of Ryegrass
(μg per g dry weight).*

		Deficiency	Normal range	Toxicity
Essential	Fe	< 50	50–240	?
trace	Mn	< 20	25–250	>500
elements	Zn	< 20	25–250	>400
	Cu	< 5	6–15	> 20
	B	< 10	18–30	>190
	Mo	< 0.1	0.5–5	?
Non-essential	Ni	–	0–8	> 80
trace	Pb	–	2–14	–
elements	Cr	–	0–0.5	>1.3
	F	–	5–8	> 50
	Co	–	0–2	>100
	Cd	–	0–0.5	>100

*After Cottenie *et al.* (1976).

Table XVIII   Yield decrease in Ryegrass with the corresponding trace element content in leaves.*

Element	Percentage yield	Leaf content (μg per g dry weight)
Zn	75	1,400
Cu		20
Cd		250
Cr^{6+}		1.8
Co		375
F		63
Zn	50	2,000 – 2,500
Cu		35
Cd		350
Cr^{6+}		2.5
Zn	25	2,000 – 2,500
Cu		55
Cd		1,250

* After Cottenie *et al.* (1976).

Analysis for V, a contaminant of petroleum, is very sensitive; so this trace element could serve as an indicator of movement of aerosols from fuel combustion (Waldron & Terry, 1975). Similarities in the patterns of Pb and As contamination around secondary Pb smelters, and the absence of As around other Pb sources, indicates that As could be used as an indicator of certain kinds of industrial Pb contamination (Temple *et al.*, 1977).

The use of biological indicators has been developed as an inexpensive alternative to sampling of air by filtration for the detection of atmospheric trace element pollution. Among bryophytes, certain mosses are particularly effective for measuring aerosol deposition (Tyler, 1972). Trace element ions are taken up passively by means of the strong ion exchange properties of moss tissue, and small amounts can be detected in wet or dry deposition. Among lichens, the correlation of the trace element content of the thallus with its rock substrate is normally poor. Instead, there is a strong relationship to the composition of the external and regional environments (Jenkins & Davies, 1966). To be useful, the elemental composition of the plant must be quantitatively related to direct measurement of the trace element content of the air (Goodman & Roberts, 1971).

The moss *Hypnum cupressiforme* has been used extensively for the indication and integration of Zn, Pb, Cu, Ni and Mg contamination of the aerial environment (Goodman & Roberts, 1971). The plants are exposed to ambient air in nylon mesh bags or in other ways for a period of weeks or months. Another method uses species of the moss genus *Sphagnum*, suspended in bags from natural vegetation (Little & Martin, 1974). Zinc, Pb and

Cd, data collected from such 'moss bags' have been used to construct maps of concentrations and distribution patterns. Species used in Scandinavia to determine gradients of Cd, Co, Cr, Cu, Ni, Pb and Zn include *Hypnum cupressiforme*, *Hylocomnium splendens*, and *Sphagnum magellanicum* (Rühling & Tyler, 1973).

Grasses, including *Festuca rubra* (Red or Creeping Fescue), can be used as indicators of trace element aerosols (Goodman & Roberts, 1971). Specific reference has also been made to the use of Sugar beets (*Beta vulgaris* cultivars) as inexpensive monitors of Hg pollution, as they are highly susceptible to Hg injury (Waldron & Terry, 1975).

SUMMARY

The term trace element pollution is used rather than 'heavy metals' because it is more inclusive and accurate. A few of these elements occur in local atmospheres in sufficient quantities to be toxic to plants. Most of these are associated with smelting operations or pesticide use; other sources include vehicle pollution, industrial activities, and coal, and petroleum and waste combustion. Arsenic, cadmium, copper and lead, are the most significant. Foliar deposits tend to be insoluble, but when accumulated in the soil may be taken up by plant roots in toxic amounts. Interactions with ozone and SO_2 may also be important.

The general modes of action of trace element pollution involve interference with stomatal function, inhibition of photosynthesis and/or respiration, and/or alteration of mitochondria membrane permeability. In regions of high incidence of copper or zinc, rapid evolution of tolerant populations of certain species has been noted.

REFERENCES

Antonovics, J., Bradshaw, A. D. & Turner, R. G. (1971). Heavy-metal tolerance in plants. *Adv. Ecol. Res.,* **7**, pp. 1–85.

Arvik, W. H. & Zimdahl, R. L. (1974). Barriers to foliar uptake of lead. *J. Environ. Qual.,* **3**, pp. 369–70.

Auermann, E., Jacobi, J., Eternach, R. & Kühn, H. (1976). Untersuchungen über den bleigehalt pflanzlicher Nahrungsmittel im Wirkungsbereich eines bleiemittierenden Betriebes. *Die Nahrung,* **20**, pp. 509–18.

Bazzaz, M. B. & Govindjee (1974). Effects of cadmium nitrate on spectral characteristics and light reactions of chloroplasts. *Environ. Lett.,* **6**, pp. 1–12.

Bazzaz, F. A., Carlson, R. W. & Rolfe, G. L. (1974*a*). The effect of heavy-metals on plants, Part 1: Inhibition of gas exchange in Sunflower by Pb, Cd, Ni, and Tl. *Environ. Pollut.,* **7**, pp. 241–6.

Bazzaz, F. A., Rolfe, G. L. & Windle, P. (1974*b*). Differing sensitivity of Corn and Soybean photosynthesis and transpiration to lead contamination. *J. Environ. Qual.,* **3**, pp. 156–8.

Beavington, F. (1973). Contamination of soil with zinc, copper, lead, and cadmium, in the Wollongong city area. *Australian J. Soil Res.,* **11**, pp. 23–31.

Beavington, F. (1975*a*). Some aspects of contamination of herbage with copper, zinc, and iron. *Environ. Pollut.,* **8**, pp. 65–71.

Beavington, F. (1975*b*). Heavy-metal contamination of vegetables and soil in domestic gardens around a smelting complex. *Environ. Pollut.,* **9**, pp. 211–7.

Bowen, H. J. M. (1977). Natural cycles of the elements and their perturbation by man. Pp. 1–37 in *The Chemical Environment: Environment and Man* (Ed. J. Lenihan & W. Fletcher). Blackie, Glasgow, Scotland, UK: Vol. 6, 163 pp., illustr.

Buchauer, M. J. (1973). Contamination of soil and vegetation near a zinc smelter by zinc, cadmium, copper, and lead. *Environ. Sci. Technol.,* **7**, pp. 131–5.

Bukavok, M. J. & Wittwer, S. H. (1957). Absorption and mobility of foliar applications of nutrients. *Plant Physiol.,* **32**, pp. 428–35.

Burton, K. W. & John, E. (1977). A study of heavy-metal contamination in the Rhondda Fawr, South Wales. *Water, Air, Soil Pollut.,* **7**, pp. 45–68.

Cannon, H. L. & Bowles, J. M. (1962). Contamination of vegetation by tetraethyl lead. *Science,* **137**, pp. 765–6.

Chow, T. J. (1970). Lead accumulation in roadside soil and grass. *Nature* (London), **225**, pp. 295–6.

Chow, T. J. & Earl, J. L. (1970). Lead aerosols in the atmosphere: Increasing concentrations. *Science,* **169**, pp. 577–80.

Cottenie, A., Dhaese, A. & Camerlynck, R. (1976). Plant quality response to the uptake of polluting elements. *Qual. Plantarum,* **26**, pp. 293–319.

Coughtrey, P. J., Jones, C. H., Martin, M. H. & Shales, S. W. (1979). Litter accumulation in woodlands contaminated by Pb, Zn, Cd, and Cu. *Oecologia* (Berlin), **39**, pp. 51–60.

Czuba, M. & Ormrod, D. P. (1974). Effects of cadmium and zinc on ozone-induced phytotoxicity in Cress and Lettuce. *Can. J. Bot.,* **52**, pp. 645–9.

Daines, R., Motto, H. & Chilko, D. M. (1970). Atmospheric lead: Its relationship to traffic volume and proximity to highways. *Environ. Sci. Technol.,* **4**, pp. 318–22.

Dedolph, R., Haar, G. Ter, Holtzman, R. & Lucas, H., jr (1970). Sources of lead in Perennial Ryegrass and Radishes. *Environ. Sci. Technol.,* **4**, pp. 217–23.

Dochinger, L. S. (1980). Interception of airborne particles by tree planting. *J. Environ. Qual.,* **9**, pp. 265–8.

Evans, D. W., Wiener, J. G. & Horton, J. H. (1980). Trace element inputs from a coal-burning power plant to adjacent terrestrial and aquatic environments. *J. Air Pollut. Contr. Assoc.,* **30**, pp. 567–73.

Fassett, D. W. (1975). Cadmium: biological effects and occurrence in the environment. *Annu. Rev. Pharmacol.,* **15**, pp. 425–35.

Flanagan, J. T., Wade, K. J., Currie, A. & Curtis, D. J. (1980). The deposition of lead and zinc from traffic pollution on two roadside shrubs. *Environ. Pollut.* (Ser. B), **1**, pp. 71–8.

Freedman, B. & Hutchinson, T. C. (1980). Effects of smelter pollutants on forest leaf litter decomposition near a nickel–copper smelter at Sudbury, Ontario. *Can. J. Bot.,* **58**, pp. 1722–36.

Gartside, D. W. & McNeilly, T. (1974). The potential for evolution of heavy metal tolerance in plants, II: Copper tolerance in normal populations of different plant species. *Heredity,* **32**, pp. 335–48.

Goldsmith, C. D., jr, Scalon, P. F. & Pirie, W. R. (1976). Lead concentrations in soil and vegetation associated with highways of different traffic densities. *Bull. Environ. Contam. Toxicol.,* **16**, pp. 66–70.

Goodman, G. T. & Roberts, T. M. (1971). Plants and soils as indicators of metals in the air. *Nature* (London), **231**, pp. 287–92.

Graham, D. L. & Kalman, S. M. (1974). Lead in forage grass from a suburban area in northern California. *Environ. Pollut.,* **7**, pp. 209–15.

Haar, G. L. Ter & Bayard, M. A. (1971). Composition of airborne lead particles. *Nature* (London), **232**, pp. 553–4.

Haar, G. L. Ter, Dedolph, R. R., Holtzman, R. B. & Lucas, H. F., jr (1969). The lead uptake by Perennial Ryegrass and Radishes from air, water, and soil. *Environ. Res.,* **2**, pp. 267–71.

Haghiri, F. (1973). Cadmium uptake by plants. *J. Environ. Qual.,* **2**, pp. 93–6.

Hallsworth, E. G. & Adams, W. A. (1973). The heavy metal content of rainfall in the East Midlands. *Environ. Pollut.,* **4**, 231–5.

Hunter, J. G. & Vergnano, O. (1952). Nickel toxicity in plants. *Ann. Appl. Biol.,* **39**, pp. 279–284.

Hutchinson, T. C. & Whitby, L. M. (1974). Heavy metal pollution in the Sudbury mining and smelting region of Canada, I: Soil and vegetation contamination by nickel, copper, and other metals. *Environmental Conservation,* **1**, pp. 123–32.

Jenkins, D. A. & Davies, R. I. (1966). Trace-element content of organic accumulations. *Nature* (London), **210**, pp. 1296–7.

Klein, D. H., Andren, A. W. & Bolton, N. E. (1975). Trace-element discharges from coal combustion for power production. *Water, Air, Soil Pollut.,* **5**, pp. 71–7.

Koslow, E. E., Smith, W. H. & Staskawicz, B. J. (1977). Lead-containing particles on urban leaf surfaces. *Environ. Sci. Technol.,* **11**, pp. 1019–21.

Kothny, E. L. (1973). The three-phase equilibrium of mercury in Nature. Pp. 48–80 in *Trace Elements in the Environment* (Ed. E. L. Kothny) (*Adv. Chem. Ser.* 123.) American Chemical Society, Washington, DC, USA: ix + 149 pp., illustr.

Krause, G. H. M. & Kaiser, H. (1977). Plant response to heavy-metals and sulphur dioxide. *Environ. Pollut.,* **12**, pp. 63–71.

Lagerwerff, J. V. (1967). Heavy metal contamination of soils. Pp. 343–64 in *Agriculture and the Quality of Our Environment* (Ed. N. C. Brady). (Amer. Assoc. Adv. Sci. Publ. 85, Washington, DC, USA: xv + 460 pp., illustr.

Lagerwerff, J. V. (1971). Uptake of cadmium, lead, and zinc, by radish from soil and air. *Soil Sci.,* **111**, pp. 129–33.

Lagerwerff, J. V. (1972). Lead, mercury and cadmium as environmental contaminants. Pp. 593–628 in *Micro-nutrients in Agriculture* (Ed. J. J. Mortvedt, P. M. Giordano & W. L. Linsey). Soil Sci. Soc. Amer., Madison, Wisconsin, USA: xviii + 666 pp., illustr.

Lagerwerff, J. V. & Specht, A. W. (1970). Contamination of roadside soil and vegetation with cadmium, nickel, lead, and zinc. *Environ. Sci. Technol.,* **4**, 583–6.

Lagerwerff, J. V. Armiger, W. H. & Specht, A. W. (1973). Uptake of lead by Alfalfa and Corn from soil and air. *Soil Sci.,* **115**, pp. 455–60.

Lakin, H.W. (1973). Selenium in our environment. Pp. 96–111 in *Trace Elements in the Environment* (Ed. E. L. Kothny). (Adv. Chem. Ser. 123.) American Chemical Society, Washington, DC, USA: ix + 149 pp., illustr.

Lamoreaux, R. J. & Chaney, W. R. (1978). Photosynthesis and transpiration of excised Silver Maple leaves exposed to cadmium and sulphur dioxide. *Environ. Pollut.,* **17**, pp. 259–68.

Lazrus, A. L., Lorange, E. & Lodge, J. P., jr (1970). Lead and other metal ions in United States precipitation. *Environ. Sci. Technol.,* **4**, 55–8.

Lee, R. E., jr & Goranson, S. (1972). National air surveillance cascade impacter network, I: Size distribution measurements of suspended particulate matter in air. *Environ. Sci. Technol.,* **6**, 1019–24.

Lee, R. E., jr, Goranson, S., Enrione, R. E. & Morgan, G. B. (1972). National air surveillance cascade impacter network, II: Size distribution measurements of trace metal components. *Environ. Sci. Technol.,* **6**, pp. 1025–30.

Little, P. (1973). A study of heavy metal contamination of leaf surfaces. *Environ. Pollut.,* **5**, pp. 159–72.

Little, P. & Martin, M. H. (1972). A survey of zinc, lead and cadmium in soil and natural vegetation around a smelting complex. *Environ. Pollut.,* **3**, pp. 241–54.

Little, P. & Martin, M. H. (1974). Biological monitoring of heavy-metal pollution. *Environ. Pollut.,* **6**, pp. 1–19.

McIlveen, W. D., Spotts, R. A. & Davis, D. D. (1975). The influence of soil zinc on nodulation, mycorrhizae, and ozone sensitivity of Pinto Bean. *Phytopathology,* **65**, pp. 645–7.

Malone, C., Koeppe, D. E. & Miller, R. J. (1974). Localization of lead accumulated by Corn plants. *Plant Physiol.,* **53**, pp. 388–94.

Marten, G. C. & Hammond P. B. (1966). Lead uptake by bromegrass from contaminated soils. *Agron. J.,* **58**, pp. 553–54.

Miles, C. D., Brandle, J. R., Daniel, D. J., Chu-Der, O., Schnare, P. O. & Uhlick, D. J. (1972). Inhibition of photosystem II in isolated chloroplasts by lead. *Plant Physiol.,* **49**, pp. 820–5.

Miller, R. J., Bittel, J. E. & Koeppe, D. E. (1973). The effects of cadmium on electron and energy transfer reactions in corn mitochondria. *Physiol. Plant.,* **28**, pp. 166–71.

Motto, H. L., Daines, R. H., Chilko, D. M. & Motto, C. K. (1970). Lead in soils and plants: Its relationship to traffic volume and proximity to highways. *Environ. Sci. Technol.,* **4**, pp. 231–7.

National Academy of Sciences (1972–77). Medical and Biologic Effects of Environmental Pollutants: Division of Medical Sciences, Washington, DC, USA:

1972 *Lead: Airborne Lead in Perspective:* xi + 330 pp., illustr.

1974 *Chromium:* ix + 155 pp.

1974 *Manganese:* vi + 191 pp.

1975 *Nickel:* vii + 277 pp.

1976 *Selenium:* xi + 203 pp., illustr.

1977 *Arsenic:* v + 332 pp., illustr.

1977 *Copper:* viii + 115 pp.

Nieboer, E. & Richardson, D. H. S. (1980). The replacement of the nondescript term 'heavy metals' by a biologically and chemically significant classification of metal ions. *Environ. Pollut.* (Ser. B), **1**, pp. 3–26.

Oberländer, H. E. & Roth, K. (1978). Die Wirkung der schwermetalle Chrom, Nickel, Kupfer, Zink, Cadmium, Quecksilber, und Blei, auf die Aufnahme und Verlagerung von Kalium and Phosphat bei jungen Gerstepflanzen. *Z. Pfl. Ernahr, Düng. Bodenk.,* **141**, pp. 107–16.

Ormrod, D. P. (1977). Cadmium and nickel effects on growth and ozone sensitivity of Pea. *Water, Air, Soil Pollut.,* **8**, pp. 263–70.

Ormrod, D. P. (1978). *Pollution in Horticulture.* Elsevier, Amsterdam, The Netherlands: xi + 260 pp., illustr.

Page, A. L. & Ganje, T. J. (1970). Accumulations of lead in soils for regions of high and low motor vehicle traffic density. *Environ. Sci. Technol.,* **4**, pp. 140–2.

Peterson, P. J. (1969). The distribution of zinc-65 in *Agrostis tenuis* Sibth. and *A. stolonifera* L. tissues. *J. Exp. Bot.,* **20**, pp. 863–75.

Pierson, D. H., Cawse, P. A., Salmon, L. & Cambray, R. S. (1973). Trace elements in the atmospheric environment. *Nature* (London), **241**, pp. 252–6.

Porter, E. K. & Peterson, P. J. (1977). Arsenic tolerance in grasses growing on mine waste. *Environ. Pollut.,* **14**, pp. 255–65.

Preer, J. R., Sekhon, H. S., Stephens, B. R. & Collins, M. S. (1980). Factors affecting heavy-metal content of garden vegetables. *Environ. Pollut.* (Ser. B), **1**, pp. 95–104.

Purves, D. (1967). Contamination of urban garden soils with copper, boron, and lead. *Pl. Soil,* **26**, pp. 380–2.

Purves, D. (1977). *Trace Element Contamination of the Environment.* Elsevier, Amsterdam, The Netherlands: xi + 260 pp., illustr.

Rühling, A. & Tyler, G. (1971). Regional differences in the deposition of heavy metals over Scandinavia. *J. Appl. Ecol.,* **8**, pp. 497–507.

Rühling, A. & Tyler, G. (1973). Heavy metal pollution and decomposition of spruce needle litter. *Oikos,* **24**, pp. 402–16.

Russo, F. & Brennan, E. (1979). Phytotoxicity and distribution of cadmium in Pin Oak seedlings determined by mode of entry. *Forest Sci.,* **25**, pp. 328–32.

Saunders, P. J. W. (1973). Effects of atmospheric pollution on leaf surface microflora. *Pestic. Sci.,* **4**, pp. 589–95.

Schlesinger, W. H., Reiners, W. A. & Dropman, D. S. (1974). Heavy metal concentrations and deposition in bulk precipitation in montane ecosystems of New Hampshire, U.S.A. *Environ. Pollut.,* **6**, pp. 39–47.

Schuck, E. A. & Locke, J. K. (1970). Relationship of automotive lead particulates to certain consumer crops. *Environ. Sci. Technol.,* **4**, pp. 324–30.

Schwela, D. H. (1979). An estimate of deposition velocities of several air pollutants on grass. *Ecotoxicol. Environ. Safety,* **3**, pp. 174–89.

Sheridan, R. P. (1979). Effects of airborne particulates on nitrogen fixation in legumes and Algae. *Phytopathology,* **69**, pp. 1011–8.

Smith, W. H. (1971). Lead contamination of roadside White Pine. *Forest Sci.,* **17**, pp. 195–8.

Smith, W. H. (1972). Lead and mercury burden of urban woody plants. *Science,* **176**, 1237–9.

Smith, W. H. (1973). Metal contamination of urban woody plants. *Environ. Sci. Technol.,* **7**, pp. 631–6.

Smith, W. H. (1976*a*). Lead contamination of the roadside ecosystem. *J. Air Pollut. Contr. Assoc.,* **26**, pp. 753–66.

Smith, W. H. (1976*b*). Air pollution—effects on the structure and function of plant–surface microbial-ecosystems. Pp. 75–105 in *Microbiology of Aerial Plant Surfaces* (Ed. C. H. Dickinson & T. F. Preece). Academic Press, New York, NY, USA: 660 pp., illustr.

Smith, W. H. (1977). Influence of heavy-metal leaf contaminants on the *in vitro* growth of urban-tree phyllophane-fungi. *Microbial Ecol.,* **3**, pp. 231–9.

Smith, W. H., Staskawicz, B. J. & Harkov, R. S. (1978). Trace-metal pollutants and urban-tree leaf pathogens. *Trans. Br. Mycol. Soc.,* **70**, pp. 29–33.

Temple, P. J., Linzon, S. N. & Chai, B. L. (1977). Contamination of vegetation and soil by arsenic emissions from secondary lead smelters. *Envron. Pollut.,* **12**, pp. 311–20.

Toivonen, P. M. A. & Hofstra, G. (1979). The interaction of copper and sulphur dioxide in plant injury. *Can. J. Plant Sci.,* **59**, pp. 475–9.

Tyler, G. (1972). Heavy metals pollute nature, may reduce productivity. *Ambio,* **1**, pp. 52–9.

Vallee, B. L. & Ulmer, D. D. (1972). Biochemical effects of mercury, cadmium, and lead. *Ann. Rev. Biochem.,* **41**, pp. 92–128.

Wainwright, S. J. & Woodhouse, H. W. (1975). Physiological mechanisms of heavy-metal tolerance in plants. Pp. 231–57, in *The Ecology of Resource Degradation and Renewal.* (Ed. M. J. Chadwick & G. J. Goodman). (15 Symp. Brit. Ecol. Soc.) Blackwell, Oxford, England, UK: xiii + 480 pp., illustr.

Waldron, L. J. & Terry, N. (1975). Effect of mercury vapor on sugar beets. *J. Environ. Qual.,* **4**, pp. 58–60.

Walley, K. W., Khan, M. S. I. & Bradshaw, A. D. (1974). The potential for evolution of heavy metal tolerance in plants. I: Copper and zinc tolerance in *Agrostis tenuis. Heredity,* **32**, pp. 309–319.

Wangen, L. E. & Turner, F. B. (1980). Trace elements in vegetation downwind of a coal-fired power plant. *Water. Air, Soil Pollut.,* **13**, pp. 99–108.

Ward, N. I., Reeves, R. D. & Brooks, R. R. (1975). Lead in soil and vegetation along a New Zealand State Highway with low traffic volume. *Environ. Pollut.,* **9**, pp. 243–51.

Wheeler, G. L. & Rolfe, G. L. (1979). The relationship between daily traffic volume and the distribution of lead in roadside soil and vegetation. *Environ. Pollut.,* **18**, pp. 265–74.

White, E. J. & Turner, F. B. (1970). A method of estimating income of nutrients in a catch of airborne particles by a woodland canopy. *J. Appl. Ecol.,* **7**, pp. 441–61.

Wickliff, C., Evans, H. J., Carter, K.R. & Russell, S. A. (1980). Cadmium effects on the nitrogen fixation system of Red Alder. *J. Environ. Qual.,* **9**, pp. 180–4.

Wu, L., Bradshaw, A. D. & Thurman, D. A. (1975). The potential for evolution of heavy metal tolerance in plants, III: The rapid evolution of copper tolerance in *Agrostis stolonifera. Heredity,* **34**, pp. 165–87.

Zimdahl, R. L. (1976). Entry and movement in vegetation of lead derived from air and soil sources. *J. Air Pollut. Control Assoc.,* **26**, pp. 655–60.

Zoller, W. H., Gordon, G. E., Gladney, E. S. & Jones, A. G. (1973). The sources and distribution of vanadium in the atmosphere. Pp. 31–47 in *Trace Elements in the Environment* (Ed. E. L. Kothny). (Adv. Chem. Ser. 123.) American Chemical Society, Washington, DC, USA: ix + 149 pp., illustr.

CHAPTER 14

Interactions of Disease and Other Stress Factors with Atmospheric Pollution

SATU HUTTUNEN

Department of Botany, University of Oulu, PO Box 191, SF-90101 Oulu 10, Finland

INTRODUCTION

Air pollutants as stress factors in the environment have many interactions with other environmental stress factors. During the past twenty years, it has been reported that air pollutants may have detrimental or beneficial effects or interactions with other stressful components, most notably diseases and insects. Air pollutants have also been reported to reduce the frost-, winter-, and drought-resistance of plants. The recent years' information has included the stress effects of increasing acidity on forest trees and forest soil biology as additional interactions.

The life of plants is normally well adapted to the environmental conditions in which they live, as the most appropriate forms have been selected under conditions of natural stress. The normal plants' response to natural stresses may be changed when an additional stress factor, such as air pollution, is imposed on it.

There are fairly numerous reports in the literature on air pollution and its effects on host–disease interactions and host–insect interactions; during recent years, an increasing proportion have focused on the possible effects of increasing acidity and the interactions with climatic and other environmental factors (Treshow, 1975; Schneider & Chłodny, 1977; Huttunen, 1979a; Shriner & Cowling, 1980).

In the future, with the increasing environmental awareness, it is to be expected that few areas will be afflicted by air pollution concentrations that cause acute damage or are toxic to plant pathogens. But larger and larger areas are being exposed to relatively low concentrations for long periods.

Furthermore, the combined effects of several air pollutants are apt to increase environmental stress-induced chronic injury and damage, which may lead to new environmental stresses with further interactions with insects, pathogenic fungi and bacteria, and non-biotic stress factors.

Air pollution induced stress makes plants more easily susceptible than otherwise to such secondary effects as attacks by viruses, bacteria, and especially fungi and insects. Air pollutants have numerous effects that can alter the interactions between a plant and its natural susceptibility to disease and insect attacks. These changes are mostly physiological, affecting the properties of the host cell; but even mechanical factors can be included. Air pollution has effects on green plant surfaces which may facilitate their penetration and infection by insects and pathogens. The altered chemical composition may attract insects, and the altered chemical *milieu* of the leaf surface may give improved possibilities for the development of diseases.

When discussing these questions, the ecophysiological perspective of interactions has been taken as the basis of consideration. The ecophysiological perspective includes the viewpoint of seasonal differences of air pollution effects and the phenology of stress occurrence.

Having read through pertinent papers on plant pathology, entomology, and plant ecology, I have found that a multidisciplinary approach is still needed in order to understand the possible interactions between air pollution and other stress factors. In some cases I have made speculative proposals on a very limited scientific basis in order to raise questions and cite investigations supporting or disproving these speculations.

INTERACTIONS WITH PATHOGENIC VIRUSES, BACTERIA, AND FUNGI

Background

Resistance can be described as the inherent ability of a plant to prevent or restrict the establishment and subsequent activities of a potential pathogen. In recent years, the terms compatible and incompatible have been used as synonyms for susceptible and resistant. The former terms have the advantage of focusing attention on the cellular interactions between the host and the pathogen (Daly, 1976).

It is well known that the overall metabolic activity of the cell is under genetic control, although it is simultaneously influenced by environmental factors. The biochemical and physiological responses of host plants to pathogens are therefore not necessarily specific. Increased cell metabolism is frequently connected with changes in the permeability of the cell walls or changes in the ultastructure of cell organelles. Among the most fundamental requirements of metabolic regulation is the maintenance of a steady state between the energy-generating catabolic processes and the myriad of energy-requiring synthetic reactions proceeding in the cell simultaneously (Kosuge & Gilchrist, 1976).

When we add some modifying environmental factors to the actuality of plant–pathogen interactions, we must consider the complex of many ecological—both biotic and non-biotic—factors. Air pollutants, as an additional environmental factor in host–pathogen interactions, have been discussed previously by Heagle (1973) and Treshow (1975). It has been suggested that air pollutants may act as an ecological factor causing stress that may predispose plants to infection by pathogens (Treshow, 1968; Parmeter & Cobb, 1972; Shriner, 1977; Huttunen, 1979a). Results of several studies suggest that there are a number of possible mechanisms that could account for the effects of air pollutants on plant diseases.

First, there may be a direct effect of the pollutant on the growth and development of the organism in question and/or the pathogen. The severity of parasitism may be increased or decreased as a result of the action of the pollutant on the susceptibility of the host or the virulence of the parasite (Treshow, 1975; Weinstein, 1977; Linzon, 1978). Another mechanism could be that the pollutant, through its effects on the structure, physiology, or metabolism, of the host, indirectly affects the pathogen. The pollutant could also alter the suitability of the host organ as a habitat for the pathogen. Moreover it could affect the progress of the disease by modifying, for instance, the external chemistry of the host or altering the quantity or quality of the host's exudates. These changes include the changes in the exterior barrier to infection, and also the changes that alter the chemical environment of the host surface.

In addition, the effects of the pollutant on the host may be intensified or reduced by the presence of a parasite. Considering the large number of host–parasite–pollutant combinations that can occur, the research required to determine the effects of these interactions is practically limitless (Linzon, 1978).

In ecophysiological research on host–pathogen interactions, increased infection as a result of heating the host tissue before inoculation has been demonstrated with several pathogens, including viruses, bacteria, Phycomycetes, Fungi Imperfecti, Basidiomycetes, and non-parasitic fungi. There are also numerous observations of plants, which had been killed or injured by frost or winter cold, being invaded by pathogens. Heat or cold are injurious mechanisms that act as predisposing agents (Goodman, 1976).

The host–pathogen interactions must be considered in more detail than by simply saying that the infection is decreasing or increasing. Relatively high concentrations of pollutants may be involved largely with directly reducing the effect of pathogens, while long-term, lower concentrations frequently have more complicated effects on the interactions between the host and the pathogen. In this connection the possibility of imbalance in forest ecosystems is the most obvious risk, and there are several reports on such possible interactions (Treshow, 1968; Lighthart *et al.,* 1971; Cowling, 1978; Shriner & Cowling, 1980). The possibility of having a detrimental effect on beneficial microorganisms, such as nitrogen-fixing bacteria and blue-green algae, is most obvious in the case of air pollution (Kallio, 1976; Huttunen *et al.,* 1981*b*). The effects of increasing acidity on soil biology and microorganisms have been discussed recently (Hutchinson & Havas, 1980; Lohm, 1980). Generally speaking, the alterations in symbiotic associations, such as changes in leaf and root exudation processes and accelerated leaching from plants, may affect the functioning of mycorrhizas, nitrogen-fixing organisms, etc. (Cowling, 1978).

Interactions with Viruses and Bacteria

Zoeten (1976) defines a plant virus as a nucleoprotein or a group of nucleoproteins containing the genetic information needed to obtain access to living cells, and to alter, generally to the detriment of the invaded cells, their metabolism in favour of its own multiplication.

The cells of plant pathogenic bacteria are essentially naked and non-spore-forming, so their persistence and growth seem to be dependent on rapidly 'finding' a conducive environment. The salient requirements of such an environment are high relative humidity, adequate nutrition, and temperatures between 24°C and 30°C.

It is obvious that soil provides a bacterial reservoir. Leben *et al.* (1968) pointed out that large populations of pathogens could be expected on leaf surfaces that remain wet for 24 hours or longer at temperatures of 24–36°C. The pathogen may be delivered in a number of ways by Man, other mammals, insects, nematodes, wind-driven rain, soil particles, or abrasion as a consequence of growth (Goodman, 1976).

The possible protective or non-protective interactions of air pollutants with plant viral and bacterial pathogens must be discussed together with other factors—such as weather conditions, rain-splash or wind-driven rain, and the factors affecting the penetration processes *via* stomata, cuticle, lenticels, and leaf-traces.

Plant viruses can be introduced by abrasion through the epidermis, by insects, by nematodes, by fungi, by grafting, by dodder (*Cuscuta* spp.), or by pollen. All of these require wounding of the host plant to establish the intimate contact between the host and the virus that is necessary for infection. Some of these 'normal infection processes', such as those by insects or fungi, can be changed by air pollutants. The wounding of the host plant can take place without any further interactions with insects or fungi in the case of air pollution through surface erosion or the damaged plant epidermis: hence the possibility of having more and more virus infections seems obvious.

The classic example of viral plant disease, Tobacco Mosaic (TMV) in tobacco leaves, was the first example described of the interactions between virus infection and air pollution. The information provided by TMV research shows that virus lesions increase when leaves are exposed to hydrogen fluoride (Dean & Treshow, 1965; Treshow *et al.*, 1967), whereas air-pollution effects decrease in the case of ozone (Yarwood & Middleton, 1959). Brennan & Leone (1969, 1970) observed systemic protection of tobacco plants from ozone when inoculated with TMV. According to a report by D. D. Davis (1978), the following virus systems have been demonstrated to protect plant foliage from subsequent ozone or ambient oxidant injury: systemic BCMV on Pinto Bean (*Phaseolus vulgaris* cultivar) plants; local lesion AMV, TRSV, TomRSV, and TMV, on Pinto Bean primary leaves; TRSV on Soybean (*Glycine max*) primary leaves: TEV and TMV on Tobacco (*Nicotiana tabacum*) in the field. The mechanism of protection has not yet been elucidated.

As for the interactions of bacterial infection with air pollution, the first information was provided by the early findings of several authors on the inactivation and sterilization of bacteria under the influence of ozone (Elford & Ende, 1942; I. Davis, 1961; Serat *et al.*, 1966). The bacteria are mostly disseminated by rain. The inability of bacteria to survive or be infective in acidic media is also well established. Shriner (1977) found that *Pseudomonas phaseolicola* (Burkh.) Dows. could not infect bean foliage after artificial acid rain treatment. The information on increased bacterial infection after abnormal abscission of leaves (J. R. Davis & English, 1969) following wind-driven rain can be modified to relate to air pollution, because air pollution injury often leads to abnormal, early abscission of affected young leaves (Guderian, 1977).

Nitrifying and nitrogen-fixing bacteria are very sensitive to acidity. Their activity declines rapidly with decreasing pH in the soil. The effects of acidifying air pollutants on symbiotic nitrifying bacteria must be considered one of the potentially most harmful effects of interactions, because the soil biology, the natural forest nitrogen-budget, and many cultivated plants, depend on this nitrifying or nitrogen-fixing capacity (Kallio, 1976).

Mechanisms of Interactions with Fungal Diseases

Pathogenic fungi can be classified into two groups, those capable of attacking completely healthy plants, and those attacking plants that have been injured in some way. In order to obtain the nutrients required for growth and development as well as reproduction, it is necessary for parasitic fungi to enter their host and establish direct contact. There are several means of penetration by which fungi breach the host. The primary ways include penetration by zoospore cysts or by non-motile propagules. The available evidence indicates that the host cell wall or cuticle is either degraded enzymatically during the penetration, or is weakened by a purely mechanical attack.

The mechanisms by which air pollutants or acidification may influence the pathologic processes are related to several stages of a disease. Fungal spores are disseminated by wind-blown rain. Free moisture on the leaf surface is essential for spore germination and external growth prior to penetration. The first stage in the formation of a new fungus colony is the germination of a spore. The infection of a plant by a parasitic fungus, and the resulting development of disease, is initiated, in most instances, by a spore germinating on, within, or near, the potential host plant. Spore germination is a crucial event for pathogenic fungi, because it is the determining factor in the onset of colonization.

There are several types of stimulants and inhibitors influencing spore germination on plant surfaces. Thus there is some evidence that the substances present on the surfaces of resistant plant species are inhibitory, while those on susceptile plants are stimulatory, to a particular pathogen (J. K. Sharma & Shinha, 1971). The substances produced by interaction with the host (phytoalexins) are particularly important in limiting the vegetative growth of the fungus, though the possible stimulatory effects of several compounds are discussed in many papers. Among these effects, there are several that can be affected by air pollutants, such as volatile aldehydes, steroid saponins, a wide range of terpenes and sesquiterpenes, etc. (Allen, 1959, 1976; Schönbeck & Schösser, 1976).

The results obtained by Kovacz & Szwöke (1956) provided some evidence about stimulants accompanying the inhibitors in water washings of leaves. Washing of surfaces by air pollutants may produce more stimulants or inhibitors that influence spore germination chemistry on wet surfaces. The germinating spore may penetrate directly through a primary barrier (the cuticle when ruptured) or through stomatal openings or wounded tissues. The pathogen is buffered by host tissue and is thus dependent on host metabolism. The initial effects of fungal infection on host cytology may be either stimulatory or degenerative.

Acidifying air pollutants may reduce pathogenic infections because of their ability to lower the ion concentration of moistened plant surfaces. In many urban and industrial areas, airborne particles of soot, dust, etc., are somewhat basic, however, and increase the pH, or neutralize the acidity, of moistened plant surfaces. In practice, the acidifying effect of air pollutants on the pathogen is only observable when no particle effect occurs on the host surfaces (Shriner & Cowling, 1980; Huttunen *et al.*, 1981a). In many polluted areas, the effect of particles has been found to increase the leaf surface pH and conductance, and to make the surfaces more basic than formerly. This is true of needles and other green surfaces having acidic reactions. The possible action of metals and toxic compounds on surfaces must also be considered.

There is evidence that increasing surface pH is connected with nitrogen dioxide and ammonia pollution, and even with sulphur dioxide pollution, when airborne particles are present (Huttunen *et al.*, 1980, 1981*a*, 1981*c*).

Several suggestions have been made concerning the importance of dry particulate deposition on foliar surfaces in relation to disease incidence. Cement kiln dust deposits have been reported to increase the incidence of fungus leaf-spot on Sugar beet (*Beta vulgaris* cultivar) leaves, and limestone processing dust appeared to stimulate leaf-spot infections on wild grape (*Vitis* sp.) and Sassafras (*Sassafras albidum*) leaves. There are several reports showing that leaf surfaces of trees growing along urban streets contain greater-than-normal quantities of certain particle-borne metal contaminants, and that these particles are effectively retained by the leaf surfaces (Shriner & Cowling, 1980).

The surface properties of tree bark are changed readily under air pollution conditions, and this may result in situations that are more suitable for some pathogens than others (Garber, 1962; Grill & Härtel, 1972; Lötschert & Kohm, 1978). The special reaction of several tree species has shown that the chemical composition of bark surfaces is easily changed, and this affects such symbiotic relations as the presence of lichens and aerophilic Algae.

The chemical composition of tree bark has been studied, and it is known that metals and other pollutants may accumulate on it (Laaksovirta *et al.*, 1976). Nitrogen oxides, ammonium compounds, and salt sprays, may all increase the basicity of the bark (Skye, 1968; Barkman, 1969; Grodzinska, 1971, 1976, 1977).

Martin & Juniper (1970), Preece & Dickinson (1971), and Dickinson & Preece (1976), have edited substantial volumes that review the role of the plant cuticle and epicuticular waxes as barriers to penetration by plant pathogens. Martin & Juniper (1970) discussed the weathering of epicuticular waxes on leaf surfaces by rain. Data suggest that the rate of weathering may be a function of the acidity of the rainfall impacting on the surface (Fowler *et al.*, 1980). Weathered or eroded plant surfaces may indeed pose a less formidable mechanical barrier than uneroded ones to direct penetration by pathogens, for weathered surfaces exhibit increased wettability, which may subsequently result in a significantly greater number of water-borne propagules being retained on the leaf surface in a position to effect successful penetration (Shriner & Cowling, 1980).

The effects of air pollution on plant cuticles and epidermal features have been studied by several authors (G. K. Sharma, 1977; Godzik, 1978; Godzik & Sassen, 1978; Huttunen & Laine, 1981; Fowler *et al.*, 1980). The degradation of the structure of the cuticular wax is most easily seen in coniferous needles. The degradation of *Pinus sylvestris* L. (Scots Pine) needle wax structure near the stomata was studied (Huttunen & Laine, 1981; Fowler *et al.*, 1980). Studies on the degradation of needle epicuticular waxes of *Pinus sylvestris* have shown that degradation of wax microfibres started soon after needle 'flushing', and clear signs of erosion and wax degradation were found in August—cf. Fig. 1(A) and 1(B).

The normal weathering of needle cuticles during the first year of growth is not very extensive, but under air pollution or acid rain conditions it may be many times faster than in unpolluted forest areas. Fig. 1(C) and 1(D) show normal and air pollution-induced erosion, respectively, during the first year of growth of Scots Pine needles (Huttunen *et al.*, 1980). The same phenomenon has been found in several forest species under air pollution conditions and in some garden species, such as *Rosa rugosa* (Huttunen *et al.*, 1980). Even mosses lacking a cuticle show typical air pollution induced surface degradation before visible injury appears (Huttunen *et al.*, 1980), though in their case the degradation is largely related to the injury or water deficit of surface cells.

The pathogen is buffered by host tissue, and is therefore dependent on the host's metabolism. Air pollution has some effect by changing and reducing the general buffering capacity of host cells, mainly through organic acid metabolism that may increase the possibility of the penetrating pathogen to colonize. Fungus and disease development have been found to change the natural buffering capacity as well against air pollutants (Grill & Härtel, 1972; Scholz & Stephan, 1974). The buffering capacity of host cells under the influence of air pollution can be considered an indicator of certain physiological conditions of the cells (Wind, 1979). The buffering capacity is species-specific and is influenced by genetic, seasonal, and environmental, factors (Wind, 1979; Huttunen *et al.*, 1981*a*, 1981*c*). The buffering capacity may be altered from being acidic in character to being more basic if the plant has been influenced by acidic substances, and the capacity may become more acidic if the plant has been influenced by basic substances.

The nutrient status and interactions with air pollution and plant pathogens have additional effects. The buffering capacity of Scots Pine needles on some test sites under sulphur dioxide pollution is schematically indicated in Fig. 2 (Huttunen *et al.*, 1981*c*).

The genetic variation of many secondary constituents that are important in host resistance, is very wide. Tobolski & Hanover (1971) found substantial genetic variation in the monoterpenes of *Pinus sylvestris* var. *lapponica*, which has a relatively high proportion of its monoterpenes in the 3-carene group. This group has been found to be influenced by air pollution in *Picea abies* (Lehtiö, 1981). The phenolic compounds and oxidative processes of higher plants have been found to increase as a consequence of air pollution (Härtel, 1972. Grill *et al.* (1975) found the total phenol content in SO_2-fumigated Norway Spruce needles to be increased. The increase was not due to picein or p-hydroxyacetophenone, two of the major phenols in Norway Spruce needles, as both of them were significantly decreased in SO_2-damaged needles. These

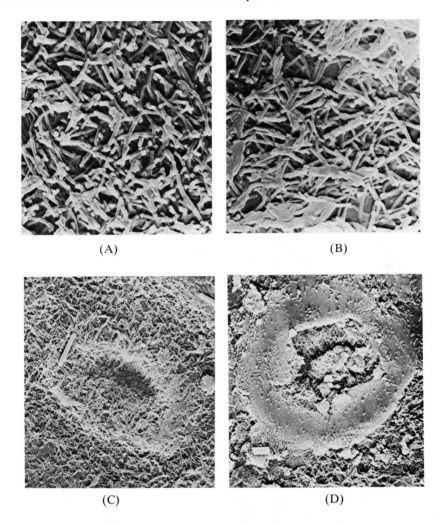

Fig. 1 The wax structure of the stomatal areas of Scots Pine (*Pinus sylvestris*) needles. (A) Stomatal wax structure of Scots Pine needles in unpolluted area ten weeks after needle 'flushing'. (B) Emerging effects of air-pollution-induced degradation with fusion of wax microfibres in the wax structure of the stomatal area ten weeks after needle 'flushing'. (C) Normal wax erosion in the stomatal area during the first year of the needles; microfibres are somewhat fused in needles eleven months old in May. (D) Air pollution induced wax erosion with severe degradation of the waxes of the stomatal area in needles eleven months old in May. (A) and (B) Magnification × 8,000. (C) and (D) Magnification × 1,600. Air pollution circumstances: (A) Clean air SO_2 < 10 μg per m^3. (B) Polluted air 30–40 μg per m^3 SO_2 and some NO_2. (C) Clean air SO_2 < 20 μg per m^3. (D) SO_2 50 μg per m^3. (Approximate results from several years' measurements.)

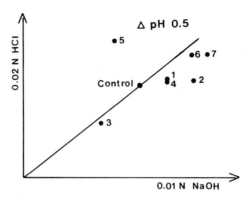

Fig. 2   The buffering capacity of *Pinus sylvestris* needles under sulphur dioxide and organic sulphur-compound pollution. Test sites 1 and 4: mean concentration annually 76 . . . 91 μg SO₂ per m³. Test sites 1 and 4: the trees show some acute needle-damage due to high short-time concentrations, but in other areas no injury except for cuticular erosion of needle surface waxes. Air pollution induced biochemical stress has been observed in all samples. Test area 5 has been fertilized by nitrogren fertilizer. Test area 3: the trees had *Lophodermium pinastri* Chev. in their needles. Jämsänkoski, Finland, 7 July 1978 (Huttunen *et al.*, 1980).

results suggest quite complex interactions, with many secondary plant products, under air pollution–pathogen interaction.

Interpretation of the changes in the oxidative enzymes of plants must be based on knowledge of the physiological role of enzymes and the regulation of their activity in an intact plant. The presence of peroxidases has been demonstrated in various subcellular components, including nuclei, mitochondria, ribosomes, cell wall, and cell membranes. The changes in the activity or the isoenzyme composition, of peroxidase are not a specific reaction of the plant to infection by the parasite; rather are they an accompanying characteristic of the altered metabolic activity of the plant cell under the various exogenous and endognous factors. In most cases it is difficult to evaluate the role of peroxidase in plant resistance. A number of investigations show that peroxidase activity in the host enhances plant resistance (Lovrekovich *et al.*, 1968a, 1968b). In numerous incompatible host–parasite combinations, peroxidase activity is often several times as high as in compatible ones (Daly *et al.*, 1971; Loon & Geelen, 1971). In some other cases, however, peroxidase activity is greater in compatible host–parasite combinations than in incompatible ones (Wood & Barbara, 1971).

The same phenomenon of increasing oxidative processes with the peroxidase and katalaze enzymes has been demonstrated under the influence of several air pollutants and in ambient air field conditions with a mixed type of air pollution, including metals (Keller & Schwager, 1971; Keller, 1974; Flückinger *et al.*, 1978; Niemtur, 1979; Schultz, 1979; Huttunen *et al.*, 1979b,

1980). Peroxidase activity has been established as one of the most sensitive indicators of air pollution effects (Horsman & Wellburn, 1975, 1977). The genetic diversity of peroxidases has been found to be quite large (Prus-Glowacki & Szweykowski, 1977). The genetic control of the peroxidase isoenzymes and their changes under phytochrome regulation has been noted (Frič, 1976).

Air pollution and biotic diseases have very similar mechanisms of action. Even with respect to oxidative processes, which may affect the host–pathogen interactions, the air pollution induced stress must be considered to be indicative of harmful effects. By the time air pollution has altered the oxidative process, the reaction against pathogens may be weakened. The same phenomenon may well be observable even in pre-infection stages, before air pollution effects become apparent.

Examples of Interactions between Air Pollution and Fungal Pathogens

The incidence of fungi infecting plant foliage has been reported to be reduced by atmospheric SO_2 in the vicinity of industries in several countries. The leaf fungi studied include *Microsphaera alni* (Wallr. ex Tul.) Wint., (Koch, 1935; Hibben & Taylor, 1975), *Hypodermella laricis, Lophodermium pinastri* Schrad. ex Fr.) Chev., and other *Hypodermella* spp. (Schaeffer & Hedgecock, 1955), *Diplocarpon rosae* (Saunders, 1970), *Venturia inaequalis* (Przybylski, 1967), *Hysterium pulicare* (Skye, 1968), and *Lophodermium juniperinum* and *Rhytisma acerinum* (Pers.) Fr. (Barkman *et al.,* 1969).

The ecology of *Rhytisma acerinum* is one of the best-known examples of air pollution and pathogen interactions. The incidence of the fungus was found to respond quantitatively to declining SO_2 levels, and the fungus has an upper tolerance limit of approximately 90 μg per m^3, above which it does not occur. Because of this definite cut-off point, the use of the fungus as a biological indicator of pollution, together with lichens, has been tested by Bevan & Greenhalgh (1976) and Bevan (1978). Similar results of bioindication with *Rhytisma acerinum* on the Dubener Heide have been reported by Dörfelt & Braun (1979). Earlier reports on *Rhytisma acerinum* have been published by Donaubauer (1966) and Barkman *et al.* (1969).

Saunders (1970) reported that Blackspot on roses (*Diplocarpon rosae* Wolf.) was checked or eliminated in areas where pollution of the atmosphere by sulphur dioxide exceeded 100 g per m^3. Dörfelt & Braun (1979) found that *Puccinia aegopodii* was more abundant in areas with heavy pollution, but the dominating factor was the nitrophilic nature of the fungus, rather than the air pollution alone.

Weinstein *et al.* (1975) studied the effects of SO_2 concentrations below 0.40 mg per m^3 on the incidence and severity of bean rust and early blight of Tomato (*Lycopersicum esculentum*). Sulphur dioxide affected bean rust, but not early blight of Tomato, under the conditions used in the experiment. The effect of SO_2 on bean rust consisted of a decrease in the incidence and severity of the disease, and in the size and percentage germination of uredospores. These effects were due to exposure of plants to SO_2 before or after inoculation with the pathogen, but exposures before inoculation were more effective than after it.

Hydrogen fluoride has been reported to decrease both the number of uredia and the rate of their development (McCune *et al.*, 1973), while exposure to ozone increased the number, but decreased the size, of uredia (Resh & Runeckles, 1973). Laurence *et al.* (1978) reported that subacute doses of sulphur dioxide applied after inoculation with *Puccinia graminis* f. sp. *tritici* caused a reduction in the number of pustules formed on Thatcher Wheat, but not on Prelude Wheat. The severity of Southern Corn leaf-blight (*Helminthosporium maydis* Race T) was affected by pre- and post-inoculation exposure.

In foresty, leaf- and needle-decaying fungi, as well as wood-decaying fungi, must all be considered. Grzywacz (1978) reported, on the basis of field and laboratory investigations, that *Pinus sylvestris* wood from stands situated within areas impacted by air pollution, mostly of SO_2, had different natural resistance against fungal decomposition in comparison with wood from stands not injured by pollution. Wood from low pollution areas was less resistant to decomposition and hence of poorer technical value. The anatomical structure of wood grown in a polluted atmosphere has been reported to change (Liese *et al.*, 1975; Liese & Eckstein, 1978). Late wood increases and early wood decreases in polluted areas, especially when nitrogen dioxides are present (Havas & Huttunen, 1972). Changes in wood composition have been revealed even by x-ray analysis (Huttunen, 1979*a*). These might have effects on the severity of decaying fungi. The possibly decreasing inhibitory effects of resins also must be borne in mind.

Under Finnish conditions, *Lophodermium pinastri* Chev. has been found to be quite common in air polluted areas (Huttunen, 1979*b*). The most severely polluted area, with SO_2 in some months exceeding 100 μg per m^3, and the areas between 30 μg and 60 μg, have shown a typical infection by *Lophodermium pinastri* Chev. after chronic air pollution stress.

The incidence of Snow Blight on Scots Pine is affected by nutrition (Kurkela, 1975), and the disease was quite abundant under mixed nitrogen dioxide and sulphur dioxide pollution (Huttunen, 1978*c*). Infection with Snow Blight correlated somewhat with the effects of air pollution on trees of different provenances. The main cause of the sensitivity was the disturbance of the nutrient conditions (Kurkela, 1975). In an industrial town on the southwestern coast of Finland, having averages of 40–50 μg per m^3 SO_2, trees were found to have been affected by recent epidemics of *Lophodermella sulcigena* Hohn (Huttunen *et al.*, 1979*b*).

Among the fungi that have their pathogenicity enhanced by industrial emissions, *Armillaria mellea* (W. Novak *et al*., 1957; Jančařik, 1961; Kudela & Novakova, 1962; Darley & Middleton, 1966; Sierpinski, 1972*a*, 1972*c*; Grzywacz, 1973), *Rhizosphaera kalkhoffii* Bubak (Chiba & Tanaka, 1968), *Lophodermium pinastri* (Schrad.) Chev. (Costonis & Sinclair, 1967), *Hirschioporus abietinus* (Dicks. ex Fr.) Donk. (W. Novak *et al.,* 1957; Jančařik, 1961; Grzywacz, 1973), *Hirschioporus fusco-violaceus* (Ehrenb. ex Fr.) Donk. (Sierpinski, 1970; Grzywacz, 1973), *Schizophyllum commune* Fr., *Nectria cinnabarina* Fr., *Stereum pini* Fr., and other species of the genus *Stereum* devouring the wood of deciduous species (Jančařik, 1961; Grzywacz, 1973), all provide good examples.

In forests suffering from chronic air pollution, the species composition of the saprophytic fungal flora on the leaves is apt to change (Manning, 1971). A number of studies, mainly experimental ones, concerning the influence of air pollution on the fungal pathogens of crop plants in agriculture and gardening, indicate that the interaction of fungi with air pollution is of synergistic character (Manning *et al.,* 1969, 1970, 1971).

Grzywacz (1973) tested *Schizophyllum commune* Fr. as being a SO_2- resistant, and *Fomes annosus* (Fr.) Cke. a SO_2-sensitive, species. He found that low concentrations, when persisting over protracted periods, stimulated mycelial growth. Sulphur dioxide stimulated linear growth of the hyphae when the concentration was low and the time was short. This phenomenon is noticeable in what is called 'escape of mycelium' when visualized in the graphic compari-

Table I   Possible interactions between plant diseases and air pollution.

Process	Mechanism of interaction
Spore germination	Surface structures and their wettability; surface chemistry
Cytology and physiology of penetration and establishment	Surface erosion, stomatal injury, etc.; surface chemistry
Physiology of host response	Biochemistry of the host Metabolic disturbances Energy-activity disturbances Disturbances of permeability Water status Protein metabolism Carbon balance Growth regulators
Modification of host response	Natural environmental factors such as temperature, humidity, light, etc.; other chemicals than air pollutants etc.

son of the colony diameter and the dry weight of mycelium. The concentrations that stimulated growth were between 0.01 and 1 mg per m^3 SO$_2$. Higher sulphur dioxide concentrations were found to produce morphological changes in the mycelial colony and to disturb fructification processes.

The sensitivity of the tested fungal species to SO$_2$ was dependent on their physiological activity. The higher the protein content in the hyphal cells, and the higher the respiration intensity, the more sensitive the species was to SO$_2$. Grzywacz (1973) pointed out that these species are sensitive to sulphur dioxide; some more recent papers have suggested this as a possible benefit of SO$_2$ pollution (OECD, 1978). But the SO$_2$-concentrations that are of practical importance are ones that stimulate the growth of the pathogen or have only minor effects on pathogen respiration. The recent information on the indirect effects of air pollution in increasing acidification suggest that even very small amounts of air pollutants and acidity may increase the susceptibility of the host to these injurious agents, alter the plant's capability to tolerate disease, or alter the virulence of pathogens (Table I).

INTERACTIONS WITH INSECTS

Research on host–insect interactions as an additional stress factor of plants in air polluted areas requires a multidisciplinary approach by geneticists, physiologists, entomologists, and meteorologists. The complex can be discussed mainly with reference to trees, because they are deeply affected by both of these stress factors. The complexity of the normal host–insect relationships has been discussed by Hanover (1975) and Philogene (1972). Host resistance has been defined in this context by Beck (1965) as the collective heritable characteristics by which a plant species, race, clone, or individual, may reduce the probability of successful utilization of the plant as a host by an insect species, biotype, or individual.

Some environmental variables may modify the expression of genetic resistance and occasionally increase host susceptibility. Variation between different provenances, for instance, has been observed in host–insect relationships. Hanover (1975) classified the imbalances in the insect–tree relationships leading to tree resistance as being due to variation in any of the four basic host characteristics: (1) morphology and anatomy of the host, (2) chemical repellents produced by the host, (3) chemical attractants produced by the host, and (4) nutritional status of the host. These four broad categories of host behaviour are also dependent upon phenological and ontogenetic patterns of variation, such as the variation associated with season and the age of the tree.

Several reports published during the past decade have contained information showing that these basic characteristics of host–insect relationships can be affected by environmental factors, such as damage of trees through mechanical injury in forest management, through chemical fertilization resulting in an imbalance in the nutritional or microelement conditions of forest soil, or through air pollution (Hanover, 1975; Smelyanets, 1977).

The effects of air pollution as an additional stress and risk-factor for insect attacks in forest and agricultural species have been reported in several countries especially in Europe for forestry (e.g. V. Novak, 1962; Templin, 1962; Schneider & Sierpinski, 1967; Sierpinski, 1968; 1972*a*, 1972*b*, 1972*c*; Schneider & Chłodny, 1977; Sierpinski, 1979, 1980; Chłodny, 1979; Przybylski, 1979). The effects of various pollutants on the occurrence of insects has been discussed for sulphur dioxide by Schneider & Chłodny (1977), for nitrogen oxides by Sierpinski (1979), and for hydrogen fluorides by Sierpinski & Szalonek (1974) and Weinstein (1977). Perhaps the most important information acquired so far has been that pertaining to forests with secondary noxious insects, which can be of great economic importance under a chronic influence of air pollution.

The possibility of increased acidity of 'acid rain' affecting host–insect relationships has been touched upon by Shriner (1977), and Cowling (1978). In many urban areas, typical urban insect attacks, especially on trees, but even on garden and agricultural species, have been reported, and the possibility of interactions with air pollution discussed.

Different Air Pollution Influenced Factors Affecting Insect Attacks

Bennett (1954) correlated the degree of insect attack by the pine needle Miner (*Exoteleia piniforiella*) with the number and size of resin ducts in several 'hard' pine species. The avoidance of the resin ducts by White pine Weevils was demonstrated by Stroh & Gerhold (1965). Plank & Gerhold (1965) reported that resistant Western White Pine (*Pinus monticola*) trees had a greater number of external cortical resin-ducts than the Eastern White Pine (*P. strobus*), which is more sensitive. The possibility of air pollutants damaging needle resin ducts has been suggested by Huttunen (1975). The damaging effects of air pollution on resin ducts were first observable in the epithelial cells of the ducts, which may have some effect on resin production (*Ibid.*).

The effects of air pollution on leaf and needle morphology, and on the cuticle and cuticular waxes, have been demonstrated in several plant species (Stewart *et al.*, 1973; G. K. Sharma, 1977; Godzik & Sassen, 1978; Huttunen *et al.*, 1981*c*). The wax layer is eroded by air pollution and acid rain, and even the chemical decomposition of waxes may be altered (Godzik, 1978; Huttunen & Laine, 1981; Fowler *et al.*, 1980). Injuries in needle and cortical resin ducts may influence the attractants of insects by altering the composition or properties of oleoresin. Even changes in the viscosity of oleoresin must be taken into account (Mergen *et al.*, 1955; Hanover, 1975).

In discussing the effects of stomatal disturbances on host–insect relationships, the effects on water balance should also be considered. Low sulphur dioxide concentrations have been observed to stimulate stomatal opening, but the effects of air pollution on stomata are modified by various environmental factors (Noland & Kozlowski, 1979).

Air pollution induced injuries and other effects often disturb the water balance of plants, especially conifers (Halbwachs, 1968; Huttunen *et al.,* 1980, 1981*a*). Air pollution may modify the morphology of the host species by fertilizing (nitrogen oxides, ammonia) or by reducing growth (sulphur dioxide, fluorides). Airborne nitrogen compounds have fertilizing effects on vegetation: conifer needles, especially, have been found to be more mesomorphic and contain more water in areas polluted by nitrogen dioxide and ammonia than in other areas (Havas, 1971; Huttunen, 1975, 1978). These features may increase the insect attacks because of the altered leaf water balance or increased fertilization (Vite, 1961; House, 1961, 1962, 1966).

Direct contact with oleoresin has been shown to be a primary factor in insect resistance by tree species (Hanover, 1975). The possible interactions between air pollution effects and the chemical repellent or attractant produced by the host is very interesting. The chemicals that have been found effectively to attract insects, and that can be qualitatively and quantitatively changed by air pollution, include monoterpenes, fatty acids, benzoic acid, and ethanol (Yasunaga *et al.*, 1962; Rudinsky, 1966; Rottink & Hanover, 1972).

Different kinds of glycocides have been found to be the main repellents in resistant trees, and direct contact with oleoresin has been reported to be the reason for host resistance (Hanover, 1975). After plant tissue injury, the release of volatile substances from the host soon provides the primary attraction for insects (Anderson & Anderson, 1968).

Air pollution has been found to affect the composition of oleoresin in conifers. Lehtiö (1981) found that the high carene group volatile oil decreased in air pollution damaged spruce needles in areas with SO_2, NO_x and HF pollution. Cobb *et al.* (1972) noted that methyl chavicol decreased in photochemical pollution injured *Pinus ponderosa* (Ponderosa Pine) trees. The terpene content of *Picea abies* (Norway Spruce) exposed to air pollution was found by Dässler (1964) to be modified. Relationships were found between the seasonal amounts of volatile substances in air-pollution-resistant trees (Pelz & Materna, 1964). Some related effects have been found in connection with forest fertilization (Hiltunen *et al.,* 1975).

Phenolic substances have been found to increase as a consequence of air pollution (Härtel, 1972; Grill *et al.,* 1975). The changes in phenolic compounds may have either positive or negative effects on host–insect relationships. Positive effects of increasing amounts of phenolic compounds in restricting insect attacks might be observable in older needles.

Amino acid metabolism, also, has been found to be affected by air pol-

Table II Physical and chemical properties of air pollution stressed hosts, possibly affecting the host–insect relationships (modified from Hanover, 1975).

Surface architecture	leaves, needles, bark
Erosion of surface waxes of hairy coat erosion	leaves, needles
pH and surface-accumulated substances	leaves, needles, bark
Surface injuries	leaves, needles, buds
Resin ducts and their injuries	needles, bark, buds
Stomata and their injuries	leaves, needles
Oleoresin composition and its changes	needles, bark, wood
Phenolic compounds and their changes	leaves, needles, bark, wood
Amino-acid content, sugar content, etc.	leaves, needles
Water-balance	leaves, needles, buds

lutants (Jäger & Pahlich, 1972; Jäger & Grill, 1975). This might be one reason for the changing status of insect nutrients. In early summer, air pollution affected plants have a higher sugar content than healthy plants, which contributes to the host–insect relationships. Table II shows the physical and chemical properties of the air pollution stressed host, possibly affecting the host–insect relationships.

The fact that oleoresin has a very wide range of variation in its chemical and physical properties, is of fundamental importance for the potential insect-resistance mechanisms of trees. Both the monoterpenes and diterpenes, and the resin acids, are under definite genetic control, and the modifying or selecting effects of air pollution may have detrimental effects on these genetic aspects of resistance (Tobolski & Hanover, 1971).

The changes in the chemical composition of coniferous bark are often related to lichen assays, but must be considered also in relation to insect attacks (Grodzinska, 1971, 1976, 1977; Lötschert & Köhm, 1978).

Some Examples of Interactions between Air Pollution and Insects

Templin (1962) reported that the mining insects *Rhyacionia buoliana* Den. & Schiff. and *Exoteleia dodecella* L., were of permanent occurrence in air pollution injured areas, but that the occurrence decreased with increasing distance from industry. The reason for the increase of these populations was the disturbance of water relations in *Pinus* forests. The cortical insects, such as *Blastophagus piniperda* L. and *Melanophila cyanea* Fr., did not increase enough to reach mass occurrence, but some intensive deaths of trees were found in air pollution areas.

In his more recent works, Sierpinski (1980) reported that air pollution exerts a great influence on the population density of various forest pests. Attention should be paid, first of all, to pests of pine plantations and thickets—especially to the butterflies *Exoteleia dodecella* L. and *Rhyacionia buoliana* Den. & Schiff., which mine the buds and thus lead to long-lasting deformation of trees. Locally, they are accompanied by *Blastethia turiionella* L., *Dioryctia mutatella* Fuchs, *Pertova resinella* L., *Aradus cinnamomeus* Panz., *Thecodiplosis brachyanthera* Schwag., and others.

Older stands stressed by air pollution occasionally display an increased appearance of foliophages, such as *Acantholyda nemoralis* Thoms., *Lumantria monacha* L., and *Diprion* spp. Insects belonging to the groups of cambiophages and xylophages, find there favourable developing conditions: *Pissodes piniphilus* Hbst., *Monochamus galloprovinsialis* Ol., *Melanophila cyanea* F., and *Plastophagus piniperda* L., feeding in the phloem and cambium of trees, and *Trypodendron lineatum* Ol., *Criocephalus rusticus* L., and *Sirex* spp., feeding in wood (Sierpinski, 1980), are of economic importance.

Differences in the host–insect relationships have been observed under different types of air pollution. Smaller populations of *Exoteleia dodecella* L., *Thecodiplosis brachyanthera* Schwag., and *Aradus cinnamomeus,* were found in young *Pinus sylvestris* stands in fluorine-polluted environments than on sites influenced by sulphur compounds. In areas of thinned forest, needles were destroyed by *Ocmerostoma piniarella* Zell. Persistent pests, such as *Rhyacionia buoliana, Petrova resinella, Brachyderes incanus* L., *Blasthethia turionella* L., etc., which are common in young pine forests that are susceptible to sulphur dioxide influence, were not significant. In industrial areas with nitrogen dioxide pollution, leaf-feeding insects are most abundant, and increase any damage caused by nitrogen dioxide. Increasing numbers of *Tortrix viridiana* L., *Operophthera brumata* L., *Erannis defoliaria*, and *Hyponomeuta cognatella,* were observed (Sierpinski, 1979). Damage induced by *Xylophagen* is of minor importance.

In areas with chronic industrial air pollution, insects classified by foresters as belonging to the group of 'secondary noxious' insects are of great economic importance. They kill trees that are weakened by industrial emissions or occasionally by parasitic fungi and feeding by *Exoteleia dodecella.* The species composition of the populations of noxious insects is largely dependent on tree stand density and the intensity of penetration by industrial gases and smoke. In thinned stands, *Melanophila cyanea, Pissodes piniphilus* Hbst., and *Siricidae* are most common. In denser and more shady stands, *Blastophagus piniperda, B. minor* Hbst., *Acanhocinus aedilis* L., *Trypodendron lineatum, Pissodes piniphilus,* and *Ips acuminatus* Gyll., are locally abundant (Sierpinski, 1968).

Ranft (1968) studied bark-beetle attacks on gas damaged pine and spruce stands, and found more severe attacks in areas with greater pollution damage. A similar relationship between attacks by *Adelges abietes* and the degree of gas damage was found in the Erzgebirge.

Air pollution has also effected the entomofauna of forest plantations in another area of heavy industrial pollution, where *Thecodiplosis brachyanth-*

era, Exoteleia dodecella, and *Brachonix pineti* Payk., were more abundant than elsewhere (Schneider & Chłodny, 1977).

Among 'primary' insects, *Acantholyda posticalis* Mats. has turned out to be most resistant to air pollution conditions. Where pollution damage was slight, *Lymantria monacha* L. sometimes showed mass occurrence (Sierpinski, 1968).

In urban areas, the occurrence of Aphilidae seems to follow air pollution induced stress. In Finland during the summer of 1978, the mass occurrence of Aphilidae on birches (*Betula verrucosa* Ehrh. and *B. pubescens* Ehrh.) and Little leaf Linden (*Tilia cordata* Mill.) leaves in the most polluted urban areas was compared with that in less polluted areas. An increasing trend of infestation with pollution was noticed by Huttunen *et al.* (1979*b*).

INTERACTIONS WITH NON-BIOTIC FACTORS

Stress has been defined as any environmental factor that is potentially unfavourable to living organisms; the stress resistance of a plant is it ability to survive such an unfavourable factor (Levitt, 1972). Among the stresses are biotic factors, such as infections and competition, and physiochemical factors, such as temperature, water relations, radiation, chemicals, and wind pressure etc. In the case of temperature, low temperature tolerance (frost and winter cold), and high temperature tolerance (heat), must be considered. As for water relations, both deficit (drought) and excess (flooding) must be included. Among the other stress factors, wind speed is one of the most important with regard to interactions.

When considering plant resistance from the viewpoint of air pollutants concentration, plants have been said to be more resistant during winter than during the vegetative growing season. But considerations of this kind have no real relevance. During winter the metabolic processes are not directly related to growth, but rather to the energy budget of wintering plant cells or to the wintering of the photosynthetic apparatus. The resistance situation is more complex than the mere combination of pollutant concentrations and exposure time. The considerations must focus on the periods of hardening, predormancy, dormancy, and postdormancy, and on all the fluctuations to which the plant may be exposed during the vegetational period—including drought, frosty nights, etc.

The modifying effects of temperature, humidity, light, and edaphic factors, have been discussed in the literature (Guderian, 1977; Weinstein, 1977). Light and stomatal behaviour, and their effects during the growth period, are very important to air pollutant uptake. Air pollution induced changes in surface features and stomatal functions are of importance to stress interactions (Unsworth *et al.,* 1976; Noland & Kozlowski, 1979). Humidity as an additional factor, can be discussed in circumstances where humidity has an increasing effect or is the maintaining factor for some limited metabolic functions. Soil moisture has been observed to have similar effects to air moisture on stomatal opening and the uptake rates of air pollutants (Guderian, 1977).

Less visible air pollutant induced injury has been found when fertilizing experiments have been carried out (Materna, 1962, 1963; Trillmich, 1969; Olszowski, 1976) than when they have not. Ashenden & Mansfield (1977) found that wind speed can modify the effects of air pollution on *Lolium perenne* (Perennial Rye grass). This shows that some of the experimental data are useless with respect to field conditions, and hence the effects of air pollution must be considered more and more under normal seasonal and field conditions.

Air pollution induced stresses are often most obvious in woody plants. Other environmental stresses, e.g. wintering, which are considered as natural, are also particularly evident in woody plants when these are subject to seasonal changes—such as when a snow cover or other protective factor is lacking. One of the early modifications of plant response to the polluted environment to be observed, was the air pollution induced reduction of frost resistance in forest trees (Munch, 1933; Huber, 1956; Wentzel, 1956, 1965; Ziegler *et al.*, 1958; Materna, 1979). These early observations pertained to the reduced frost resistance in areas of marginal air pollution damage (Lux, 1964; Materna, 1979). In the nordic countries, observations on increasing injury to conifers during the wintering periods were made by Havas (1971), Havas & Huttunen (1972), and Huttunen (1973, 1974, 1975). The possible effects of air pollution on winter crop plants have been discussed by Guderian & Stratmann (1968), Guderian (1977), and the OECD report (1978).

Winter crop plants have appeared to be more sensitive to SO_2 during winter than during summer (OECD, 1978). A reduction in resistance to water deficit and winter drought has been noted under air pollution conditions (Halbwachs, 1968; Huttunen *et al.*, 1980, 1981*a*). The possibility of increased microscopic and ultrastructural cell injury in conifers is considerable when the temperatures are over $0°C$—especially during the long late-winter period when light intensity, and occasionally daytime temperatures, are high (Börtitz, 1969; Havas, 1971; Huttunen, 1979*c*; Huttunen *et al.*, 1980).

The role of air pollutants as additional stress factors is particularly important under the long-term effects of low concentrations. The earlier findings on reduced frost resistance in pollutant damaged trees must focus on ambient-air concentrations. Badly injured plant cells cannot respond to environmental changes, and the cold is only the final killing agent. The total amount of tree damage increases because of the secondary injury induced by cold.

There are several possibilities of air pollution modifying or disturbing the normal seasonal rhythm of plants, depending on the severity of air pollution induced injuries or stresses. These possibilities are discussed at three different 'stress doses' (invisible metabolic injuries) and possible pathways of additional interactions (in Table III) (Huttunen, 1979a; Huttunen, 1982).

The cold resistance of woody plants is not constant in winter but changes with temperature. In autumn, plants that are capable of hardening start their hardening period as the length of day and the temperature begin to change. The quality and intensity of light also have an important influence. The hardening period can be explained in terms of having two different stages, the first stage being with biochemical changes and the second more with physical changes. After the wintering period, the postdormancy period retains the hardening mechanisms for the growth period. Under the influence of air pollution, all these stages can be affected, either alone or separately–depending on the intensity of the air pollution induced stress.

Acute injury, at either a visible or microscopic level, has severe consequences, due to cell damage, on water deficiency and cold resistance. Disturbances in wintering or frost resistance can be found (Havas, 1971; Huttunen, 1974, 1975, 1978). Chronic injury, which is produced gradually as membrane injury during unfavourable periods, can be increased, the marginal environmental conditions being the injuring agent. The effects on ultrastructure indicate, however, that the toxic effect of different pollutants has different types of injury patterns (Soikkeli & Tuovinen, 1979), but the relationships between the biochemical and ultrastructural injuries are not fully known. The metabolic stress effects, e.g. oxidative processes, seem to proceed at the same time as the injuries start to develop. The oxidative processes and the accumulation of air pollutants, have clear environmental connections (Fig. 3). Depending on the duration of the stress, new injuries can develop late in the period of wintering (Huttunen *et al.*, 1981c).

The question of the nature of stress is very interesting. The total activity of peroxidases can be taken as an indicator of the stress metabolism of needles. Both qualitative and quantitative changes have been found in peroxidases during the hardening to withstand cold temperatures. It seems that oxidative processes change markedly during the hardening period. At the first stage, marked changes were found in the oxidative processes related to air pollution in winter (Keller, 1974; Huttunen, 1981). Further information has been to the effect that the increase in oxidative processes is especially correlated to air pollution conditions in autumn and winter, and that stress is most common under midwinter conditions (Huttunen *et al.*, 1979a, 1979b, 1980).

Another air pollution induced change is in the sugar content of leaves and needles. Materna (1972) found that glucose and sucrose concentrations decreased in spruce needles exposed to air pollutants. Huttunen *et al.* (1980, 1981c) found that the total sugar content was higher in polluted than in unpolluted needles, and the amount of sugars in autumn was less in polluted than in unpolluted needles of Norway Spruce (*Picea abies*). But no similar results could be verified for Scots Pine (*Pinus sylvestris*). Air pollution

Table III Stress effects of airborne pollutants on forest trees adjusted for different dose relationships (modified according to Huttunen, 1979 and 1982).

Season	Acute effects	Chronic effects	Stress effects (invisible effects)
	Stress derived from preceding year		
Growth period	Delay of growth period. High risk of acute effects on current-year needles during the flushing and elongation period. Effects during leaf development and bud flushing.	Buds and needles transfer the stress from the previous season. Biological stress effects (derived even from previous seasons) changes, or 'strains' of inhibited and delayed growth.	Development of stress effects/stress (reversible, irreversible) in some weeks.
Vegetational period	Risk of acute effects during the period of physiological differentiation in conifers; marked effects on leaves of broad-leafed trees; blocking of photosynthesis.	Accumulation of toxic compounds on the green surface and in the needles and leaves; stress effects e.g. increasing peroxidase activity or increasing respiration activity.	Heavy environmental stresses; accumulation of toxic substances in leaves and needles; stress effects e.g. increasing peroxidase activity.
Hardening period	Disturbances in timing the process of hardening; severe stress in wintering buds.		

Predormancy	Severe injuries due to heavy cold; timing disturbances.		Breakdown of stress tolerance (buffering capacity, oxidative processes).
Dormancy	Primarily membrane structure injuries; stomatal response injuries; energy disbalance.	Development of heavy stress effects. Disturbances of energy flow. Development of typical needle symptoms (not always). Bud deaths.	
Postdormancy period	Winter water stress due to radiation conditions. Typical post effects of winter injury; inhibition of photosynthesis; stress respiration; compensatory effects; yield losses; delay of next growth period.	Stress effects even on the current-year needles and in opening buds. Yield losses after some years.	Post-wintering stress changes water deficiency etc. Stress effects. Changes after some years.

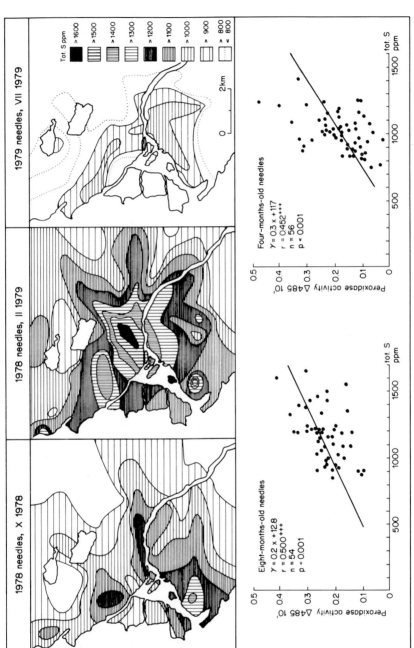

Fig. 3   The dispersion of airborne sulphur compounds in urban forests of Oulu (65°N). The mean total sulphur content of *Pinus sylvestris* needles in October 1978 (X), February 1979 (II), and July 1979 (VII). The correlation is between sulphur content and peroxidase activity in February 1979 (eight-months-old needles) and October (four-months-old needles). The correlation indicates the greatest stress under winter circumstances (in July 1979, four-weeks-old needles have no significant correlation with the stress) (Huttunen *et al.*, 1980)

induced stress seemed to delay the growth period, so that the changes in sugar concentrations were postponed compared with normal conditions. The increase in sugars was less, and occurred later, than normally, so that the timing of the processes was altered. A similar observation on timing disturbances was made for transpiration from needles; the deep dormancy stage was not fully developed, the polluted needles being ready to transpire in January (Huttunen *et al.*, 1981*a*).

The wintertime erosion of the green-needle surface, and its possible effects on the winter water economy, have been discussed. It is obvious that accumulation of pollutants on green surfaces in the winter causes degradation. This was most rapid during January and February (Huttunen & Laine, 1981). Some of the *Pinus sylvestris* needles died because of stomatal wax erosion.

In the case of Norway Spruce, frost resistance is reduced because of air pollution. In winter, coniferous trees usually tolerate much colder temperatures than normally occur (Larcher, 1975), but wintering disturbances may be associated with winter drought (Huttunen *et al.*, 1981*a*), or with the toxic properties of accumulated pollutants that affect the membranes of the photosynthesizing apparatus (Soikkeli & Tuovinen, 1979) or the energy metabolism (Keller & Schwager, 1977; Huttunen, 1982).

The increase in oxidative processes to approaching the toxic level must be one of the explanations for the possibility of having injuries without any measurable metabolic stresses during the growth period. The water deficit, and metabolically coupled buffering capacity, are the main ways that the plant cells can compensate or defend against the pollutant stress. During the winter, this capacity is very limited because it is the biologically non-active period (Huttunen *et al.*, 1980).

Investigations into the biochemical background of air pollution induced winter injury has shown that changes in cold tolerance, cellular osmotic values, sugar content, etc., are common in the case of acute injury and can occur in cases of chronic injuries. Whenever such injuries develop, the following winter becomes critical for the coniferous needles (Huttunen, 1979*c*; Huttunen *et al.*, 1980). In the presence of nitrogen compounds, cold-tolerance seems to be disturbed in many ways (Huttunen *et al.*, 1980), while in the case of sulphur dioxide the situation is more clearly connected with the toxicity of pollutant concentrations (Huttunen *et al.*, 1980).

The winter pollutant content of buds and needles, and the winter uptake of pollutants, have been discussed recently, and the question of using the pollutant content as an indicator has been raised again (Huttunen, 1979*c*,1981; Keller, 1978*a*; Keller & Jäger, 1980). In recent papers by Keller (1978*b*), Keller & Jäger (1980), Huttunen (1979*c*), and Huttunen *et al.* (1981*a*), the sulphur content has been studied on a seasonal basis, and the earlier information on the winter uptake of pollutants has been confirmed (Materna & Kohout, 1963; Materna, 1974, 1979; Huttunen, 1975).

As a measure of the stress caused by a given dose of pollutants, and especially as a measure of more extensive exposures to air pollutants, the sulphur compound accumulation has proven particularly useful. Such a relationship between sulphur accumulation and stress has been demonstrated (Huttunen *et al.*, 1980; Huttunen, 1981, 1982). In these cases the role of increasing environmental stress cannot be separated from the increase in air pollution. The increase in air pollution during the winter months is one of the interacting factors in the environment (Huttunen, 1982). This has been established as one of the most important factors pertaining to the possible effects of air pollution on coniferous forests under nordic conditions (Huttunen *et al.*, 1981c). The most unfavourable pollution situation coincides with the onset of autumn, whereupon an increasing chronic or subchronic stress development can be seen.

SUMMARY

Plants are adapted to the environmental conditions in which they have evolved, including many natural stresses; but air pollutants, as well as acting directly, may act indirectly by altering plant tolerance and response to such stresses. They may also influence interactions with the plant diseases caused by viruses, bacteria, and fungi. Any stage of parasitism may be affected, and the effect may inhibit or enhance disease expression—depending on the pathogen, the host, and the pollutant. The interaction depends upon the aspect on which the pollutant exerts the strongest effect. Pollutants, most simply, may weaken plants to the point at which they become more sensitive then formerly to a pathogen. More subtly, interactions may involve altering the acidity or otherwise affecting the plant's chemistry to the extent of altering normal mechanisms of tolerance, or disrupting nutrient or water balance relationships.

Pollutant–insect interaction may be attributable either to the altered metabolic products of the host, or to direct effects on the insects. The outcome tends to be most serious where secondary noxious insects occur in areas of chronic pollution and the weakened trees may be killed.

Non-biotic factors are also influenced by pollutants—notably winter injury, which appears to be accentuated in the presence of air pollution. They are associated with cold tolerance, cuticle integrity and water-balance—or, in the case of sulphur deposition, with the toxicity of sulphur accumulation.

REFERENCES

Allen, P. J. (1959). Physiology and biochemistry of defense. Pp. 435–67, in *Plant Pathology*, Vol. I (Ed. J. G. Horsfall & A. E. Diamond), Academic Press, New York & London: xii + 674 pp., illustr.

Allen, P. J. (1976). Control of spore germination and infection structure formation in the Fungi. Pp. 51–85 in *Encyclopaedia of Plant Physiology, Vol. 4, Physiological Plant Pathology* (Ed. R. Heitefuss & P. H. Williams). Springer Verlag, Berlin–Heidelberg–New York: 890 pp.

Anderson, N. H. & Anderson, D. B. (1968). Ips bark-beetle attacks and brood development on a lightening-struck pine in relation to its physiological decline. *Florida Entomol.*, **51**, pp. 23–30.

Ashenden, T. W. & Mansfield, T. A. (1977). Influence of wind-speed on the sensitivity of Ryegrass to SO_2. *J. of Exp. Bot.*, **28**(104), pp. 729–35.

Barkman, J. J. (1969). The influence of air pollution on bryophytes and lichens. Pp. 197–209 in *Air Pollution* (Proc. First European Congress on the Influence of Air Pollution on Plants and Animals, Wageningen 1968.) Wageningen, The Netherlands: 419 pp., illustr.

Barkman, J. J., Rose, F. & Westhoff, V. (1969). Discussion in Section 5: The effects of air pollution on non-vascular plants. Pp. in *Air Pollution* (Proc. First European Congress on the Influence of Air Pollution on Plants and Animals, Wageningen 1968.) Wageningen, The Netherlands: 419 pp., illustr.

Beck, S. D. (1965). Resistance of plants to insects. *Ann. Rev. Entomol.*, **10**, pp. 207–32.

Bennett, W. H. (1954). The effect of needle structure upon the susceptibility of hosts of Pine-needle Miner (*Exoteleia piniforiella* Chamb.). *Can. Entomol.*, **86**, pp. 49–54.

Bevan, R. J. (1978). The ecology of *Rhytisma acerinum* in relation to atmospheric pollution. P. 351 in *Abstracts of Papers, 3rd International Congress of Plant Pathology, München, 16–23 August 1978*. Paul Parey, Berlin & Hamburg: 435 pp.

Bevan, R. J. & Greenhalgh, G. N. (1976). *Rhytisma acerinum* as a biological indicator of pollution. *Environ Pollut.*, **10**, pp. 271–85, 3 figs.

Brennan, E. & Leone, I. A. (1969). Suppression of ozone toxicity symptoms in virus-infected tobacco. *Phytopathology*, **59**, pp. 263–4.

Brennan, E. & Leone, I. A. (1970). Interaction of tobacco mosaic virus and ozone in *Nicotiana sylvestris*. *J. Air Pollut. Contr. Ass.*, **20**, p. 470.

Börtitz, S. (1969). Physiologische und biochemische Beiträge zur Rauchschadenforschung, 8 Mitteilung: Physiologische Untersuchungen über die Wirkung von SO_2 auf den Stoffwechsel von Koniferennadeln im Winter. *Biol. Zbl.*, **87**, pp. 489–506.

Cape, J. N. & Fowler, D. (1981). Changes to epicuticular wax of *Pinus sylvestris* exposed to polluted air. *Silva Fennica*, **15**(4), pp. 457–58.

Chiba, O. & Tanaka, K. (1968). The effect of sulphur dioxide on the development of pine needle blight caused by *Rhizosphera kalkhoffii* (L.) Bubak. *J. Jap. Forest. Soc.*, **50**, pp. 135–9.

Chłodny, J. (1979). The effect of industrial pollution on the number of insects invading *Betula verrucosa*. Poster presentation abstract, *Proceedings of the Symposium on the Effects of Airborne Pollution on Vegetation*, UN/ECE, Warszawa, Poland, 20–24 August, 1979.

Cobb, F. W., Zavarin, E. & Bergot, J. (1972). Effect of air pollution on the volatile oil from leaves of *Pinus ponderosa*. *Phytochemistry*, **11**, pp. 1815–8.

Costonis, A. C. & Sinclair, W. A. (1967). Effects of *Lophodermium pinastri* and *Pullularia pullulans* on healthy and ozone-injured needles of *Pinus strobus*. *Phytopathology*, **57**(8), p. 807.

Cowling, E. B. (1978). Effects of acid precipitation and atmospheric deposition on terrestrial vegetation. Pp. 46–63, in *A National Program for Assessing the Problem of Atmospheric deposition*: A Report to the Council on Environmental Quality. NC–141.

Daly, J. M. (1976). Some aspects of host–pathogen interactions. Pp. 27–50 in *Physiological Plant Pathology* (Ed. R. Heitefuss & P. H. Williams). Springer-Verlag, Berlin–Heidelberg–New York: xx + 890 pp., illustr.

Daly, J. M., Ludden, P. & Seevers, P. M. (1971). Biochemical comparisons of resistance to wheat stem-rust diseases controlled by Sr 11 alleles. *Physiol. Plant Pathol.*, **1**, pp. 397–407.

Darley, E. F. & Middleton, J. T. (1966). Problems of air pollution in plant pathology. *Ann. Rev. Phytopath.*, **4**, pp. 103–18.

Dässler, H.-G. (1964). Der Einfluss des Schwefeldioxids auf den Terpengehalt von Fichtennadeln. *Flora*, **154**, pp. 376–82.

Davis, D. D. (1978). Interactions between oxidants and virus infections. P. 351 in *Abstracts of Papers, 3rd International Congress of Plant Pathology, München, 16–23 August 1978*. Paul Parey, Berlin & Hamburg.

Davis, I. (1961). Microscopic studies with ozone-quantitative lethality of ozone for *Escherichia coli*. *U.S. Air Force Aerospl. Med. Cent. Rep.*, 61–54, pp.

Davis, J. R. & English, H. (1969). Factors related to the development of bacterial canker in peach. *Phytopathology*, **59**, pp. 588–95.

Dean, G. & Treshow, M. (1965). Effects of fluoride on the virulence of tobacco mosaic virus *in vitro*. *Utah Acad. Sci. Arts. Lett. Proc.*, **42**, pp. 236–9.

Dickinson, C. H. & Preece, T. F. (1976). *Microbiology of Aerial Plant Surfaces*. Academic Press, New York, NY, USA: 669 pp.

Donaubauer, E. (1966). Durch Industieabegase gedingte Sekundärschäden am Wald. *Mitt. Forstl. Versuchsanst. Mariabrunn*, **73**, pp. 101–10.

Dörfelt, H. & Braun, U. (1979). Untersuchungen zur Bioindikation durch Pilze in der Dübener Heide (DDR). Pp. 131–2 in *Abstracts of the International Workshop on Problems of Bioindication* (Ed. R. Schubert & E. Weinert), Halle (Saale), 27–31 August, 1979.

Elford, W. J. & Ende, J. van den (1942). An investigation of the merits of ozone as an aerial disinfectant. *J. Hyg.*, **42**, pp. 240–65.

Flückinger, W., Flückinger-Keller, H. & Oertli, J. J. (1978). Der Einfluss verkehrsbedingter Luftverunreinigungen auf die Peroxydaseaktivität, das ATP-Bildungvermögen isolierter Chloroplasten und das Lägenwachstum von Mais. *Z. Pfl. Krankh.*, **85**(1), pp. 41–7.

Fowler, D., Cape, J. N., Nicholson, I. A., Kinnaird, J. W. & Paterson, I. S. (1980). The influence of a polluted atmosphere on cuticle degradation in Scots Pine (*Pinus sylvestris*), P. 146, in *Proceedings of the International Conference on Ecological Impact of Acid Precipitation*, Norway, 11–14 March, 1980. 383 pp.

Frič, F. (1976). Oxidative enzymes. Pp. 617–31, in *Physiological Plant Pathology* (Ed. R. Heitefuss & P. H. Williams). Springer-Verlag, Berlin–Heidelberg–New York: xx + 890 pp., illustr.

Garber, K. (1962). Über die Aufnahme von Schadstoffen durch die Rinde der Bäume. *Viww. Z. TU Dresden*, **11**, pp. 549–52.

Godzik, S. (1978). External needle waxes of *Pinus sylvestris* and their modification by air pollution. P. 349, in *Abstracts of Papers, 3rd International Congress of Plant Pathology, München, 16–23 August, 1978*. Paul Parey, Berlin & Hamburg: 435 pp.

Godzik, S. & Sassen, M. M. A. (1978). A scanning electron microscope examination on *Aesculus hippocastanum* L. leaves from control and air polluted areas. *Environ. Pollut.*, **17**, pp. 13–18.

Goodman, R. N. (1976). Physiological and cytological aspects of bacterial infection process. Pp. 172–96 in *Physiological Plant Pathology* (Ed. R. Heitefuss & P. H. Williams). Springer-Verlag, Berlin–Heidelberg–New York: xx + 890 pp., illustr.

Grill, D., Esterbauer, H. & Beck, G. (1975). Untersuchungen an phenolischen Substanzen und Glucose in SO$_2$-geschädigten Fichtennadeln. *Phytopath. Z.*, **82**, pp. 182–4.

Grill, D. & Härtel, O. (1972). Zellphysiologische und biochemische Untersuchungen an SO$_2$-begasten Fichtennadeln: Resistenz und Pufferkapazität. *Mitt. Forstl. Bundes-Versuchsanst.* (Wein), **97**, pp. 367–84.

Grodzinska, K. (1971). Acidification of tree-bark as a measure of air pollution in southern Poland, *Bull. Acad. Polon. Sci. Ser. Biol. Cl II*, **19**, pp. 189–95.

Grodzinska, K. (1976). Acidity of tree-bark as a bioindicator of forest pollution in southern Poland. Pp. 905–11 in *Proceedings of the First International Symposium on Acid Precipitation and Forest Ecosystems*, USDA Forest Service General Technical Report NE-23, 1074 pp.

Grodzinska, K. (1977). Acidity of tree-bark as a bioindicator of forest pollution in southern Poland. *Water, Air and Soil Pollut.*, **8**, pp. 3–7.

Grzywacz, A. P. (1973). Sensitivity of *Fomes annosus* Fr. Cooke and *Schizophyllum commune* Fr. to air pollution with sulphur dioxide. *Acta Soc. Bot. Poloniae*, **XLII** (3), pp. 347–60.

Grzywacz, A. P. (1978). Interactions between air pollutants and decaying Fungi of pine wood. P. 351 in *Abstracts of Papers, 3rd International Congress of Plant Pathology, München, 16–23 August, 1978*. Paul Parey, Berlin & Hamburg: 435 pp.

Guderian, R. (1977). Air pollution. P. 127 in *Ecological Studies 22*. Springer-Verlag, Berlin–Heidelberg–New York: 127 pp.

Guderian, R. & Stratmann, H. (1968). Freilandversuche zur Ermittlung von Schwefeldioxidwirkungen auf die Vegetation, III Teil: Grenzwerte schädlicher SO$_2$-Immissioned für Obst- und Forstkulturen sowie fur landschaftliche und gärtnerische Pflanzenarten. *Westdeutscher Verlag. Forsch. Ber. d. Landes Nordrhein-Westfalen* 1920.

Halbwachs, G. (1968). Untersuchungen über den Wasserhaushalt rauchgeschädigter Forstgehölze. Pp. 209–18 in *Referaten der VI Internationalen Arbeitstagung forstlicher Rauchschaden-sachverstängiger*. Polskiej Akademii Nauk. Katowice, 9–14 Sept. 1968.

Hanover, J. W. (1975). Physiology of tree resistance to insects. *Ann. Rev. Entomol.*, **20**, pp. 75–95.

Härtel, O. (1972). Langjährige Messreichen mit dem Trübungstest an abgasgeschädigten Fichten. *Oecologia*, **9**, pp. 103–11.

Havas, P. (1971). Injury to pines in the vicinity of a chemical processing plant in northern Finland. *Acta Forestalia Fennica*, **121**, pp. 1–21.

Havas, P. & Huttunen, S. (1972). The effect of air pollution on the radial growth of Scots Pine (*Pinus sylvestris* L.). *Biol. Conservation*, **4** (5), pp. 361–8, illustr.

Heagle, A. S. (1973). Interactions between air pollutants and plant parasites. *Ann. Rev. Phytopathol.*, **11**, pp. 365–88.

Hibben, C. R. & Taylor, M. P. (1975). Ozone and sulphur dioxide effects on the Lilac Powdery-mildew Fungus. *Environ. Pollut.*, **9**, pp. 107–14.

Hiltunen, R., Schantz, M. v. & Löyttyniemi, K. (1975). The effect of nitrogen fertilization on the composition and the quantity of volatile oil in Scots Pine (*Pinus sylvestris* L.). *Comm. Inst. For. Fenn.*, **85**(1), pp. 1–14.

Horsman, D. C. & Wellburn, A. R. (1975). Synergistic effects of SO$_2$ and NO$_2$ polluted air upon enzyme activity of pea seedlings. *Environ. Pollut.*, **8**, pp. 123–33.

Horsman, D. C. & Wellburn, A. R. (1977). Effect of SO$_2$ polluted air upon enzyme activity in plants originating from areas with different annual mean atmospheric SO$_2$ concentrations. *Environ. Pollut.*, **13**, pp. 33–9.

House, H. L. (1961). Insect nutrition, *Ann. Rev. Entomol.*, **6**, pp. 13–26.

House, H. L. (1962). Insect nutrition. *Ann. Rev. Entomol.*, **7**, pp. 653–72.

House, H. L. (1966). Effects of varying the ration between the amino-acids and the other nutrients in conjunction with salt mixture on the fly *Agria affinis* Fall. *J. Insect Physiol.*, **12**, pp. 299–310.

Huber, B. (1956). Winterfrost 1956 und Rauchschäden. *Allg. Forstztschrift*, **11**, pp. 609–10.

Hutchinson, T. C. & Havas, M. (eds) (1980). Effects of acid precipitation on terrestrial ecosystems. *Nato Conference Series I Ecology*, pp. 1–654.

Huttunen, S. (1973). Studies on tree damage due to air pollution in Oulu. *Aquilo Ser. Bot.*, **12**, pp. 1–11.

Huttunen, S. (1974). A preliminary monitoring survey on a test field near a chemical processing plant. *Aquilo Ser. Bot.*, **13**, pp. 23–34.

Huttunen, S. (1975). The influence of air pollution on the forest vegetation around Oulu. *Acta Univ. Oul. Series A 33, Biol.*, **2**, p. 73.

Huttunen, S. (1978). The effects of air pollution on provenances of Scots Pine and Norway Spruce in northern Finland. *Silva Fennica*, **12**(1), pp. 1–16.

Huttunen, S. (1979a). The integrative effects of airborne pollutants on boreal forest ecosystems. Pp. 111–132 in *Proceedings of the Symposium on the Effects of Airborne Pollution on Vegetation, Warsaw, 20–24 August 1979*. UN/ECE. 410 pp.

Huttunen, S. (1979b). [*Report on the Effects of Air Pollution in Jämsänkoski*—Unpublished report in Finnish.] 7 pp. (typescript obtainable from Author).

Huttunen, S. (1979c). Winter injuries of coniferous trees and the accumulation of sulphur compounds in pine needles. (Tagungsberichte der X. internationalen Tagung der IUFRO-Fachgruppe 2.09 'Luftverunreinigung'.) *Institut za Gozdno in Lesho Gospodarstvo, Ljubljana, Yugoslavia, Zbornik* (Mitteilungen), **1**, pp. 103–14.

Huttunen, S. (1981). Seasonal variation of air pollution stresses in conifers. *Mitt forstl. VersAnst. Wien*, **137**(II), pp. 103–113.

Huttunen, S. (1982). Air-pollution-induced stresses in forest ecosystems. Pp. 1392–400 in *Proceedings of the 5th International Clean Air Congress, Buenos Aires, 20–25 October, 1980*. Buenos Aires, Argentina: 1498 pp.

Huttunen, S. & Laine, K. (1981). The structure of pine needle surface (*Pinus sylvestris* L.) and the deposition of airborne pollutants. *Archiwum Ochromy Srodowiska 24*, Polish Academy of Sciences, pp. 29–38.

Huttunen, S., Havas, P. & Laine, K. (1981a). Effects of air pollutants on wintertime water economy of the Scots Pine (*Pinus Sylvestris* L.) *Holarctic Ecology*, **4**, pp. 94–101.

Huttunen, S., Karhu, M. & Kallio, S. (1981b). The effect of air pollution on transplanted mosses. *Silva Fennica*, **15**(4), pp. 495–504.

Huttunen, S., Karenlampi, L. & Kolari, K. (1981c). Changes of osmotic values and some related physiological variables in polluted coniferous needles'. *Ann. Bot. Fenn.*, **18**, pp. 63–71.

Huttunen, S., Karenlampi, L., Laine, K., Soikkeli, S., Pakonen, T., Karhu, M. & Törmälehto, H. (1980). [Dispersion of airborne pollutants and their effects in forest environments—Research Report for the Academy of Finland 1980-06-30, in Finnish.] *Oulun yliopiston kasvitieen laitoksen monisteita no. 12*, pp. 1–153.

Huttunen, S., Laine, K., Ohvo, S., Pakonen, T. and Törmälehto, H. (1979a). Ilman epäpuhtauksien leviäminen ja vaikutus kasvillisuuteen Kemissä (Dispersion of sulphur compounds and their effects on vegetation in Kemi, northern Finland) [*Finnish Report to the Municipality of the City of Kemi*] 14 pp. (typescript).

Huttunen, S. Manninen, J., Laine, K., Forsten, P., Pakonen, T. & Törmälehto, H. (1979b). Rikkiyhdisteiden leviäminen ja vaikutus kasvillisuuteen Porissa vuonna

1979. [In Finnish: Dispersion of sulphur compounds and their effects on vegetation in Pori, in 1979]. *Porin kaupungin tutkimuksia*, **32**, pp. 1–38.

Jančařik, V. (1961). Vyskyt drevokaznych hub u kouřem poškozovane oblasti Krušnych hor. *Lesnictvi*, **7**, pp. 667–92.

Jäger, H. J. & Grill, D. (1975). Effect of SO_2 fumigation on the activity of enzymes of amino-acid metabolism and free amino-acid contents in plants of different resistance. *PlKrankh.*, **82**, pp. 139–48.

Jäger, H. J. & Pahlich, E. (1972). Einfluss von SO_2 auf den Aminosäurestoffwechsel von Erbsenkeimlingen. *Oecologia*, **9**, pp. 135–40.

Kallio, S. (1976). *Studies on Elemental Nitrogen Fixation in Lichens in North Finland.* Academic Dissertation, Department of Biology, University of Turku, Turku, Finland: 112 pp. (mimeogr).

Keller, T. (1974). The use of peroxidase activity for monitoring and mapping air pollution areas. *Eur. J. For. Path.*, **4**(1), pp. 11–19.

Keller, T. (1978*a*). Wintertime atmospheric pollutants—Do they affect the performance of deciduous trees in the ensuing growing-season? *Environ. Pollut.*, **16**, pp. 243–7. fig.

Keller. T. (1978*b*). Frostschäden als Folge einer 'latenten' Immissionsschädigung. *Staub-Reinhalt. Luft*, **28**(1), pp. 24–6.

Keller, T. & Jäger, H.-J. (1980). Der Einfluss bodenburtiger Sulfationen auf den Schwefelgehalt SO_2-begaster Assimilationsorgane von Waldbaumarten. *Angew. Botanik.*, **54**, pp. 77–89.

Keller, T. & Schwager, H. (1971). Der Nachweis unsichtbarer Fluor-Immissionsscädigungen an Waldbäumen durch eine einfache kolorimetrische Bestimmung der Peroxidase Aktivität. *Eur. J. For. Path.*, **1** (1), pp. 6–18.

Keller, T. & Schwager, H. (1977). Air pollution and ascorbic acid. *Eur. J. For. Path.*, **7**(6), pp. 338–50.

Koch, G. (1935). Eichenmehltau und Rauchgasschaden. *Z. PfKrank.*, **14**, pp. 44–5.

Kosuge, T. & Gilchrist, D. G. (1976). Metabolic regulation in host–parasite Interactions. Pp. 679–702 in *Physiological Plant Pathology* (Ed. R. Heitefuss & P. H. Williams). Springer-Verlag, Berlin–Heidelberg–New York: 890 pp.

Kovacz, A. & Szwöke, E. (1956). Die phytopathologische Bedeutung der kutikularen Exkretion. *Phytopathol. Z.*, **27**, pp. 335–49.

Kudela, M. & Novakova, E. (1962). Lešni skudci a škody zvěři u lesich poškozovanych kouřem. *Lesnictvi*, **8**(6), pp. 493–502.

Kurkela, T. (1975). Incidence of snow-blight on Scots Pine as affected by fertilization and some environmental factors. *Comm. Inst. For. Fenn.*, **85** (2), pp. 1–35.

Laaksovirta, K., Olkkonen, H. & Alakuijala, P. (1976). Observations on the lead content of lichen and bark adjacent to a highway in southern Finland. *Environ. Pollut.*, **11**, pp. 247–55.

Larcher, W. (1975). Pp. 199–213 in *Physiological Plant Ecology*. Springer-Verlag, Berlin–Heidelberg–New York: xx + 890 pp.

Laurence, J. A., Weinstein, L. H., McCune, D. C. & Aluisio, A. L. (1978). Effects of sulphur dioxide on disease of wheat and maize. P. 352 in *Abstracts of Papers, 3rd International Congress of Plant Pathology, München, 16–23 August 1978*. Paul Parey Berlin & Hamburg.

Leben, C., Dafr, G. C. & Schmitthenner, A. F. (1968). Bacterial blight of Soybeans: Population levels of *Pseudomonas clycinea* in relation to symptom development. *Phytopathology*, **58**, pp. 1143–6.

Lehtiö, H. (1981). Effect of air pollution on the volatile oil from needles of Scots Pine (*Pinus sylvestris* L.). *Silva Fennica*, **15**(2), pp. 122–9.

Levitt, J. (1972). *Responses of Plants to Environmental Stresses*. Academic Press, New York, NY, USA: ix + 697 pp., illustr.

Liese, W. & Eckstein, D. (1978). Anatomical depositions and responses of trees to air pollution effects. P. 354 in *Abstracts of papers, 3rd International Congress of Plant Pathology, München, 16–23 August 1978*. Paul Parey, Berlin & Hamburg.

Liese, W., Schneider, M. & Eckstein, D. (1975). Histometrische Untersuchungen am Holz einer rauchgeschädigten Fichte. *Eur. Jour. For. Path.*, **5** (3), pp. 152–61.

Lighthart, B., Hiatt, V. E. & Rossano, A. T., jr (1971). The survival of airborne *Serratia marcescens* in urban concentration of sulphur dioxide. *J. Air Pollut. Contr. Ass.*, **21**, pp. 638–42.

Linzon, S. (1978). Effects of airborne sulfur pollutants on plants. Pp. 110–62 in *Sulfur in the Environment, Part II: Ecological Impacts* (Ed. J. O. Nriagu). John Wiley & Sons, New York: xii + 482 pp. illustr.

Lohm, U. (1980). Effects of experimental acidification on soil organism populations, decomposition, and turnover of nitrogen and carbon. P. 43 in *Abstracts of Voluntary Contributions, International Conference on the Ecological Impact of Acid Precipitation, Sadefjord, Norway, 11–14 March 1980.*, Vol. I, 383 pp.

Loon, L. C., Geelen, J. L. M. (1971). The relation of polyphenoloxidase and peroxidase to symptom expression in tobacco var. Samsun NN after infection with tobacco mosaic virus. *Acta Phytopathol. Acad. Sci. Hung.*, **6**, pp. 9–20.

Lötschert, W. & Köhm, H.-J. (1978). Characteristics of tree-bark as an indicator in high-emission areas. *Oecologia*, **27**(1), pp. 47–64.

Lovrekovich, L., Lovrekovich, H. & Stahmann, M. A. (1968*a*). The importance of peroxidase in the wildfire disease. *Phytopathology*, **58**, 193–98.

Lovrekovich, L., Lovrekovich, H. & Stahmann, M. A. (1968*b*). Tobacco mosaic virus-induced resistance to *Pseudomonas tabaci* in tobacco. *Phytopathology*, **58**, pp. 1034–5.

Lux, H. (1964). Der Lesitungsverfall der Kiefer und seine Bedeutung in industrienahen Gebieten. *Soz. Forstwirtchaft*, **1**, pp. 19–21.

McCune, D. C., Weinstein, L. H., Mancini, J. F. & Leuken, P. van (1973). Effects of hydrogen fluoride on plant–pathogen interactions. Pp. A146–9 in *Proc. Third International Clean Air Congress, Düsseldorf*. VDI-Verlag.

Manning, W. J., Feder, W. A., Perkins, J. & Glickman, H. (1969). Ozone injury and infection of potato leaves by *Botrytis cinerea*. *Plant Dis. Rep.*, **53**(9), pp. 691–3.

Manning, W. J., Feder, W. A., Perkins, J. & Glickman, H. (1969). Ozone injury and infection of potato leaves by *Botrytis cinerea*. *Plant Dis. Rep.*, **53**(9), pp. 691–3.

Manning, W. J., Feder, W. A. & Perkins, J. (1970). Ozone injury increases infection of geranium leaves by *Botrytis cinerea*. *Phytopathology*, **60**(4), pp. 669–70.

Manning, W. J., Feder, W. A., Popia, P. M. & Perkins, J. (1971). Effect of low levels of ozone on growth and susceptibility of cabbage plants to *Fusarium oxysporum* f. sp. *conglutinans*. *Plant Dis. Rep.*, **54**(1), pp. 47–9.

Martin, J. T. & Juniper, B. E. (1970). *The Cuticles of Plants*. St Martin's Press, New York, NY, USA: 347 pp.

Materna, J. (1962). Die Schwefeldioxideinwirkung auf die mineralische Zusammensetzung von Fichtennadeln. *Arb. forstl. Forschungsanstalt Tschechoslowakei*, **24**, pp. 7–36.

Materna, J. (1963). Steigerung der Wiedestansfähigkeit von Holzarten gegen Rauchgaseinwirkung durch Düngung. *Prace, Vyzkumnych Ustav Lesnickych CSSR*, *Svazek*, **26**, pp. 209–35.

Materna, J. (1972). Einfluss niedriger Schwefeldioxydkonzentrationen auf die Fichte. *Mitt. Forst. Bundes Anstalt* (Wien), **97**, pp. 219–32.

Materna, J. (1974). Einfluss der SO_2 Immission auf Fichtenpflanzen in Wintermonaten. Pp. 107–14 in *Tagungsbericht IX. Internationale Tagung uber die Luftverunreinigung und Forstwirtschaft, Ministerium für Forst und Holzwirtschaft der CSR. Mariánské Lázně*, Tschechoslowakei den 15 bis 18 Oktober 1974. Vytiskl Tomos, Praha: 436 pp.

Materna, J. (1979). Frostschäden in Fictenbeständen in Abhäugigkeit von der Immiszionseinwirkung. Pp. 34–352 in *Tagungsberichte der X. internationalen Tagung der IUFRO-Fachgruppe 2.09 'Luftverunreinigung'*. Zbornik (Mitteleilung) Vol. 1–390. Institut za Gozdno in lesno Gospodarstvo, Ljubljana, Yugoslavia.

Materna, J. & Kohout, R. (1963). Die Absorption des Schwefeldioxids durch die Fichte. *Naturwissenschaften*, **50**, pp. 407–8.

Mergen, F., Hoekstra, P. E. & Echols, R. (1955). Genetic control of oleoresin yield and viscosity in Slash Pine. *Forest Sci.*, **1**, pp. 19–30.

Munch, E. (1933). Winterschäden an immergrünen Gehölzen. *Ber. Deutsch. Bot. Ges.*, **51**, p. 21.

Mikkonen, H. & Huttunen, S. (1981). Dwarf shrubs as bioindicators. *Silva Fennica*, **15**(4), pp. 475–480.

Niemtur, S. (1979). Influence of zinc smelter emissions on peroxidase activity in Scots Pine needles of various families. *Eur. J. For. Path.*, **9**, pp. 142–7.

Noland, T. L. & Kozlowski, T. T. (1979). Effect of SO_2 on stomatal aperature and sulphur uptake of woody angiosperm seedlings. *Can. J. For. Res.*, **9**(1), pp. 57–62.

Novak, V. (1962). Vyzkum sukcese podkorniho hmyzy na stromech chradnoucich vliv-m prumyslovych exhalaci v Krusnych horach. (Chechoslovenska Akademie Zemendelskych Ved Sbornik.) *Lesnictvi*, **35**, pp. 329–42.

Novak, W. Jancarik, V. & Jermanova, H. (1957). Hlavni zivocisni skudci a hubove choroby v oblasti Krusnych hor. *Zpravy VULH*, **1**, pp. 44–6.

OECD, (1978). *Plant Damage Caused by SO_2*. Report of the workshop held at the OECD, Paris, 7–8 June 1978: ENV/AIR/78, 15, pp. 1–30.

Olszowski, J. (1976). The effect of fertilization on a pine forest ecosystem in an industrial region, VII: Summary of the studies. *Ekol. Pol.*, **24**(3), pp. 359–63.

Parmeter, J. R. & Cobb, F. W., jr (1972). Long-term impingement of aerobiology systems on plant production systems. Pp. 61–8 in *US/IBP Aerobiology Program Handbook No. 2*.

Pelz, E. & Materna, J. (1964). Beiträge zur Problem der Individuellen Rauchhärte von Fichte. *Arch. Forstwes.*, **13**(2), pp. 177–210.

Philogene, B. J. R. (1972). Physiological studies and pest control. *BioScience*, **22**, pp. 715–18

Plank, G. H. & Gerhold, H. D. (1965). Evaluating host resistance to the White-pine Weevil, *Pissodes strobi*, using feeding preference tests. *Ann. Entomol. Soc. Am.*, **58**, pp. 527–32.

Preece, T. F. & Dickinson, C. H. (1971). *Ecology of Leaf-surface Microorganisms*. Academic Press, New York, NY, USA: xvii + 640 pp., illustr.

Prus-Glowacki, W. & Szweykowski, J. (1977). Studies on isoenzyme variability in Scotch Pine (*Pinus sylvestris* L.) and Mountain Dwarf Pine (*Pinus mugo*) populations. *Bull. Soc. Amis. Scient. Lettr. Poznan, Serie D*, **17**, pp. 15–27.

Przybylski, Z. (1967). Wyniki obserwacji nad dzalaniem gasow i par SO_2, SO_3 i H_2SO_4 na drzewa owocowe i niektore szkodliwe owady w rejonie kopalni i zakladow przetworczych siarki w machowie k/Tarnobrzega. *Postepy Nauk Roln*, **2**, 111–18.

Przybylski, Z. (1979). The effects of automobile exhaust gases on the arthropods of cultivated plants, meadows, and orchards. *Environ. Pollut.*, **19**, pp. 157–61.

Ranft, H. (1968). Zur Bewirtschaftung rauchgeschädigter Fichtenjungbestände. *Sozial. Forstw.* (Berlin), **18**(10), pp. 299–301.

Resh, H. M. & Runeckles, V. C. (1973). Effects of ozone on Bean Rust *Uromyces phaseoli. Can. J. Bot.*, **51**(4), pp. 725–7.

Rottink, B. A. & Hanover, J. W. (1972). Identification of Blue Spruce cultivars by analysis of cortical oleoresin monoterpenes. *Phytochemistry*, **11**, pp. 3255–7.

Rudinsky, J. A. (1966). Scolytid beetles associated with Douglas Fir: Response to terpenes. *Science*, **152**, pp. 218–19.

Saunders, P. J. W. (1970). The toxicity of sulphur dioxide to *Diplocarpon rosae* Wolf. causing Blackspot of roses. *Ann. Appl. Biol.*, **58**(1), pp. 103–14.

Schaeffer, T. C. & Hedgecock, G. G. (1955). Injury to northwestern forest trees by sulphur dioxide from smelters. *U.S. Dep. Agr., Tech. Bull.*, **1117**, pp. 1–49.

Schneider, Z. & Chłodny, J. (1977). Entomofauna of forest plantations in the zone of disastrous industrial pollution. Pp. 81–108 in *Relationships Between Increase in Air Pollution Toxicity and Elevation above ground (Janusz Wolak)*. Forest Research Institute, Warszawa, Poland.

Schneider, Z. & Sierpinski, Z. (1967). Stan zagrozenia przez owady niektorych gatunkow drzew lesnych w okolicach przemyslowych Slaska (Gefährdungszustand mancher forstlichen Holzarten inindustriegebieten Schlesiens durch Insekten). *Institut Badawczy Lesnictwa*, **316**, 114–50.

Scholz, F. & Stephan, B. R. (1974). Physiologische Untersuchungen über die untersiedliche Resistenz von *Pinus sylvestris* gegen *Lophodermium pinastri*. I. Die Pufferkapazität in Nadeln. *Eur. J. For. Path.*, **4**, pp. 118–26.

Schönbeck, F. & Schlösser, E. (1976). Performed substances as potential protectants. Pp. 653–78 in *Physiological Plant Pathology* (Ed. R. Heitefuss & P. H. Williams.) Springer-Verlag, Berlin–Heidelberg–New York: xx + 890 pp., illustr.

Schultz, H. (1979). Methodik und erste Ergebnisse zur Peroxydaseaktivitätsbestimmung als Indikation der experimentellen Störung eines Ökosystemes. In *Abstracts of International Workshop on Problems of Bioindication, Halle (Saale), 27–31 August 1979*. (Ed. R. Schubert & E. Weinert).

Serat, W. F., Budinger, F. E. & Muller, P. K. (1966). Toxicity evaluation of air pollutants by use of luminescent Bacteria. *Atmos. Environ.*, **1**, pp. 21–32.

Sharma, G. K. (1977). Cuticular features as indicators of environmental pollution. *Water, Air and Soil Pollution*, **8**, pp. 15–19.

Sharma, J. K. & Shinha, K. (1971). Effect of leaf exudates of *Sorghum* varieties varying in susceptibility and maturity on the germination of conidia of *Colletotrichum graminicola* (Ces.) Wilson. Pp. 597–601 in *Ecology of Leaf-surface Microorganisms* (Ed. T. F. Preece & G. H. Dickinson). Academic Press, London & New York: xvii + 640 pp., illustr.

Shriner, D. S. (1977). Effects of simulated rain acidified with sulfuric acid on host–parasite interactions. *Water, Air and Soil Pollution*, **8**, pp. 9–14.

Shriner, D. S. & Cowling, E. B. (1980). Effects of rainfall acidification on plant pathogens. Pp. 435–42 in *Effects of Acid Precipitation on Terrestrial Ecosystems* Ed. T. C. Hutchinson & M. Havas.) Nato Conference Series I Ecology: xi + 654 pp. illustr.

Sierpinski, Z. (1968). Wplyw gazow i dymow przemyslowych na dynamike populacji niektorych szkodnikow pierwotnych sosny (Einfluss von industriellen Luftverunreinigungen auf die Populationsdynamik einiger primärer Kiefernscädlinge). *Prace, Instytutu Badawczego Lesnictwa* (Warszawa), Nr. 365, **13**, pp. 139–50.

Sierpinski, Z. (1970). Gospodarcze znaczenie szkodliwych owadow w drzewostanach sosnowych objetych chronicznym dzialaniem przemyslowych zanieczyszczen powietrza. *Sylwan*, **6**, pp. 59–71.

Sierpinski, Z. (1972*a*). Szkodniki wtorne sosny na tle zmian zachodzacych w drzewostanach znajdujacych sie w zasiegu dzialania emisji azotowych (Secondary injurious insects of pine on the background of changes occurring in stands in the range of the influence of nitrogenous air pollution). *Prace, Instytutu Badawczego Lesnictwa* (Warszaiva), pp. 433–4, pp. 55–99.

Sierpinski, Z. (1972*b*). Wystepowanie przedziorka sosnowka (*Paratetranychus ununguis* Jacoby) na osnie pospolitej w zasiegu dzialania emisji przemyslowych. (The occurrence of the spruce spider (*Paratetranychus ununguis* Jacoby) on Scotch Pine in the range of the influence of industrial air pollution). *Prace, Instytutu Badawczego Lesnictwa* (Warszawa), **434**, pp. 101–9.

Sierpinski, Z. (1972*c*). Die Bedeutung der sekundären Kieferschädlinge in Gebieten chronischer einwirkung industrieller Luftverunreinigungen. *Mitt. Forstlichen Bundes-Versuchsanstalt Wien*, **97**, pp. 609–15.

Sierpinski, Z. (1979). *Schädinsekten auf Laubfaumarten alse Folgevon Stickstoffhaltigen Immissionen*, Zbornik, Ljubljana, Yugoslavia, pp. 225–30.

Sierpinski, Z. (1980). Noxious insects in pine stands in industrial areas in Poland. Pp. 241–9 in *Proceedings of the Symposium of the Effects of Airborne Pollution on Vegetation, 20–24 August, Warsaw, Poland*. UN/ECE: 440 pp.

Sierpinski, Z. & Szalonek, I. (1974). Kiefernschädlinge in einwirkungsraum von Fluorverbindungen. Pp. 315–22 in *IX Internationale Tagung über die Luftverunreinigung und Forstwirtschaft, den 15 bis 18 Oktober 1974*. Marianske Lazne, Tsechoslowakei, Ministerium für Forst und Wasserwirtschaft der CSR.

Skye, E. (1968). Lichens and air pollution. *Acta Phytogeogr. Suec.*, **52**, pp. 1–123.

Smelyanets, V. P. (1977). Mechanisms of plant resistance in Scots Pine (*Pinus sylvestris* L.). *Z. Ang. Ent.*, **84**, pp. 113–23.

Soikkeli, S. & Tuovinen, T. (1979). Damage in mesophyll ultrastructure of needles of Norway Spruce in two industrial environments in central Finland. *Ann. Bot. Fenn.*, **16**(1), pp. 50–64.

Stewart, D. M., Treshow, M. & Harner, F. (1973). Pathological anatomy of conifer needle necrosis. *Can. J. Bot.*, **51**(5), pp. 983–8.

Stroh, R. C. & Gerhold, H. D. (1965). Eastern White Pine characteristics related to weevil feeding. *Silvae Genet.*, **14**, pp. 160–9.

Templin, E. (1962). Zur Populationsdynamik einiger Kiefernschadeinsekten in rauchgeschädigten Beständen. *Wiss. Z. TU Dresden*, **11**(3), pp. 631–7..

Tobolski, J. J. & Hanover, J. W. (1971). Genetic variation in the monoterpenes of Scotch Pine. *Forest Sci.*, **17**(3), pp. 293–9.

Treshow, M. (1968). Impact of air pollutants on plant populations. *Phytopathology*, **58**, pp. 1103–13.

Treshow, M. (1975). Interaction of air pollutants and plant diseases. Pp. 307–34 in *Responses of Plants to Air Pollution* (Ed. J. B. Mudd & T. T. Kozlowski). Academic Press, London & New York: xii + 383 pp. illustr.

Treshow, M., Dean, G. & Harner, F. M. (1967). Stimulation of tobacco mosaic virus-induced lesions on bean by fluoride. *Phytopathology*, **57**, pp. 756–8.

Trillmich, H. D. (1969). Düngung von Mischbeständen in einem Rauchschadengebiet des Erzebirges. *Wiss. Z. TU Dresden*, **18**, 807–16.

Unsworth, M. H., Biscoe, P. V. & Black, V. (1976). Analysis of gas exchange between plants and polluted atmospheres. Pp. 4–16 in *Effects of Air Pollutants on Plants* (Ed. T. A. Mansfield). Cambridge University Press, Cambridge, England, UK: 209 pp.

Vite, J. P. (1961). The influence of water supply on oleoresin exudation pressure and resistane to bark-beetle attack in *Pinus ponderosa*. *Contrib. Boyce Thompson Inst. Plant Res. Prof. Pap.*, **21**, pp. 37–66.

Weinstein, L. H. (1977). Fluoride and plant life. *J. of Occup. Medicine*, **19**(1), pp. 49–78.
Weinstein, L. H., McCune, D. C., Aluisio, A. L. & Leuken, P. Van (1975). The effect of sulphur dioxide on the incidence and severity of bean rust and early blight of Tomato. *Environ. Pollut.*, **9**, pp. 145–55.
Wenzel, K. F. (1956). Winterfrost 1956 und Rauchschäden. *Allg. Forstztschr.*, **11**(32), pp. 541–3.
Wenzel, K. F. (1965). Die Winterfrostschäden 1962–1963 in Koniferenkulturen des Ruhrgebietes un ihre vehrmütlichen Ursachen. *Forstarchiv*, **36**(3), pp. 49–59.
Wind, Eva (1979). Pufferkapazität in Koniferennadeln. *Phyton* (Austria), **3–4**, pp. 197–215.
Wood, K. R. & Barbara, D. J. (1971). Virus multiplication and peroxidase activity in leaves of Cucumber (*Cucumis sativus* L.) cultivars systemically infected with W strain of cucumber mosaic virus. *Physiol. Plant Pathol.*, **1**, pp. 73–81.
Yasunaga, K., Oshima, Y. & Kuwatsuka, S. (1962). Attractants of pine-bark beetles, I: Isolation of an attractant, benzoic acid, from Red Pine bark. *J. Agr. Chem. Soc. Jap.*, **36**, pp. 802–4.
Yarwood, C. E. & Middleton, J. T. (1959). Virus infection and heating reduce smog damage. *Plant Dis. Rep.*, **43**, pp. 129–30.
Ziegler, E., Pelz, E. & Horning, W. (1958). Ergebnisse einer Umfrage über Umgang und Art der Frostschäden des Winters 1955/56 in den Staatlichen Forstwirtschaftbetrieben der DDR. *Archiv f. Forstw.*, **7**, pp. 316–29.
Zocien, G. A. De (1976). Cytology of virus infection and virus transport. Pp. 129–49 in *Physiological Plant pathology* (Ed. R. Heitefuss & P. H. Williams), Springer-Verlag, Berlin–Heidelberg–New York: xx + 890 pp., illustr.

Air Pollution and Plant Life
Edited by M. Treshow
© 1984 John Wiley & Sons Ltd.

CHAPTER 15

Impact of Atmospheric Pollution on Agriculture

HOWARD E. HEGGESTAD & JESSE H. BENNETT

Plant Stress Laboratory, Agricultural Research Service, US Department of Agriculture, Beltsville, Maryland 20705, USA

INTRODUCTION

Concern about the impact of phytotoxic air pollutants on agricultural productivity has increased markedly in the United States during the past several decades. Vegetation injury by the most ubiquitous and damaging class of pollutants—constituents of photochemical smog—was first recognized in the mid-1940s, when stippling and glazing or bronzing of the leaves of vegetables were discovered in the Los Angeles basin (Middleton *et al.*, 1950). Ozone (O_3) and associated photochemical oxidants which caused the injury are formed by the action of sunlight on nitrogen oxides and hydrocarbons. Many sources of these precursor chemicals exist, but motor vehicle emissions are the main source.

Prior to the recognition of atmospheric oxidants as plant-damaging agents, sulphur dioxide and fluorides were of the greatest interest. The latter phytotoxicants are both 'point-source pollutants', emitted primarily by large industries such as smelters, refineries, and electric generating plants that consume huge amounts of sulphur-containing fossil fuels or process sulphur-and fluoride-bearing raw materials. According to Thomas (1961), by the middle of the present century SO_2 had been studied longer and more intensively than any other pollutant. Although damage from hydrogen fluoride was reported in Europe as early as 1920 (Weinstein, 1977), it was not of much concern in the United States until about 1940—only a few years before the photochemical oxidant problem was first perceived as such.

The significance of O_3 as a crop-damaging air pollutant was not appreciated until 1958, when it was shown to be the primary cause of grape (*Vitus vinifera*) injury in California (Richards *et al.*, 1958) and of Tobacco (*Nicotiana tabacum*) damage in the Eastern United States (Heggestad & Middleton,

357

1959). Later, peroxyacetyl nitrate (PAN) and closely related chemicals were identified as the causative agents of the undersurface leaf glazing observed in the Los Angeles Basin (Stephens *et al.,* 1961). Although it is of importance in the Los Angeles Basin, there is no evidence that PAN is a significant phytotoxicant in the Eastern United States. Research with gas mixtures was stimulated by the discovery that O_3 and SO_2, at concentrations that were too low to be injurious by themselves, could act synergistically to injure leaves of Tobacco when present as mixtures (Menser & Heggestad, 1966).

Economic Assessment of Crop Loss from Air Pollution

Economic assessments of crop losses in the United States caused by air pollutants became the subject of much interest after the Stanford Research Institute (SRI) estimated that photochemical oxidants caused $121 millions worth of annual loss of a proposed $135.5 millions of total annual loss in the US due to air pollution (Benedict *et al.,* 1971). Of the $121 millions, a total of $76 millions was reported as damage to crops. A decade later, in 1981, SRI estimated crop losses from air pollutants to be $1.8 billion, with $1.7 billion (almost 95%) due to oxidants and $34 millions to SO_2. They estimated that 41% of the crop loss occurred in the eastern North-Central region of the US and 22% each occurred in the Middle Atlantic and Pacific regions. The remainder (15%) was attributed to 6 other regions (SRI, 1981).

The order-of-magnitude increase in the loss estimate from ozone (SRI, 1981) was due in part to the subsequent availability of crop yield research data, previous estimates having relied on leaf injury appraisals. The 1981 SRI estimate and other recent estimates may be too high, because the early research tended to involve sensitive crops and inadequate dose–response information was available. Even now there is very little monitoring in rural areas. Much of the existing monitoring data do not include 7- or 8-hr averages that are needed to compare probable exposures with oxidant doses used in research studies. The US Environmental Protection Agency (EPA) in 1980 organized a National Crop Loss Assessment Network (NCLAN) to determine crop-losses in field situations due to ozone, SO_2, and their mixtures (R. G. Wilhour, pers. comm. 1980). Research planning and cooperation stresses interdisciplinary approaches. Specialists engaged in the field studies are located in different regions of the country. Primarily open-top field chambers are used as the investigative tool. Some exposures with SO_2 are planned, using ambient field plots without chambers (cf. J. J. Lee & Lewis, 1978; Miller *et al.,* 1980).

Any assessment of the impact of air pollution on crop yields is inevitably complicated, owing to pollutant variations that are found in different air-sheds, their concentration and fluctuation patterns, the different responses of

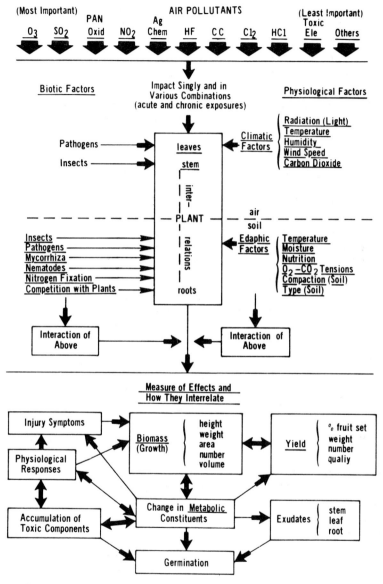

Fig. 1 Air pollution effects on plants: Conceptual interrelationships. Source: Heck *et al.* (1973).

the plant species and varieties present, and yet other interacting environmental stresses and parameters as shown in Fig. 1 (Heck, *et al.*, 1973). Comprehensive reviews dealing with the major individual pollutants have been compiled recently, of which the following are recommended: Mansfield

(1976), Guderian (1977), Heck *et al*. (1977*a*), Weinstein (1977), Linzon (1978), McCune & Weidensaul (1978), Ormrod (1978), and Jacobson (1982).

Methods of Assessing Air Pollution Impact on Agriculture

Methods for assessing the impact of air pollution on crop yields in field situations have been described in recent review papers (Heck *et al.*, 1977*b*; Heggestad, 1980; Reinert, 1980; Heagle & Heck, 1980). The available methodologies currently employed in field studies include: (1) using natural concentration-gradients present in ambient air, (2) studies employing closed chambers. (3) field-grown plants investigated in open-top chambers, (4) open-air fumigations, and (5) the use of chemical protectants.

In this chapter we discuss the four most important classes of air pollutants affecting agriculture, and their impacts on the growth and yields of agricultural crops with special reference to the USA.

OZONE

Atmospheric Ozone Concentrations

Background O_3 concentrations at the Earth's surface are generally about 0.02 or 0.03 ppm. Values ten times greater than these, however, can result in polluted urban atmospheres due to O_3-generating photochemical reactions. Ozone concentrations are highest during the summer months. They tend to peak around midday and become depleted at night as O_3 destruction by dust, reactive hydrocarbons, nitric oxide, and other atmospheric pollutants, occurs. Meteorological parameters associated with high O_3 days in the Eastern United States are: (1) ambient temperatures in the range 25–35°C, (2) high light intensities, (3) low relative humidity with little or no precipitation, and (4) calm days with low to moderate (0–15 km per hr) wind-speeds (Wanta *et al.*, 1961).

Maximum O_3 generation may be found long distances downwind from metropolitan source-areas, depending upon solar intensities, time of day, and other factors (Westberg *et al.*, 1974). Ozone and its precursors, transported hundreds of kilometres from cities, can overlap and affect large areas of the country (Cleveland *et al.*, 1976). Occasionally, cyclonic storms bring ozone to the lower atmosphere, but elevated ozone pollution is essentially due to concentrated traffic emissions from heavily populated areas. In the densely populated northeastern US, megalopolitan airsheds merge—leading to elevated oxidant concentrations throughout the entire Atlantic seaboard region. The highest ozone concentrations in the country are found, however, in the Los Angeles Basin, where the pollution becomes trapped and is often intense.

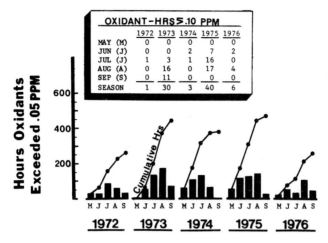

OXIDANT-HRS ≥ .10 PPM					
	1972	1973	1974	1975	1976
MAY (M)	0	0	0	0	0
JUN (J)	0	0	2	7	2
JUL (J)	1	3	1	16	0
AUG (A)	0	16	0	17	4
SEP (S)	0	11	0	0	0
SEASON	1	30	3	40	6

Fig. 2   Number of hours that ambient oxidants exceeded 0.05 (graphs) and 0.10 (upper table) ppm at Beltsville, Maryland, during May–September 1972–76.

Oxidant (mostly ozone) data in Fig. 2 illustrate the variation in elevated ozone concentrations found for Beltsville, Maryland, in different years over a five-years' period. The data were taken to relate the effects of ozone exposure doses on the yields of early and late Snap (Green) Beans (*Phaseolus vulgaris*) grown in each of the five years (Heggestad *et al.*, 1980). The greatest yield loss, determined by comparing the productivity in open-top chambers with that in charcoal filtered and non-filtered air, occurred in the early crop of 1972, when there was one day in July on which oxidants exceeded 10 pphm (parts per hundred million). The Snap Bean plants were in the flowering and early fruiting stages. Leaf damage was severe in the chambers of non-filtered air.

Although the total dose of pollutant that plants receive is important, a on plant productivity are determined also by genetic and environmental factors. As compared with 1972, yield losses due to ozone were less in 1973 and 1975—the years with the highest seasonal ozone levels for the five years' period. In 1973 the plants were less sensitive, owing to inadequate soil moisture, while in 1975 it was necessary to harvest early—when pods were immature—because of excessive rain. Many years of observation on Tobacco indicate that the greatest amounts of leaf injury tend to occur when the plants have optimum soil moisture and grow rapidly.

Ozone Effects

Chambers with closed tops, resembling small greenhouses, were built around 16-years old, producing *citrus* trees in the Los Angeles Basin, to investigate the effects of ozone and other oxidants on yields of *Citrus* fruits (Thompson & Taylor, 1969). In these studies, photochemical oxidants reduced Orange (*Citrus aurantium*) and Lemon (*C. limonum*) yields by as much as 50%, due to premature leaf- and fruit-drop. In other experiments, Zinfandel Grape (*Vitis vinifera* cultivar) yields were found to be reduced 12% the first year and 61% the second year, due to reduced vitality of the vines after one year of exposure to oxidant (Thompson *et al.*, 1969). Similar investigations with cotton *(Gossypium* spp.) in the San Joaquin Valley, California, indicated 5–29% less yield in non-filtered air than in carbon filtered air (Brewer & Ferry, 1974).

Oxidants reduce the yields of Rice (*Oryza sativa*) in Japan, according to studies using closed chambers over field-grown Rice (Nakamura & Ota, 1978). Yields in non-filtered air compared with charcoal-filtered air were reduced 13%, 16%, and 17%, by oxidants at Konosu, Tachikawa, and Chiba, respectively. Leaf injury caused by oxidants has been observed in many rice-growing areas. Ozone concentrations exceeds 0.11 ppm on the day prior to the appearance of new injury. At Konosu the ozone concentrations during one growing season averaged 0.039 ppm. Peak hourly concentrations were 0.27 in 1974 and 0.24 in 1975. As with other crop species, rice cultivars show a wide range of sensitivity to oxidant injury.

Oshima *et al.* (1975) utilized ozone concentration-gradients in the Los Angeles Basin to assess the impact on crop yields. Plants grown in containers with a standard soil mix, were located at sites ranging from the ocean (with least ozone) to the mountains (with highest ozone levels). Losses of yield for Alfalfa (*Medicago sativa*) (Oshima *et al.*, 1976) and Tomatoes (*Lycopersicum esculentum*) (Oshima *et al.*, 1977) were related to the ozone dose. Based on regression analyses, it was predicted that Alfalfa yields would be reduced by 53% in the area with the highest ozone dose (5,750 pphm-hour $\geq$ 10 pphm).

Oshima (1978) also predicted a 23% reduction in the yield of Tomatoes for an ozone exposure dose of 1,000 pphm-hour, and a 50% reduction at about 2,000 pphm-hour. The highest ozone dose in the Los Angeles basin in 1975 was 3,085 pphm-hours, compared with 5,750 in 1974. These exposure doses were *very high* compared with those measured in Eastern US. For example at Yonkers, NY, which may have the greatest ozone concentrations in the East, 493 hours above 8 pphm O_3 and 144 hours exceeding 12 pphm O_3 occurred in 1977—the year with the highest ozone for the period 1970 to 1978 (Jacobson, 1982). At Beltsville, Maryland, records for 1968–76 indicate a maximum for this period (occurring in 1975) of 40 hours with ozone concentrations equal to or greater than 10 pphm.

The development of open-top chambers (Heagle *et al.*, 1973; Mandl *et al.*, 1973), supplied with either non-filtered ambient air or charcoal-filtered air, made it possible to study yield losses due to photochemical oxidants with minimum chamber interference. A 5 years' (1972–76) study at Beltsville—which is fairly representative of the mid-Atlantic seaboard—investigating two

Snap Bean crops each year, showed that fresh weights of the pods were reduced an average of 4.5% (Heggestad *et al.*, 1980). However, two of the four cultivars used in the research had sufficient natural tolerance to ozone to prevent yield losses. One cultivar, the most sensitive one, showed an average yield-loss of 14% for the five-years' period. A three-years' study with Soybeans (*Glycine max*) showed an average yield reduction, for the four cultivars tested, of 20% (Howell *et al.*, 1979). The authors believed that reduced yield was due in part to an interaction of chamber effects and air pollutant effects.

A summary describing results of field research at Beltsville from 1972 to 1980, using open-top chambers with either charcoal-filtered or non-filtered air, is presented in Table I. Crop yields in open-top chambers with charcoal-filtered air was used as the standard for comparison. The indicated yearly mean loss for five crop species examined: Potatoes, Tomatoes, Snap Beans, Soybeans, and Sweet Corn, was 12% (Heggestad *et al.*, 1977; Heggestad, 1980). Whether charcoal-filtered air is a realistic standard may be questioned. But, in open-top chambers there is some mixing of non-filtered air with filtered air owing to ambient air entering through the chamber top when wind speeds increase above 3–5 km per hour. Relatively little ingress of ambient air occurs, however, during periods of air stagnation, when ozone levels are highest (Heagle *et al.*, 1973; Heggestad *et al.*, 1980).

Chemical protectants have been used to assess the effects of O_3 on crop yields. The fungicide Benomyl (methyl 1-(butylcarbamoyl)-2-benzimidazole-carbamate), applied as a foliar spray to protect Snap Beans in Massachusetts from ozone injury, indicated that ambient O_3 reduced

Table I   Effects of photochemical oxidants on growth and yield of 5 crops species grown at Beltsville, Maryland, and on the Delmarva peninsula.

Species	Cultivars tested each year	Years*	Yields as compared to chambers with charcoal-filtered air	
			Non-filtered chambers	No chambers
Potatoes[†]	4	1973–77	90	112[¶]
Soybeans[‡]	4	1973–75	80	101[¶]
Sweet Corn[§]	4	1977–78	91	93
Snap Beans[§]	4	1972–76	96	96
Tomatoes[§]	1	1979–80	83	64[¶]

* Four replications each year except for Tomatoes (three replications).
† Grown at Painter, Virginia, 1973–75, cooperation of R. E. Baldwin & Boyette Graves, Virginia Truck and Ornamentals Research Station; Georgetown, Delaware, 1976–77, cooperation of D. Fieldhouse, University of Delaware, Newark; and Beltsville, Maryland, 1974.
‡ From R. K. Howell *et al.*, 1979; grown at Queenstown, Maryland.
§ Grown at Beltsville, Maryland.
¶ Higher and lower values than in non-filtered chambers may indicate chamber effects.

yields about 33% on two susceptible variaties, but a more tolerant variety showed no yield reduction (Manning *et al.*, 1974). In Canada, foliar applications of N-(2-(2-oxo-1-imidazolindinyl)ethyl)-N′-phenylurea (EDU) to White Beans increased yields 36%, presumably because of EDU-induced tissue-tolerance to ozone (Hofstra *et al.*, 1978). Treatment with EDU (Carnahan *et al.*, 1978) gave more protection than did Benomyl or Carboxin (5,6-dihydro-2-methyl-1,4-oxathiin-3-carboxanilide). Chemical protectants must be applied at the stage of maturity when the plants are most sensitive to yield loss effects; this is thought to be at the flowering and fruiting stages for most species. Sprays may need to be applied several times during the growing period—for example at 2-weeks' intervals—to obtain maximum protection. None of the above chemical protectants is used commercially at the present time to reduce plant injury and yield-losses caused by ozone.

Much of the available dose–response information assessing the impact of ozone on crop yields comes from studies conducted in North Carolina by Heagle & Heck (e.g. 1980). Increasing ozone concentrations were added for 7 hours per day throughout the growing season to open-top chambers set over field-grown crops. Seasonal mean ozone concentrations added to the chambers are illustrated by the data in Fig 3. Regression curves showing

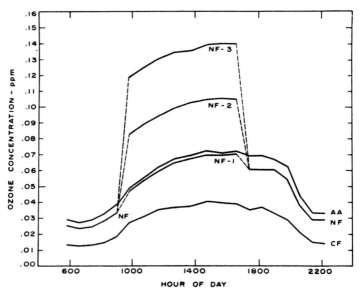

Fig. 3   Seasonal mean ozone concentrations for various hours of the day for the experimental period: 9 April through 31 May 1977. Symbols: AA = Ambient Air; CF = Chambers with filtered air; NF–1, NF–2, NF–3 = Non-filtered chambers supplied with ambient air or two levels of O_3 added for 7 hours per day (0930–1630). Source: Heagle & Heck (1980).

exposure-dose yield effects for Field Corn (*Zea mays*), Winter Wheat (*Triticum aestivum*), Soybean, and Spinach (*Spinacia oleracea*), are given in Fig. 4. Heagle & Heck utilized these data to estimate annual crop losses for different seasonal 7 hr per day ozone concentrations. Based on the US Department of Agriculture value of $54 billion ($1 \times 10^9$) for the nation's 1978 crop harvest, they estimated that a seasonal 7 hour per day mean ozone concentration of 0.06 ppm for the year could reduce crop yields by 5.6% and result in a $3 billion loss. A 10% loss in crop value was indicated if the mean 7-hour daily ozone concentration was 0.08 ppm, and a 15% loss if ozone levels were to reach 0.10 ppm. The $3 billion loss in crop value attributed to 0.06 ppm daytime O_3 means, can be compared with the 1.8 billion loss estimated by the Stanford Research Institute in 1981.

A major problem of the assessment by Heagle & Heck (1980) is the lack of adequate ambient air monitoring data for ozone to compare with 7-hour means. Also, in Nature, ozone concentrations vary from day to day, whereas those supplied by Heagle & Heck (1980) were the same each day throughout

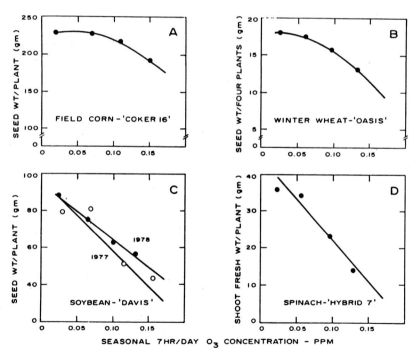

Fig. 4 Measured values and regression curves showing the yields of selected varieties of Field Corn, Winter Wheat, Soybean, and Spinach, exposed to different (7 hours per day) seasonal mean ozone concentrations. Source: Heagle & Heck (1980).

Table II   Distribution of maximum two-months' means of daily maximum 7 hour
values for US sites during 1975, 1976, and 1977.

Mean ppm 2-months' ozone values at sites in 1975, 1976, and 1977	Representative locations	Number of sites with mean ozone values in one or more years
0–0.019	Honolulu, HI	3
0.02–0.029	Palm Beach, FL	3
0.03–0.039	Spokane, WA; Salem, OR	7
0.04–0.049	Peoria, IL; Chattanooga, TN	26
0.05–0.059	Syracuse, NY; Omaha, NE	40
0.06–0.069	Charlotte, NC; Louisville, KY	42
0.07–0.079	Dallas, TX; Tulsa, OK	27
0.08–0.089	Fresno, CA; Dayton, OH	11
0.09–0.099	Southeast Desert, CA	8
0.10–0.109	Bethlehem, PA	2
0.11–0.119	———	0
0.12–0.129	———	0
0.13–0.139	Philadelphia, PA	1
0.14–0.149	———	0
0.15–0.159	———	0
0.16–0.169	Los Angeles, CA	1

171

Source: Data from Environmental Protection Agency, (1979).

the growing-season—except that such 'fumigations' were not conducted on
rainy days.

Data on 7-hour daytime means for a two-months' period in 1975, 1976 and
1977, were summarized by the Environmental Protection Agency (1979) for
171 sites in the US (Table II). Two-thirds of the sites had mean values
between 0.05 and 0.08 ppm. The lowest value was from Hawaii. The two
highest values were from metropolitan Los Angeles and Philadelphia. Nine
locations identified as rural averaged 0.069 ppm, ranging from 0.053 ppm at
Waltham, Massachusetts, to 0.086 ppm on the San Joaquin Valley, California. The two consecutive months with the highest value were July and August,
which totalled 34% of the high values. Twenty-six per cent of the high values
were found in the two bimonthly periods: May–June, and June–July. August
and September were high for only 14% of the sites.

Cultivar Tolerance to Ozone

Crop cultivars developed in regions with elevated pollution-levels tend to be
more tolerant to pollutants than cultivars developed in relatively clean areas
(Reinert *et al*, 1982). Documented cases of the above include cultivars of
Potatoes, Cotton, Alfalfa, Sugar Beets, and Dry Beans.

Table III   Yield of Potato cultivars in greenhouses with charcoal-filtered and non-filtered air at Beltsville, Maryland in 1971.*

Cultivar	State where developed	Yield		NF/CF × 100
		Charcoal-filtered air (CF)	Unfiltered air (NF)	
Katahdin	Maryland (USDA)	218	279	128
Penn 71	Pennsylvania	251	269	107
Pungo	Virginia & Maryland (USDA)	282	287	102
Norgold Russett	North Dakota	251	252	100
Superior	Wisconsin	318	304	96
Kennebec	Maine & Maryland (USDA)	369	339	92
Wauseon	Maine & Maryland (USDA)	273	218	80
Norchip	North Dakota	466	346	74
LaChipper	Louisiana	380	276	73
Alamo	Texas & Maryland (USDA)	410	272	66
Haig	Nebraska	295	187	63
Norland	North Dakota	401	199	50

* LSD 0.05 = 71 g for cultivars within an environment, and 81 g for cultivars between the two environments. Data from Heggestad (1973).

Breeders of cigar-wrapper Tobacco, the first crop identified as having ozone injury in the Eastern United States (Heggestad & Middleton, 1959), have been able to eliminate losses due to ozone by plant-breeding procedures. Based on cultivar screening tests involving ozone-induced foliar injury, cultivars within most crop and horticultural species vary considerably in their resistance to ozone in the seedling stage (Reinert *et al.*, 1982). Much less is known about cultivar variation in resistance to ozone based upon yield response. Foliar injury, especially in the seedling stage, may not predict resistance revealed by yield response in field studies (Heggestad & Bennett, 1981).

To illustrate species-response to ozone, yield information on Potato (*Solanum tuberosum*) cultivars is provided (Table III). Kennebec, Katahdin, and Pungo, which originated from research in Maryland and Virginia, are believed to be more tolerant to ozone air pollution than those originating in less-polluted areas of the country. They were selected under pollution-stress conditions (Heggestad, 1973). The five most sensitive cultivars were developed in the Midwest and South, which have less of an ozone problem than the urban–industrial states of the Atlantic Coastal Region. Potato cultivars with the greatest ozone tolerance (Penn 71, Pungo, and Norgold Russett) though, tended to have lower yielding-ability in charcoal-filtered air than some of the more sensitive cultivars.

In 1971, growers on the Eastern Shore of Virginia changed from Pungo to the higher-yielding cultivars: Norchip, LaChipper, and Alamo. The combination of (1) a serious pollution episode, and (2) unusual environmental conditions for two weeks prior to that episode which increased plant sensitivity to ozone, resulted in severe foliar injury and a reported 50% yield-reduction on some farms. The following year, the growers returned to the more tolerant cultivars, Pungo and Superior. Only relatively minor losses have occurred since the 1971 season.

Ozone monitoring on the Eastern Shore of Virginia in 1972, 1973, and 1974, revealed that days with the highest ozone concentrations occurred when winds were light and from the North, suggesting that large metropolitan areas in the Washington to New York City corridor were the sources of precursor chemicals for the photochemical generation of ozone. The US Department of Agriculture and Virginia Agriculture Experiment Station continue to utilize the Research Station at Painter, Virginia, to screen new breeding lines and cultivars for their tolerance to ozone. A possible mechanism involved in the resistance of plants to ozone is discussed in a later section.

SULPHUR DIOXIDE

Effects of SO_2 on Foliar Injury and Yield

Sulphur dioxide (SO_2) and acid rain are becoming of more and more concern as energy demands increase and we switch to the use of coal to generate electricity. About three-quarters of the coal that is currently used in the United States is burned by electric utilities (Abelson, 1981). Although SO_2 emissions only sporadically cause acute enough foliar injury to damage crops, there is mounting evidence that prolonged exposure to low concentrations—either as a single pollutant or in combination with others, such as O_3 and NO_2—can reduce crop productivity. As low concentrations of atmospheric sulphur may also provide fertilizer benefits to sulphur-deficient crops, dose–response field studies are needed to quantify the impact of SO_2 on agriculture in different regions. Crops grown under the humid conditions that normally occur in Eastern United States, are more sensitive to pollutants than are those grown in the arid and semi-arid Western United States.

The effects of SO_2 impingement downwind from a major polluting source are likely to be quite different from the effects induced in areas with many merging sources. Plant responses to short-term, high-level SO_2 exposures may result in acute foliar injury, whereas long-term, low-level exposures in some instances can be beneficial—especially if the soil is deficient in sulphur. Conversely, as will be discussed later, long-term, subacute exposures can also result in crop yield losses—sometimes in the absence of recognizable foliar injury.

Relationships between peak SO_2 concentrations and induced injury are better known than those between foliar injury and yield loss. If the damaged leaf area is expressed as a percentage of the total foliar area produced during the growing season, useful estimates of crop yield losses can be predicted from the leaf injury data (Haase *et al.*, 1980). These authors, nevertheless, state that 'leaf injury from SO_2 does not always result in crop yield losses.'

Effects, on crop yields, of SO_2 exposures that simulate frequency distributions found in multiple-source areas, have been the subject of relatively few studies. In multiple-source areas, other phytotoxic gases also are likely to play an important interacting role in determining impact on the crop. There have been very few controlled experiments with mixed gases, especially in field situations; consequently the analysis of available results involving mixtures of SO_2 and other pollutants is tenuous. Some data suggest that mixtures of SO_2 and O_3 are likely to increase losses—especially when the pollutants are at or near the thresholds for causing yield losses by the individual pollutants (Oshima, 1978; Heggestad & Bennett, 1981).

The impact of SO_2-induced foliar injury on the growth and yield of Alfalfa (*Medicago sativa*), could be simulated by clipping leaf tissue (equivalent to the amount of damaged foliar area) from control plants at the times when SO_2-damage occurred (G. R. Hill & Thomas, 1933). According to Haase *et al.* (1980), foliar injury caused by a single SO_2 exposure does not vary greatly with growth stage, whereas crop yield losses due to foliar damage do depend upon the stage of maturity.

The most extensive data relating foliar injury occurring in the field to continuously monitored ambient SO_2 concentrations are those reported by

Dreisinger & McGovern (1970) from studies around smelters near Sudbury in Ontario, Canada, and by H. C. Jones *et al.* (1979) obtained around electric power-generating plants in the Tennessee Valley. Both investigations demonstrate that considerable variation exists among species in their tolerance to SO_2. Leaf injury was related to peak concentrations in the Sudbury area. Injury occurred most commonly in June and July. To prevent SO_2 injury to most species, the authors concluded that SO_2 concentrations should not exceed 0.70 ppm for 1 hour, 0.40 ppm for 2 hours, 0.26 for 4 hours, or 0.10 ppm for 8 hours. Some particularly sensitive species may be injured by concentrations slightly below the above recommended 1- and 2-hour maximum means.

H. C. Jones *et al.* (1979) studied foliar effects caused by ambient SO_2-levels on native plants as well as crops. After 6,500 field inspections, they concluded that the threshold dose for foliar injury on sensitive species was 0.32 ppm for 1 hour or 0.17 ppm for 3 hours. The probability that foliar effects would occur on any species examined, or that yields of Soybean would be reduced, was less than 50% for 3-hours' exposures to concentrations less than 0.50 ppm. Corn (*Zea Mays*), Cotton (*Gossypium* sp.), Wheat (*Triticum* sp.), and Tobacco, were much more resistant than Soybeans or the southern pines (*Pinus taeda* and *P. virginiana*). Some field studies showed an 8.8 kg per ha reduction in Soybean yield for each 1% of foliar injury. This amounted to one-half of 1% of the normal Soybean yield.

Crop Yield Related to SO_2 Gradient at Varying Distances from a Source

In 1959 and 1960, plants of several species grown in containers were placed at varying distances from an iron-ore roasting-plant near Biersdorf, Germany (Guderian & Stratmann, 1962, 1968). The results are summarized in Table IV. Among the cereals, Winter Wheat and Winter Rye were more sensitive than Spring Wheat and Oats. In general, the 10 crop species tested were more tolerant than 8 woody species included in the study. Two species of *Ribes* (currants and gooseberries) showed reduced fruit yield at all locations. Since episodes below 0.10 ppm were discounted, the exposure doses shown in Table IV are underestimated. Also, hydrogen fluoride may have been a complicating factor in the study.

Open Air Fumigations with SO_2

To eliminate chamber microclimate effects on plant responses to SO_2, open-air SO_2 fumigation systems have been used. Beginning in 1972, in France, de Cormis *et al.* (1975), and Bonte (1977), released pollutants at 4 heights (0.5 to 2.8 m), utilizing 128 vertical pipes spaced in a grid over a 60 m × 35 m plot. Trees were planted at the mid-points between diffuser pipes. An automated

Table IV   SO$_2$ concentrations and crop yields given as percentages of the control for varying distances from an iron ore roasting-plant.*

	6,000	Sites, SO$_2$ dose, and yields compared with control (%)				
Distance (m)	6,000	1,900	1,350	725	600	325
Concentration (ppm)†		0.15	0.18	0.24	0.34	0.45
% Time†		8	12	20	26	30
1/2 hour max. conc. (ppm)		1.3	1.6	2.3	5.1	6.0
Species						
Lycopersicum esculentum (Tomato)	100	100	104	102	90	91
Brassica rapa (Turnip)	100	90	90	90	87	89
Beta vulgaris (Garden Beet)	100	98	99	91	83	68
Medicago sativa (Alfalfa)	100	96	97	93	81	64
Triticum aestivum (Spring Wheat)	100	100	100	88	74	64
Avena sativa (Oats)	100	100	95	84	76	62
Secale cereale (Winter Rye)	100	99	97	85	58	48
Solanum tuberosum (Potato)	100	98	90	83	68	44
Triticum aestivum (Winter Wheat)	100	100	92	41	56	43
Trifolium repens (Red Clover)	100	97	99	90	64	30

* Source: Guderian & Stratmann (1962, 1968).
† Concentration at times that SO$_2$ concentrations exceeded 0.10 ppm.

system produced concentrations between 0 and 0.075 ppm (mean 0.05 ppm). A control plot of comparable size was located 100 metres away. Results, compiled after 3 years of continuous exposure, indicated that growth in circumference was reduced in 5 of 7 conifers. Apples (*Malus* sp.) and Pears (*Pyrus* sp.) showed reduced growth and yield losses without visible symptoms of injury. SO$_2$ exposure predisposed some trees to frost damage.

Another open-air study involving horizontal pipes set out over a prairie grassland in Montana was initiated in 1975 (J. J. Lee & Lewis, 1978). A similar fumigation system established in 1977 was used over Soybeans in Illinois (Miller *et al.,* 1980). In Montana, SO$_2$ was released continuously over the prairie grassland plots, generating average concentrations of 0.01, 0.02, 0.04 and 0.07 ppm. Foliage of the dominant species, Western Wheat-grass

(*Agropyron smithii*), senesced more rapidly due to the SO_2 fumigation, but no significant reduction in net above-ground production was detected (Heitschmidt *et al.*, 1978).

In the Illinois Soybean studies (Miller *et al.*, 1980; Sprugel *et al.*, 1980), the SO_2 exposures caused some acute injury. The fumigations were for 4 to 5 hours a day but conducted only on days when the wind direction was favourable in keeping SO_2 away from control plots. Mean plot concentrations during the fumigations were 0.12, 0.30, and 0.79, ppm in 1977, and 0.11, 0.19, 0.26, and 0.36, ppm in 1978. Yield reductions were observed with all SO_2 exposure-doses. The yield effects observed with low SO_2-concentrations may have been due in part to ozone. Background ozone concentrations averaged 0.05 ppm during the fumigations. Ozone levels periodically reached 0.10 ppm. Although these studies lacked statistical replication of the fumigated plots, which is a major concern, two distinct advantages with such open-air systems are immediately evident: (a) the plants are grown under ambient field conditions, and (b) chamber effects are avoided.

At four locations, long-term effects of fluctuating ambient SO_2 concentrations on crop yields were assessed by comparing yields in charcoal-filtered and non-filtered air in either closed or open-top chambers. Bleasdale (1973) found significant yield reductions in Rye grass (*Lolium multiflorum*) at 0.05–0.07 ppm SO_2 under both winter and summer conditions. Crittenden & Read (1978, 1979) observed effects on Ryegrass at mean levels as low as 0.02 ppm with daily peaks less than 0.10 ppm. Navara (1975) also reported that Buckwheat (*Fagopyrum esculentum*) and Rye (*Secale cereale*) grew better in charcoal-filtered air. These studies were all in urban areas where other pollutants could have been present. Unfortunately, plot replications to test for inherent experimental variation were not obtained for the above chamber treatments.

Brough *et al.* (1978), using replicated open-top and closed-chamber experiments conducted over three seasons, found that mean SO_2 levels of about 0.02 ppm during the growing season reduced Barley (*Hordeum vulgare*) yields by 14–30%. Interference due to HF could have complicated the study due to emissions from a nearby brickworks.

Interactive Effects of SO_2 and O_3 on Yield

Oshima (1978) employed continuously stirred, closed chambers in a field study near Riverside, California, to assess the effects of 0.10 ppm SO_2 and various O_3 mixtures on Red Kidney-bean (*Phaseolus vulgaris* cultivar) yields. SO_2 was added for 6 hours per day for a total of 335 hours from August to October. The SO_2 was mixed with O, 25%, 50%, 75%, or 100%, carbon-filtered air. Bean yields were significantly reduced by SO_2 mixed with 50% carbon-filtered air, which provided an ozone dose just below the threshold required to reduce Bean yield without added SO_2. In this experiment, 50%

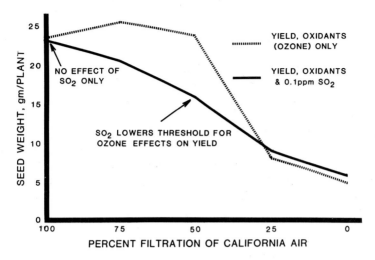

Fig. 5   Red Kidney-bean yields for plants exposed to 0.1 ppm SO_2 added to open-top chambers supplied with charcoal-filtered (oxidant-cleansed) California air or to filtered : unfiltered air mixed in the proportions: 3 : 1, 1 : 1, 1 : 3, and 0 : 1. Source: Adapted from Oshima (1978).

carbon-filtered air provided a total of 95 hours above 0.10 ppm ozone. Data summarizing the consequences of the O_3–SO_2 mixtures are arrayed in Fig. 5.

Heggestad & Bennett (1981) showed that ambient ozone levels at Beltsville, Maryland, portended yield losses in Snap Beans exposed to SO_2. Three Snap Bean cultivars, grown in the field in open-top chambers, received recurrent midday treatments with 0, 0.06, 0.12, and 0.30, ppm SO_2 added to non-filtered air, and (in separate plots) 0 and 0.30 ppm SO_2 added to charcoal-filtered air. When compared with bean yields in chambers with charcoal-filtered air, yields were reduced 43% in the non-filtered air with 0.30 ppm SO_2 added. They were reduced only 16% when 0.30 ppm SO_2 was added to charcoal-filtered air. However, ambient oxidant (ozone) levels in the non-filtered air, but no SO_2 added, did not cause a measurable yield reduction in the 1979 study. A significant negative linear regression (Fig. 6) of Bean yields with increasing SO_2 doses was indicated ($r = -0.99$). One cultivar, Astro, which exhibited no visible pollutant-induced foliar injury, exhibited as much yield loss as two other cultivars showing foliar injury symptoms.

Variation in Plant Resistance to SO_2

Threshold concentrations required to produce foliar injury on resistant crop-plants can be threefold greater than those that injure susceptible plants (H. C. Jones *et al.*, 1979). Desert species show even greater tolerance to SO_2 (A. C.

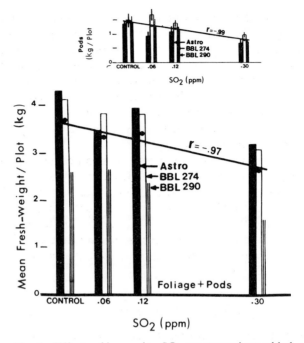

Fig. 6   Effects of increasing SO$_2$ concentrations added
to open-top chambers on Snap Bean yield and biomass
production.

Hill *et al.*, 1974; Thompson *et al.*, 1980). This may be due in part to inherent
genetic control, but also to environmental stress factors that increase plant
tolerance. For example, SO$_2$ uptake is reduced by low humidities (Thomas,
1935; Black & Unsworth, 1980) and low soil-moisture (A. C. Hill *et al.*,
1974). Humidities below 40% are particularly protective (McLaughlin &
Taylor, 1980). Variation in SO$_2$ resistance among Rye grass cultivars apparently originated by selection pressures in a polluted area (Mansfield, 1976)
similar to that discussed above for cultivar resistance to ozone.

SO$_2$–Plant Disease Interactions

The literature describing the effects of sulphur dioxide on plant–pathogen
relations has been reviewed by Heagle (1973), Treshow (1975), and Laurence (1978). Sulphur dioxide, which was one of the first fumigants to be used
to control transit and storage diseases, can have fungicidal properties at high
concentrations, but the effective exposure-doses required range from about
100 to 2,500 ppm applied for several minutes to a few hours. Schaeffer &
Hedgecock (1955) observed the reduced prevalence of certain fungi attack-

ing trees in polluted air downwind from a smelter, in contrast to their prevalence on trees in clean air control areas. Heagle (1973) concluded that documented cases of SO_2 causing the inhibition of disease development are more common than the converse. Reduced disease development appears to occur primarily with obligate fungal parasites such as rust fungi (*Puccinia* spp.) (Weinstein *et al.*, 1975) and powdery mildews (Erysiphales) (Hibben & Taylor, 1975). Decreased incidence and severity of rust diseases was correlated with smaller sizes of the uredospores and reduced spore germination, suggesting diminished viability.

Disease development can be sensitive to small changes in SO_2 concentrations as shown by Saunders (1966) for Black-spot disease of roses (*Rosa* sp), caused by *Diplocarpon rosae* Wolf. Exposure to 0.04 ppm SO_2 for 2 or 14 days reduced infection, but exposure to 0.02 ppm for 14 days increased it. Saunders reported that the disease was checked or eliminated in the urban areas studied where mean SO_2 concentrations exceeded 0.04 ppm. Apparently, disease suppression occurs only at concentrations that are relatively rare today (Treshow, 1975).

Beneficial Effects of Atmospheric Sulphur

Vegetation can obtain significant amounts of nutrient sulphur (*see* below) from polluted atmospheres. Downwind from a source, sulphur uptake due to dry deposition often exceeds that introduced in rain (Nyborg, 1978). In remote areas, rainfall contributes as little as 1 kg sulphur per ha per year, compared with over 100 kg sulphur per ha per year near some urban–industrial areas producing SO_2. Emitted SO_2 may be transported several hundred kilometres, though that which remains in the atmosphere is eventually oxidized to sulphate and removed by acid rain or other particulate deposition.

Atmospheric sulphur, absorbed in small doses, is known to be beneficial to plants—especially when they are grown on sulphur-deficient soils (Cowling *et al.*, 1973; Cowling & Lockyer, 1978; Noggle & Jones 1979). Sulphur deficiency occurs in 29 states of the US, 5 provinces of Canada (Beaton *et al.*, 1971), many parts of Australia and New Zealand (R. K. Jones *et al.*, 1975), and in parts of Africa and Europe (Brogan, 1978; Terman, 1978). In recent years, sulphur deficiency has become more common than formerly on cropped lands, due to the replacement of ammonium sulphate by ammonium nitrate in high-analysis fertilizers. Recycled sulphur is made available from decaying plant and other organic matter and manures, while to some extent pesticides and herbicides add sulphur to the plant–soil environment.

Sulphur deficiencies frequently occur in sandy soils, such as those found on the Coastal Plains of southeastern US. If SO_2 levels remained near background levels (about 2 to 4 ppb), annual sulphur deposition could be expected to be less than 10 kg per ha. Good growth of many crops requires about

25–75 kg of S per ha per year. In regions with multiple SO_2 sources where concentrations exceed background levels, atmospheric sulphur could supply all of the sulphur that crops need. Potted Cotton plants placed downwind of coal-fired power-plants, were shown to grow better than plants in sulphur-free areas (Noggle & Jones, 1979). Cotton was more efficient than fescue-grass (*Festuca* sp.) in accumulating sulphur. Sulphur accumulation can be due to both foliar SO_2 uptake through stomata and to root uptake of sulphur deposited on the soil.

SO_2 Summary

Most agricultural areas of the world are only exposed to SO_2 concentrations that are well below injury thresholds. In these rural areas, SO_2 hourly averages never exceed 0.06 ppm and average concentrations, especially in summer months, are less than 0.01 ppm—even in the United States. At the latter concentrations, SO_2 either has no effect on plant growth or it may be beneficial if soil sulphur is inadequate. Elevated and potentially injurious concentrations of SO_2 may occur near large urban–industrial areas or downwind of a significant point-source. SO_2 pollution was estimated recently to cause only 2% as much crop loss as ozone pollution in the US (Stanford Research Institute, 1981). All other atmospheric pollutants combined contribute towards the remaining 3% annual crop losses estimated for the US.

PESTICIDES

Need for Pesticides and Pesticide Regulation

Pesticides and other agricultural chemicals are vital to the intensive agriculture that permits our modern way of life. Pesticides are employed to increase crop productivity by controlling destructive insects and pathogens, checking the growth of competing weeds, and regulating other unwanted biotic interferences. Success in combating agricultural pests and diseases is necessary to maximize agricultural productivity and lower production costs. The introduction of pesticides into the environment for this purpose, however, presents hazards of its own that require solution. The benefits from increased food production and pest control nevertheless far outweigh the hazards involved, provided proper control measures are taken.

Physical processes and biological activity in the environment degrade many applied pesticides in a relatively short period of time. Others can persist for years or decades, accumulating at times to significant levels in soils and water or by concentrating in the food-chain. Concern about the protection of public health and a nation's resources has led to widespread legislation to regulate the marketing and use of pesticides. In the United States, the US Department of Agriculture (USDA), the Food and Drug Administration (FDA), and the Environmental Protection Agency (EPA), have specific control functions relating to pesticide registration, marketing, safety, and environmental monitoring. Regulation of pesticide users, and in some cases licensing, is conducted at the State level. The United States is among the strictest countries in the world in protecting the public from harmful pesticides. Many under-developed countries, in contrast, do not have effective control measures in force.

Pest control and food production are major world-wide problems. The World Health Organization (WHO) and the Food and Agriculture Organization (FAO) of the United Nations have pressed for expanded use of pesticides throughout the world in order to raise agricultural efficiency and increase food supplies for the planet's growing population. Particularly in the tropics, food production would be seriously impaired without the use of pesticides. The control of environmental contamination by pesticides, though not an insurmountable problem, requires an adequate knowledge of the dispersal, fates, and effects, of toxic chemicals. Regulation is needed at both local and national levels to minimize the potential hazards to human health and to the environment.

Pesticide Use in the US

'Pesticide' is a general term covering herbicides, fungicides, insecticides, fumigants, and rodenticides. Chemical agents that modify and regulate plant growth (defoliants, dessicants, etc.) are also included in the definition under the US Federal Insecticide, Fungicide and Rodenticide Act (FIFRA). Synthetic organic pesticides are the most important in terms of annual production. Table V gives the US production of synthetic herbicides, insecticides, and fungicides, for 1976, and the amounts used on crop lands. Of the 2.3 billion (1×10^9) acres (nearly 1 billion hectares) of land in the United States, farmlands planted to major crops account for approximately 17% of the total. Well over 500 million kilograms of pesticides are produced annually in the United States, involving approximately 40,000 pesticide formulations (Council on Environmental Quality, 1979). Approximately half of the annual pesticide production is used on crop lands. Herbicides, and selective and non-selective weed killers, constitute about two-thirds of this applied volume. Table VI lists the major herbicides and insecticides applied by US farmers on croplands. The relative amounts of each pesticide included in the table, given as a percentage of the total pesticide usage for its class, is also shown.

Pesticides in the US are also applied each year to some forest stands, pasture lands, and rangelands, which between them cover the major portion of the country. In addition, pesticides are employed as public health measures

Table V US production of synthetic pesticides for 1976, and their consumption by farmers on crop lands.*

Pesticide class	Use	US production (million kg)	Consumption on crop lands	
			(million kg)	(millions of hectares treated)†
Herbicides	Control of weeds	298	179 (60%)	79 (53%)
Insecticides‡	Control of insects, mites, and related organisms; also of disease vectors	257	74 (29%)	30 (20%)
Fungicides	Control of moulds, rots, blights, wilts, and other fungal disease	65	data unavailable	data unavailable
	Total	620		

* Values in parentheses give percentages of the total US pesticide production by pesticide class for 1976 or percentage of total farm croplands treated.
† Total US 1976 farm crop-land = 150 million hectares (1 ha = 2.471 acres).
‡ Insecticide class includes fumigants, rodenticides, and soil conditioners.

Source: Council on Environmental Quality (1979).

Table VI   Major herbicides and insecticides used by US farmers on crops, 1976.*

	% of pesticide by class
Herbicide	
Atrazine	23
Alachlor/Propachlor	25
2,4-D	10
Trifluralin	7
All others	35
	100
Insecticides	
Toxaphene	19
Methyl/Ethyl Parathion	18
Carbofuran	7
Aldrin/Dieldrin	1
All Others	55
	100

* Average application rate: Herbicides (0.36 kg per ha); Insecticides (0.4 kg per ha). Source: US Department of Agriculture.

to control mosquitoes and other disease-vectors on marshlands and in cities. Finally, pesticides are used in the management of nuisance or obstructive vegetation around waterways, roadways, and railroads, parks and lawns, and utility rights-of-way.

Off-target Contamination and Effects

Pesticide drift from target areas occurs when spray or dust formulations are applied by aircraft or by equipment on the ground, polluting the surrounding environment. Some pesticide formulations remain airborne long enough to be transported over vast distances, contaminating The Biosphere in low concentrations. Even after deposition on plant, soil, or water, surfaces volatilization and/or the re-entrainment of pesticide-treated soil particles into the atmosphere owing to strong winds (dust storms), there may be redistribution of pesticides in the environment. Moreover, they can enter the air in smokes from burning wastes and in vapours lost during their manufacture, processing, and use. Environmental factors (temperature, humidity, rainfall, light, soil particle-composition and size, etc.) and the pesticides' chemical and physical properties, affect their eventual fates and persistence-times in the environment.

Off-target drift of herbicides can cause losses in the yields or aesthetic value of neighbouring crops and vegetation. Pollution engendered by some insecticides has the potential for affecting livestock grazing in the area, birds and aquatic life, and other desirable biota—notably pollinating bees and natural insect predators. The highest concentrations of airborne pesticides are found near to their points of application, and persons applying the chemicals or working in the area, receive the greatest exposures—requiring special precautions and protective measures to minimize detrimental health-hazards.

The physical form, and the droplet or particle size, of the pesticide material used, affects the distance it will drift downwind after application. Most insecticides are formulated as dusts, sprays, or granules, varying in active concentrations between 0.1% and 95% by weight. In some cases unformulated materials are applied. The drift potential is greatest for aerosols and dusts, and least for sprays and large granules. In the 1940s, dusts were widely used in aerial applications of insecticides and fungicides in the United States, but since then they have been progressively curbed over the years because of their drifting potential. Sprays are more convenient because the droplet size can be controlled within certain limits by the type of nozzle employed, the pressure used, and the orientation-direction of the nozzle. Surfactants can also be added to facilitate foliar absorption. Large granules have little tendency to drift, so their use is restricted to soil applications. Approximate horizontal drift predicted in a steady 5 km/h wind is shown in Table VII as a function of particle-size for applications 3 m above the target surface (Brooks, 1947).

Early research into the loss of pesticides from target areas owing to drift and volatilization, suggested high losses in certain cases. Better understanding of pesticide formulation properties affecting loss-rates, improvements in equipment and application techniques, and concern about meteorological parameters, have reduced nontarget contamination in recent years. Maybank *et al.* (1978) reported that off-target drift of 2,4-D esters sprayed from ground-rigs on Canadian prairies, varied between 1% and 8%—depending upon nozzle type and wind-speed. From 20% to 35% drifted off-target when

Table VII   Approximate distances that particles of various sizes will drift horizontally in a 5 km wind while falling 3 m vertically.*

Particle	Particle size (μm)	Horizontal drift
Sprays		
coarse	400	2.5 m
medium	150	7 m
fine	20–100	15–300 m
Mists	50	50 m
Dusts	10–20	0.3–1.3 km
Aerosols	2	35 km

* Source: Brooks (1947).

sprayed from aircraft. When deposited as the butyl ester, 30–40% evaporated and drifted downwind within 2 hours of spraying. Use of the less-volatile octyl ester resulted in 10–15% loss during this period. Vapourization was the major contributor to target-area loss. However, this was minimized by phasing out the use of 2,4-D high-volatile esters such as the butyl ester.

For many years, 2,4-D herbicides have been extensively used in the wheat-growing region of central Washington State. Over the past several decades, widespread damage to vineyards, due to 2,4-D drift, has occurred periodically (Robinson & Fox, 1978), grapes being among the most sensitive of plants to 2,4-D. Monitoring programmes in the area showed that high-volatile types of 2,4-D accounted for three-quarters of the dispersed herbicide found at sampling stations, even though the use of high-volatile esters was prohibited in the area. Non-volatile formulations (2,4-D salts and amine compounds) represented only 3% of the total. The source of high-volatile 2,4-D esters was linked to southerly winds transporting the compounds into the region from the State of Oregon more than 30 kilometres away (where they could be legally used). Atmospheric monitoring data and weather patterns in relation to observed vineyard damage indicated that transport up to 80 kilometres was an important factor in the 2,4-D distribution pattern.

Pesticide Removal from the Atmosphere

Pesticides are removed from the atmosphere by (i) surface deposition, (ii) rainout, and (iii) chemical or photochemical reactions. Relatively little is known about the fates of atmospheric pesticides, but photochemical conversion and reactions with atmospheric oxidants are considered to be principal reaction mechanisms. The atmosphere contains numerous absorptive (and potentially catalytic) aerosol particles which facilitate conversion processes in association with water and organic matter. Sunlight provides photochemical reaction energy. For these reasons, the atmosphere is often a sink rather than a source for some classes of pesticides. Nevertheless, atmospheric pesticide reactions can produce secondary compounds that have biological activity. Fig. 7 summarizes, in schematic form, pesticide inputs to and removal processes from the atmosphere.

Surface deposition processes involve gravitational settling of particles, rainout, and the deposition of vapours or particulates on vegetation, soils, and other surfaces. Rain and surface impaction/interception cleanse particles from the air more efficiently than they remove vapours. Consequently, particles may have short residence-times in the atmosphere—usually measured in days or a few weeks. Vapours are more readily photooxidized, though, and may also have short lifetimes. Vapour–solution partitioning and photochemical/oxidative reactions are major processes removing vapours from the

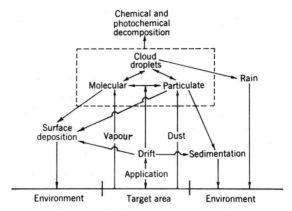

Fig. 7 Scheme illustrating pesticide inputs into the atmosphere and general removal processes. Source: Glotfelty (1978).

atmosphere. The longer the pesticide residue remains airborne, the greater is the probability that even slow reactions will result in chemical changes before the residual material returns to Earth.

The pesticide for which the most data are available concerning atmospheric biogeochemical cycling is DDT. The chemical reaction-rate of DDT in the atmosphere is not known, but the half-life for its physical removal is calculated to be about 7–10 weeks (Glotfelty, 1978). General atmospheric reactions for this relatively stable pesticide are believed to proceed at a rate of less than 1–10% conversion per week. DDT continues to command considerable attention as an environmental contaminant, even though it can no longer be legally used in the United States and some other countries. Because of its persistence in the environment and tendency to accumulate in food-chains, DDT was banned from general use in 1972 in the United States, where it was the first pesticide substance to be categorically prohibited. Banning of DDT, however, was a political decision that is not always well-supported by benefit–risk assessments.

Pesticide Interactions

Pesticides, used as plant and animal poisons, growth regulators, and surface contaminants, interact in The Biosphere to influence organisms' responses to other environmental stresses. Both negative and positive interactions have been reported, relating to such aspects as crop host–parasite relations as well as a wide range of—often subtle—pathophysiological and competitive processes. For example, certain fungicides and related chemicals are known to influence plant responses to phytotoxic air pollutants, offering protection in some cases to plants against ozone injury (Rich, 1975; E. H. Lee & Bennett, 1981). This has provided the impetus to investigate selected fungicidal and growth regulator chemicals for their potential to increase plant tolerance to oxidants.

In this search a whole host of substances have been studied, of which dithiocarbamates, thiophanates, and benzimidazole compounds, are noteworthy. One of the most effective plant systemic compounds found to date, which markedly increases the foliar resistance of ozone-susceptible plants, is the imidazolidinyl/phenylurea derivative, N-[2-(2-oxo-1-imidazolidinyl)ethyl]-N′-phenylurea (EDU) (Carnahan *et al.*, 1978). This plant regulating chemical also retards senescence in leaves and shows other cytokinen-like activity, including the enhancement of RNA and protein synthesis (E. H. Lee *et al.*, 1981). A major difference between EDU and kinetin, however, is that kinetin is not effective in transforming oxidant-sensitive plants into highly resistant ones. EDU-induced O_3 resistance is thought to be mediated through the induction of specific oxy-radical scavenging enzymes increasing the basic 'aerobic' nature of the cells, and tissues (Bennett *et al.*, 1981; E. H. Lee *et al.*, 1981). This is believed to be a natural phenomenon by which tolerant plant cultivars gain improved resistance to oxidants and ageing.

Certain growth-regulating herbicides cause symptoms of injury that may be mistaken for damage due to other air pollutants or environmental stresses. 2, 4-D is of interest, as it causes 'suture red-spot' on peaches whereby the fruits develop hypertrophy along the suture. The affected area may then soften and not before the rest of the peach is ripe. A similar pathological response is caused by fluoride injury to peaches. This crop damage response is discussed in more detail in the following section on fluoride pollution and affects.

FLUORIDE

The Fluoride Problem

Fluoride is rated as the fourth most important class of air pollutants affecting agriculture in the United States (Heck *et al.*, 1973). Historical references to fluoride effects on agricultural plants and animals resulting from volcanic eruptions and anthropogenic pollution date back ten centuries. As was indicated previously, fluoride became recognized as an agricultural problem in the United States in the 1940's following the rapid expansion of steel and aluminium industries during and after World War II. Prior to the installation of modern pollution control equipment in major fluoride-emitting industries, atmospheric fluoride concentrations near large polluting sources at times ranged upwards from 10 to 50 parts per billion (ppb). Now, atmospheric concentrations rarely exceed 1 to a few ppb for short durations. This has reduced the potential fluoride problem in the United States and some other countries.

Although fluoride ranks behind some other, more ubiquitous atmospheric pollutants in general importance, it is the most phytotoxic of known air-pollutants on the basis of atmospheric concentrations required to injure plants. This is largely due to the tendency of fluoride to accumulate in plant foliage. Leaves are extremely efficient absorbers of gaseous fluorides entering through the stomata. When once it is inside the leaf, little fluoride is translocated out of the organ by way of the conducting tissues. Acropetal diffusion does occur, causing characteristic 'marginal necrosis' and 'tip burn' of foliage when it attains toxic concentrations. Soluble particulate fluoride deposited on leaf surfaces may penetrate into the mesophyll to some degree through the cuticle, but particulate forms are also subject to removal by rain, so ending up in the soil. Studies with a number of crop species indicate that particulate fluoride on leaves do not typically result in significant leaf injury.

Fluoride is widespread in soils and rocks, so that background levels are found in plants in the absence of polluting atmospheric sources. Fluoride comprises 0.06–0.09% of the Earth's crust, and soil concentrations of it range from a few ppm to several thousand ppm. The mean fluoride concentration in US soils is reported to be about 200 ppm (Committee on Biologic Effects of Atmospheric Pollutants, 1971). Major fluoride-containing minerals include calcium fluoride, calcium fluorophosphate, and aluminium fluoride compounds. Natural fluoride in waters varies widely in different regions, depending upon the sources. Western US waters are higher in fluoride (above 0.2 ppm) than are those in the Northeast (0.1 ppm). Where the water contains several ppm of fluoride, mottled tooth enamel is often observed in the animal populations.

Major crop plants usually take up relatively little fluoride from the soil, even when the soil content of it is high. Roots contain more soil-derived fluoride than foliage, while atmospherically absorbed fluoride concentrates in the leaves. Background fluoride in foliage typically results in leaf concentrations between 0.1 and 20 ppm of the dry weight. This, however, varies with plant age, species and variety, soil pH, environmental conditions, and other factors. Fluoride-tolerant 'accumulator plants', such as certain members of the Tea family (*Theaceae*) and some poisonous plants of South America, Africa, and Australia, build up high concentrations of soil-absorbed fluoride in the foliage (50 to several hundred ppm). *Camellia* (*Theaceae*) plants fertilized with superphosphate have been shown to accumulate fluoride levels to over a thousand ppm in the dry matter. Most agricultural crop plants tend to discriminate against excessive fluoride uptake from the soil under managed agronomic conditions. The hazard to agriculture by fluoride is regarded as essentially an air pollution problem arising from the absorption of gaseous fluoride compounds by foliage.

As fluoride is always present to some extent in plant tissues, and can apparently induce the formation or activation of certain enzymes in cells, stimulating growth at subinjurious concentrations, a number of investigators have questioned whether some fluoride may be beneficial or even essential to good plant growth. Although fluoride is generally considered to be non-essential, it does appear to benefit plants growing in aluminium-rich media. Aluminium phytotoxicity is reported (Foy *et al.,* 1978) to be a major problem in acid soils of the Eastern United States, restricting root growth. It is not known to what extent fluoride can mitigate the phytotoxic impact of available aluminium in the ambient environment.

Fluoride Effects on Crop Growth

Effects of fluoride on physiobiochemical processes are discussed in earlier chapters. It should be pointed out here, though, that chloroplasts are major sites of fluoride accumulation—accounting for as much as three-fifths of the elevated fluoride in some leaves (Chang & Thompson, 1966). High fluoride concentrations inhibit many enzymes involved in photosynthesis and inter-mediary metabolism; thus, numerous metabolic processes are potentially affected at elevated levels. A special challenge is presented to young develop-ing cells that rely on primary metabolism for cell formation and development. Cell compartmentalization in mature cells provide a means by which older cells can mitigate the impact of toxic substances. This may influence the relative impact of fluoride on young developing cells as compared with older, more structured, cells.

Reversible effects of HF on crop canopy photosynthesis rates correlate more closely with short-term acute fluoride exposures than with chronic low-level exposures that typify the polluted ambient environment (Thomas & Hendricks, 1956; A. C. Hill, 1969). HF exposures for several hours above 10 ppb can measurably depress CO_2 exchange-rates of Alfalfa and Barley (*Hordeum vulgare*) canopies (Bennett & Hill, 1973). Approximately 15.0–20.0 ppb HF for 2 hours is sufficient to cause a trace of leaf necrosis. Slow recovery after termination of the treatments occurs in non-necrotic tissues.

Chronic fluoride exposures over extended durations at concentrations near 1 ppb reduce photosynthesis in relation to the amount of tissue necrosis or chlorotic mottling induced. This is, of course, irreversible and results largely from fluoride redistribution to the leaf extremities, causing damage to these parts. Crop yield reductions are more likely in areas where chronic exposures to HF occur. Older (lower-canopy) leaves accumulate more total fluoride from extended low-level exposures than new (upper) leaves, as they are exposed over a longer period of time. But, as the oldest leaves of the canopy senesce and drop off, foliar fluoride is shed. This tends to reduce the range of concentrations that otherwise would be found in canopy foliage. Leaching of fluoride from tissues by rain water also makes the amount of fluoride measured in leaves decreasingly coupled to the amount absorbed from the air.

Impact of Fluoride on Agricultural Crops and Domestic Animals

The vegetation's fluoride content, exposure, and cultural conditions, and any significant interacting biological factors that affect plant concentrations and tissue sensitivity, provide the diagnostic criteria for assessing the impact of fluoride on crops. Foliar levels are also used to evaluate the potential hazard to livestock feeding on the vegetation. Some investigators feel that the hazard of fluorosis to ruminants transcends the hazard to crop production in importance (Weinstein, 1977). Cattle have been the mammals most often affected by fluoride pollution (Suttie, 1977). Part of the domestic animal problem can be traced to high fluoride in irrigation water and occurring as contaminants in commercial feed supplements.

Fig. 8 summarizes fluoride exposure dose *versus* incipient injury response relationships for a number of plant species and groups (McCune, 1969). The series of curves demarcate threshold doses required to cause injury. The most sensitive plants (conifers, *Gladiolus,* and *Sorghum*) were injured by extended (10–30 days) exposures to 0.5 ppb. Tomato plants required 3 to 6 ppb. Exposures of less than a day required air concentrations of about 3 ppb for conifers and more than 10–30 ppb for Corn or Tomato. McCune (1974) proposed general limiting values designed to protect most vegetation, those given being in the range of 5–10 ppb for 2- to 4-hours' exposures (peak concentrations), or 30–60 days' exposure to concentrations between 0.3–0.6

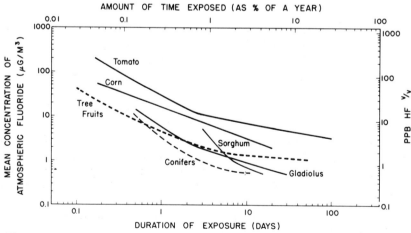

Fig. 8   Foliar injury and fluoride dose–response relationships for different plant species and groups. Source: McCune (1969).

ppb. Daily mean limits proposed were 1–4 ppb, depending upon diurnal fluctuations and peak concentrations in the average. A. C. Hill (1969) believes that as the fluoride content in leaves can fluctuate due to rain and other factors, and tissue sensitivity varies not only with the species but with tissue age and environmental circumstances, fluoride air quality standards based on atmospheric concentrations are not reliable. He recommends that the most practical approach to air standards involving chronic fluoride exposures would be to base the standard on the actual fluoride content measured in leaf tissues. The advantage of leaf analyses over air analyses is that tissue concentrations more closely relate to the potential for plant and animal effects.

Depending to a large extent upon whether 'crop yield' is measured in terms of vegetative biomass (as for Alfalfa) or production of useful fruiting structures, foliar symptoms of injury may or may not reflect yield reductions—particularly when injury is light. Long-term studies with Alfalfa indicate that growth- and yield-reductions caused by fluoride can be accounted for by leaf chlorosis and necrosis (Thomas & Hendricks, 1956). When fruit or seed production is involved, the relationship becomes more complex. This is further complicated by difficulty in determining the statistical significance of yield differences of less than 10% due to natural variation.

Valencia Orange trees (*Citrus aurantium* cultivar), grown in greenhouses with filtered or unfiltered air and located in a high fluoride area in Florida, showed yield losses in these studies which were only partly attributable to fluoride-induced leaf chlorosis. Chlorosis occurred when fluoride leaf levels were above 20 ppm (Leonard & Graves, 1966). Reduced leaf size was thought to lower the amount of photosynthate available for partitioning. The extent to which flower development or fruit drop affected yields was not known. The data indicated that one-quarter of the crop was lost for each 60 ppm increase in the fluoride measured in 5-months-old current-year leaves. These results can be contrasted with those observed for Bush Bean (*Phaseolus vulgaris* cultivar), which is more resistant to fluoride, and of which the growth was reduced only when foliar concentrations exceeded 300 ppm (Treshow & Harner, 1968).

Work with sensitive plant cultivars under hydroponic culture suggests that threshold levels for injurious effects may be in the range of 20–100 ppm fluoride in the leaves if conditions are right (A. C. Hill, 1969). Tissue age is an important consideration, and whereas 100 ppm in expanded new-flush leaves can reduce growth, mature leaves may require 200 ppm or more. Plant tissues that are injured by leaf concentrations below 200 ppm are classified as susceptible, while highly susceptible tissues show injury at concentrations below 50 ppm. Plant tissues which are regarded as intermediate or resistant, can tolerate 200 ppm or more of fluoride.

The impact of fluoride on the quality of crop plants and ornamentals may be obvious or subtle. Foliar necrosis in ornamental *Gladiolus* and conifers devalue them aesthetically. In contrast to leaves, petals of flowers are rarely injured by fluoride, although a few floral species (e.g. of Petunia (*Petunia hybrida*) and Cyclamen (*Cyclamen* sp.)) may show fluoride injury. Fluoride damage to stone fruits (Peach (*Prunus persica*), Apricot (*Prunus armeniaca*), and cherry (*Prunus* sp.)), result in unmarketable fruits. This can occur even though the total fluoride concentrations in the fruits rarely exceed a few ppm. Fluoride-injured Peaches, the classic example, ripen prematurely along the suture line near the distal end of the fruits. Affected areas become overripe or rot before the rest of the fruit is ripe. This disease was called 'fluoride suture' in the early part of the century when fluoride sprays were being tested to control insects and bacterial diseases. Today, it is known as 'soft suture' or 'suture red-spot'. Water-stress and poor mineral nutrition aggravate the condition, but spraying the trees with calcium chloride during the pit-hardening stage can economically prevent the disease from attaining expression (Benson, 1959).

Fluoride-stress and injury to crop plants can be minimized by substituting fluoride resistant plants for sensitive ones when annuals or biennials are involved. This is often done unknowingly by farmers and gardeners, who are reluctant to grow unhealthy plant varieties a second year. Replanting to other desirable species and cultivars that are known to grow well in the area results in the substitution of susceptible plants by more tolerant ones.

The fluoride content of forage varies markedly during the growing season. During rapid plant growth in the spring and early summer, foliar levels are lowest (Suttie, 1977), whereas autumn forage contains more fluoride. Early symptoms of excessive fluoride intake by young grazing animals appear as mottling and abnormalities of the teeth, while extended ingestion of toxic amounts can cause lesions to the bones as fluoride ions replace hydroxyl groups, so converting bone apatite to fluoroapatite. The joints may also be affected. Severe fluorosis may result in systemic effects accompanied by loss of appetite, reduced weight gain, and lower milk-yields of cows. Often the animals become lame and stiff, and death may result in the most severe cases.

Table VIII gives the maximum fluoride concentrations recommended for feedstuffs fed to young and mature cattle as well as for several other important domestic animals. The recommended tolerance-limits are designed to protect against significant interference with normal growth and performance. Air quality standards adopted for the protection of domestic animals requires that fluoride emissions be regulated so that the yearly *average* fluoride content of forage does not exceed 40 ppm (dry weight). The standard further stipulates that the fluoride content of forage must not be greater than 60 ppm for more than two consecutive months. Eighty ppm fluoride must not be exceeded in any one month period. Lower fluoride levels may cause pathological changes in some animals, resulting in slight dental mottling or periosteal

Table VIII Dietary fluoride tolerance for domestic animals.*

Animal	Fluoride[†] concentrations (ppm)
Beef or dairy heifers	40
Mature beef or dairy cattle	50
Finishing cattle	100
Feeder lambs	150
Breeding ewes	60
Horses	60
Finishing pigs	150
Breeding sows	150
Growing or broiler chickens	300
Laying or breeding hens	400
Turkeys	400

* Source: Suttie (1977).

† The values are presented as ppm in dietary dry-matter and assume the ingestion of a soluble fluoride such as NaF.

hyperostosis, but these changes have not been shown to affect significantly the animals' growth and performance in clinical trials.

SUMMARY

Concern about the impact of air pollutants on agriculture has led to economic assessments of crop losses. In 1981 this was estimated to be $1.8 billion in the USA. The impact has been assessed by field studies using natural concentration-gradients, closed or open-top chambers, open-air fumigations, and chemical protectants. The greatest losses are associated with ozone which, in the Los Angeles basin, may reduce *Citrus* yield by as much as 50%. Production losses for Grape, Cotton, Alfalfa, and Tomatoes, may be only slightly less. Over the years, losses have been somewhat minimized, but not eliminated, by the selection by most astute growers of more tolerant varieties. Tobacco and Potato provide the best examples of this.

Sulphur dioxide may also cause significant losses, particularly in the eastern USA and in Europe where ambient concentrations often remain high. In eastern USA, Soybeans are notably sensitive, and in Europe cereals and forest crops appear to be the most affected. Where both O_3 and SO_2 are present, yield losses are apt to be compounded.

Pesticides also may have adverse effects, but such are far outweighed by the benefits from increased production associated with their use. Finally, fluoride pollution is unlikely to impact crop yields even in local areas, although a few examples exist of reduced forest quality. The more serious effect is in the accumulation of fluoride in forage crops, and resultant problems with grazing domestic animals.

References

Abelson, P. H. (1981). Coal research. *Science,* **212,** p. 1341.
Beaton, J. D., Tisdale, S. L. & Platou, J. (1971). Crop responses to sulfur in North America. *The Sulphur Institute, Tech. Bul.,* **18,** 38 pp.
Benedict, H. M., Miller, C. J. & Olson, R. E. (1971). *Economic Impact of Air Pollutants on Plants in the United States.* Final Report, SRI Project LSD-1056, Stanford Research Institute, Menlo Park, California, USA: 77 pp.
Bennett, J. H. & Hill, A. C. (1973). Inhibition of apparent photosynthesis by air pollutants. *J. Environ. Qual.,* **2,** pp. 526–30.
Bennett, J. H., Lee, E. H., Heggestad, H. E., Olsen, R. A. & Brown, J. C. (1981). Ozone injury and aging in leaves: Protection by EDU. Pp. 604–5 in *Oxygen and Oxy-radicals in Chemistry and Biology* (Ed. M. A. J. Rodgers & E. L. Powers). Academic Press, New York, NY, USA: xxx + 808 pp.
Benson, N. R. (1959). Fluoride injury in soft suture and splitting of peaches. *Proc. Amer. Soc. Hort. Sci.,* **74,** pp. 184–98.
Black, V. J. & Unsworth, M. H. (1980). Stomatal responses to sulphur dioxide and vapour pressure deficit. *J. Exp. Bot.,* **31,** pp. 667–77.
Bleasdale, J. K. A. (1973). Effects of coal-smoke pollution gases on the growth of Ryegrass (*Lolium perenne* L.). *Environ. Pollut.,* **5,** pp. 275–85.
Bonte, J. (1977). *Effets du Dioxyde de Soufre Vegetaux en Plein Champ, à Faible Concentration et Appliqué d'une Façon Permanente.* Ministère de l'Agriculture, Institut National de la Recherche Agronomique, Domaine Saint Paul, 84140 Montfavet, France: 10 pp., illustr. (mimeogr.).
Brewer, R. F. & Ferry, G. (1974). Effects of air pollution on cotton in the San Joaquin Valley. *Calif. Agr.,* **28,** pp. 6–7.
Brogan, J. C. (Ed.) (1978). Sulphur in forages. *Proc. Symposium, Wexford Inc., 3–4 October 1978* (Sponsored by Agricultural Institute, Ireland, and Sulphur Institute, Paris). An Foras, Taluntais, Dublin, Ireland: 262 pp., illustr.
Brooks, F. A. (1947). The drifting of poisonous dusts applied by airplanes and land-rigs. *J. Agr. Eng.,* **28,** pp. 233–9.
Brough, A., Parry, M. & Whittingham, C. P. (1978). The influence of aerial pollution on crop growth. *Chemistry and Industry,* **21,** pp. 51–3.
Carnahan, J. E., Jenner, E. L. & Wats, E. K. W. (1978). Prevention of ozone injury to plants by a new protectant chemical. *Phytopathology,* **68,** pp. 1225–9.
Chang, C. W. & Thompson, C. R. (1966). Site of fluoride accumulation in navel orange leaves. *Plant Physiol.,* **42,** pp. 211–3.
Cleveland, W. S., Kleiner, B., McRae, J. E. & Warner, J. L. (1976). Photochemical air pollution: Transport from the New York City area into Connecticut and Massachusetts. *Science,* **191,** pp. 179–81.
Committee on Biologic Effects of Atmospheric Pollutants (1971). *Fluorides.* National Academy of Sciences, Washington, DC, USA: x + 295 pp.
de Cormıs, L., Bonte, J & Tisne, A. (1975). Experimental technique for determining the effect on vegetation of SO_2 pollutants applied continuously in subnecrotic doses. *Pollution Atmosphérique,* **66,** pp. 103–7.
Council on Environmental Quality (1979). *Environmental Quality—1979: Tenth Annual Report.* US Government Printing Office No. 041-011-0047-5, Washington, DC, USA: xlii + 816 pp.

Cowling, D. W., Jones, L. H. P. & Lockyer, D. R. (1973). Increase yield through correcting sulphur deficiency in Rye-grass exposed to sulphur dioxide. *Nature*, (London), **243**, pp. 479–80.

Cowling, D. W. & Lockyer, D. R. (1978). The effect of SO_2 on *Lolium perenne* L. grown at different levels of S and N nutrition. *J. Exp. Bot.*, **29**, pp. 257–65.

Crittenden, P. D. & Read, D. J. (1978). The effects of air pollution on plant growth with special reference to sulphur dioxide, II: Growth studies with *Lolium perenne* L. *New Phytol.*, **80**, pp. 49–62.

Crittenden, P. D. & Read, D. J. (1979). The effects of air pollution (SO_2) on plant growth—III. Growth studies with *Lolium multiflorum* Lam. and *Dactylis glomerata* L. *New Phytol.*, **83**, pp. 645–651.

Dreisinger, B. R. & McGovern, P. C. (1970). Monitoring SO_2 and correlating its effects on crops and forestry in the Sudbury area. Pp. 12–28 in *Impact of Air Pollution on Vegetation Speciality Conference, Toronto* (Ed. S. N. Linzon). Air Pollution Control Association, Pittsburgh, Pennsylvania, USA: 122 pp., illustr.

Environmental Protection Agency (EPA) (1979). *Evaluation of Alternative Secondary Ozone Air Quality Standards.* Strategies and Air Standards Division, Office of Air Quality Planning and Standards, OAQPS78-8, IV-A3, US Environmental Protection Agency, Research Triangle Park, North Carolina, USA: 36 pp., Attachments 3, Appendices 3 (mimeogr.).

Foy, C. D., Chaney, R. L. & White, M. C. (1978). The physiology of metal toxicities in plants. *Ann. Rev. Plant Physiol.*, **29**, pp. 511–66.

Glotfelty, D. E. (1978). The atmosphere as a sink for applied pesticides. *J. Air Pollut. Control Assoc.*, **28**, pp. 917–21.

Guderian, R. (1977). *Air Pollution: Phytoxicity of Acidic Gases and its Significance in Air Pollution Control.* (Ecological Studies 22.) Springer-Verlag., Berlin & New York: 127 pp., illustr.

Guderian, R. & Stratmann, H. (1962). *Freilandversuche zur Ermittlung von Schwefeldioxidwirkungen auf die Vegetation. I. Teil: Übersicht zur Versuchsmethodik und Versuchsauswertung*, Nr. 1118. Westdeutscher Verlag, Forsch. Ber. d. Landes Nordrhein-Westfalen, West Germany: 102 pp., illustr.

Guderian, R. & Stratmann, H. (1968). *Freilandversche zur Ermittlung von Schwefeldioxidwirkungen auf die Vegetation. III. Teil: Grenzwerte schädlicher SO_2 Immissionen fur Obst- und Forstkulturen sowie für landwirtschaftliche und gärtnerische Pflanzenarten*, Nr. 1920 Köln und Opladen. Westdeutscher Verlag, Forsch. Ber. d. Landes Nordrhein-Westfalen, West Germany: 114 pp. illustr.

Haase, E. F., Morgan, G. W. & Salem, J. A. (1980). Field surveys of SO_2 injury to crops and assessment of economic damage. *Air Pollution Control Association, Pittsburgh, Pennsylvania, 73rd Annual Meeting, Montreal, Canada:* Preprint No. 80-62-2, 22 pp.

Heagle, A. S. (1973). Interactions between air pollutants and plant parasites. *Ann. Rev. Phytopath.*, **11**, 365–88.

Heagle, A. S., Body, D. E. & Heck, W. W. (1973). An open-top field chamber to assess the impact of air pollution on plants. *J. Environ. Quality*, **2**, pp. 365–8.

Heagle, A. S. and Heck, W. W. (1980). Field methods to assess crop losses due to oxidant air pollutants. Pp. 296–305 in *Proceeding of the E. C. Stakman Commemorative Symposium—Assessment of losses which constrain production and crop improvement in agriculture and forestry.* University of Minnesota, Miscel. Publ. 7-1980, Agric. Exp. Sta., St. Paul, Minnesota, USA: 327 pp.

Heck, W. W., Taylor, O. C. & Heggestad, H. E. (1973). Air pollution research needs: Herbaceous and ornamental plants and agriculturally generated pollutants. *J. Air Pollut. Control Assoc.*, **32**, pp. 267–76.

Heck, W. W., Mudd, J. B. & Miller, P. R. (1977*a*). Plants and microorganisms. Pp. 437–585 in *Ozone and Other Photochemical Oxidants,* Vol. 2. National Academy of Sciences, Washington, DC, USA: vii + 719 pp., illustr.

Heck, W. W., Heagle, A. S. & Cowling, E. B. (1977*b*). Air pollution impact on vegetation. Pp. 193–203 in *New Directions in Century 3: Strategies for Land and Water Use.* Proceedings of 32nd Annual Meeting, Soil Conservation Society of America, Ankeny, Iowa, USA: 279 pp., illustr.

Heggestad, H. E. (1973). Photochemical air pollution injury to Potatoes in the Atlantic Coastal States. *American Potato Jour.*, **50,** pp. 315–28.

Heggestad, H. E. (1980). Field assessment of air-pollution impacts on growth and productivity of crop species. *Air Pollution Control Association, Pittsburgh, 73rd Annual Meeting, Montreal, Canada: Preprint No. 80-26-1.*

Heggestad, H. E. & Bennett, J. H. (1981). Photochemical oxidants potentiate yield-losses in Snap Beans attributable to SO_2. *Science,* **213,** pp. 1008–10.

Heggestad, H. E. & Middleton, J. T. (1959). Ozone in high concentrations as cause of Tobacco leaf injury. *Science,* **129,** pp. 208–10.

Heggestad, H. E., Howell, R. K. & Bennett, J. H. (1977). The effects of oxidant air pollutants on soybeans, snap beans, and potatoes. EPA *Ecological Research Series, EPA-600/3-77-128,* 38 pp.

Heggestad, H. E., Heagle, A. S., Bennett, J. H. & Koch, E. J. (1980). The effects of photochemical oxidants on Snap Beans. *Atmospheric Environment,* **14,** pp. 317–26.

Heitschmidt, R. K., Lauenroth, W. K. & Dodd, J. L. (1978). Effects of controlled levels of SO_2 on Western Wheat-grass in southeastern Montana grassland. *J. Applied Ecol.,* **14,** pp. 859–68.

Hibben, C. R. & Taylor, M. P. (1975). Ozone and sulphur dioxide effects on the Lilac powdery Mildew Fungus. *Environ. Pollut.,* **9,** pp. 107–14.

Hill, A. C. (1969). Air Quality standards for fluoride vegetation effects. *J. Air Pollut. Control Assoc.,* **19,** pp. 331–36.

Hill, A. C., Hill, S., Lamb, C. & Barrett, T. W. (1974). Sensitivity of native desert vegetation to SO_2 and to SO_2 and NO_2 combined. *J. Air Pollut. Control Assoc.,* **24,** pp. 153–7.

Hill, G. R. & Thomas, M. D. (1933). Influence of leaf distribution by sulphur dioxide and by clipping on yield of Alfalfa. *Plant Physiol.,* **8,** pp. 223–45.

Hofstra, G., Littlejohns, D. A. & Wukasch, R. T. (1978). The efficacy of the antioxidant ethylenediurea (EDU) compared to carboxin and benomyl in reducing yield-losses from ozone in Navy Beans. *Plant Dis. Reptr,* **62,** pp. 350–2.

Howell, R. K., Koch, E. J. & Rose, L. P. (1979). Field assessment of air pollution induced Soybean yield losses. *Agronomy Jour.,* **71,** pp. 285–8.

Jacobson, J. S. (1982). Ozone and the growth and productivity of agricultural crops. Pp. 293–304 In *Effects of Gaseous Air Pollution in Agriculture and Horticulture* (Ed. M. H. Unsworth & D. P. Ormrod). Butterworth Scientific, London, England, UK: 532 pp., illustr.

Jones, H. C., Weatherford, F. P., Noggle, J. C., Lee, N. T. & Cunningham, J. R. (1979). Power-plant siting: Assessing risks of SO_2 effects on agriculture. *Air Pollution Control Association, Pittsburgh, 72nd Annual Meeting, Cincinnati, USA:* Preprint No. 79-13.5.

Jones, R. K., Robert, M. E. & Crack, B. J. (1975). Pp. 127–35 in *Sulphur in Australasian Agriculture* (Ed. K. D. McLachlan). Sydney University Press, Sydney, Australia: 261 pp., illustr.

Laurence, J. A. (1978). Effects of air pollutants on plant–pathogen interactions. *Air Pollution Control Association, Pittsburgh, 71st Annual Meeting, Houston, USA: Preprint No. 78-44.5.*

Lee, E. H. & Bennett, J. H. (1981). Chemical modification of oxidant-sensitive plant cultivars: Relationship between senescence regulation and O_3 tolerance. *Plant Physiol.* Supplement 567 **67**, p. 68.

Lee, E. H., Bennett, J. H. & Heggestad, H. E. (1981). Retardation of senescence in Red Clover leaf discs by a new antiozonant N-[2-(2-oxo-1-imidazolidinyl)-ethyl]-N'-phenylurea. *Plant Physiol.*, **67**, pp. 347–50.

Lee, J. J. and Lewis, R. A. (1978). Zonal air pollution system design and performance: Sulphur dioxide impact on grassland. Pp. 322–44 in *Bioenvironmental Impact of Coal-fired Power Plant*. Third Interim Report, U.S. Environmental Protection Agency, Ecol. Res. Ser. 600/3-78-021. Corvallis, Oregon USA:

Leonard, C. D. & Graves, H. B. (1966). Effect of airborne fluoride in Valencia Orange yields. *Proc. Fla State Hort. Soc.*, **79**, pp. 79–86.

Linzon, S. N. (1978). Effects of airborne sulphur pollutants on plants. Pp. 109–62 in *Sulphur in the Environment, Part II: Ecological Impact* (Ed. J. O. Nriagu). John Wiley & Sons, New York, NY, USA: xii + 482 pp., illustr.

McCune, D. C. (1969). *On the Establishment of Air Quality Criteria with Reference to the Effects of Atmospheric Fluorine in Vegetation*. Air Quality Monograph 69–3; American Petroleum Institute, New York, NY, USA: 33 pp.

McCune, D. C. (1974). Acceptable limits for air pollution dosage and vegetation effects: Fluoride. *Air Pollution Control Association, Pittsburgh, 67th Annual Meeting, Denver, USA:* Preprint No. 74-226.

McCune, D. C. & Weidensaul, T. C. (1978). Effects of atmospheric sulfur oxides and related compounds on vegetation. Pp. 80–129 in *Sulphur Oxides*. National Academy of Sciences, Washington, DC, USA: ix + 209 pp.

McLaughlin, S. B. & Taylor, G. E. (1980). Relative humidity important modifier of pollutant uptake by plants. *Science*, **211**, pp. 167–9.

Mandl, R. H., Weinstein, L. H., McCune D. C. & Keveny M. (1973). A cylindrical, open-top chamber for the exposure of plants to air pollutants in the field. *J. Environ. Quality*, **2**, pp. 371–6.

Manning, W. J., Feder, W. A. & Vardaro, P. M. (1974). Suppression of oxidant injury by Benomyl: Effects on yields of bean cultivars in the field. *J. Environ. Quality*, **3**, pp. 1–3.

Mansfield, T. A. (1976). *Effects of Air Pollutants on Plants*. Cambridge University Press, Cambridge, England, UK: 209 pp., illustr.

Maybank, J., Yoshida, K. & Grover, R. (1978). Spray drift from agricultural pesticide applications. *J. Air Pollut. Control Assoc.*, **28**, pp. 1009–14.

Menser, H. A. & Heggestad, H. E. (1966). Ozone and sulfur dioxide synergism: Injury to Tobacco plants. *Science*, **153**, pp. 424–5.

Middleton, J. T., Kendrick, J. B., jr & Schwalm, H. W. (1950). Injury to herbaceous plants by smog or air pollution. *Plant Dis. Rept*, **34**, pp. 245–52.

Miller, J. E., Sprugel, D. G., Muller, R. N., Smith, H. J. & Xerikos, P. B. (1980). Open-air fumigation system for investigating sulfur dioxide effects on crop. *Phytopathology*, **70**, pp. 1124–8.

Nakamura, H. & Ota, Y. (1978). An injury to rice plants caused by photochemical oxidants in Japan. *Japanese Agricultural Research Quarterly*, **12**, pp. 69–73.

Navara, J. (1975). The effects of emissions on winter crops in different phases of ontogeny. *Biologia* (Bratislava), **33**, pp. 757–66.

Noggle, J. C. & Jones, H. C. (1979). *Accumulation of Atmospheric Sulfur by Plants and Sulfur-supplying Capacity of Soil*. Ecological Research Series EPA-600/7-79-109, 37 pp.

Nyborg, M. (1978). Sulfur pollution and soils. Pp. 359–90 in *Sulfur in the Environment, Part II* (Ed. J. O. Nriagu). John Wiley & Sons, New York, NY, USA: xii + 482 pp., illustr.

Ormrod, D. P. (1978). *Pollution in Horticulture.* Elsevier Scientific Pub. Co., Amsterdam–Oxford–New York: xi + 260 pp., illustr.

Oshima, R. J. (1978). *The Impact of Sulphur Dioxide on Vegetation; a Sulphur Dioxide–Ozone Response Model.* Statewide Air Pollution Research Center, University of California at Riverside, Report A6-162-30: 91 pp. (mimeogr.).

Oshima, R. J., Taylor, O. C., Braegelmann, R. K. & Baldwin, D. W. (1975). Effect of ozone on the yield and plant biomass of a commercial variety of Tomato. *J. Environ. Quality,* **4,** pp. 463–4.

Oshima, R. J., Poe, M. P., Braegelmann, P. K., Baldwin, D. W. & Way, V. Van (1976). Ozone dosage–crop loss function for Alfalfa: A standardized method for assessing crop losses from air pollutants. *J. Air Pollut. Control Assoc.,* **26,** pp. 861–5.

Oshima, R. J., Braegelmann, P. K., Baldwin, D. W., Way, V. Van & Taylor O. C. (1977). Reduction of Tomato fruit-size and yield by ozone. *J. Amer. Soc. Hort. Sci.,* **102,** pp. 289–93.

Reinert, R. A. (1980). Assessment of crop productivity after chronic exposure to ozone or pollutant combinations. *Air Pollution Control Association, Pittsburgh, 73rd Annual Meeting, Montreal, Canada:* Preprint No. 80-26.4.

Reinert, R. A., Heggestad, H. E. & Heck W. W. (1982). Response and genetic modification of plants for tolerance to air pollution. Pp. 259–92 in *Breeding Plants for Marginal Environments* (Ed. M. N. Christiansen & C. F. Lewis). Wiley–Interscience; New York, NY, USA: viii + 459 pp.

Rich, S. (1975). Interactions of air pollution and agricultural practices. Pp. 335–60 in *Responses of Plants to Air Pollution* (Ed. J. B. Mudd & T. T. Kozlowski). Academic Press, New York, NY, USA: xii + 383 pp.

Richards, B. L., Middleton, J. T. & Hewitt, W. B. (1958). Air pollution with relation to agronomic crops, V: Oxidant stipple of Grape. *Agronomy J.,* **50,** pp. 559–61.

Robinson, E. & Fox, L. L. (1978). 2,4-D herbicides in Central Washington, *J. Air Pollut. Control Assoc.,* **28,** pp. 1015–20.

Saunders, P. J. W. (1966). The toxicity of sulfur dioxide to *Diplocarpon rosae* Wolf causing Blackspot of roses. *Ann. Appl. Biol.,* **58,** pp. 103–14.

Schaeffer, T. C. & Hedgecock, G. G. (1955). Injury to Northwestern forest trees by sulfur dioxide from smelters. *US Dept of Agric., Tech. Bull.* **1117,** 49 pp.

Sprugel, D. G., Miller, J. E., Muller, R. N., Smith, H. J. & Xerikos, P. B. (1980). Sulfur dioxide effects on yield and seed quality in field-grown Soybeans. *Phytopathology,* **70,** pp. 1129–33.

Stanford Research Institute (SRI) (1981). *An Estimate of the Non-health Benefits of Meeting the Secondary National Ambient Air Quality Standards.* (Final Report, National Commission on Air Quality, Washington, DC.) SRI Project 2094, Menlo Park, California, USA: vii + 53pp.

Stephens, E. R., Darley, E. F., Taylor, O. C. & Scott, W. E. (1961). Photochemical reaction products in air pollution. *Int. J. Air and Water Pollut.,* **4,** pp. 79–100.

Suttie, J. W. (1977). Effects of fluoride on livestock. *J. Occup. Med.,* **19,** pp. 40–8.

Terman, G. L. (1978). Atmospheric sulphur–the agronomic aspects. *The Sulphur Institute Bull.,* Washington, DC, **23,** 15 pp.

Thomas, M. D. (1935). Absorption of sulphur dioxide by Alfalfa and its relation to leaf injury. *Plant Physiol.,* **10,** pp. 291–307.

Thomas, M. D. (1961). Effects of air pollution on plants. Pp. 233–78 in *Air Pollution.* (World Health Organization, Monograph 46.) Columbia University Press, New York, NY, USA: 442 pp., illustr.

Thomas, M. D. & Hendricks, R. H. (1956). Effects of air pollution on plants. Section 9 in *Air Pollution Handbook* (Ed. R. L. Magill, F. R. Holden & C. Ackley). McGraw-Hill, New York, NY, USA: x + 15 sections.

Thompson, C. R. & Taylor, O. C. (1969). Effects of air pollutants on growth, leaf-drop, fruit-drop, and yield, of Citrus trees. *Environ. Sci. & Technol.*, **3**, pp. 934–40.

Thompson, C. R., Hensel, E. G. & Kats, G. (1969). Effects of photochemical oxidants on Zinfandel Grapes. *Hort. Sci.*, **4**, pp. 222–4.

Thompson, C. R., Kats. G., & Lennox, R. W. (1980). Effects of SO_2 and/or NO_2 on native plants of the Mojave Desert and Eastern Mojave–Colorado Desert. *J. Air Pollut. Control Assoc.*, **30**, pp. 1304–9.

Treshow, M. (1975). Interactions of air pollutants and plant diseases. Pp. 309–34 in *Responses of Plants to Air Pollution* (Ed. J. B. Mudd & T. T. Kozlowski). Academic Press, New York, NY, USA: xii + 383 pp., illustr.

Treshow, M. & Harner, F. M. (1968). Growth responses of Pinto Bean and Alfalfa to sublethal fluoride concentrations. *Can. J. Bot.*, **46**, pp. 1207–10.

Wanta, R. C., Moreland, W. B. & Heggestad, H. E. (1961). Tropospheric ozone: An air pollution problem arising in the Washington, DC, metropolitan area. *Monthly Weather Review*, **89**, pp. 289–96.

Weinstein, L. H. (1977) Fluoride and plant life. *J. Occup. Med.*, **19**, pp. 49–78.

Weinstein, L. H., McCune, D. C., Alusio, A. L. & Leuken, P. van (1975). The effect of sulphur dioxide on the incidence and severity of bean rust and early blight of Tomato. *Environ. Pollut.*, **9**, pp. 145–55, fig.

Westberg, H. H., Robinson, E. & Zimmerman, P. (1974). Ozone and light hydrocarbon measurements in Phoenix, Arizona. *Air Pollution Control Association, Pittsburgh, 67th Annual Meeting, Denver, USA:* Preprint No. 74-54.

Air Pollution and Plant Life
Edited by M. Treshow
© 1984 John Wiley & Sons Ltd.

CHAPTER 16

Impact of Atmospheric Pollution on Natural Ecosystems

JAN MATERNA

Institute of Forestry, Zbraslav, Prague, Czechoslavakia

BACKGROUND

In categorizing the effects of air pollutants on ecosystems, Guderian (1977) presents the type and extent of effects on various levels of ecosystem organization as follows:

(1) Accumulation of pollutants in the plant and other ecosystem components (such as soil and surface- and ground-water).
(2) Damage to consumers as a result of pollutant accumulation (e.g. fluorosis).
(3) Changes in species diversity due to shifts in competition.
(4) Disruption of biogeochemical cycles.
(5) Disruption of stability and reduction in the ability of self-regulation.
(6) Breakdown of stands and associations.
(7) Expansion of denuded zones.

There are many reports illustrating the processes in ecosystems that are influenced by air pollution in this general way. But there are also many reports that demonstrate only the end result where pollution has been extreme. This transition of the declining landscape from one stage to another can last a very long time. Consequently, one can find regions where air pollution has persisted for 100 to 200 years, and where well organized ecosystems, dominated by relatively sensitive plant species of fairly high productivity, still remain.

If we assume that we may have to live and produce under the influence of a fair amount of air pollution in future decades, it is necessary to analyse carefully the effects of air pollution at various concentrations, and to appraise their importance to the structure, function, and stability, of important ecosystems. However, the information available on ecosystem–pollutant inter-actions is limited, despite such contributions as those of Miller & McBride (1975), Bordeau & Treshow (1978), McClenahan (1978), and Treshow (1980).

AIR POLLUTION AS AN ECOLOGICAL FACTOR

Air pollution, owing to its regional character and sometimes global extent, has become a relatively newly recognized but important ecological factor (Treshow, 1968). It is scarcely possible, in ecological studies, to consider a single air pollutant alone, any more than it is usual to have it acting alone. We must keep in mind the influence of both individual pollutants and their mix-tures. Some of the reasons for this are:

(1) Several of the pollutants (*sensu latissimo*) occur naturally in the atmos-phere—for example CO_2, SO_2, nitrogen oxides, and O_3—and to some extent natural mechanisms compensate for their influence.
(2) Each pollutant has its own relationship to the reaction of components of an ecosystem, and therefore its own threshold at which it begins to act in the ecosystem.
(3) There are distinct differences in the mechanism of action of some of these components; especially must the differences in persistence be considered.

Special Features of Air Pollution as an Ecological Factor

A distinct feature of air pollution is that only seldom does a single pollutant act alone, and another is that even pollutants from a single source, such as those resulting from burning coal, can have a relatively great variability of individual components; and these can vary during a relatively short time. The relationship between the concentration of SO_2, HF, and nitrogen oxides (NO_x), provides one example. This factor must be considered even with respect to such short-lived components of ecosystems as annuals. When we consider such long-living ecosystems as forests, and certainly during the development of a generation of trees, the quality of air pollution will change to such a degree that it cannot possibly be neglected.

Naturally the changes in air quality will be greater in cases where industrial production and technology (or at least the quality of raw products) are altered. Not only are the amounts of emissions from various sources decisive, but dispersion conditions are critical to the pollutant concentration. There-fore, air pollution can be adequately characterized only following long-term measurements. Although changes in emissions may be decisive to air pollution concentrations and their effects on ecosystems, it is not sufficiently meaningful only to measure pollutant concentrations or their activity over short intervals of time. One example, provided by our measurements of SO_2

concentrations at a point surrounded by a damaged Norway Spruce (*Picea abies*) forest in the outskirts of an industrial region, showed that whereas emission sources were relatively stable over the last decade, the pollutant concentrations varied with the dispersion of air as related to meteorological factors.

Besides the long-term changes over a period of years, there are changes with a regular seasonal character (e.g. occurrence of higher concentrations in winter) that have a regular rhythm. Variations can also occur during a day due to changes in the temperature stratification between day and night. The variation during a very short period of time (e.g. an hour) can be especially significant. Such changes in concentrations, during shorter and longer periods, are important with all pollutants.

While physiological changes in individual plant reactions can be demonstrated following brief exposures to varied concentration of air pollutants, far longer periods are necessary to detect any ecosystem changes. The dependence of the occurrence and concentration of pollutants on meteorological factors and climatic conditions critically influence the impact of the pollutants. These variables, influencing natural as well as artificial ecosystems, make a qualitative description of phenomena very difficult, and render the term 'air pollution' highly mutable.

Mechanism of Air Pollution Influence on Ecosystems

The long-term influence of air pollution on an ecosystem is a result of:

— Direct influence of gaseous pollutants, acid deposition, and particulates—especially on the assimilatory apparatus of plants and the respiratory organs of animals, as well as on soils and even on climatic conditions.
— Indirect influence on the individual components of ecosystems due to alteration by air pollution of the stage or function of these components.

The combined sensitivity of individual components of the ecosystem to air pollution and other stress factors is of great importance to an ecosystem's development. In this case, great differences among the plant species, as well as individual and clonal variability in their reactions, also can be observed. Knowledge about the individual processes of this mechanism is very unbalanced, but it is sometimes possible to apply existing knowledge about the influence of other factors on ecosystems.

The direct influence of air pollution on individual components of ecosystems: Our understanding of the direct influence of pollutants on ecosystems is based on numerous experiments and observations characterized by a number of physiological and biochemical criteria (Knabe, 1970a; Materna,

1973b; Guderian, 1977). But from the point of view of an ecosystem, the integrated influence of all the variables must be considered. Owing to the great variation in the sensitivity of plants, there are numerous tables dividing plant species into various degrees of sensitivity (e.g. Davis & Wilhour, 1976). Such information is gained mainly from two sources.

(1) Short-term experiments with more or less precisely defined conditions of air pollution (concentration and duration of pollutants or their mixtures).
(2) Observations in polluted areas where, in most instances, information about air pollution concentrations is not available, or where information is limited only to short-term measurements—instead of the long periods during which air pollution damage develops.

In the first case, (1) the results can be used for annuals, and using them we can characterize the reactions of such plants. For perennial plant species exposed to air pollution for a long time, the results of experiments on annuals can be misleading. The response of long-lived plants depends on far more than the influence of air pollution on leaves; rather does it involve interactions of all other stress-factors in combination with air pollution.

Observations in polluted areas offer a very useful basis for the evaluation of plant sensitivity, provided reliable information is available about the concentrations of air pollutants during the period of development. Though unfortunately few such observations are available, there appear to be three main criteria for the success of individual plant species in ecosystems in polluted areas:

—adequate biomass formation and suitable structure;
—ability to survive the lasting impact of pollutants; and
—ability to reproduce under the pollution stress.

These criteria are not always related to the amount of foliar injury, especially from a short-term fumigation. They determine the competitive ability of the individual plant species in the ecosystem, and particularly the stability of the principal plant species. This is naturally valid only within distinct limits, not exceeding the ability of any organism to survive.

A number of observations and studies in industrial areas of North America and some regions of Europe concern the susceptibility of forest trees to (and sometimes mortality caused by) SO_2, HF, and oxidants (Miller & McBride, 1975). In the Ruhr area of Europe, the long-lasting influence of relatively high concentrations of SO_2 has influenced the distribution of Scots Pine (*Pinus sylvestris* L.) (Knabe, 1970a, 1970b). It has also been possible to correlate the scarcity of this tree species with SO_2 concentrations in the industrial Peninnes of England (Farrer *et al.*, 1977).

Perhaps the most sensitive tree species to have suffered severe losses in European forests in the last decades is the European Silver Fir (*Abies alba*,

Mill.). Air pollution is the main cause of its increased mortality, at least in some regions (Wentzel, 1979). Our observations made in Beskydy (CSSR), show that the dieback of this species begins at a mean, long-lasting concentration of about 20 μg per m^3 SO$_2$. Also, exposure to such concentrations increases the susceptibility of the tree to frost damage. In a mountain forest where Silver Fir, Norway Spruce, and Common Beech (*Fagus sylvatica* L.) associations are dominant, the Silver Fir is especially important owing to its deep rooting system and high biomass production which together make it vital to the stability and productivity of the whole ecosystem; but it is the first to be injured. A further increase of SO$_2$ concentrations, to about 50–70 μg per m^3 as a long-term average, also affects the Norway Spruce, causing its gradual dieback (Materna, 1973*b*). The stability of the resulting Beech forest becomes endangered because of the reduced reproductive capacity of this last tree species.

In the mountain spruce forests, where the Norway Spruce is of great economic importance, the influence of air pollution is most pronounced. Despite its competitive ability in clean air, elevated SO$_2$ concentrations have caused the disappearance of this species in some regions of Central Europe. Thus the stability of the spruce ecosystem has been destroyed. Such admixed tree species as Silver Birch (*Betula verrucosa* Ehrh.) and Mountain Ash or Rowan (*Sorbus aucuparia* L.), which are more tolerant than the above-mentioned species to SO$_2$, have become the dominant, stable trees of new ecosystems. They are able to produce a great amount of biomass in a relatively short time, and can reproduce under the canopy of spruce stands.

The influence of natural ecological conditions on the degree of susceptibility of plants to air pollution is very important. Factors such as radiation, temperature, humidity of air and soil, nutrient availability, and air movement, can very substantially modify the response of plants to individual pollutants (Guderian, 1977).

There are only limited possibilities to analysing the influence of individual factors on the susceptibility of ecosystems. But it is possible to demonstrate the total importance of the combined effects of radiation, temperature, wind velocity, etc., on the ability of Norway Spruce forests to compensate for the effects of air pollution. The example given in Table I demonstrates the intensity of damage to Norway Spruce forest on an elevation-profile in the Ore mountains of Czechoslavakia.

At even higher elevations, especially in the upper area of the forest, heavy air pollution damage of trees can be observed where SO$_2$ concentrations have ranged between 20 and 30 μg per m^3 for a long period. It is a characteristic effect of air pollution at lower concentrations that it can last for decades or more, and severe damage in forests surrounding some industrial plants has been recognized since production began in the last century.

Such zonation, of increasing sensitivity of forest tree species with elevation,

Table I   Degree of damage to spruce forest after the same duration of air pollution exposure.

Locality: elevation	Concentration of SO_2 as a mean of an 8-years' period in μg SO_2 per m^{-3}	Degree of damage
440 m	79	Only light visible symptoms.
680 m	51	Distinct reduction of assimilatory apparatus.
1,200 m	38	Heavily damaged forest stands, high level of dieback of individual trees and groups of trees.

confirms the fact that the resultant injury of trees is an integrated effect of air pollution and other stress factors. The same conclusion is drawn in the boreal forest ecosystems by Linzon (1978) and Huttunen (1980) and in other cases by Treshow (1980).

The final stages of ecosystem disruption—complete destruction of the vegetation cover—is not common, but is still occurring where there are very high levels of SO_2, HF (Miller & McBride, 1975), or nitrogen compounds (Sokolowski, 1971), or magnesite dust exists (Hajdúk, 1965).

The direct influence on animals: There is no information known to the author about any direct influence of air pollution on animals in natural ecosystems. But Nature does not differentiate between the *direct* influence of such pollutants acting through the respiratory tract of the animals, and the *indirect* action for example of changes in food composition due to the influence of pollutants on the vegetation. Natural plant–animal interactions would assure the existance of indirect effects. Some indirect effects on insects are dealt with in the preceding chapter.

Influence of air pollution on soil: Besides the direct influence on the plants and animals in an ecosystem, it is also necessary to consider the influence on soil—especially on its chemistry and biology. The direct influence of pollutants on soil is most significant when the vegetation cover is destroyed. When once the soil has been exposed directly to the pollutants, the major buffering action of the cover is lost. In contrast to the regeneration of damaged vegetation that occurs as air pollution concentrations are reduced, or favourable growth conditions allow an increase of plant vitality, changes in the soil are distinctly cumulative.

Air pollution has a regional or sometimes global distribution, and can be responsible for soil changes over large areas. Locally, there are some examples of total disturbance of soil fertility and the vegetation cover by basic dust, such as magnesite or cement kiln dust. Distinct changes can be caused by fly-ash that has a high calcium content (Enderlein & Stein, 1964).

Fallout of magnesite dust is extraordinarily dangerous—not only for its impact on some soil characteristics, but also for its direct toxicity. A change of pH from 4.7 to 9.3, a significant decrease in exchangeable K, Ca, and Mg, as well as decrease in the content of available P and K to the critical level for plant nutrition, was demonstrated in the soil profile in such a case by Löffler (1974). In these circumstances, the vegetation cover was totally destroyed.

The fall-out of heavy metals in some regions in the vicinity of smelters can be very dangerous to the vegetation, and their content in the soil makes severe problems for the regeneration of the vegetation cover. The ecological consequences of their accumulation in the soil are serious, for some of them are very persistent in the soil (Whitby & Hutchinson, 1974).

The extent of changes in ecosystems caused by such an impact of particulate materials on the soil, or the total destruction of ecosystems, is limited by the rather rapid deposition of basic materials near the surrounding sources. Gaseous pollutants are far more widely dispersed.

Acidic Precipitation

Interest in the influence of acid deposition on ecosystems is stimulated by the fact that it can act far beyond the areas influenced by gaseous pollutants. In areas surrounding sources of pollutants, it is very difficult to separate the effects of acid deposition in ecosystems from the direct influence of noxious gases. With an increasing background of individual pollutants in the air, mainly SO_2, this combined effect also can have a regional character, if we consider the reaction of such sensitive ecosystems as coniferous forests.

While the influence of acid deposition on forest trees is not yet sufficiently elucidated, the effect on aquatic ecosystems and on soils has been relatively deeply examined (Overrein, 1980; Smith, 1980). Studies of the influence of acid deposition on soils can be traced to much older studies from early in the century (e.g. Wieler, 1912). But there are also very interesting, older complex studies of air pollution effects on forests based on analysis of the direct influence of air pollution on trees and soils, and of indirect effects of changes in soil properties—acidification, higher availability of both heavy metals and aluminium, difficulties in mineral nutrition of forest trees, etc. (Nemec, 1958). These studies were concentrated in localities with heavy air pollution.

Measurements have shown that precipitation water contains various elements derived from the surface of the sea or from wind erosion of the Earth's surface from volcanic action as well as from processes passing in the atmosphere. Substantial changes are caused by human activity. Upon contacting gaseous emissions and particulates, precipitation water is enriched by sulphate, nitrate, and ammonia, ions *plus* any other elements that might be present. The abundance of these elements is increasing to such an extent that the quality of surface water can be influenced and, through this, the freshwater ecosystems and the chemistry of the soil. At a comparable level of deposition, more distinct changes can be expected in uncultivated soils—especially as the water composition is further changed when passing through a forest stand (Cowling & Dochinger, 1980).

A detailed analysis of processes leading to the acidification of soils due to human activity has been made on the IBP plot in Solling. This is a region of Norway Spruce and Silver Beech stands that are subjected to deposition of impurities from particulate air pollution but with no direct influence from gaseous pollutants. It was concluded that the input of substances (derived from the combustion of fossil fuels) into the ecosystem, represented an extreme interruption of the cycle between plant uptake and mineralization. This would lead to the formation of sulphuric and nitric acids in the water present in air, plants, soils, and water bodies. Comparison of data between 1966–67 and 1973 showed that soil acidification had increased during that period. Aluminium and manganese were leached from the soils, and exchangeable calcium was lost completely.

The intensity of these processes depends on the composition of plants in the ecosystem, though there are distinct differences between tree species. A mean annual wet deposition was observed to be 0.8 kmol H^+ per ha. Interception under the Beech canopy was 0.84 km and under the Spruce it was 2.61, while production in the soil was *ca.* 2.5. Thus the deposition of air pollutants can be either a minor or a major source of Man-made acidification.

There are no great problems in estimating the amounts of substances reaching the soil surface under a forest stand. But the problems lie in estimating the amount of accumulation of gaseous substances through the direct absorption or adsorption on or by soil surfaces, which might explain relatively great differences in the published data of the deposition velocity (the velocity of absorption of gases) (Schwela, 1977). The mean velocity of deposition of 1 cm per sec at an average concentration of 100 μg SO_2 corresponds to an annual input of S of 15.75 t per km^2. A lower deposition velocity and a corresponding lower input of S was estimated on the surface of organic layer under the forest stands. The deposition velocity in this case depends on the amount of mulch substances in the humus layer: undecomposed organic material can absorb only very small amounts of SO_2, but the capacity for SO_2 absorption and adsorption is far greater in more decomposed humus.

Naturally for the overall calculations, it is also necessary to consider the influence of snow cover in winter and of dry periods; in both cases the absorption capacity of the soil surface is lowered. Therefore, it is probable that the total amount of sulphur actually entering humus-covered soils as a result of this type of deposition, is substantially lower than the amount previously calculated for mineral soils (Materna & Kohout, 1980).

As for the amounts of other ecologically important air pollutants, information about their absorption *plus* adsorption velocities is too scarce to make any generalizations.

Finally, the soil can be enriched by organic materials such as leaves and other organs of the trees or other plants which had been contaminated with pollutants during their active function in the forest ecosystem. The amount of sorbed substances depends on the sorption velocity, the character of the leaves and other organs, and conditions for the absorption and adsorption as well as the capacity of the leaves to accumulate the pollutants or substances derived from them.

The accumulation rates of individual pollutants vary considerably. The increase of sulphur content in comparison with natural content does not amount to an accumulation rate of more than 1 : 10. The amount of accumulated fluorine can be 10^3 times higher under HF influence than in a clean atmosphere. This may influence the rate of decomposition of organic matter.

Soil Changes Due to 'Acid Deposition'

From an ecological point of view, the problems resulting from acid deposition on ecosystems can be very serious over large areas. In forest stands suffering major injury, it is especially difficult to separate the soil changes caused by the input of pollutants from secondary changes resulting from alterations in the vegetation cover—such as disturbance of the canopy's closure and hastened decomposition of organic matter, changes in stand composition, etc. But there are criteria that make it possible to demonstrate the influence of acid deposition.

In the first place there are changes in soil reaction. According to a study in a heavily polluted area it was observed that during the period between 1956 and 1977 the pH in the humus layer dropped from 3.54 to 3.07, and in mineral soil—to a depth of 30 cm—from the pH 3.9 to 3.57. But it is possible to detect distinct changes in shorter periods of time, too. Ulrich has demonstrated pH changes under spruce forest in soil to a depth of 60 cm in a few years (Ulrich *et al.*, 1979). In addition to alterations of the soil pH, changes can occur in the concentrations of available nutrients in the soil, as well as the amount of sorbed substances depends in the concentrations of elements with possible toxic effects, such as aluminium (Overrein, 1972; Mayer & Ulrich, 1976).

SECONDARY EFFECTS OF AIR POLLUTION IN AN ECOSYSTEM

The consequences of the direct influence of pollutants on individual components of an ecosystem are extremely complex, and there is little information regarding secondary effects.

Under air pollution stress, competition between plant species due to decreased vitality or dieback of sensitive plants is altered (Guderian, 1966). But there are also changes caused by such factors as the lack of shade under injured coniferous trees, leading to disappearance of mosses and other shade-loving plants. Secondary ecosystem responses may also include altered host–parasite relationships (Treshow, 1975), but it is scarcely possible to define the importance of such changes to the stability of natural ecosystems in polluted areas.

Some observations indicate that considerable changes occur in the distribution of various species of insects in damaged forest stands (Pfeffer, 1963; Sierpinski, 1972). It is suspected that the greatest insect damage to coniferous forests in Norway by *Exteleia dodecellar* has been indirectly due to acid rain (Huttunen, 1980). Likewise, the explosive reproduction of *Semasia diniana* Gen, and heavy damage to spruce in some mountain areas in Middle Europe in recent years, occurred only in forests that had been damaged by air pollution (Edmunds & Alstad, 1982).

Reproduction and dispersal of many plants depend on insect pollinators. Alteration of the pollinator complex, such as injury to Honey Bees (*Apis mellifera* L.) due to pollution by fluorides or SO_2, can have a detrimental effect on fruit tree production (Carlson & Dewey, 1971).

As secondary effects occur in the soil, alterations in plant production can be expected. Changes in distribution of plant species can also result from changes in soil characteristics, such as are caused by magnesite or cement kiln dust. Some of the secondary changes caused by the impact of pollution are of the same nature as those following some forest management practices, being related to an increased access of precipitation and radiation to the soil surface, coupled with decreased consumption by the stand. Because of an accelerated decomposition of the humus layer, increased amounts of soil nutrients are set free, and this can improve the growing conditions. On the other hand, decreased consumption of water by trees in a humid climate can cause waterlogging of the soil, leading to a deterioration in growing conditions. Increased moisture in the soil can also change the sensitivity of the stand to the impact of air pollution. Such changes naturally have a very important influence on the stability of an ecosystem. When they occur over large areas, they are important to the entire landscape.

Consequences on Ecosystem Functions of Changes in the Environment

The direct influence of pollutants on the individual components of an ecosystem, and indirect influences such as those on the soil or changes in the microclimate, affect the processes that proceed in the ecosystem. This influence is dynamic, so that it is not reasonable to expect the effect of a particular pollutant to remain the same throughout the development of an ecosystem.

It is also clear that changes in any process in the ecosystem tend to become more and more evident as air pollutant concentrations increase. Also, alteration of other ecological factors may increase the sensitivity of individual compartments of the ecosystem to pollution. In extreme cases, biomass production or mineral cycling in the ecosystem can be fully stopped, and a total destruction of the vegetation cover may then take place. If changes prevail that injure a sensitive ecosystem and diminish its stability, it is replaced gradually by another, more resistant one.

Biomass Production

It is curious that there appears to be no information available on regional or global losses or changes in biomass in natural ecosystems that could be attributed to air pollution. There are, however, calculations and information showing a possible positive effect of increased CO_2 production from human activity on the biomass production in temperate zone forests (Armentano & Ralston, 1980). Also, Ulrich (1979) considers that the emission of nitrogen oxides from industrial sources and transportation is to some extent favourable to the growth of forest species.

But if we consider smaller areas exposed to elevated concentrations of SO_2, HF, O_3, and other pollutants, there is considerable information about increment decreases in tree growth, and in plant production in general. However, these are mostly the result of studies concerned with estimating production-losses of economically important products, notably wood. Information seems to be lacking on changing biomass production in an ecosystem as a whole.

Except for the extreme cases mentioned, the loss in biomass production of sensitive species may be substituted by the biomass production of resistant species (Guderian, 1966). The loss of biomass production of certain plant species is a result of: (i) Decreased absorption and assimilation of CO_2, which will subsequently influence other physiological or biochemical processes; (ii) Reduction of the leaf area by necrosis, premature leaf abscission, or the leaf's reduced development; and/or (iii) Increased mortality of the most sensitive individuals.

Studies of changes in the CO_2 absorption associated with air pollutants, combined with a study of their effects on increment and yield, are rare (T. Keller, 1979). It is possible, however, to consider that air pollution concentrations which cause only decreased assimilation, without visible symptoms of leaf injury, cause losses in organic matter production that may not exceed losses from a very dry or very cold year. This would not seriously destabilize an ecosystem, and alterations at this level should have no consequences on the functions of a given ecosystem in the landscape. It has, however, been demonstrated that long-lasting exposure to very low concentrations of SO_2 on a sensitive plant species, Norway Spruce, can cause relatively significant changes in the content of various substances in the needles (Materna, 1972). Therefore, it seems quite possible that such changes could accompany a decreased resistance of the tree to both biotic pathogens and abiotic factors, as well as cause losses in the biomass of the needles and even increased mortality.

The most pronounced changes are evident when high concentrations of pollutants cause acute leaf damage. But if these are only unique incidents, their influence on biomass production need not be very important. The long lasting influence of lower concentrations can be more significant in limiting the development of twigs and leaves and even causing their premature abscission.

Under air pollution stress, a fluent transition exists between fully developed, healthy leaf area and its total destruction. There are distinct, individual differences in the reactions of leaves to pollutants. These are genetically determined, but their expression is influenced by interaction with such ecological factors in 'microspace' as soil, humidity, and nutrient supply. The individual variability of the reaction of the leaf to increasing air pollution becomes more and more distinct, and extends to individual differences in the ability to adjust (Materna, 1964; Pelz, 1964).

In the early stages of air pollution influence on forest stands, it is difficult to distinguish the mortality caused by natural processes from that caused by the impact of pollutants, as the pollutant action can destroy the same individuals in a stand that would later decline because of natural processes.

The mortality of weakened individuals in forest stands at low degrees of injury mostly is not only the result of a direct influence of air pollution but the trend is also determined by other factors, especially by such stress factors as frost, drought, waterlogging, or insect damage. Especially severe frost can cause an increasing dieback of trees in forest stands weakened by air pollution.

The influence of air pollution on biomass production depends to a large degree on damage to the main component of the ecosystem. In forests this involves injury to the main tree species—the dominant or co-dominants—whose sensitivity to air pollution, and the rate at which injury develops, determines whether the result is a decrease of total biomass production or, alternatively, a decrease of biomass production of one component which is replaced by the development of another plant species. If biomass formation depends on a very sensitive plant species, then after it has been destroyed, space is liberated for the development of another, more resistant plant species whose organic matter production can be the same as, or more than, that of the original plant species.

An example of this is the destruction of very sensitive mountain Spruce forests (Norway Spruce) which can be, after their destruction, replaced by stands of Silver Birch (*Betula verrucosa*) and Mountain Ash (*Sorbus aucuparia*) with undergrowth of herbs. The total biomass formation of such stands can be in the first decades higher than occurred in the corresponding spruce forests. But if the main plant species in an ecosystem is relatively resistant, then a loss of its biomass production due to air pollution determines the degree of loss in the whole ecosystem. Changes in the decomposition of organic matter can be accelerated under the destroyed canopy of forest stands owing to increased access of light and precipitation. This can accelerate the mobilization of nutrients. But at higher levels of air pollution influence, causing acidification of soil or enriching humus layer with toxic elements, the decomposition of organic matter can be restricted.

Changes in biomass production are connected with the economic interests of society in two respects, namely (1) Changes in the production of economically available products such as wood or fruits, and (2) Changes in the further functions of the vegetation cover in the landscape.

Naturally, great changes in a forest ecosystem can influence the production of forest fruits either positively or negatively, but the losses in wood production are generally far more important. Their estimation has been an object of considerable study, which confirms the great losses that occurred in sensitive coniferous forest owing to the decrease in increment of individual trees and also increased mortality. Correlation between these last two criteria exists as long as an interaction with additional stresses does not interfere.

From studies of increment-losses in forest stands exposed to pollution, the following conclusions can be drawn:

(1) It is possible that increment reduction in stands can occur without any external signs of injury.
(2) Under more severe conditions, a correlation exists between the external markings or degree of injury and the extent of production losses.
(3) Increment development also reflects the influence of individual differences in sensitivity to pollution and to a whole complex of other possible factors.
(4) The greater the impact of pollutants is, the less are the stands able to react to favourable ecological conditions such as higher precipitation, temperature, etc.

These changes can be fully explained in terms of the direct influence of air pollution on stands, and eventually of the interacting soil changes (Vins & Pollanschütz, 1977).

Verifying the influence of soil changes due to acid deposition on the increment to forests is very complicated. It is based on comparisons between regions that are differently stressed by acid deposition, and sites that are supposed to differ in sensitivity to acidification (Abrahamson *et al.*, 1975). Therefore, it is chiefly speculation that decreases in increment occur (Abrahamson *et al.*, 1975; Johnsson, 1976).

Cycling of Mineral Substances

The impact of air pollutants on the cycling of elements is an important part of their influence on ecosystems. Pollutants can have a broadened significance in the landscape by influencing the quality of water and the character of the soil. Modifications in element cycling in affected ecosystems are caused by:

  (i) Amount and quality of input;
 (ii) Changes in the nutrient level and composition, especially of plant leaves;
(iii) Changes in the availability of nutrients in the soil;
 (iv) Changes in the dynamics of humus layer decomposition; and
  (v) Changes in the participation of plant species of varying sensitivity, and their nutrition.

The amounts of substances from a polluted atmosphere that are included in the mineral cycling in the ecosystem are a function not only of the input, but, indeed, their availability after they have settled on the soil surface is critical. Sulphur and nitrogen compounds from air pollution are readily included in the cycling, but fluorine and many of the heavy metals are absorbed by plant roots from the enriched soil in limited amounts (Ulrich *et al.*, 1979).

Some components of pollution, such as SO_2, can also influence the concentration of other elements in assimilating organs. It is known that there is a distinct increase of silicic acid, as well as of the contents of Mg, K, and Ca, in the needles of Scots Pine and Norway Spruce (Themlitz, 1960; Materna, 1972). Changes in the concentration of N and P in the needles of Big Cone Spruce (*Pseudotsuga macrocarpa*) (Mayer & Ulrich, 1976; Zinke, 1980), and of N, P, K, Ca, Mg, Al, Mn, and Zn, in the needles of Jack Pine × Lodgepole Pine (*Pinus banksiana* × *P. contorta*) were also observed (Legge, 1980).

Further changes in the soil result from the availability of nutrients or toxic heavy metals induced by changes in soil reaction, such as a decrease in the calcium content of plants growing on soils with deposits of magnesite. Clear deficiences of manganese in pines growing on soil enriched by cement kiln dust, and the higher availability of aluminium, reaching toxic amounts in acidified soils, are evident. These changes can develop over a long period of time without disturbing the stability of the ecosystem. Changes in the decomposition of organic matter can be accelerated under the destroyed canopy of forest stands owing to increased access of light and precipitation. This can accelerate the mobilization of nutrients. But at higher levels of air pollution influence, causing acidification of soil or enriching the humus layer with toxic elements, the decomposition of organic matter can be restricted.

One of the most important changes in an ecosytem related to nutrient cycling is the change in plant species. The substitution of relatively sensitive coniferous tree species by more resistant, broadleafed species represents a basic change not only through the composition of leaves and other organs, but also through the utilization of nutrient reserves in various parts of the soil profile. Elements are bound in the stable and partly dead portion of the biomass—the wood—and accumulate in the soil differently in different cases because of changes in the rate of decomposition of organic matter from the different plant species involved. Some of these processes are able to compensate for the unfavourable effect of acid precipitation for a transitional period or even for a long time, thus slowing down depauperization of the site.

Changes in Other Functions of the Vegetation Cover

If the vegetation cover is totally destroyed, all of its desirable attributes in the landscape will disappear. But before such an extreme condition is reached, it is generally possible to recognize some reduction in plant function relative to the degree of disturbance by air pollution (Knabe, 1976; Materna, 1980).

The impact of air pollutants on an ecosystem depends on the amount of pollutants entering the system. As the vegetation cover itself functions as a filter, the amount of leaf area, or absorptive surface, and its stratification, is critical. Thus a closed spruce stand can provide an effective filter. But a broken stand can be more effective in separating off dust particles, especially if a rich understorey of shrubs and herbs develops beneath the canopy of the main forest species. The higher turbulence of the air in a broken stand also causes a higher sedimentation of particles. In such cases, an injured ecosystem can fulfill this distinct function better than an intact stand.

More complicated relations occur between the changes induced by air pollution on the vegetation cover and the runoff from affected watersheds. There are observations of a greater runoff from such watersheds (Kemel, 1979). After the forest under study was injured by air pollution, water consumption was reduced in the ecosystem, owing to decreased interception. This effect in some regions can be viewed as positive if there is a need for increased water delivery to inhabitants and industry. But it is also related to the decreased ability of forests to balance the runoff, and especially to protection of the snow reserves in the spring and slowing down their melting (Běle, 1979). Further erosion of soils under the damaged forests can be accelerated, especially if the stands are dying off quickly. The new, more pollutant-tolerant cover that then develops may be more or less efficient in controlling soil erosion.

This view is valid only at elevations where the amount of water from precipitation is reduced by interception of forest stands. At higher elevations, especially under the crowns of coniferous tree species, more water is coming to the soil surface than in an open space (H. M. Keller, 1968). In this case a reduction of needle biomass caused by air pollution has a negative influence upon the amount of water.

The hydrological function of forests that are disturbed by air pollution cannot be seen only from the point of view of runoff. It is also necessary to consider the quality of the water. Not only must the increase in elements from the air pollution be considered, but also the changes in the humus layer, with the fast decomposition of the organic matter and increased nitrification. Also, changes in vegetation cover, such as that from a coniferous to a broadleafed forest, can have a negative influence on water quality. Further negative consequences on aesthetic value, microclimatic function, and recreational use, of ecosystems, must also be considered (Wentzel, 1965).

It is possible to compare the losses caused by the direct and indirect effects of air pollution on the productivity of ecosystems, with the losses resulting from the influence on their other functions. The increased risk of high water-table, soil erosion, and decreased quality of water, can be substantially more serious than the losses from reduced quantity and quality of the biomass formed.

SUMMARY

Pollutants may affect ecosystems at various levels: accumulation in the plants, soil, or ground water; damage to consumers; altering competition; disrupting biogeochemical cycles; disrupting stability; breakdown of stands and associations; and expanding denuded zones. These changes may originate as consequence of pollutant exposures for periods ranging from a few hours to a year or more. In the last decades the character of air pollution influences on ecosystems has been significantly altered. While in time past the impact of high concentrations of pollutants was limited to small surrounding areas in the vicinity of the pollution sources, today large areas with low concentrations of pollutants at great distances from the emitting sources can be found; there the character of pollutant influence is partially changed, and the impact of acid precipitation takes places in a more expressive way. Conifer forests provide the most seriously affected examples. Great differences exist between various ecosystems mainly because of the different sensitivity of plant species, as well as of individuals and clones.

Also, at low concentrations of pollutants, the total biomass production of natural ecosystems can be altered. If the main compartment of the ecosystem is a very sensitive one, e.g. Norway Spruce, its biomass production, reduced on account of pollution, can be replaced by that of one of a more resistant tree species or plant species of the ground vegetation. But even in this case the reduction in the production of, for example, economically significant forest products, might be a serious and continuing impact on forest management. Besides, the damaging of the natural ecosystems, especially forests, over large areas, might also seriously affect the other functions of the forests in the landscape, for example, their ability to protect the soil, their effect on water-flow from the forest watersheds, and their suitability for recreation purposes.

REFERENCES

Abrahamson, G., Horntvedt, R. & Tveite, B. (1975). Impact of acid precipitation on coniferous forest ecosystems. Pp. 1–15 in *Acid Precipitation — Effects on Forests and Fish. Research* Rept., 2/75—SNSF Project.
Armentano, J. V. & Ralston, C. W. (1980). The role of temperate-zone forests in the global carbon cycle. *Can. J. For. Res.,* **10**, 53–60.
Běle, J. (1979). [Influence in the changes of forest's composition on their hydrological effects.—in Czech.] *CSVTS Hydrologicka Problematiká pri Úpravách Toků,* Karlovy Vary, **22**, pp. 99–107.
Bordeau, P. & Treshow, M. (1978). Ecosystem response to pollution. Pp. 313–30 in *Principles of Exotoxicology* (Ed. G. C. Butler). SCOPE: xxii + 350 pp., illustr.
Carlson, C. E. & Dewey, J. E. (1971). Environmental pollution by fluorides. Flathead National Forest and Glacier National Park, Division of State and Private Forestry. *U.S. Dept. Agr. Forest Service, Northern Region Headquart.,* **10**, 57 pp.
Cowling, E. B. & Dochinger, L. S. (1980). Effects of acidic precipitation on health and the productivity of forests. Pp. 165–75 in *Proc. Symp. Effects of Air Pollutants on Mediterranean and Temporate Forest Ecosystems,* USDA for. Serv., Berkeley, California, USA: 256 pp.
Davis, D. D. & Wilhour, R. G. (1976). Susceptibility of woody plants to sulfur dioxide and photochemical oxidants. *Ecol. Res. Stud.,* EPA 600/3-76-102, pp. 1–65.
Edmunds, G. F. & Alstad, D. A. (1982) Effects of air pollutants on insect populations. *Annual Review of Entomology,* **27**, 369–84.
Enderlein, H. & Stein, G. (1964). Der Säurezustand der Humusauflage in den rauch-geschädigten Kiefernbeständen des Staatlichen Forstwirtschftsbetriebes Dübener Heide. *Arch. Forstw.,* **13**, pp. 1181–91.
Farrar, J. F., Relton, J. & Rutter, A. J. (1977). Sulphur dioxide and the scarcity of *Pinus sylvestris* in the industrial Pennines. *Environ. Pollut.,* **14**, pp. 63–8, illustr.
Guderian, R. (1966). Reaktionen von Pflanzengemeinschaften des Feldfutterbaues auf Schwefeldioxideinwirkungen. *Schrift. Landesanst. Immis. Bodennutzungsschutz NRW,* **4**, pp. 80–100.
Guderian, R. (1977). *Air Pollution, Phytotoxicity of Acidic Gases and its Significance in Air Pollution Control.* Springer-Verlag, New York, NY, USA: 122 pp.
Hajdúk, J. (1965). The influence of magnezite dust on vegetation and soil. Pp. 31–8 in *Problems of Air Pollution.* SAV, Bratislava.
Huttunen, S. (1980). The integrative effects of airborne pollutants on boreal forest ecosystems. Pp. 111–34 in *Effects of Airborne Pollution on Vegetation,* UN Econ. Comm. Europe, Warsaw, Poland.
Johnsson, B. (1976). Soil acidification by atmospheric pollution and forest growth. Pp. 837–45 in *Proc. 1st Int. Symp. Acid Precipitation and the Forest Ecosystems.* USDA Forest Service.
Keller, H. M. (1968). Der heutige Stand des Forschung über den Einfluss der Waldes auf den Wasserhaushalt. *Schweiz Ztschr. f. d. Forstw.,* Nr **4/5**, pp. 364–79.
Keller, T. (1979). Der Einfluss mehrwöchiger neidriger SO_2 Konzentrationen auf CO_2 und Jahrringbau der Fichte. *Bericht X Fachtagung IUFRO S 2.09* (Ljubljana), **1**, pp. 73–89.

Kemel, M. (1979). [Evaluation of future changes of culminating flow-off on watersheds in Ore mountains.—in Czech.] *CSVTS Hydrologicka Problematiká pri Úpravách Toků*, pp. 84–98.

Knabe, W. (1970*a*). Air quality criteria and their importance for forests. *Mitt. Forstl. Bundesversuchsanst* (Wien), **92**, pp. 129–50.

Knabe, W. (1970*b*). Kiefernwaldverbreitung und Schwefeldioxid Immisionen im Ruhrgebiet. *Staub,* **30**(1), pp. 32–5.

Knabe, W. (1976). Effects of sulfur dioxide on terrestrial vegetation. *Ambio,* **5**, pp. 213–8.

Legge, A. H. (1980). Primary productivity, sulfur dioxide, and the forest ecosystem; an overview of a case-study. Pp. 51–62 in *Proc. Symp. Effects of Air Pollutants on Mediterranean and Temperate Forest Ecosystems,* USDA For. Serv., Berkeley, California, USA: 256 pp.

Linzon, S. N. (1978). Effects of airborne sulfur pollutants on plants. Pp. 110–62 in *Sulphur in the Environment, Part II.* John Wiley & Sons, New York, NY, USA.

Löffler, A. (1974). Research on the influence of magnezite and fluorine pollution on the soil. *Report Forest Res. Inst., Zvolen* (CSSR).

McClenahan, J. R. (1978). Community changes in a deciduous forest exposed to air pollution. *Can. J. For. Res.,* **8**, pp. 432–8.

Materna, J. (1972). Einfluss niedriger SO_2 konsentrationen auf die Fichte. *Mitt. Forstl. Bundesversuchsanst* (Wien), **97**, pp. 219–31.

Materna, J. (1973*a*). Kriterien zur Kennzeichnung der Immissionseinwirkung auf Waldbestände. Pp. A121–3 in *Proc. III Int Clean Air Congr., Düsseldorf.*

Materna, J. (1973*b*). Relationship between SO_2 concentration and damage of forest trees in the region of the Slavkov forest. *Práce VÚLHM,* **43**, pp. 169–80.

Materna, J. (1979). Effects of pollution on the capacity of vegetation to perform such functions as water retention, soil protection, wildlife habitat, etc. *Symp. on the Effects of Airborne Pollution on Vegetation,* United Nations Economic Commission for Europe, 20–24 August, Warsaw, Poland.

Materna, J. (1980). Effects of pollution on the capacity of vegetation to perform such functions as water retention, soil protection, wildlife habitat, etc. Pp. 242–53 in *Effects of Airborne Pollution on Vegetation.* UN Econ. Comm. Europe, Warsaw, Poland.

Materna, J. & Kohout, R. (1980). [Sorption of sulfur dioxide by forest humus—in Czech]. *Práce VÚLHM,* **56**, pp. 111–27.

Mayer, R. & Ulrich, B. (1976). Acidity of precipitation as influenced by the filtering of atmospheric sulphur and nitrogen compounds—its role in the element-balance and effect on soil. Pp 737–45 in *Proc. 1st Int. Symp. on Acid Precipitation and the Forest Ecosystem. (Eds L. S. Dochinger and T. A. Seliga)* USDA Forest Service, Columbus, Ohio: 1074 pp.

Miller, P. R. & McBride, J. R. (1975). Effects of air pollutants in forests. Pp. 195–236 in *Responses of Plants to Air Pollution* (Eds J. B. Mudd & T. T. Kozlowskij). Academic Press, New York–San Francisco–London: xii + 388 pp., illustr.

Nemec, A. (1958). Air pollution and its influence on soil and the decline of forest tree species in Ore Mountains region. *Vedecké Práce V.Ú. lesa a Myslivosti,* **1**, pp. 143–76.

Ottar, B. (1980). Air pollution, a survey of sources and dispersion modelling. Pp 65–80 in *Symp. on the Effects of Airborne Pollution on Vegetation,* United Nations Economic Commission for Europe, 20–24 August, Warsaw, Poland.

Overrein, L. N. (1972). Sulphur pollution patterns observed: Leaching of calcium in forest soil determined. *Ambio*, **1**, pp. 145–7.

Overrein, L. N. (1980). Acid precipitation impact on terrestrial and aquatic systems in Norway. Pp. 145–52 in *Proc. Symp. Effects of Air Pollutants on Mediterranean and Temperate Forest Ecosystems*. USDA For. Serv., Berkeley, California, USA.

Pelz, E. (1964). Poskeneni lesů kourem a prachem. *Studijni informace lesnictví*, **1–2**, 164 pp.

Pelz, E. & Materna, J. (1964). Beiträge zum Problem der individuellen Ranchhärte der Fichte. *Arch. Forster.*, **13**(2), pp. 177–210.

Pfeffer, A. (1963). Insektenschädling an Tannen-in Bereich der Gasexhalationen. *Ztschr. Angew Entomol. St.*, (Praha), **51**, pp. 203–7.

Schwela, D. (1977). Die trockene Deposition gasförmiger Luftverunreinigungen. *Schriftenreihe der LIB Essen*, **42**, pp. 46–85.

Sierpinski, Z. (1972). The significance of secondary pine insects in areas of chronic exposure to industrial pollution. *Mitt. Forst. Bundesvers*, **97**, pp. 609–15.

Skelly, J. M. (1980). Photochemical oxidant impact on Mediterranean and temperate forest ecosystems: Real and potential effects. Pp. 38–50 in *Proc. Symp. Effects of Air Pollutants on Mediterranean and Temperate Forest Ecosystems*. USDA For. Serv., Berkeley, California, USA: 256 pp.

Smith, L. (1980). The acidity problem—its nature, causes, and possible solution. Pp. 136–44 in *Proc. Symp. Effects of Air Pollutants on Mediterranean and Temperate Forest Ecosystems*. USDA For. Serv., Berkeley, California, USA: 256 pp.

Sokolowski, A. (1971). [The influence of forest plants of emissions from the nitrogen industrial plants of Pulawy— in Polish.] *Sylwan*, **3**, pp. 47–56.

Themlitz, R. (1960). Die individuelle Schwankung des Schwefelgehaltes gesunder und rauchgeschädigter Kiefern und seine Beziehung zum Gehalt an der übrigen Hauptnährstoffen. *Allg. Forst-Jagdztg.*, **131**, pp. 261–4.

Treshow, M. (1968). The impact of air pollutants on plant populations. *Phytopathology*, **58**, pp. 1108–13.

Treshow, M. (1975). Interactions of air pollutants and plant disease. Pp. 307–34 in *Responses of Plants to Air Pollution* (Eds J. B. Mudd & T. T. Kozlowski). Academic Press. New York, NY, USA: xii + 383 pp., illustr.

Treshow, M. (1980). Pollution effects on plant distribution. *Environmental Conservation*, **79**, pp. 279–86, 5 figs.

Ulrich, B. (1979). Stoffhaushalt von Wald Ökosystemen, II: Bioelement Haushalt. *Schriftenreihe Inst. Bodenk. u. Waldernährung*, Univ. Göttingen, II Aufl.

Ulrich, B., Mayer, R. & Khanna, K. (1979). Deposition von Luftverunreinigungen und ihre Auswirkungen in Walkdökosystemen im Solling. *Schriftenreihe Forstl. Fak. Univ. Göttingen*, **58**, Sauerländer.

Vins, B. & Pollanschütz, J. (1977). Erkennung und Beurteilung immissionsgeschädigter Wälder an Handvon Jahrringanalysen. *Allg. Forstztg.*, **6**, pp.

Wentzel, K. F. (1965). Immissionschäden und Erholungswert des Waldes in der Industrielandschaft. *Forst u. Holzwirt*, **20**, pp. 378–81.

Wentzel, K. F. (1979). Waldbauliche Erfahrungen über die Immissions-Empfindlichkeit der Tanne (*Abies alba* Mill.). *Bericht. X Fachagung IUFRO S 2.09* (Ljubljana), **1**, pp. 199–211.

Whitby, L. M. & Hutchinson, T. C. (1974). Heavy-metal pollution in the Sudbury mining and smelting region of Canada. *Environmental Conservation*, **1**(2), pp. 123–31.

Wieler, W. (1912). Untersuchungen über den Einfluss der Entkalkung des Bodens durch Hüttenrauch und über die giftige Wirkung von Schwermetallen auf das Pflanzenwachstum. *Pflanzenwachstum und Kakmangel im Boden* (Berlin).

Zinke, P. J. (1980). Influence of chronic air pollution on mineral cycling in forests. Pp. 88–109 in *Proc. Symp. Effects of Air Pollutants on Mediterranean and Temperate Forest Ecosystems,* USDA For. Serv., Berkeley, California, USA: 256 pp.

Air Pollution and Plant Life
Edited by M. Treshow
© 1984 John Wiley & Sons Ltd.

CHAPTER 17

Pollutant Uptake by Plants

WILLIAM H. SMITH

*School of Forestry and Environmental Studies, Yale University, New Haven,
Connecticut 06511, USA*

BACKGROUND

Pollutant transfer from the atmosphere to natural or artificial (e.g. agricultural) plant communities is a very complex and incompletely understood process. Atmospheric contaminants may be removed by both the soil and vegetative compartments of an ecosystem through a variety of mechanisms. The primary processes are precipitation scavenging, chemical reaction, dry deposition (sedimentation), and absorption (impaction) (Rasmussen *et al.,* 1974). Loss *via* precipitation may occur as 'rainout', which involves both absorption and the capture of particles by falling raindrops. Primary and secondary contaminants are subject to a large number of chemical reactions in the atmosphere, that may ultimately transform them into an aerosol or oxidized or reduced product. Attachment by aerosols and subsequent deposition on the surface of the earth is termed 'dry deposition'. Absorption by water bodies, soils, or vegetation, at the surface of the earth, is an additional extremely important removal process.

Components of the ecosystems of the earth that remove pollutants from the atmospheric compartment and store, metabolize, or transfer them, may conveniently be termed 'sinks' (Warren, 1973). Forest ecosystems in general, and temperate forest ecosystems in particular, are locally, regionally, and globally, important sinks for a wide variety of atmospheric contaminants. Soil and vegetation surfaces represent the major sink for pollutants introduced into terrestrial ecosystems (Little, 1977).

The transfer of contaminants from the atmospheric compartment to the surface of soil or vegetation is expressed as a flux (pollutant uptake) rate and is given as a weight of pollutant removed by a given surface area per unit of time. Actual determinations of flux rates (sink strengths) are extremely complex, and involve an appreciation of atmospheric conditions (wind, turbulence, temperature, and humidity), pollutant nature and concentration, sink surface conditions (geometry and presence or absence of moisture), and other parameters. Awareness of the following simple relationship between flux-rate and pollutant concentration, will assist the reader in evaluating the literature dealing with sink function:

$$F \text{ (pollutant uptake, or flux)} = v \text{ (proportionality constant or deposition velocity)} \times C \text{ (pollutant concentration)}$$

$$\text{typical units: } F, \text{ mg per cm}^2 \text{ per sec}$$
$$v, \text{ cm per sec}$$
$$C, \text{ mg per cm}^3$$

The deposition velocity parameter is generally a function of the surface of the sink (soil or vegetation) whether it is wet or dry. It may be thought of as the rate at which an absorbing surface 'cleans' a pollutant from the air. If the deposition velocity of a pollutant is 1.0 cm per sec, it suggests that the surface is completely removing the pollutant from a layer of air 1.0 cm thick each second, with the 'clean' layer immediately replaced by a 'new' contaminated layer. Keep in mind that both the proportionality constant and pollutant concentration must be known in order to estimate pollutant uptake. This summary relationship between flux-rate and pollutant concentration, and associated jargon, has been developed by those primarily concerned with the transfer of particles from the atmosphere to natural or other surfaces. The concept, however, is widely employed in air pollution literature to express the transfer of particulate and gaseous pollutants from the atmosphere to various terrestrial and aquatic sinks.

The principal repository in terrestrial ecosystems for air contaminants of anthropogenic origin is the soil. By virtue of their distribution and physical and chemical characteristics, forest soils may be particularly efficient short- and long-term sinks. For selected trace metals and gases, the evidence supporting the importance of soils is considerable.

FOREST SOILS AS PARTICULATE SINKS

Particles are transferred from the atmosphere to forest soils directly by dry deposition and precipitation scavenging and indirectly *via* leaf and twig fall. A very large number of human activities generate small particles (0.1–5 μm) with high concentrations of trace metals. Depending on weather conditions, these particles may remain airborne for days or weeks and be transported hundreds or thousands of km from their source. The evidence that forest soils may be the ultimate or temporary repository for the trace elements associated with these particles is substantial. Soils have a very high affinity for heavy metals, particularly the clay and organic colloidal components (Lagerwerff,

1967; John *et al.*, 1972; Stevenson, 1972; Korte *et al.*, 1976; Zunino & Martin, 1977*a*, 1977*b*; McBride, 1978; Petruzelli *et al.*, 1978; Somers, 1978).

Lead

Lead is naturally present, in small amounts, in soil, rocks, surface waters, and the atmosphere. Owing to its unique properties, it has been an element widely useful to humans. This utility has resulted in greatly elevated lead concentrations in certain ecosystems. In locations where lead is being mined, smelted, and refined, where industries are consuming lead, and in urban–suburban complexes, the environmental lead level is greatly elevated. It is widely agreed that a primary source in these latter sites is the combustion of gasoline containing lead additives. Other important sources include coal combustion, refuse and sludge incineration, burning or attrition of lead-painted surfaces, and industrial processes. As the vast majority of atmospheric lead particles are less than 0.5 μm in diameter (W. H. Smith, 1976), they have been widely distributed to all parts of the Earth.

The input of lead, its cycling within forest ecosystems, its transfer in foodchains, its rate of loss to downstream aquatic systems, and its residence in soil, have received considerable research attention.

The lead concentration of the upper soil horizons of unmineralized and uncontaminated areas ('baseline' level) is generally given as approximately 10–20 μg per g of dry soil. Analyses of forest soils from throughout temperate zone forests, however, frequently show elevated lead amounts associated with the soil compartment. Lead may be added to the soil as a component of organic compounds in plant debris. The divalent cationic nature of lead added *via* precipitation, dry fallout, throughfall, or stem-flow, may cause the lead to be bound to organic exchange surfaces (Zimdahl & Skogerboe, 1977) such as are abundant in the forest floor. Subsequent reaction of lead with sulphate, phosphate, or carbonate, anions may reduce its solubility and impede its downward migration in forest soil profiles. Lead deposited from the atmosphere may have a residence time approximating 5,000 yr in the surface organic soil horizons (Benninger *et al.*, 1975), and long term concentration increases can be predicted as long as inputs to forest soils exceed outputs.

The author and colleague T. G. Siccama have been involved in an intensive biogeochemical study of several trace metals, including lead, in the northern hardwood forest (Siccama & Smith, 1978; W. H. Smith & Siccama, 1980). The study is being conducted on the Hubbard Brook Experimental Forest in central New Hampshire (elevation 230–1010 m). This forest is typified by an unbroken canopy of second growth northern hardwood (Sugar Maple [*Acer saccharum*], American Beech [*Fagus grandifolia*], and Yellow Birch [*Betula lutea*]), with patches of Red Spruce (*Picea rubens*) and Balsam

Fir (*Abies balsamea*) particularly at higher elevations and along the valley bottoms. The annual lead flux to the forest is 266 g per ha per yr. The lead output from the system in streamwater approximates 6 g per ha per yr (W. H. Smith & Siccama, 1980). The extraordinary disparity in these input–output figures is accounted for by the accumulation of lead in the soil compartment. The average lead concentration in the forest floor was determined to be 89 μg per g, with total lead of the forest floor averaging 9 kg per ha.

Reiners *et al.* (1975) also have investigated lead retention by forest soils in the White Mountains of New Hampshire in an area that they estimated had a lead flux rate of approximately 200 g per ha per yr. Samples were collected from Mt Moosilauke that represented various vegetative zones from the northern hardwood forest (to 700 m) to the alpine tundra (to 1400 m). Cores were removed from the organic layer of the forest floor and analysed for lead. Lead concentrations ranged from 11 to 336 μg per g dry weight, with the highest values occurring in the fir forests and especially in the litter layer.

Hook *et al.* (1977) have examined lead cycling in eastern Tennessee's oak forest in the vicinity (14 km) of coal-fired electric generating facilities. The estimated lead input to the forest was given as 286 g per ha per yr. For four forest types examined, the litter layer was fractioned into two horizons: original leaf form still discernible (O_1) and original form lost (O_2). This investigation involved calculation of the standing pools of lead in the vegetative components of the forests that were studied. From these data it was observed that the litter (O_1 and O_2 horizons), which constitued 13% of the total forest organic matter, contained 71% of the lead in the ecosystem. Movement of lead from the O_2 litter to the underlying soil horizons was concluded to be high, and estimated at 182 g per ha per yr.

As part of the International Biological Programme, Heinrichs & Mayer (1977) gathered soil lead data from beech and spruce forests in Germany that are representative of ecosystems that are widely distributed in central Europe. These authors suggested that the forests sampled were in a 'relatively unpolluted' environment, yet the measured lead flux in precipitation below the beech canopy was 365 g per ha per yr and below the spruce canopy was 756 g per ha per yr (adjacent open field lead input in precipitation was given as 405 g per ha per yr). All studies summarized to this point were conducted in rural sites, in locations that were considered by their authors to be relatively remote from excessive lead contamination. Despite this fact, it is obvious that considerable quantities of lead are accumulating in soils, and that this compartment is servicing as an important sink for this trace metal. For locations that are subject to excessive lead input—for example, urban, industrial, or roadside situations—the accumulation of lead in the soil sink is even more impressive.

Parker *et al.* (1978) have examined lead distributions in an oak forest in a heavily urbanized and industrialized section of East Chicago, Indiana. Lead flux *via* precipitation and dry fallout was determined to be 815 g per ha annually.

Soil samples analysed from sites 1–2 km from a lead smelter in Kellogg, Idaho, averaged 4,640 µg per g at the surface (0–2 cm) (Ragaini *et al.*, 1977). Soil samples collected from 65 major street intersections in urban locations in southern Ontario, revealed a high of 21,000 µg per g lead in the upper 5 cm of soil (Linzon *et al.*, 1976). These soils were not collected from forest sites, but woody vegetation did occur in the general area.

Roadside environments are grossly contaminated with lead, as motor vehicles combusting gasoline containing lead alkyls release approximately 80 mg of lead per km driven. The total area of roadside ecosystems approximates 3.04×10^7 ha (118,000 square miles) in the United States, and so it is important to consider the roadside soils' lead burden. If 20 µg per g lead is accepted as a baseline lead concentration for uncontaminated soils, it can be seen that soil samples taken within a few metres of the road surface of a heavily-travelled highway may range to more than 30 times the baseline. At 10 m distance from the roadway, however, the lead level is typically only 5–15 times the baseline. At approximately 20 m distance, several studies suggest that a constant level of soil lead is achieved, and the influence of the roadway is lost (W. H. Smith, 1976).

When a trace metal such as lead is added to soil it may be (1) absorbed on soil particle exchange sites, (2) precipitated as an insoluble compound, (3) leached to lower depths in the soil profile, (4) lost to the atmosphere, (5) metabolized by soil fauna or microbes, or (6) absorbed by plant roots. In the case of lead and forest soils, it is apparent that mechanisms (3) through (6) are generally unimportant, that mechanisms (1) and (2) prevail, and as a result the soil compartment, particularly the organic forest floor, is an important sink for atmospheric lead. The flux of lead from the atmosphere to the forest floor may approximate 200–400 g per ha per yr in the forest soils of the eastern United States.

The threshold concentration of soil lead for phytotoxicity approximates 600 µg per g dry-weight (Linzon *et al.*, 1976). As most forest ecosystems are well below this threshold, except those in selected urban, industrial, and roadside sites, and as evidence for tree root uptake of lead is meagre, this biologically non-essential element is judged not to be responsible for direct impairment of tree health.

Other Trace Elements

Forest soils may constitute a sink for a variety of trace elements in addition to lead. Elements judged to have particular potential for biological significance if accumulated in terrestrial ecosystems include cadmium, nickel, thallium, copper, fluorine, vanadium, zinc, cobalt, molybdenum, tungsten, mercury, and selenium (Hook & Shults, 1977). Under certain conditions manganese, chlorine, chromium, and iron, might be appropriate additions to this list. Numerous of these elements differ substantially from lead in their interaction with plants. Some, for example iron, manganese, zinc, copper, and molybdenum, are required by vegetation in small amounts for normal growth and development. Others, for example cadmium, nickel, thallium, and tungsten, are very mobile in plants. Cadmium, nickel, fluorine, thallium, vanadium, mercury, and copper, have high potentials for phytotoxicity.

Information concerning the ability of forest soils to act as a sink for these elements is very limited. A comparison of trace element concentrations for selected temperate forest soils, with baseline trace element concentrations for a variety of soils in relatively unpolluted areas and from regions lacking major mineral deposits, suggests that certain forest soils may be accumulating elevated levels of zinc, cadmium, copper, nickel, manganese, or iron. This is obviously the case for forests within several kilometres of an urban area—for example, the East Chicago situation—or of metal smelters, as in the case of the Palmerton, Pennsylvania environment. It is most significant, however, that it also appears that forest soils are accumulating elevated concentrations of these trace elements even when located many kilometres from a primary source.

The residence time of heavy metals in forest soils is of fundamental importance in judging sink efficiency. Tyler (1978) has examined the leachability of manganese, zinc, cadmium, nickel, vanadium, copper, chromium, and lead, from two organic spruce forest soils—one of them from a site that was grossly polluted by copper and zinc from a metal smelter, and the other from an unpolluted forest location. Both soils were treated in lysimeters with artificial rain water, acidified to pH 4.2, 3.2, and 2.8. At pH 4.2—relatively commonly measured in the northeastern United States and in northern Europe—the residence times are impressive, particularly in the polluted soil. The high buffer capacity of the latter soil is cited by Tyler as conferring protection against leaching. While extrapolation of lysimeter data to natural soils can only be made with considerable risk, Tyler's observation is very informative.

It is concluded that, in addition to lead, forest soils—in particular the organic forest floor—function with variable efficiency as a sink for zinc, cadmium, copper, nickel, manganese, vanadium, and chromium, associated with particulate inputs from the atmosphere. Judgements concerning the relative efficiency and importance of the retention of these various metals, must await greater understanding of their dynamics in a larger range of soil types than has yet been investigated.

Other Particulates

For several forest ecosystems in the United States, the annual deposition of sulphate, expressed as sulphur (SO_4^-S), is between 10 and 20 kg per ha

(Likens *et al.,* 1977; Swank & Douglass, 1977; Shriner & Henderson, 1978). While extremely limited information is available on the biogeochemical cycle of sulphur in forest ecosystems, Shriner & Henderson (1978) have examined sulphur cycling in various eastern Tennessee forest types in a region that they estimated had an annual average deposition of 18.1 kg sulphate per ha. Eighteen per cent of the deposition occurred as dry particulate fallout, while the bulk (82%) was dissolved in rainfall. Of this input, 64% or 11.5 kg per ha was lost from the forest in stream flow. The retention of 6.6 kg per ha (net annual accumulation) in this Tennessee forest greatly exceeds the sulphate retention estimated for New Hampshire forests (1.1 kg per ha; Likens *et al.,* 1977) but approximates the estimate for North Carolina forests (7.6 kg per ha; Swank & Douglass, 1977).

Shriner & Henderson (1978) estimated that 65% of the net annual sulphur accumulation is located in the soil. Unlike trace metal accumulation, these authors conclude that 92% of the sulphur is located in mineral soil horizons, with only 3% contained in the organic forest floor.

FOREST SOILS AS SINKS FOR ATMOSPHERIC GASES

Our understanding of the capacity of various soils to function as sinks for gaseous air contaminants is very limited. This general topic has received only meagre research attention over the last ten years. Obviously, soils have considerable capacity to absorb a variety of gases from the atmosphere, and to incorporate and transform them in or on the soil through a large number of microbial, other biological, physical, and chemical, processes. Specific information is lacking, however, on the relative importance of source *versus* sink function, the capacities and rates of various soils for absorption, residence, and reaction rate times, the influence of soil physical (mineral and organic matter content, structure, and porosity) and chemical (pH, moisture content, and exchange capacity) properties, climate on removal rates, and the significance of soil management practices on removal rates.

Bohn (1972) reviewed the literature on soil sink function and presented some generalizations. He concluded that soils will absorb organic gases more rapidly and in larger quantities with increasing molecular weight and with greater numbers of nitrogen, phosphorus, oxygen, sulphur, and other functional group substitutions in the compound. Absorption of low-molecular weight and less substituted organic gases was judged to be dependent on the development of an appropriate microbial population. Soil removal of inorganic gases was concluded to involve primarily chemical and physical processes. The author observed that the literature regarding soil removal of reducing gases (oxides of carbon, sulphur, and nitrogen, hydrocarbons, and aldehydes) was modest, while the information regarding oxidizing gases (ozone, peroxy compounds, and chlorine) was non-existent.

Carbon Monoxide

Carbon monoxide is formed in all combustion processes as a result of the incomplete oxidation of carbon; as a result, anthropogenic production is locally and regionally enormous. In excess of 6×10^{14} g of carbon monoxide is annually discharged into the world's atmosphere (Seiler, 1974). The primary contributors of combustion carbon monoxide are the United States, Europe, and Japan. As a result, most of the anthropogenic emissions are concentrated in the temperate latitudes of the northern hemisphere. Consequently, a key feature of the global distribution of carbon monoxide is a higher concentration in the Northern than in the Southern Hemisphere.

Atmospheric carbon monoxide contents are at a maximum during winter and spring and at a minimum during the summer. Over the past two decades, winter concentrations have tended to increase while summer levels have remained constant. Despite the geographic and seasonal variations, the available information supports the conclusion that global carbon monoxide concentrations have remained relatively constant in 'clean' atmospheres. Available data, further, support the conclusion that, during the warm season, atmospheric levels of carbon monoxide are regulated by intense natural sink function (Nozhevnikova & Yurganov, 1978).

A large number of potential natural sinks have been proposed for carbon monoxide: (1) absorption by oceans, (2) oxidation to carbon dioxide by OH^- in the troposphere, (3) migration to the stratosphere, followed by photochemical reaction, (4) reaction with animal hamemoprotein, (5) fixation by higher plants, and (6) absorption by soil. In the light of available evidence, only the latter two hypotheses can be said to be of general importance.

Vegetative oxidation of carbon monoxide to carbon dioxide (Ducet & Rosenberg, 1962), and fixation as serine (Chappelle & Krall, 1961), have been described. Employing ^{14}CO, Bidwell & Fraser (1972) observed uptake by leaves of bean plants under both light and dark conditions. Numerous other non-tree species were tested for carbon monoxide uptake at low gas concentration in the light. Using their bean plant data, these authors estimated a summer removal capacity of 12–120 kg per km^2 per day, or globally $3–30 \times 10^8$ tons per yr (six months' growing season).

With the risks of extrapolation from *in vitro* work with greenhouse plants to the global scene left aside, judgements concerning the significance of vegetation as a carbon monoxide sink remain difficult to make. Inman & Ingersoll (1971) failed to observe any capability of several plants, including seedlings of Monterey and Knobcone pines (*Pinus* spp.) and Mimosa (*Mimosa* sp.), to remove carbon monoxide from the atmosphere. Also, the role of plants as producers of carbon monoxide (Nozhevnikova & Yurganov, 1978) must be more accurately understood before the role of vegetation as a sink for carbon monoxide can be judged appropriately. While vegetation may play some role in the maintenance of carbon monoxide in the natural atmosphere, soils are concluded to be the most important removal agent.

The first evidence supporting the importance of soil as a sink for carbon monoxide was presented in 1926, and this study, along with many that have followed, indicate that soil microorganisms are responsible for the removal.

Inman & Ingersoll (1971) conducted preliminary experiments with a greenhouse potting mixture and found that the test soil could deplete carbon monoxide in an experimental atmosphere (containing 120 ppm [13.8×10^4 μg per m^3] CO) to near zero within 3 h. Treatment of the soil with steam sterilization, antibiotics, salt, and anaerobic conditions, all prevented carbon monoxide uptake and indicated the importance of biological processes. A variety of soils from California, Hawaii, and Florida, were brought into the laboratory and tested for their ability to remove carbon monoxide. Inman & Ingersoll generalized from their results that cultivated soils were less active than natural soils, and that higher organic matter and lower pH soils were the most active. These observations clearly support the potential importance of forest soils!

In order to improve the confidence of their observations, Inman and others (Ingersoll *et al.*, 1974) outfitted a mobile laboratory and field tested soils in most major vegetative regions of the United States. Field testing was accomplished by covering a square metre of undisturbed soil and vegetation with a gas-tight chamber. The carbon monoxide uptake rate showed considerable variation in the field, ranging from 7.5 to 109 mg per hr per m^2. As in the previous work, cultivated soils were invariably of less effectiveness than natural soils. Ingersoll *et al.* (1974) concluded that the potential rates of carbon monoxide uptake by the soils of the United States and the world were 505 million and 14.3 billion metric tons per year, respectively. Forest soils are indicated to be of particular significance. If these authors' estimates approximate to the natural condition, it must be concluded that temperate and tropical forests play extraordinarily important roles as sinks for global carbon monoxide.

Seiler (1974) has determined carbon monoxide uptake rates for several European soils (location and vegetation unspecified) and has calculated an average flux-rate of 1.5×10^{-11} g per cm^2 per sec at 15°C. This rate is very approximately an order of magnitude less than the Ingersoll *et al.* average. Seiler explained the discrepancy by indicating that the latter group had employed initial carbon monoxide concentrations of 100 ppm (11.5×10^4 μg per m^3), while his laboratory had employed 0.20 ppm (230 μg per m^3) which was judged to be the normal ambient concentration in unpolluted areas of Europe. Seiler's estimate for global soil removal was 4.5×10^8 tons per year.

The question of inadequate account of carbon monoxide evolution by soils has been raised by K. A. Smith *et al.* (1973). This deficiency could contribute to important overestimations of sink function. Perhaps the actual capacity for carbon monoxide removal lies between the estimates of Inman's and Seiler's groups, or perhaps it approximates one or the other. Whatever the case, it can be concluded that forest soils play a role of very significant importance as a repository for atmospheric carbon monoxide. As the soil removal rate increases with increasing levels of carbon monoxide, forest ecosystems in and around urban and industrial areas may be especially important sinks.

Sulphur Dioxide and Hydrogen Sulphide

The sulphur containing gases that are currently recognized as important components of tropospheric air include sulphur dioxide (SO_2), hydrogen sulphide (H_2S), carbonyl sulphide (COS), carbon disulphide (CS_2), dimethyl sulphide (CH_3SCH_3), and sulphur hexafluoride (SF_6) (Bremner & Steele, 1978). Until very recently it was generally assumed that the primary and most important forms of sulphur in the atmosphere were in hydrogen sulphide, sulphur dioxide, and sulphates. The information concerning soil as a sink for atmospheric sulphur has dealt almost exclusively with the latter two.

As previously discussed in this chapter, sulphate is added to soil largely by precipitation, and becomes part of the soluble sulphur content held by soil colloids. Soils also have a large capacity to absorb sulphur dioxide quickly from the atmosphere. Factors that tend to increase the soil uptake of sulphur dioxide include fine texture, high soil organic matter content, high pH, presence of free $CaCO_3$, high soil moisture content, and the presence of soil microorganisms (Nyborg, 1978).

Since Alway *et al.* (1937) presented the first evidence that soils can absorb sulphur dioxide, a large number of studies have followed. K. A. Smith *et al.* (1973) studied the capability of six soils with variable chemical and physical properties, from Oregon, Iowa, and Saskatchewan, and found that removal by them of sulphur dioxide and hydrogen sulphide from the atmosphere was much more rapid than the removal of carbon monoxide. It is not clear if any of the soils were from forest ecosystems, but the pH (4.8) and organic carbon percentage (9.38) suggest that the Astoria, Oregon, soil may have been. Clearly soil moisture favours uptake of sulphur dioxide—presumably owing to the high solubility of this gas in water. The Astoria soil appears quite average in its ability to remove sulphur dioxide and hydrogen sulphide. Bohn (1972) has suggested that the absorption rate of hydrogen sulphide slows with high soil moisture contents, perhaps due to slow diffusion rates in water filled pores.

The fact that hydrogen sulphide absorption capacity increases with higher soil pH, may reduce the importance of forest ecosystems in removing this gas from the atmosphere. Despite the relatively large number of papers addressing the soil removal of this gas and sulphur dioxide (e.g. Moss, 1975; Bremner & Steele, 1978), estimates of regional and global removal amounts by soil are relatively few (Rasmussen *et al.,* 1974). Available data suggest that deposition velocities (v) for sulphur dioxide are generally in the range of 0.2 to 0.7 cm per sec (Rasmussen *et al.*, 1974), which is less than those for vegetation (*see* Chapter 5). Eriksson (1963) has estimated that the global removal of sulphur dioxide by soil equals 25×10^9 kg of sulphur dioxide-sulphur annually. Abeles *et al.* (1971) examined the capacity of soil collected from Waltham, Massachusetts, to remove sulphur dioxide under laboratory conditions (100 ppm [26.2×10^4 μg per m^3] gas) and extrapolated their results to suggest that United States soils may be capable of removing 4×10^{13} kg of sulphur dioxide (2×10^{13} kg sulphur dioxide-sulphur) per year.

The gas chromatographic studies of Bremner & Banwart (1976) showed that air dry and moist soils had the capacity to sorb dimethyl sulphide, dimethyl disulphide, carbonyl sulphide, and carbon disulphide, but did not sorb sulphur hexafluoride. The first four gases were removed more efficiently by moist soils than dry ones. Soil sterilization indicated that soil microorganisms were partially responsible for the removal of these gases. As the rates of removal were substantially less than those for sulphur dioxide or hydrogen sulphide, however, the Authors concluded that, while soils may constitute a sink for low levels of dimethyl sulphide, dimethyl disulphide, carbonyl sulphide, and carbon disulphide, they would not effectively reduce elevated levels of these gases in areas of high anthropogenic emission.

Nitrogen Oxides and Other Nitrogen-containing Gases

In the atmosphere, nitric oxide is either oxidized to nitrogen dioxide or photolysed to nitrogen gas. Nitrogen dioxide reacts photochemically or is removed by precipitation, primarily in the form of nitric acid. Nitric oxide and nitrogen dioxide may also be removed by soils. Nitric oxide is oxidized to nitrogen dioxide in soil, but the former gas does not persist in acid soils as long as it does in basic soils (Bohn, 1972).

Working with Waltham, Massachusetts, soil, Abeles *et al.* (1971) found that the removal rate for nitrogen dioxide was slower than the removal rate for sulphur dioxide. Twenty-four hours were required to reduce nitrogen dioxide concentrations from 100 to 3 ppm (18.8×10^4 to 56.4×10^2 μg per m^3) in test atmospheres. Extrapolation of these laboratory experiments allowed these authors to suggest that the soils of the United States may be capable of removing 6×10^{11} kg of nitrogen dioxide per year.

Both Ghiorse & Alexander (1976) and E. A. Smith & Mayfield (1978) have documented rapid absorption of nitrogen dioxide by both sterile and non-sterile soil. In the latter study, soil from an uncultivated grassland in Ontario absorbed 99% of the nitrogen dioxide introduced into a test vessel at 25°C in 15 min.

Mechanisms of nitrogen dioxide uptake may involve reaction with soil cations to form $NaNO_2$ or KNO_2, reaction with soil water to form HNO_2 and HNO_3, binding with organic matter, or persistence as a gas in interparticle soil spaces (E. A. Smith & Mayfield, 1978).

As ammonia would probably be present in the atmosphere in the form of $(NH_4)_2SO_4$ rather than NH_3 in all but the most unusual environments, direct soil absorption of gaseous NH_3 is probably not important. Where ambient conditions might expose soils to high ammonia levels, however, evidence suggests that acidic soils are particularly efficient removal agents (Rasmussen *et al.*, 1974).

The specific capabilities that forest soils may have to remove nitrogenous gases must await further experimentation.

Hydrocarbons

Hydrocarbons are generally not soluble in water, and, as a result, soil uptake where it is important, is concluded to be primarily microbial. The light hydrocarbon from motor vehicles that is most actively removed by soil, is ethylene (Zimmerman & Rasmussen, 1975). Abeles *et al.* (1971) observed that Maryland soil samples removed ethylene more slowly than other soils removed sulphur and nitrogen dioxides. Soil removal of ethylene, mediated by various microorganisms, was calculated by Abeles's group to approximate 7×10^9 kg of ethylene annually in the United States.

K. A. Smith *et al.* (1973) determined that the soil flux-rate for acetylene was from 0.24 to 3.12 per 10 mole per g per day. Their test soil with the lowest pH and highest organic matter (forest soil) was the most active of all soils tested for acetylene removal.

Oxidants

There is limited evidence that soils function as a sink for atmospheric ozone (Rasmussen *et al.*, 1974). Aldaz (1969) concluded that the soil and vegetation of the surface of the Earth represent a major sink for this gas, and estimated the capacity of this sink to be within the range of 1.3 to 2.1×10^{12} kg ozone per yr.

Most reports of ozone removal have examined plant uptake or plant and soil uptake combined. Turner *et al.* (1973) have tested the sink capacity of a freshly cultivated sandy loam devoid of vegetation. Their results, which were recorded under field conditions, showed that the flux rate of ozone removal varied from 3 to 12×10^{11} mole per cm^2 per sec, making bare soil in these authors' judgement an important sink for ozone.

Other Gases

Fang (1978) examined the uptake of mercury vapour by five Montana soils by exposing them to a test atmosphere containing 75.9 μg metallic [203]Hg vapour per m^3 for 24 h. The soil with the highest organic matter content showed the highest mercury uptake. While mercury vapour is currently only an extremely localized problem, more than 90% of the mercury contained in coal is vapourized during combustion. Even with relatively low mercury concentrations in coal, widespread or large volume coal combustion may increase the significance of soil retention of mercury vapour.

Summary of Removal of Atmospheric Substances by Soils

Forest soils are important sinks for a variety of air contaminants. Retention of particulate lead by organic materials in the forest floor is a most dramatic example. The flux of lead to temperate forest ecosystems downwind of industrial, urban, or roadside, souces may approximate 200–400 g per ha per yr, with much of this lead accumulating in the forest floor. Certain forest soils may also serve as a sink for additional trace metals, including zinc, cadmium, copper, nickel, manganese, vanadium, and chromium. The efficiency of sink functioning for the latter metals is generally substantially less than for lead, but may be important—particularly in forest systems close to primary sources.

Forest soils remove pollutant gases from the atmosphere *via* several microbial, chemical, and physical, processes. Forest soils function as an especially efficient sink for carbon monoxide, and may play a dominant role in regulating the concentration of this gas in the atmosphere. Other gases that may be significantly removed by forest soils include sulphur dioxide, ammonia, some hydrocarbons, and mercury vapour.

FOREST VEGETATION AS A SINK FOR PARTICULATE CONTAMINANTS

The vegetative compartment of forest ecosystems, as well as the soil compartment, functions as a sink for atmospheric contaminants. As in the case of soils, a complex range of biological, chemical, and physical, processes are involved in the transfer of pollutants from the air to the surfaces of vegetation. For certain contaminants—for example, persistent heavy metal particles —the repository functions of vegetation and soils are intimately linked, as a portion of the heavy metals input to the soil is derived from vegetative sources contributing litter to the forest floor.

Interest in the ability of plants to remove pollutants from the air has grown considerably in recent years, as individuals have become increasingly aware of the amenity functions (W. H. Smith, 1970a; Heisler, 1975) of woody plants—particularly in urban and suburban areas. The capability of plants to act as a sink for air contaminants has been addressed by a variety of recent reviews—for example, Hill (1971), Hanson & Thorne (1972), Warren (1973), Bennett & Hill (1975), Environmental Health Science Center (1975), W. H. Smith & Dochinger (1976), US Environmental Protection

Agency (1976), and Keller (1978). These papers indicate that the surfaces of vegetation provide a major filtration and reaction surface to the atmosphere and function importantly to transfer pollutants from the atmosphere to the living components of The Biosphere.

Much of the understanding of the mechanics of deposition of particles on natural surfaces has been gleaned from studies with particles in the size range 1–50 μm, and is reviewed in the excellent works of Chamberlain (1967, 1970, 1975). Ingold (1971), Gregory (1971, 1973), and Slinn (1976).

The physics and theory of interception and retention of fine particles by vegetation are well beyond the scope of this chapter. It is useful for the interpretation of the evidence to follow, however, to have an introduction to some basic observations.

Particulates are deposited on plant surfaces by three processes: sedimentation under the influence of gravity, impaction under the influence of eddy currents, and deposition under the influence of precipitation. *Sedimentation* usually results in the deposition of particles on the upper surfaces of plant parts and is most important with large particles. Sedimentation velocity varies with particle density, shape, and other factors. *Impaction* occurs when air flows past an obstacle and the airstream divides, but particles in the air tend to continue on a straight path, due to their momentum, until they strike the nearest obstacle. The efficiency of collection *via* impaction increases with decreasing diameter of the collection obstacle and increasing diameter of the particle. Chamberlain (1967) suggested that impaction is the principal means of deposition if (1) particle size is of the order of tens of micrometres or greater, (2) obstacle size is of the order of centimetres or less, (3) approach velocity is of the order of metres per second or more, and (4) the collecting surface is wet, sticky, hairy, or otherwise retentive. Ingold (1971) presented data indicating that leaf petioles are considerably more efficient particulate impactors than either twigs (stems) or leaf lamina. For particles of dimensions 1–5 μm, impaction is not efficient and interception by fine hairs on vegetation is possibly the most efficient retentive mechanism. The efficiency of washout of particles by rain is high for particles approximately 20–30 μm in size. The capturing efficiency of raindrops falls off very sharply for particles of 5 μm or less.

Following *deposition,* particles may be retained on vegetative surfaces, may rebound from such surfaces, or they may be temporarily retained and subsequently removed. If either the particles or the tree surface is wet or sticky, deposited particles are generally retained. Superficial salt accumulation by plants in marine coastal or north temperate roadside environments (where de-icing chemicals are employed) results as vegetation 'acts' to trap salt particles. Fluorine, sulphate, and nitrate, molecules associated with moisture droplets (fog) in the atmosphere may be distributed to vegetative surfaces with great efficiency (Chamberlain, 1975).

The transfer of particles from the atmosphere to natural surfaces is commonly expressed *via* deposition velocity. For small particles—for example, condensation aerosols of less than 1 μm—deposition velocities are much less than for large particles, such as spores and pollen grains 20–40 μm in diameter.

In addition to spores and pollen, particles in the atmosphere larger than 10 μm are frequently the result of mechanical processes—for example, wind erosion, grinding, or spraying. Soil particles, process dust, industrial combustion products, and marine salt particles, are typically between 1 and 10 μm in diameter. Particles in the 0.1 to 1 μm range frequently represent gases that have condensed to form non-volatile products.

The interaction of these variously sized particles with exceedingly diverse vegetative surfaces under conditions of extremely variable microclimate and particle source characteristics, suggests an enormously complex relationship. Since this is indeed the case, field evidence to quantify the amounts of natural or anthropogenic particles removed by trees is very sparse. Numerous investigations have studied detached plant parts, small plants, or seedlings under wind tunnel, growth chamber, or greenhouse, conditions. This is an appropriate and necessary initial step, and these studies have yielded considerable qualitative perspective on the capacity of plants to filter air. The studies reviewed in the following sections were typically not conceived nor conducted specifically to evaluate the role of plants as repositories of atmospheric contaminants. Nevertheless, the hypothesis that trees are important particulate sinks is supported by evidence obtained from studies dealing with diverse particulates including radioactive, trace element, pollen, spore, salt, precipitation, dust, and other unspecified particles.

Radioactive Particles

Because of the considerable interest in the distribution of radioactive materials following the use of nuclear weapons or nuclear accidents, and because of the ease of counting, several investigations have examined the ability of above ground plant parts to intercept radioactive aerosols (Oak Ridge National Laboratory, 1969; Chamberlain, 1970).

Witherspoon & Taylor (1969) treated potted, seedling White Pine (*Pinus strobus*) and Red Oak (*Quercus rubra*) with 88–175-μm-diameter quartz particles tagged with ^{134}Cs under field conditions. Initial particle retention by the Oak foliage was 35% while for White Pine only 24%. After 1 hour, however, the Oak leaves lost 91% of their initial concentrations of particles while the pines lost only 10% of theirs. On the Pine, particles were trapped at the base of the needle-bundles situated around branch termini, while particles on the Oak were retained in small, hairy recesses along leaf veins. Particle half-lives in the 'traps' were calculated at intervals of 0–1 day, 1–7 days, and

7–33 days. For Pine the values were 0.25, 5, and 21, days respectively. For Oak they were 0.12, 1, and 25, days respectively. Particulate loss was primarily attributed to the action of wind and rain.

Trace Metal Particles

Trace metals, especially heavy metals, are most commonly associated with fine particles in contaminated atmospheres. Trace element investigations conducted in roadside, industrial, and urban, environments have dramatically demonstrated the impressive burdens of particulate heavy metals that can accumulate on vegetative surfaces. In the case of lead in roadside ecosystems, for example, the increased lead burden of plants, largely due to surface deposition, may be 5–20, 50–200, and even 100–200 times baseline (non-roadside environment) lead levels for unwashed agricultural crops, grasses, and trees, respectively (W. H. Smith, 1976).

Eastern White Pine (*P. strobus*) is widely planted in the roadside environment in New England, and its capacity to accumulate fine particles (~ 7 μm diameter; Heichel & Hankin, 1972) has been shown to be substantial (W. H. Smith, 1971). Heichel & Hankin (1976) have investigated the distribution of lead deposited on this species in roadside situations and have advanced several important observations. The lead burden of older needles and twigs was consistently greater than that of younger organs, and was greater in samples taken adjacent to rather than far from the road. These are consistent with observations we made with the same species (W. H. Smith, 1971), and are important as the former study indicates that lead accumulates over time on the trees while the latter argues against the importance of soil uptake as a mechanism of lead acquisition. Heichel & Hankin further concluded that twigs retained particles more effectively than needles throughout the season. This was judged to be due to the roughness of twigs relative to needles. The authors observed that a 12-m high White Pine, growing in a dense planting, would have about 15×10^4 cm^2 of woody surface and about 15×10^5 cm^2 of foliage surface. Although White Pine exposed approximately 10-fold more foliage than woody surface, the woody surfaces retained about 20-fold the lead burden of foliage per unit area.

The literature is replete with studies demonstrating significant trace metal particle accumulation on trees in roadside, urban, and industrial situations. Industrial regions—particularly those with metal smelters—may excessively contaminate surrounding woody vegetation with particles containing trace metals. A representative study was that conducted by Little & Martin (1972) in the Avonmouth industrial complex, Severnside, England. Close to this complex, Elm (*Ulmus* sp.) leaves exhibited 8,000, 5,000, and 50 μg per g zinc, lead, and cadmium, respectively.

Data provided, however, rarely permit a quantitative estimate of the sink

Table I Calculated particulate metal sink capacity of the leaves and current twigs of a single, 30-cm (12-inch) trunk-diameter urban Sugar Maple during the course of a growing season.*

Metal contaminant	Growing-season removal (mg per tree)
Lead	5,800
Nickel	820
Chromium	140
Cadmium	60

* Source: W. H. Smith (1974).

capability of the woody plants. The required mass or vegetative surface area calculations are not presented, as the purpose of the studies typically did not include sink function assessment. When quantitative estimates of sink capacity are made, the results can be impressive. During the course of interpretation of some of our urban lead data, we combined Sugar Maple (*Acer saccharum*) dimension analysis information from New Hampshire with contamination data from New Haven, Connecticut in order to speculate on the sink capacity of a single urban Sugar Maple tree (Table I).

Pollen and Spores

Pollen studies have provided important evidence of vegetative interception of large particles. G. S. Raynor of the Brookhaven National Laboratory, Upton, Long Island, NY, has conducted a series of dispersion experiments employing Ragweed (*Ambrosia* sp. or spp.) pollen emitted from sources at various distances and heights upwind of a forest edge. Pollen loss from the plume occurred in two stages and by two mechanisms—impaction near the forest edge, and deposition well within the forest. Pollen lost to the forest was considerably greater than the loss over open terrain (Raynor *et al.,* 1966; Raynor, 1967). Interception of ragweed pollen by a Pennsylvania forest canopy reduced pollen concentration in the forest atmosphere to only 70% of the concentration in a nearby open field (Elder & Hosler, 1954). Neuberger *et al.* (1967) measured ragweed pollen concentrations in and out of forests, and found that 100 m inside a dense coniferous forest over 80% of the pollen had been substracted from the atmosphere. Data indicate that deciduous species are less effective than conifers for filtration of pollen (Neuberger *et al.,* 1967; Steubing & Klee, 1970). For these large ($\sim$20 μm) particles, the dominant transfer to vegetative surfaces is *via* sedimentation and not impaction (Aylor, 1975).

Fungal spore (size-range 1.5–30 μm) interception studies have provided important evidence for understanding particulate capture (Gregory, 1971).

Ingold (1971) has concluded that the most efficient plant parts for spore collection are petioles, twigs, and leaf lamina (in descending order). Efficiency of spore collection increases with decreasing diameter of the collecting cylinder. Observations of basidiomycete spores in Washington Douglas Fir (*Pseudotsuga menziesii*) forests have emphasized the extraordinary importance of microclimate and forest stand structure on the distribution and deposition of these particles in the forest. Wind speed, air temperature, inversions, cloud cover, and forest openings, all influenced particle movement (Fritschen *et al.,* 1970; Edmonds & Driver, 1974).

Salt Particles

Vegetative interception of saline aerosol (primarily NaCl) and nutrient particles has also contributed to our understanding of plant sink function. Particulate deposition of salt particles occurs in roadside environments where deicing salts are employed, in the vicinity of cooling towers, and in maritime regions (Eaton, 1979; Moser, 1979). Conifers planted close to roads receiving deicing salt applications frequently exhibit needle necrosis due to the accumulation of toxic levels of salt transported from the road to the leaves *via* the atmosphere (W. H. Smith, 1970*b*; Hofstra & Hall, 1971; Constantini & Rich, 1973). In coastal ecosystems that are subject to airborne marine salt, accumulation of salt particles by above ground plant parts, injures foliage and twigs (Wells & Shunk, 1938; Oosting & Billings, 1942; Oosting, 1945; Boyce, 1954) and may control species success or failure depending on tolerance to salt loading (W. E. Martin, 1959). Clayton (1972) described the trapping of particulate salts by *Baccharis* brushlands in coastal California, while Woodcock (1953) provided evidence that the shape of plant leaves influences the amount of salt deposited. By employing plates of various shapes, he found that long narrow plates accumulated more salt per unit area than did circular plates. Edwards & Claxton (1964) found over four times the deposition of salt on the windward side of a hedgerow compared with the leeward side.

　Where foliar capture of marine particulates is below the threshold of foliar injury, particle accumulation may be an important mechanism for nutrient acquisition (Art, 1971; Art *et al.,* 1974). Numerous investigations, reviewed by White & Turner (1970), have indicated that non-maritime trees also catch airborne nutrient particles. These authors found that a mixed deciduous forest was capable of removing annually from the atmosphere 125 kg per ha of sodium, 6 kg per ha of potassium, 4 kg per ha of calcium, 16 kg per ha of magnesium, and 0.1 kg per ha of phosphorus. The degree of leaf hairiness was inversely correlated with particle retention. Apparently the small droplets employed had insufficient inertia to penetrate the stable boundary layer created by the hairy leaves. Small diameter branches were more efficient particle collectors than large diameter branches in all species examined.

In their examination of the impact of saline aerosols of cooling tower origin, McCune *et al.* (1977) emphasized the importance of particle wetness in causing damage to surrounding trees. Dry particles appeared less toxic than hydrated particles. This supports the contention that moist particles are more effectively retained by vegetative surfaces than dry ones.

Precipitation, Dust, and Other Particles

Foliar interception of precipitation has been intensively investigated (Zinke, 1967), but the relatively large size of the particles (range 50–700 μm) makes these data of limited application for considerations of fine particle retention The enormous importance of rainout in transferring fine particles from the atmosphere to vegetation is recognized. Several precipitation studies support the general observation that conifers intercept more particles than deciduous species; for example, Helvey (1971), who reported loss by canopy-interception to be greatest in a spruce–fir–hemlock forest type, intermediate in pine, and least in broad-leafed deciduous forests.

Numerous additional studies employing dust—synthetic or un-specified—particles have contributed to our understanding of particulate capture by vegetation. Rosinki & Nagamoto (1965) investigated the deposition of 2 μm particles on Rocky Mountain Juniper (*Juniperus scopulorum*) and Douglas Fir. At low dosages, particles preferentially accumulated on the windward leaf edge. Eventually a new layer was formed on the previously deposited layer, where thickness increased until an equilibrium was reached. Total deposition was increased when wind exposed different leaf surfaces for deposition. Langer (1965) concluded that dust deposition on coniferous leaves was not significantly influenced by electrostatic effects. Podgorow (1967) investigated the relative effectiveness of pine, birch, and aspen in filtering dust particulates. Pine proved most effective, and its interior crown needles accumulated more and retained more dust than exterior needles. Bach (1972) also presented evidence supporting the superior collecting capacity of pines relative to deciduous species. In an Ohio study, Dochinger (1972) examined dustfall and suspended particulate matter in three areas—treeless, deciduous canopy, and conifer canopy—and concluded that trees have the capacity to reduce particulate pollutants in the ambient atmosphere.

Wedding *et al.* (1975) found, under controlled wind tunnel conditions, that particulate retention by rough, pubescent Sunflower (*Helianthus annuus*) leaves was 10 times as great as it was by smooth, waxy Tulip-poplar (*Liriodendron tulipifera*) leaves. In a unique study, Graustein (1978) employed the ratio of strontium isotopes in soil dust to determine strontium input to forested watersheds in New Mexico. Most of the atmospherically transported strontium entered the watershed by impaction of soluble particles on spruce foliage. Quaking Aspen (*Populus tremuloides*), also present in the ecosystem, was judged to trap little if any, dust. The flux of dust-derived strontium to the forest floor was four times as great as the flux to an unforested area.

In an extremely informative set of experiments, Little (1977) exposed freshly collected leaves of several tree species in a wind tunnel to various sizes of polystyrene aerosols labelled with technetium. Particles sized 2.75, 5.0, and 8.5, μm were tested with leaves from European Beech (*Fagus sylvatica*), White Poplar (*Populus alba*), and Stinging Nettle (*Urtica dioica*). Surface texture was critical in determining capture efficiency, the rough and hairy leaves of the Nettle proving more effective than the densely tomentose leaves of the poplar or the smooth surfaces of those of the Beech. For each species there was a strong negative linear correlation between leaf area and deposition velocity, the latter being smallest for the largest leaves. Deposition was heaviest at the leaf tip and along leaf margins, where a turbulent boundary layer was present. Leaves with complex shapes and the largest circumference-to-area ratio were the most efficient collectors. Increased wind speed and particle size were both reflected in increased deposition velocities. Deposition velocities to petioles and stems were many times greater than deposition velocities to leaf aminas, even though the majority of particles of the total catch were intercepted by the leaf lamina. The non laminar catch was significant, however, and little (1977) suggested that this may cause deposition of atmospheric particles on trees to be relatively high even during winter when deciduous species are devoid of leaves.

FOREST VEGETATION AS A SINK FOR GASEOUS CONTAMINANTS

Substantial evidence is available to support the potential that plants in general Hill, 1971; Bennett & Hill, 1975; Rasmussen *et al.*, 1975), and trees in particular (Roberts, 1974; Warren, 1973; W. H. Smith & Dochinger, 1975, 1976; W. H. Smith, 1979), function as sinks for gaseous pollutants. The latter are transferred from the atmosphere to vegetation by the combined forces of diffusion and flowing air movement. When once in contact with plants, gases may be bound or dissolved on exterior surfaces or taken up by the plants *via* stomata. If the surface of the plant is wet, and if the gas is water soluble, the former process can be very important. When the plant is dry, or in the case of gases with relatively low water solubilities, the latter mechanism is assumed to be the more important.

Stomatal Uptake

Stomatal pores are small openings, typically approximately 10 μm in length and 2 and 7 μm in width when open, in the epidermal surface of leaves through which plants naturally exchange carbon dioxide, oxygen, and water vapour with the atmosphere. The waxy cuticle of leaf surfaces restricts diffusion, so that essentially all gas exchange carried out by leaves is *via* stomatal openings. Even though these openings make up only approximately 1% of the leaf's surface area, their orientation and mechanics prove to be nearly optimal for maximum gas diffusion in and out of the leaf (Salisbury & Ross, 1978). Stomata undergo diurnal opening and closing, with the pores of most plants opened within an hour of sunrise and closed by dark. The timing and degree of opening of stomatal apertures, and hence gas diffusion to and from leaves, is strongly influenced by a number of complex environmental factors.

During daylight periods, when plants' leaves are releasing water vapour and taking in carbon dioxide, other gases, including trace pollutant gases in the vicinity of the leaf, will also be taken in through the stomata. When once they are inside the leaf, these gases will diffuse into intercellular spaces and be absorbed on or in the surfaces of palisade or spongy parenchyma cell walls.

The rate of pollutant gas transfer from the atmosphere to interior leaf cells is regulated by a series of resistances. Factors controlling atmospheric resistance include wind speed, leaf size and geometry, and gas viscosity and diffusivity. Stomatal resistance is regulated by stomatal aperature, which is influenced by water deficit, carbon dioxide concentration, and light intensity. Mesophyllic resistance is regulated by gas solubility in water, gas–liquid diffusion, and leaf metabolism (Kabel *et al.*, 1976). Because the rate of pollutant uptake is regulated by numerous forces and conditions, the rate of removal under field conditions is highly variable. If leaf characteristics, wind speed, atmospheric moisture, temperature, and light intensity, are quantified, however, the pollutant uptake rate can be estimated (Bennett & Hill, 1973; Kabel *et al.*, 1976).

General Plant Uptake

The fundamental investigations of A. Clyde Hill and Jesse H. Bennett, of the University of Utah, allow several general conclusions to be made concerning gaseous pollutant uptake (Hill, 1971; Bennett & Hill, 1973, 1975). Their studies have concentrated on Alfalfa (*Medicago sativa*), Oats (*Avena sativa*), Barley (*Hordeum vulgare*), and other grasses. Standard Alfalfa canopies removed gaseous pollutants from the atmosphere at rates in the following order of descent: hydrogen fluoride > sulphur dioxide > chlorine > nitrogen dioxide > ozone > peroxyacetylnitrate > nitric oxide > carbon monoxide. In general, plant uptake rates increased as the solubility of the pollutant in water increased. Hydrogen fluoride, sulphur dioxide, nitrogen dioxide, and ozone, which are soluble and reactive, were readily absorbed. Nitrogen dioxide and carbon monoxide, which are very insoluble, were absorbed relatively slowly or not at all. The rate of pollutant removal was found to increase linearly as the concentration of the pollutant was increased over the ranges of concentration that are encountered in ambient air and that were low enough not to cause stomatal closure.

Under growth chamber conditions, wind velocity, canopy height, and light intensity, were shown to affect the rate of pollutant removal by vegetation. As previously stressed, light plays a critical role in determining physiological activities of the leaf and stomatal opening and, as such, exerts great influence on foliar removal of pollutants. Under conditions of adequate soil moisture, however, pollutant uptake by vegetation was judged almost constant throughout the day, as the stomata were fully open. Pollutants were absorbed most efficiently by plant foliage near the canopy surface where light-mediated metabolic and pollutant diffusivity rates were greatest. Sulphur and nitrogen dioxides were taken up by respiring leaves in the dark, but uptake rates were greatly reduced relative to rates in the light.

Tree Uptake

Sulphur dioxide: Because of its high solubility in water, large amounts of sulphur dioxide are absorbed on external tree surfaces when they are wet. In the dry condition, sulphur dioxide is readily absorbed by trees' leaves and rapidly oxidized to sulphate in their mesophyll cells. At low uptake rates, sulphur dioxide is presumed to be oxidized about as rapidly as it is absorbed (Bennett & Hill, 1975).

Roberts (1974) measured sulphur dioxide sorption by single leaves or shoots of several one-year-old seedlings of numerous woody species. All species examined were capable of reducing high ambient levels within his test chambers. Because of the large dose employed, 1 ppm (2,620 μg per m^3) for 1 hour, Roberts reduced the concentration in subsequent trials and examined uptake at concentrations of 0.2 and 0.5 ppm (524 and 1,310 μg per m^3). At the lower concentration, uptake by birch (*Betula* sp.) and Firethorn (*Cotoneaster pyracantha* Spach.) was significantly reduced. It was speculated that higher concentrations of sulphur dioxide may maintain stomatal opening. Under controlled environmental conditions, comparable to those employed by Roberts, Jensen (1975) fumigated hybrid poplar (*Populus*) cuttings with sulphur dioxide ranging in concentration from 0.1 to 5 ppm (262 to 13.1 $\times$ 10 μg per m^3) for periods of 5 to 80 hr. Uptake was determined by measuring total sulphur content of the leaves. At low levels of fumigation (0.1 and 0.25 ppm [262 and 655 μm per m^3]) leaf sulphur initially increased but then declined to unfumigated levels as fumigation continued. This reduction was judged by the author to be due to one or more of the following: reduction in absorption rates, translocation of sulphur out of the leaves, leaching of sulphate from the roots, or release of hydrogen sulphide by the leaves.

Jensen & Kozlowski (1975) have provided additional perspectives on trees seedling uptake of sulphur dioxide in their experiments that exposed one-year-old Sugar Maple (*Acer saccharum*), Bigtooth Aspen (*Populus gran-didentata*), White Ash (*Fraximus americana*), and Yellow Birch (*Betula lutea*), to 2.75 ppm (7,205 μg per m³) for 2 hours. Prefumigation with 0.75 ppm (1,965 μg per m³) sulphur dioxide for 20 hours or more reduced the rate of absorption in all species except White Ash. The authors speculated that tolerance to sulphur dioxide injury following uptake may be related to the rate at which accumulated sulphur can be moved out of the leaves. Roberts & Krause (1976) monitored sulphur dioxide uptake of intact plants of rhododendron (*Rhododendron* sp.) (three-months-old) and Firethorn (12-months-old), and suggested that the great uptake of the latter may have been partially due to its abundant trichomes (leaf hairs).

These various controlled environment studies are important as they qualify and caution our efforts to extrapolate the sink function to trees in natural environments. Uptake under ambient conditions may be less than under experimental conditions, as the latter frequently employ unnaturally high concentrations of sulphur dioxide. Prefumigation, common in natural situations, may further reduce natural uptake. Uptake rates in the field may decline over time as pollution episodes continue. Even though maintained by several investigators, the seedling studies have not addressed two very important questions concerning uptake, namely the relative importance of stomatal uptake *versus* adsorption to the surface of dry plant parts, and the relative uptake of dry plants *versus* plants with moisture films on their surfaces. Garland & Branson (1977), however, have recently provided evidence supporting the importance of stomata and wet surfaces in uptake. These investigators determined the rates of water vapour and sulphur dioxide conductance in detached Scots Pine (*Pinus sylvestris*) needles collected from a 45-years-old stand and in intact trees in a 10-years-old plantation. The similarity of the conductances observed for these two cases led the authors to conclude that they were both controlled by diffusion through the stomata. By analogy with transpiration, the authors further estimated that the deposition velocity of sulphur dioxide to a dry pine canopy would vary from 0.2 to 0.6 cm per sec during daytime (0.05–0.1 cm per sec at night) but might be 10 times this rate if the canopy was wet from precipitation or dew.

This deposition velocity appears slightly conservative when compared with the dry deposition rates provided by other investigators (Sheih, 1977; Sheih *et al.*, 1979). A. E. Martin & Barber (1971) monitored the sulphur dioxide loss in the immediate vicinity of a large hawthorn (*Crataegus* sp.) hedge (approximately 4 m high × 3 m wide) that was subject to effluent from an electricity generating facility. With ambient concentrations generally less than 10 pphm (262 μg per m³), the authors observed significant loss of sulphur dioxide near ($\approx$ 150 mm) the foliage. The greatest loss was during periods of rain or dew when the hedge foliage was wet.

Fluoride: Plants take up fluoride more effectively than any other pollutant (Bennett & Hill, 1975). Hydrogen fluoride can be taken up effectively by both the leaf mesophyll and exposed plant surfaces. Several studies have shown that more than 50% of the fluoride taken up was deposited externally on the foliage. This is influenced in part by the same principles described previously. The great absorption into the leaf is due in large measure to the solubility and reactivity of the fluorides with water. Plant leaves are extremely effective in concentrating as well as accumulating fluorides from the atmosphere (Hill, 1971). Leaves exposed to HF can accumulate as much as a million times the concentration to which they are exposed. Much of the fluoride moves in the transpiration stream and becomes concentrated towards the leaf margins or tips, where evaporation and transpiration is greatest—hence the deposition 'externally on the foliage'. Precipitation, when sufficiently intense, not only washes much of the fluorides from the leaf surfaces, but can also leach some within the leaf.

Oxidants: Ozone is relatively insoluble in water (0.052 g per 100 g H_2O at 20°C) but readily diffuses into stomatal cavities (Wood & Davis, 1969; Rich *et al.*, 1970; Rich & Turner, 1972; Thorne & Hanson, 1972). The very reactive nature of this gas undoubtedly causes it to react rapidly on the surface of leaf mesophyll cells.

Under controlled environmental conditions, Townsend (1974) has monitored ozone uptake by a variety of seedlings of tree species. Ozone sorption exhibited a linear increase up to 0.5 ppm (980 μg per m^3) for both White Birch (*Betula populifolia*) and Red Maple (*Acer rubrum*). These two species were also capable of reducing ambient ozone throughout a prolonged 8-hours' exposure.

While tree removal of atmospheric peroxyacetylnitrate has not been reported, Garland & Penkett (1976) have suggested that the deposition velocity of this gas to grass was approximately 0.25 cm per sec, which is lower than the value for ozone (0.8 cm per sec) or sulphur dioxide (1 cm per sec).

Nitrogen dioxide dissolved in water yields nitrite and nitrate ions in solution. The latter can be reduced to ammonia in leaf cells (Bennett & Hill, 1975). Rogers *et al.* (1979) have provided nitrogen uptake rates for Loblolly Pine (*Pinus taeda*) and White Oak (*Quercus alba*).

Models of Forest Gas-sink Function

Efforts to estimate the sink capability of forest vegetation under natural conditions must consider a complex set of variables—including pollutant concentration and deposition velocity, meteorological parameters, and dimensions (leaf or canopy area, dry weight, etc.) and conditions of the trees. The systematic approach of model development is desirable and may even be necessary, despite the obvious deficiencies in field data.

Murphy *et al.* (1977) have modelled the sulphur dioxide uptake of a simulated Loblolly Pine forest exposed to 50 ppb (131 μg per m^3) sulphur dioxide on two clear days in January and June, using climate data from a station near Aiken, South Carolina. The simulated uptake compared favourably, but was smaller than the seedling uptake rates reported by Roberts (1974). Murphy *et al.* (1977) also applied their model to regions where forest vegetation was dominant and where actual frequency-distributions of sulphur dioxide concentrations were known. At a site of the Savannah River Laboratory having an average sulphur dioxide concentration of 8 ppb (21 μg per m^3) during the spring, the model predicted an uptake of 11 metric tons per day over the 778 km^2 area of the southern pine forest site. For Long Island, New York, over an area of 1,723 km^2 in June, the model predicted a sulphur dioxide uptake of 103 metric tons per day (SO$_2$ 32 ppb, or 84 μg per m^3) for a west wind condition. According to the authors, the New York estimate was larger owing to the larger land area, higher level of ambient sulphur dioxide, and greater leaf area employed.

Kabel *et al.* (1976) have argued, and appropriately so, that the uptake rate of sulphur dioxide on a leaf area basis must be extrapolated to a ground area basis in order to predict large area pollutant removal. In addition, deposition velocities, or mass-transfer coefficients as these authors prefer (compare Kabel, 1976), must be given for uptake by stems and branches as well as leaves. Fortunately, Whittaker & Woodwell (1967) have provided generalized area ratios for temperate forest communities. Kabel *et al.* (1976) calculated the following deposition velocities, using the appropriate ratios: 0.015 m per sec for dry conditions (stomatal *plus* soil) and 0.21 m per sec for damp canopy conditions, yielding a total deposition velocity of 0.23 m per sec for a moist forest canopy. Uptake rates were calculated for a model forest (dry condition) downwind from a sulphur dioxide source.

The radioactive measurements of Garland (1977) propose that the deposition velocity for a dry forest canopy varies from 0.001 to 0.006 m per sec. The Kabel *et al.* (1976) figure is presumably higher than this owing to the inclusion of soil uptake in the latter.

Unfortunately, models describing sink capabilities of forests for other gaseous contaminants are not readily available. Waggoner (1971, 1975) has made some preliminary observations with ozone, but no formal model has been proposed.

SUMMARY

Our examination of the literature permits us to conclude that abundant field data are available to support the suggestion that plant surfaces accumulate a variety of natural and anthropogenic particles from the atmosphere and that controlled environment and wind-tunnel studies allow the following generalizations:

1. The interception and retention of atmospheric particles by plants is highly variable and primarily dependent on:
   (a) size, shape, wetness, and surface texture, of the particles;
   (b) size, shape, wetness, and surface texture, of the intercepting plant-part; and
   (c) micro- and ultramicroclimatic conditions surrounding the plant.
2. More is known concerning the physical–mechanical aspects of particle deposition under controlled conditions than is known about the relative capture and retention efficiencies of various plants and parts of plants under natural conditions.
3. Generally, greater leaf surface roughness increases particle capture efficiency for particles approximately 5 μm (and less) in diameter. Smooth-leaved species (for example, Horse Chestnut [*Aesculus hippocastanum*] and Yellow Poplar [*Liriodendron tulipifera*]) are less efficient than rough-leaved species (for example, elm [*Ulmus* spp.] and hazel [*Corylus* spp.]).
4. Surface roughness acts to decrease the stability of the boundary layer (region of retarded air flow) surrounding the leaf, and thus acts to increase particle impaction. Leaf hairs and leaf veins are principal contributors to surface roughness.
5. Smaller leaves are generally more efficient particle collectors than larger leaves.
6. Particle deposition (but probably not retention) is heaviest at the leaf tip and along leaf-margins, where a turbulent boundary layer is present. Leaves with complex shape and large circumference-to-area ratio collect particles most efficiently.
7. Increased wind speeds and increased particle size typically increase particulate deposition velocities.
8. Deposition velocities to petioles and stems are generally many times greater than deposition velocities to leaf laminas. Collection of atmospheric particles by leafless deciduous species in winter may remain quite high due to twig- and shoot-impaction.
9. Conifers are generally more effective particulate sinks than deciduous species.
10. Mechanisms by which particles are resuspended or otherwise removed from tree surfaces need to be investigated more thoroughly.

Unfortunately, only a modest portion of all experiments conducted with artificially generated particles has been conducted in natural environments, and very few models have been developed to estimate the removal capacity of groups of trees. Hosker (1973) applied a standard plume diffusion model to hypothetical sources located at several heights above homogeneous stretches of grassland and forest. He concluded that the amount of effluent to be physically deposited on the foliage would be significantly larger for the forest than for the field. Slinn (1975) has provided the best analysis of the problems associated with developing a model for dry deposition in a plant canopy. He stresses the importance of considering 'resuspension' of particles from vegetation in model design. Bache (1979) stresses the importance of considering all trapping mechanisms, including sedimentation and impaction, when designing particle trapping models.

It is tempting to conclude that trees may be especially efficient filters of airborne particles because of their large size, high surface-to-volume ratio of foliage, petioles, and twigs, and frequently hairy or rough leaf, twig, or bark, surfaces. Because the interior portions of forest stands act to still the air, mean wind speeds are reduced and particle sedimentation will be augmented. We are unable to quantify the filtration capacity of forests at the present time, however, because of deficient field data. Many of the data on particle loading of trees are expressed in a μg per g (ppm) dry weight basis. Little (1977) has very appropriately indicated that judgements regarding particle collection efficiencies, which are expressed as a function of plant dry weight, must be used with caution and consider differences in area:dry weight ratios if they are to be meaningful. Furthermore, unless accurate dimension-analysis data (leaf and twig dry weight, leaf and trunk surface area) are available along with ambient microclimatological information, tree or stand loading can only be speculated. Much of the information on deposition velocities available for particle transfer to natural surfaces was not accumulated from experiments designed to assess sink capability, and did not employ particles small enough ($< 1 \mu$m) to be particularly relevant to individuals primarily interested in air quality.

We conclude that there is also substantial evidence that trees remove gaseous contaminants from the atmosphere. Experiments, again largely performed under controlled environmental circumstances and with seedling or young plants, allow the following generalizations:

1. Plant uptake rates increase as the solubility of the pollutant in water increases. Hydrogen fluoride, sulphur dioxide, nitrogen dioxide, and ozone, which are soluble and reactive, are readily sorbed pollutants. Nitric oxide and carbon monoxide, which are very insoluble, are absorbed relatively slowly or not at all by vegetation.
2. When vegetative surfaces are wet (or at least damp), the pollutant removal rate may increase up to 10-fold. Under damp conditions, the entire above-ground plant surface—leaves, twigs, branches, stems—is available for uptake.
3. Light plays a critical role in determining physiological activities of the leaf and stomatal opening, and as such exerts great influence on foliar removal of pollutants. Under conditions of adequate soil moisture, pollutant

uptake by vegetation is almost constant throughout the day, as the stomata are fully open. Moisture stress sufficient to limit stomatal opening, and relatively common in various urban environments, would severely restrict gaseous pollutant uptake.

4. Pollutants are absorbed most efficiently by plant foliage near the canopy surface, where light-mediated metabolic and pollutant diffusivity rates are greatest.

5. Sulphur and nitrogen dioxides are taken up by respiring leaves in the dark, but uptake rates are greatly reduced relative to rates in the light.

REFERENCES

Abeles, F. B., Craker, L. E., Forrence L. E. & Leather G. R. (1971). Fate of air pollutants: Removal of ethylene, sulphur dioxide, and nitrogen dioxide, by soil. *Science,* **173**, pp. 914–6.

Aldaz, L. (1969). Flux measurements of atmospheric ozone over land and water. *J. Geophys. Res.,* **74**, pp. 943–6.

Alway, F. J., Marsh, A. W. & Methley, W. J. (1937). Sufficiency of atmospheric sulphur for maximum crop yields. *Soil Sci. Soc. Am. Proc.,* **2**, pp. 229–38.

Art, H. W. (1971). Atmospheric salts in the functioning of a maritime forest ecosystem. Ph.D. thesis, School of Forestry and Environmental Studies, Yale University, New Haven, Connecticut, USA: 135 pp. (typescript).

Art, H. W., Bormann, F. H., Voigt, G. K. & Woodwell, G. M. (1974). Barrier island forest ecosystem: Role of meteorologic nutrient inputs. *Science,* **184**, pp. 60–62.

Aylor, D. E. (1975). Deposition of particles of ragweed pollen in a plant canopy. *J. Appl. Meteorol.,* **14**, pp. 52–7.

Bach, W. (1972). *Atmospheric Pollution.* McGraw-Hill, New York, NY, USA: 144 pp.

Bache, D.H. (1979). Particle transport within plant canopies—I: A framework for analysis. *Atmos. Environ.,* **13**, pp. 1257–62.

Bennett, J. H. & Hill, A. C. (1973). Absorption of gaseous air pollutants by a standardized plant canopy. *J. Air Pollut. Control Assoc.,* **23**, pp. 203–6.

Bennett, J. H. & Hill, A. C. (1975). Interactions of air pollutants with canopies of vegetation. Pp. 273–306 in *Responses of Plants to Air Pollution* (Eds J. B. Mudd & T. T. Kozlowski), Academic Press, New York–San Francisco–London: xii + 388 pp., illustr.

Benninger, L. K., Lewis, D. M. & Turekian, K. K. (1975). The use of natural Pb-210 as a heavy metal tracer in the river-estuarine system. Pp. 201–10 in *Marine Chemistry and Coastal Environment* (Ed. T. M. Church). American Chemical Society Symposium Series No. 18, American Chemical Society, Washington, DC, USA.

Bidwell, R. G. S. & Fraser, D. E. (1972). Carbon monoxide uptake and metabolism by leaves. *Can. J. Bot.,* **50**, pp. 1435–9.

Bohn, H. L. (1972). Soil absorption of air pollutants. *J. Environ. Qual.,* **1**, pp. 372–7.

Boyce, S. G. (1954). The salt spray community. *Ecol. Monogr.,* **24**, pp. 29–67.

Bremner, J. M. & Banwart, W. L. (1976). Sorption of sulphur gases by soils. *Soil Biol. Biochem.,* **8**, pp. 79–83.

Bremner, J. M. & Steele, C. G. (1978). Role of microorganisms in the atmospheric sulphur cycle. *Adv. Microb. Ecol.,* **2**, pp. 155–201.

Chamberlain, A. C. (1967). Deposition of particles to natural surfaces. Pp. 138–64 in

Airborne Microbes (Eds P. H. Gregory & J. L. Monteith). 17th Symp. Soc. Gen. Microbiol. Cambridge University Press, London.

Chamberlain, A. C. (1970). Interception and retention of radioactive aerosols by vegetation. *Atmos. Environ.,* **4**, pp. 57–78.

Chamberlain, A. C. (1975). The movement of particles in plant communities. Pp. 155–203 in *Vegetation and the Atmosphere,* Vol. I (Ed. J. L. Monteith). Academic Press, New York, NY, USA: xviii + 278 pp., illustr.

Chappelle, E. W. & Krall, A. R. (1961). Carbon monoxide fixation by cell-free extracts of green plants. *Biochem. Biophys. Acta,* **49**, pp. 578–80.

Clayton, J. L. (1972). Salt spray and mineral cycling in two California coastal ecosystems. *Ecology,* **53**, pp. 74–81.

Constantini, A. & Rich, A. E. (1973). Comparison of salt injury to four species of coniferous tree seedlings when salt was applied to the potting medium and to the needles with or without an anti-transpirant. *Phytopathology,* **63**, p. 200.

Dochinger, L. S. (1972). Can trees cleanse the air of particulate pollutants? *Int. Shade Tree Conf. Proc.,* **48**, 45–48.

Ducet, G. & Rosenberg, A. I. (1962). Leaf respiration. *Ann. Rev. Plant Physiol.,* **13**, pp. 71–200.

Eaton, T. E. (1979). Natural and artificially altered patterns of salt spray across a forested barrier island. *Atmos. Environ.,* **13**, pp. 705–9.

Edmonds, R. L. & Driver, C. H. (1974). Dispersion and deposition of spores of *Fomes annosus* and fluorescent particles. *Phytopathology,* **64**, pp. 1313–21.

Edwards, R. S. & Claxton, S. M. (1964). The distribution of airborne salt of marine origin in the Aberystwyth area. *J. Appl. Ecol.,* **1**, pp. 253–63.

Elder, F. & Hosler, C. (1954). *Ragweed Pollen in the Atmosphere.* Report, Department of Meteorology, Pennsylvania State University, University Park, Pennsylvania USA:

Environmental Health Science Center (1975). *The Role of Plants in Environmental Purification.* Environmental Health Science Center, Oregon State University Corvallis, Oregon, USA: 34 pp. (mimeogr.).

Eriksson, E. (1963). The yearly circulation of sulphur in Nature. *J. Geophys. Res.,* **68**, pp. 4001–8.

Fang, S. C. (1978). Sorption and transformation of mercury vapor by dry soil *Environ. Sci. Technol.,* **12**, pp. 285–8.

Fritschen, L. J., Driver, C. H., Avery, C., Buffo, J., Edmonds, R., Kinerson, R. & Schiess, P. (1970). *Dispersion of Air Traces into and within a Forested Area (3).* Report No. OSD01366, College of Forest Resources, Washington University, Seattle, Washington, USA: 53 pp. (Mimeogr.).

Garland, J. A. (1977). The dry deposition of sulphur dioxide to land and water surfaces. *Proc. Royal Soc. London,* **354**, pp. 245–68.

Garland, J. A. & Branson, R. (1977). The deposition of sulphur dioxide to pine forest assessed by a radioactive tracer method. *Tellus,* **29**, pp. 445–54.

Garland, J. A. & Penkett, S. A. (1976). Absorption of peroxyacetylnitrate and ozone by natural surfaces. *Atmos. Environ.,* **10**, pp. 1127–31.

Ghiorse, W. C. & Alexander, M. (1976). Effect of microorganisms on the sorption and fate of sulphur dioxide and nitrogen dioxide in soil. *J. Environ. Qual.,* **5**, pp. 227–30.

Graustein, W. D. (1978). Measurement of dust input to a forested watershed using $^{87}Sr/^{86}Sr$ ratios. *Geol. Soc. Am. Abst.,* **10**, p. 411.

Gregory, P. H. (1971). The leaf as a spore trap. Pp. 239–43 in *Ecology of Leaf Surface Microorganisms* (Ed. T. F. Preece & C. H. Dickinson). Academic Press, New York, NY, USA: xvii + 640 pp., illustr.

Gregory, P. H. (1973). *The Microbiology of the Atmosphere*, 2nd Edn. (Plant Science Monographs Gen. Ed. Nicholas Polunin.) Leonard Hill–intertext, 24 Market Place, Aylesbury, England, UK: pp. xxi + 377 pp. illustr.

Hanson, G. P. & Thorne, L. (1972). Vegetation to reduce air pollution. *Lasca Leaves*, **20**, pp. 60–5.

Heichel, G. H. & Hankin, L. (1972). Particles containing lead, chlorine and bromine detected on trees with an electron microprobe. *Environ. Sci. Technol.*, **6**, pp. 1121–2.

Heichel, G. H. & Hankin, L. (1976). Roadside coniferous windbreaks as sinks for vehicular lead emissions. *J. Air Pollut. Control Assoc.*, **26**, pp. 767–70.

Heinrichs, H. & Mayer, R. (1977). Distribution and cycling of major and trace elements in two central European forest ecosystems. *J. Environ. Qual.*, **6**, pp. 402–7.

Heisler, G. M. (1975). How trees modify metropolitan climate and noise. Pp. 103–12 in *Forestry Issues in Urban America*. (Proc. 1974 National Convention Society of American Foresters, New York.)

Helvey, J. D. (1971). A summary of rainfall interception by certain conifers of North America. Pp. 103–13 in *Proc. Biological Effects in the Hydrological Cycle*. USDA Forest Service, Washington, DC, USA.

Hill, A. C. (1971). Vegetation: A sink for atmospheric pollutants. *J. Air Pollut. Control Assoc.*, **21**, pp. 341–6.

Hofstra, G. & Hall, R. (1971). Injury on roadside trees: Leaf injury on pine and White Cedar in relation to foliar levels of sodium chloride. *Can. J. Bot.*, **49**, pp. 613–22.

Hook, R., I. Van & Shults, W. D. (1977). *Effects of Trace Contaminants from Coal Combustion*. (Proc. Workshop, 2–6 August, 1976, Knoxville, Tennessee, US ERDA Publication No. 77–64, U.S. Energy Research and Development Administration, Washington, DC, USA: 79 pp.

Hook, R. I. Van, Harris, W. F. & Henderson G. S. (1977). Cadmium, lead and zinc distributions and cycling in a mixed deciduous forest. *Ambio*, **6**, pp. 281–6.

Hosker, R. P., jr (1973). Estimates of dry deposition and plume depletion over forests and grassland. Pp. 250–65 in *Proc. IAEA Symposium on the Physical Behaviour of Radioactive Contaminants in the Atmosphere*. (International Atomic Energy Authority, Vienna, Austria, 12–16, November 1973.)

Ingersoll, R. B., Inman, R. E. & Fisher, W. R. (1974). Soils potential as a sink for atmospheric carbon monoxide. *Tellus*, **26**, pp. 151–8.

Ingold, C. T. (1971). *Fungal Spores*. Clarendon Press, Oxford, England, UK: 302 pp., illustr.

Inman, R. E. & Ingersoll, R. B. (1971). Uptake of carbon monoxide by soil Fungi. *J. Air Pollut. Control Assoc.*, **21**, pp. 646–57.

Jensen, K. F. (1975). Sulfur content of hybrid poplar cuttings fumigated with sulfur dioxide. *USDA Forest Service, Res. Note No NE-209*, Upper Darby, Pennsylvania, USA: 4 pp.

Jensen, K. F. & Kozlowski, T. T. (1975). Absorption and translocation of sulphur dioxide by seedlings of four forest tree species. *J. Environ. Qual.*, **4**, pp. 379–82.

John, M. K., Chuah, H. H. & Vandaerhoven, C. J. (1972). Cadmium and its uptake by oats. *Environ. Sci. Technol.*, **6**, pp. 555–7.

Kabel, R. L. (1976). Natural removal of gaseous pollutants. Pp. 26–36 in *3rd Symposium on Atmospheric Turbulance, Diffusion and Air Quality*. (American Meteorological Society, 19–22 October, 1976, Raleigh, North Carolina.)

Kabel, R. L., O'Dell, R. A., Taheri, M. & Davis, D. D. (1976). *A Preliminary Model of Gaseous Pollutant Uptake by Vegetation*. Center for Air Environment Studies, Publ.

No. 455–76, Pennsylvania State University, University Park, Pennsylvania, USA: 96 pp.

Keller, T. (1978). *How Effective are Forests in Improving Air Quality?* Eighth World Forestry Conference, Jakarta, Indonesia, 16–28 October 1968, 9 pp.

Korte, N. E., Skopp, J., Fuller, W. H., Niebla, E. E. & Alessii, B. A. (1976). Trace-element movement in soils: Influence of soil physical and chemical properties. *Soil Sci.,* **122**, pp. 350–9.

Lagerwerff, J. V. (1967). Heavy-metal contamination of soils. Pp. 343–64 in *Agriculture and the Quality of our Environment* (Ed. N. C. Brady). Amer. Assoc. Adv. Sci., Publ. No. 85, Washington, DC, USA: xv + 460 pp.

Langer, G. (1965). Particle deposition and re-entrainment from coniferous trees, Part II: Experiments with individual leaves. *Kolloid Z. Z. Polym.,* **204**, pp. 119–24.

Likens, G. E., Bormann, F. H., Pierce, R. S., Eaton, J. S. & Johnson, N. M. (1977). *Biogeochemistry of a Forested Ecosystem.* Springer-Verlag, New York, NY, USA: 146 pp.

Linzon, S. N., Chai, B. L. Temple, P. J. Pearson, R. G. & Smith, M. L. (1976). Lead contamination of urban soils and vegetation by emissions from secondary lead industries. *J. Air Pollut. Control Assoc.,* **26**, pp. 650–4.

Little, P. (1977). Deposition of 2.75, 5.0 and 8.5 μm particles to plant and soil surfaces. *Environ. Pollut,* **12**, pp. 293–305.

Little, P. & Martin, M. H. (1972). A survey of zinc, lead and cadmium in soil and natural vegetation around a smelting complex. *Environ. Pollut.,* **3**, pp. 241–54.

McBride, M. B. (1978). Transition metal bonding in humic acid: An ESR study. *Soil Sci.,* **126**, pp. 200–9.

McCune, D. C., Silberman, D. H., Mandl, R. H., Weinstein, L. H., Freudenthal, P. C. & Giardina, P. A. (1977). Studies on the effects of saline aerosols of cooling-tower origin on plants. *J. Air Pollut, Control Assoc.,* **27**, pp. 319–24.

Martin, A. & Barber, F. R. (1971). Some measurements of loss of atmospheric sulfur dioxide near foliage. *Atmos. Environ.,* **5**, pp. 345–52.

Martin, W. E. (1959). The vegetation of Island Beach State Park, New Jersey. *Ecol. Monogr.,* **29**, pp. 1–46.

Moser, B. C. (1979). Airborne salt spray techniques for experimentation and its effects on vegetation. *Phytopathology,* **69**, pp. 1002–6.

Moss. M. R. (1975). Spatial patterns of sulphur accumulation by vegetation and soils around industrial centres. *J. Biogeography,* **2**, pp. 205–22.

Murphy, C. E., jr, Sinclair, T. R. & Knoerr, K. R. (1977). An assessment of the use of forests as sinks for the removal of atmospheric sulphur dioxide. *J. Environ. Qual.,* **6**, pp. 388–96.

Neuberger, H., Hosler, C. C. & Koemond, C. (1967). Vegetation as an aerosol filter. Pp. 693–702 in *Biometeorology 2* (Ed. S. W. Tromp & W. H. Weihe). Pergamon Press, New York, NY, USA:

Nozhevnikova, A. N. & Yurganov, L. N. (1978). Microbiological aspects of regulating the carbon monoxide content in the Earth's atmosphere. *Adv. Microbial. Ecol.,* **2**, pp. 203–44.

Nyborg, M. (1978). Sulphur pollution and soils. Pp. 359–90 in *Sulphur in the Environment, Part II: Ecological Impacts* (Ed. J. O. Nriagu). Wiley, New York, NY, USA: xii + 482 pp., illustr.

Oak Ridge National Laboratory (1969). *Progress Report in Postattack Ecology.* Interim Progress Report No. ORNL-TM-2466, Oak Ridge, Tennessee, USA: 60 pp.

Oosting, H. J. (1945). Tolerance to salt spray of plants of coastal dunes. *Ecology,* **26**, pp. 85–9.

Oosting, H. J. & Billings, W. D. (1942). Factors affecting vegetational zonation on coastal dunes. *Ecology,* **23**, pp. 131–42.

Parker, G. R., McFel, W. W. & Kelly, J. M. (1978). Metal distribution in forested ecosystems in urban and rural northwestern Indiana. *J. Environ. Qual.,* **7**, pp. 337–42.

Petruzelli, G., Guidi G. & Lubrano, L. (1978). Organic matter as an influencing factor on copper and cadmium adsorption by soils. *Water, Air, Soil Pollut.,* **9**, pp. 263–9.

Podgorow, N. W. (1967). Plantings as dust filters. *Les. Khoz.,* **20**, pp. 39–40.

Ragaini, R. C., Ralston, H. R. & Roberts, N. (1977). Environmental trace-metal contamination in Kellogg, Idaho, near a lead-smelting complex. *Environ. Sci. Technol.,* **11**, pp. 773–81.

Rasmussen, K. H., Taheri, M. & Kabel, R. L. (1974). *Sources and Natural Removal Processes for Some Atmospheric Pollutants.* U.S. Environmental Protection Agency, Publication No. EPA-650/4-74-032, Washington, DC, USA: 121 pp.

Rasmussen, K. H., Taheri, M. & Kable, R. L. (1975). Global emissions and natural processes for removal of gaseous pollutants. *Water, Air, Soil Pollut.,* **4**, pp. 33–64.

Raynor, G. S. (1967). Effects of a forest on particulate dispersion. Pp. 581–6 in *Proc. USAEC Meteorological Information Meeting* (Ed. C. A. Mawson). (Chalk River Nuclear Laboratories, Chalk River, Ontario, Canada, 11–14 September 1967.)

Raynor, G. S., Smith, M. E., Singer, I. A., Cohen, L. A. & Hayes, J. V. (1966). The dispersion of ragweed pollen into a forest. Pp. 2–6 in *Proc. 7th National Conf. Agricultural Meteorology,* 29 August–1 September 1966, Rutgers University, New Brunswick, New Jersey, USA.

Reiners, W. A., Marks, R. H. & Vitousek, P. M. (1975). Heavy-metals in subalpine and alpine soils of New Hampshire. *Oikos,* **26**, pp. 264–75.

Rich, S. & Turner, N. C. (1972). Importance of moisture on stomatal behaviour of plants subjected to ozone. *J. Air Pollut. Control Assoc.,* **22**, pp. 369–71.

Rich, S., Waggoner, P. E. & Tomlinson, H. (1970). Ozone uptake by bean leaves. *Science,* **169**, pp. 79–80.

Roberts, B. R. (1974). Foliar sorption of atmospheric sulphur dioxide by woody plants. *Environ. Pollut.,* **7**, pp. 133–140, illustr.

Roberts, B. R. & Krause, C. R. (1976). Changes in ambient SO_2 by rhododendron and pyracantha. *Hort. Sci.,* **11**, pp. 111–2.

Rogers, H. H., Jeffries, H. E. & Witherspoon, A. M. (1979). Measuring air pollutant uptake by plants: Nitrogen dioxide. *J. Environ. Qual.,* **8**, 551–7.

Rosinki, J. & Nagamoto, C. T. (1965). Particle deposition on and re-entrainment from coniferous trees, Part I: Experiments with trees. *Kolloid Z. Z. Polym.,* **204**, pp. 111–9.

Salisbury, F. B. & Ross, C. W. (1978). *Plant Physiology.* Wadsworth, Belmont, California, USA: 422 pp.

Seiler, W. (1974). The cycle of atmospheric CO. *Tellus,* **26**, pp. 116–35.

Sheih, C. M. (1977). Application of a statistical trajectory model to the simulation of sulphur pollution over northeastern United States. *Atmos. Environ.,* **11**, pp. 173–8.

Sheih, C. M., Wesely, M. L. & Hicks, B. B. (1979). *A Guide for Estimating Dry Deposition Velocities of Sulfur Over the Eastern United States and Surrounding Regions.* Argonne National Laboratory Report No. ANL-RER-79-2, 55 pp.

Shriner, D. S. & Henderson, G. S. (1978). Sulfur distribution and cycling in a deciduous forest watershed. *J. Environ. Qual.,* **7**, pp. 392–7.

Siccama, T. G. & Smith, W. H. (1978). Lead accumulation in a northern hardwood forest. *Environ. Sci. Technol.,* **12**, pp. 593–4.

Slinn, W. G. N. (1975). Dry deposition and resuspension of aerosol particles—A new look at some old problems. Pp. 1–40 in *Proc. Conf. Atmosphere–Surface Exchange of Particles and Gases*. ERDA Conf. Series, No. CONF-740921, Washington, DC, USA:

Slinn, W. G. N. (1976). *Some Approximations for the Wet and Dry Removal of Particles and Gases from the Atmosphere*. Atmosphere Sciences Department, Battelle Memorial Institute, Pacific Northwest Laboratory, Richland, Washington, USA.

Smith, E. A. & Mayfield, C. I. (1978). Effects of nitrogen dioxide on selected soil processes. *Water, Air, Soil Pollut.*, **9**, pp. 33–43.

Smith, K. A., Bremner, J. M. & Tabatabai, M. A. (1973). Sorption of gaseous atmospheric pollutants by soils. *Soil Sci.*, **116**, pp. 313–19.

Smith, W. H. (1970*a*). Technical review: Trees in the city. *J. Am. Inst. Planners*, **36**, pp. 429–36.

Smith, W. H. (1970*b*). Salt contamination of White Pine planted adjacent to an interstate highway. *Plant Dis. Reptr,* **54**, pp. 1021–5.

Smith, W. H. (1971). Lead contamination of roadside White Pine. *For. Sci.*, **17**, pp. 195–8.

Smith, W. H. (1974). Air pollution—Effects on the structure and function of the temperate forest ecosystem. *Environ. Pollut.*, **6**, 111–29.

Smith, W. H. (1976). Lead contamination of the roadside ecosystem. *J. Air Pollut. Control Assoc.*, **26**, pp. 753–66.

Smith, W. H. (1979). Urban vegetation and air quality. Pp. 284–305 in *Proc. National Urban Forestry Conference, Washington, DC, 13–16 November 1978*. USDA Forest Service, Washington, DC, and State University of New York, Publication No. 80–003, Syracuse, New York, USA.

Smith, W. H. & Dochinger, L. S. (1975). *Air Pollution and Metropolitan Woody Vegetation*. (Pinchot Institute, Consortium for Environmental Forestry Research, Publication No. PIEFR-PA-1.) USDA Forest Service, Upper Darby, Pennsylvania, USA: 74 pp.

Smith, W. H. & Dochinger, L. S. (1976). Capability of metropolitan trees to reduce atmospheric contaminants. Pp. 49–59 in *Proc. Better Trees for Metropolitan Landscapes*, (Ed. H. Gerhold, F. Santamor & S. Little). U.S.D.A. Forest Service, Gen. Tech. Report No. NE-22, Upper Darby, Pennsylvania, USA:

Smith, W. H. & Siccama, T. G. (1980). The Hubbard Brook Ecosystem study: Biogeochemistry of lead in the northern hardwood forest. *J. Environ. Qual.*,

Somers, G. F. (1978). The role of plant residues in the retention of cadmium in ecosystems. *Environ. Pollut.*, **17**, pp. 287–95.

Steubing, L. & Klee, R. (1970). Comparative investigations into the dust-filtering effects of broad leaved and coniferous woody vegetation. *Angew. Bot.*, **4**, pp. 73–85.

Stevenson, F. J. (1972). Role and function of humus in soil with emphasis on adsorption of herbicides and chelation of micronutrients. *BioScience*, **22**, pp. 643–50.

Swank, W. T. & Douglas, J. E. (1977). Nutrient budgets for undisturbed and manipulated hardwood forest ecosystems in the mountains of North Carolina. Pp. 343–64 in *Watershed Research in Eastern North America*, Smithsonian Institution, Edgewater, Maryland, USA.

Thorne, L. & Hanson, G P. (1972). Species differences in rates of vegetal ozone absorption. *Environ. Pollut.*, **3**, pp. 303–12, illustr.

Townsend, A. M. (1974). Sorption of ozone by nine shade-tree species. *J. Am. Soc. Hort. Sci.*, **99**, pp. 206–8.

Turner, N. C., Rich, S. & Waggoner, P. E. (1973). Removal of ozone by soil. *J. Environ. Qual.*, **2**, pp. 259–64.

Tyler, G. (1978). Leaching rates of heavy-metal ions in forest soil. *Water, Air, Soil Pollut.,* **9**, pp. 137–48.

US Environmental Protection Agency (1976). *Open Space as an Air Resource Management Measure, Vol. I: Sink Factors.* U.S.E.P.A. Publication No. EPA-450/3/76-028a, Research Triangle Park, North Carolina, USA.

Waggoner, P. E. (1971). Plants and polluted air. *BioScience,* **21**, pp. 455–9.

Waggoner, P. E. (1975). Micrometeorological models. Pp. 205–28 in *Vegetation and the Atmosphere,* Vol. I (Ed. J. L. Monteith). Academic Press, New York, NY, USA: xviii + 278 pp., illustr.

Warren, J. L. (1973). *Green Space for Air Pollution Control.* School of Forest Resources, Technical Report No. 50, North Carolina State University, Raleigh, North Carolina, USA: 118 pp. (mimeogr.).

Wedding, J. B., Carlson, R. W., Stukel, J. J. & Bazzaz, F. A. (1975). Aerosol deposition on plant leaves. *Environ. Sci. Tech.,* **9**, pp. 151–3.

Wells, B. W. & Shunk, I. V. (1938). Salt spray: An important factor in coastal ecology. *Bull. Torr. Bot. Club* **65**, pp. 485–92.

White, E. J. & Turner, F. (1970). Method of estimating income of nutrients in catch of airborne particles by a woodland canopy. *J. Appl. Ecol.,* **7**, pp. 441–61.

Witherspoon, J. P. & Taylor, F. G. jr (1969). Retention of a fallout simulant containing ^{134}Cs by pine and oak trees. *Health Phys.,* **17**, pp. 825–9.

Whittaker, R. H. & Woodwell, G. M. (1967). Surface area relations of woody plants and forest communities. *Am. J. Bot.,* **8**, pp. 931–9.

Wood, F. A. & Davis, D. D. (1969). Sensitivity to ozone determined for trees. *Pennsylvania State Univ., Sci. Agr.,* **17**, pp. 4–5.

Woodcock, A. H. (1953). Salt nuclei in marine air as a function of altitude and wind-force. *J. Meteorol.,* **10**, pp. 362–71.

Zimdahl, R. L. & Skogerboe, R. K. (1977). Behaviour of lead in soil. *Environ. Sci. Technol.,* **11**, pp. 1202–7.

Zimmerman, P. & Rasmussen, R. (1975). Identification of soil denitrification peak as N_2O. *Environ. Sci. Technol.,* **9**, pp. 1077–9.

Zinke, P. J. (1967). Forest interception studies in the United States. Pp. 137–160 in *Forest Hydrology*, Pergamon Press, Oxford, England, UK.

Zunino, H. & Martin J. P. (1977*a*). Metal-binding organic macromolecules in soil, 1: Hypothesis interpreting the role of soil organic matter in the translocation of metal ions from rocks to biological systems. *Soil Sci.,* **123**, pp. 65–76.

Zunino, H. & Martin, J. P. (1977*b*). Metal-binding organic macromolecules in soil, 2: Characterization of the maximum binding ability of the macromolecules. *Soil Sci.,* **123**, pp. 188–202.

Air Pollution and Plant Life
Edited by M. Treshow
© 1984 John Wiley & Sons Ltd.

CHAPTER 18

Controlling Atmospheric Pollution

STEFAN BIAŁOBOK

Polish Academy of Sciences, Institute of Dendrology, 62–035 Kórnik, Poland

---

INTRODUCTION

The economic activities of Man often lead to changes in the environment. The nature and intensity of such changes depend on the sources and methods of producing energy, the quality of available raw materials, the type of industry and its localization, and on the scale of productivity. The danger to the environment that is posed by airborne pollution is caused by many operations of heavy industry—primarily in metals and the chemistry of energy production (whether based on brown or black coal, peat, or oil).

On the basis of the estimation of injuries to plants that are occurring in natural or artificial plant communities under the influence of chronic air pollution, we attempt to analyse the changes taking place in their functioning, in the production of biomass, and also to evaluate the economic losses involved (Miller, 1973; Miller & McBride, 1975; Heck & Brandt, 1977; Linzon, 1978; Knabe, 1980).

DANGER TO THE ENVIRONMENT FROM AIR POLLUTION

The degree to which vegetation is endangered by airborne pollution is dependent on the degree of industrialization in the given country. This condition can be expressed as the unitary consumption of primary energy per inhabitant per year. This index is usually given in metric tons of a conventional fuel yielding 7000 kcal per kg. If we express the mean danger to the environment with this index in relation to the density of populations per year, as well as per km^2 per year, we shall obtain very characteristic figures of the index for Poland, France, West Germany, and USA (Juda, 1978).

The unitary consumption of primary energy for 1980 was 5.5 in Poland, 8.3 in West Germany, and 6.1 in France. All are lower than the 12.3 value for USA. However, this is on a *per capita* basis. When the index of primary energy consumption takes into consideration the density of the population, and is expressed *per capita* per km^2, then it will be 114 for Poland, 250 for Germany, and 100 for France—all much higher than the corresponding 24 for the USA. If we calculate the unitary consumption of primary energy per km^2 per year, we then find that the danger to the environment is greatest in countries with small areas: 630 for Poland, 2,100 for West Germany, 610 for France, and 290 for the USA.

At present, without a system-simulating modelling process (Grodziński & Lesiński, 1978; Juda, 1978; Kickert, 1980), it is not possible to protect fully the natural and artificial ecosystems, nor to search for the most satisfactory economic solutions for reconciling the welfare of plants and animals with industrial production. For this purpose it is necessary to determine which factors demand long-term observation, and which are sufficiently sensitive to provide the good early data needed to determine the effects of long-term air pollution, without having to wait for the consequences of these emissions in the environment (Kickert, 1980).

The socio-economic response is also an important factor in management principles of this system so as to balance losses and gains in the systems: people–natural resources–food production–industry.

The most important prerequisites for the simulation model for estimating ecosystem losses are, first, to determine the levels of air pollutions that are not harmful, and, secondly, to manage industry, energy production, and goods manufacturing, in such a fashion as to be less injurious to the environment than they formerly were.

CONTROL OF ATMOSPHERIC POLLUTION AT THE SOURCE

The basic factor in protecting plants is to reduce emissions to non-injurious levels. This is a complex problem; but it can be solved by various means, such as using the most appropriate raw materials for industry, optimizing site locations of industrial plants, and using the most modern technologies for efficient reduction of emissions.

It is generally easier to provide the necessary equipment for pollution control when industrial plants are localized in specific places: to reduce pollutant concentrations from mobile sources, such as automobile engines and air transport, is far more difficult.

Oxides of Sulphur

Currently, SO_2 in the air is reduced in either of two ways: (a) removal of sulphur from fumes, and (b) removal of sulphur from fuels (Magill *et al.*, 1956; Juda & Chróściel, 1974). There are about one hundred methods of purifying fumes, but because of the high costs of installation of purifying equipment, and the large quantities of wastes, only a few are practicable. In removing SO_2 from fumes, a wet absorption method has been widely used. One was perfected in the Battersea power-plant, London, England, using an appliance produced by Gottfried Bischoff, of Essen, W. Germany. The efficiency of sulphur removal obtained was 80–90% (Juda & Chróściel, 1974).

Another wet absorption method was developed by Combustion Engineering Inc. The value of this method lies in the use of cheap materials such as limestone or dolomite. The efficiency of SO_2 removal is about 85–95%, and this method is used widely.

The DAP-Mn Mitsubishi method, employing oxides of manganese, is interesting and has a potential future in the opinion of Juda & Chróściel (1974). Its efficiency of SO_2 removal is of the order of 90%, this technology having been first employed in 1971 in the Yokkaichi Power Station. Together with the Bischoff method, it belongs to the simplest and technically most reliable group of methods.

The Reinluft, dry adsorption method of removing SO_2 from fumes is based on activated carbon and ensures a 90% efficiency. The by-product in this method is H_2SO_4. An installation working on a semi-commercial basis in Lünen Steag-Kraftwerk Kellermann, attains an efficiency of 65–70% (Juda & Chróściel, 1974).

Research on the removal of sulphur from fumes is being conducted by the Industrial Research Institute of the Tokyo Electric Power Company, Hitachi Ltd. Their method ensures an efficiency of 90%, and in 1970 an installation of this sort was set up in the Kashime Power Station in Ibaragi (Juda & Chróściel, 1974).

The greatest hopes for the reduction of SO_2 emissions are placed in the removal of sulphur from solid fuels and in improved technology. Work in this latter field is being conducted intensively in the UK and USA. The principle of improved technology known as FBC (Fluidized Bed Combustion) was developed in the UK in 1972 (Juda & Chróściel, 1974).

Another method of reducing air pollution when obtaining energy from coal is COGAS (Combined Gas–Steam Turbine System), proposed by Lurgi Gesellschaft für Wärme and Chemotechnik m.b.H. Frankfurt. This method uses oxygen, producing NO_x (Juda & Chróściel, 1974). It seems safe to predict that the future will bring even more satisfactory solutions, and a more rational protection of the environment, than the methods discussed above.

Reduction of Photochemical Oxidant Emission

The combustion of fluid fuels in gasoline and diesel engines leads to the emission of such substances as CO_2, CO, O_2, SO_2, NO_x, H_2, hydrocarbons, and aldehydes. As a result of photochemical reactions, ozone and peroxyacetyl nitrate (PAN) are formed, which are extremely injurious to plants.

The possibility of eliminating these pollutants from the air is hindered by the multitude of emission sources. Various possibilities have therefore been proposed for reducing these injurious emissions, particularly in the USA and Japan. However, many technical problems are being encountered in finding the most economic solutions. In this case the greatest possibilities of reducing emissions lie in the construction of 'cleaner' engines and in the changing of the basic fuel.

Reduction of Hydrogen Fluoride Emission by Industrial Plants

Hydrogen fluoride is a gas that is readily soluble in water, so the use of towers with water sprinklers (scrubbers) reduces the level of HF in fumes by more than 97%. The passing of fumes containing HF through a layer of calcium carbonate is an even more efficient method. Using the equipment described by T. P. Hignett & M. R. Siegel (in Magill *et al.*, 1956) removes 99.9% of HF from the fumes. Use of wet-cell washers in a system with two such cells, the level of HF is reduced by more than 99% from fumes initially containing 25–4,000 mg per m^3 of the gas. For the absorption of HF, solutions of sodium hydroxide and calcium hydroxide are also used, and the resultant calcium fluoride is easy to remove (Magill *et al.*, 1956).

Subsequently, improved control technology has made it economically feasible and practicable to remove over 99% of the potential HF emissions from most industries, and toxic thresholds of this pollutant are now rare.

Reduction of Oxides of Nitrogen in Industrial Emissions

Oxides of nitrogen are most commonly removed by their absorption in water despite the fact that these methods are not very efficient. Much greater efficiency of absorption is obtained by misting under pressure in a Venturi injector. Packed towers and spray towers operating with countercurrent water- and gas flow are characterized by a very high efficiency when there is a low concentration of oxides of nitrogen. Adsorption on silica gels is also used, and such gels, containing adsorbent nitrogen dioxide, are employed to catalyse the oxidation of nitric oxide (Magill *et al.*, 1956).

Protection of ecosystems against injurious influences of industrial air pollution is dependent on the development of technology for the reduction of emissions of pollutants into the atmosphere, and on the expansion of a functional monitoring system that would constitute a factor in the control of air cleanliness.

DETERMINATION OF AIR POLLUTANT CONCENTRATIONS

One of the important factors in the control of atmospheric pollution is the determination of official norms for concentrations in air. Most commonly a maximal concentration of toxic substances is adopted, which may not be exceeded in a given time-period.

Thus the relationship has to be established between the dose (concentration × duration) of gaseous pollutant and the plant's or ecosystem's response. In order to determine this exposure level, a response has to be selected that best characterizes the reaction of a plant to the action of the pollutant. Visible injuries to plant organs most commonly provide such criteria. Inhibition of growth and development, foliage losses, or crop losses in terms of biomass production, are estimated. Physiological effects of air pollution are also estimated in situations where no visible injuries to plants occur.

The magnitude of the dose, and the reaction of plants to it, are modified by edaphic and climatic conditions, by the genetic properties of the plants (such as heterogeneity of the population), by gaseous components of any air pollution, and by the duration of action of various concentrations (Heck & Brandt, 1977; Linzon, 1978; Guderian & Küppers, 1979; Białobok, 1980; Jeffree, 1980).

Various strategies exist for coping with this problem (Hartogensis, 1980). This latter author outlines the following ways of regulating concentrations of air pollution: (a) 'The emission standard strategy', which depends on the regulation of concentrations of pollution levels to determine the maximal permissible ones. This requires the use of the best available technology and its wide application in an effort to restrict the amounts of pollutants emitted into air (Hartogensis, 1980). (b) Another approach to air purity protection is the 'economic strategy', which depends on penalizing industrial enterprises that emit too much pollution into the atmosphere with monetary fines or by making them pay special taxes. When using this strategy, protection of air purity requires the existence of a functional pollution monitoring system. As a rule, the effects of this strategy are low, because it will cover only a part of the damage done to Nature, and this is difficult to determine. (c) The most efficient approach is the 'total emission strategy' for air purity protection. It started to develop only a few years ago, involving the establishment of regulations concerning airborne pollutants (Hartogensis, 1980).

Injurious emissions circulate around the globe that may endanger vegetation at considerable distances from the source. The process of self-purification of poisonous emissions is very slow, and so it is necessary to eliminate poisonous emissions at the source, before they reach the atmosphere.

The determination of criteria for the maximal permissible concentrations of injurious substances (i.e. air quality standards) requires improved knowledge of the critical levels for various wild and cultivated plants. Determination of principles on which the protection of air purity has to be based, requires a compromise between results of scientific investigations on the effects of emissions on people and plants, the level of industrial technology, and the economy of plants' and industrial production.

POLLUTANT CONCENTRATIONS INJURIOUS TO PLANTS

Two types of damage are recognized, acute and chronic.

Acute damage occurs as a result of the action of high concentrations of gases over a short period of time. When the poisonous gases reach the plant cells, the latters' natural physiological and biochemical processes are altered. Accumulation of large quantities of poisonous substances leads to injuries to leaf blades, the formation of necroses, and eventually to the rapid dying of the plant (Taylor, 1973).

Chronic damage occurs when low concentrations of polluting gases act for a longer period of time. This leads to reduction in growth and development of plants and consequently of the crop, without any initially visible injuries (Feder, 1973; Linzon, 1978).

Sulphur Dioxide

It is not always fully possible to determine the losses to biomass production in many crop and wild plants. This topic has been very widely reviewed in the literature (Linzon, 1972, 1978; Kozlowski & Mudd, 1975; Guderian, 1977; Heck & Brandt, 1977; Jeffree, 1980). For this reason, it is only possible to discuss results of selected experiments, as the aim here is to outline the scope of the problem rather than to exhaust it in detail.

In order to determine the critical concentrations of SO_2 for trees, detailed data were collected for Scots Pine (*Pinus sylvestris*) and Norway Spruce (*Picea abies*) (Knabe, 1970, 1972; Jeffree, 1980). Knabe has shown that the cultivation of Scots Pine in the Ruhr, FRG, region is possible only if the emission of SO_2 during the vegetative period is lower than 0.08 mg per m^3. J. F. Farrar *et al.* (in Jeffree, 1980) have found a small number of mature trees in a natural pine stand where the mean concentration of SO_2 during the investigation period was equal to 0.250 mg per m^3. These authors have observed a similar situation at a concentration of SO_2 equal to 0.1 mg per m^3. Jeffree (1980) reports that, at SO_2 concentrations of 0.070 mg per m^3, losses to timber production are of the order of 40%, and when the concentration is 0.180 mg per m^3, the losses rise to 80%. Growth of Norway Spruce, which is more sensitive to SO_2 than is Scots Pine, is reduced at 0.025–0.035 mg per m^3, while at SO_2 concentrations of 0.030–0.040 mg per m^3, some trees die and volume production drops 50%. At 0.070–0.090 mg per m^3 SO_2, whole stands die (Knabe, 1973; cf. Jeffree, 1980). The peak concentrations, which may be more critical, are not given.

Knabe (1971), in cooperation with H. G. Dässler & H. Stein, H. Stein & H. G. Dässler, J. Materna, and J. Materna *et al.*, reported the following data on maximal doses of SO_2 pollution tolerated by Norway Spruce (Table I).

The selective action of SO_2 on plant crops is very strong. The concentration of gas and the duration of its action has to be characterized. Guderian (1977)

Table I   Information on air quality criteria.[*]

	Mean emission criteria Concentration (mg SO_2 per m^3 per year) Injury		
Measurement interval	severe	medium	slight
20 minutes 3 samples per day	0.16	0.06−0.13	0.07−0.09
30 minutes continuous monitoring	0.12	0.06−0.09	−
24 hours	0.53	0.07−0.24	0.09−0.10

[*] As estimated by the degree of injury to Norway Spruce (*Picea abies* L. Karst) needles following exposure to specified concentrations of SO_2 over specified periods of time.

describes the action of SO_2 on such plants as *Helianthus annuus* (Sunflower), *Zea mays* (maize), *Pisum sativum* (Garden Pea), and *Vicia sativa* (Spring Vetch). A concentration of SO_2 of 1 mg per m^3 for a period of 48 hours reduces the crop on individual plants while the total crop is reduced by 14%. A concentration of 2 mg per m^3 per 24 hours reduces the crop by 26% (*Ibid.*).

Perennial Ryegrass (*Lolium perenne* main cultivar S 23) has been the object of many studies relating crop reduction with concentrations of SO_2 and the time of year (Jeffree, 1980). Table II summarizes some of these data for the summer period.

Davis & Wilhour (1976), and Hicks (1978), quoting the data of P. J. O'Gara, report that Alfalfa (*Medicago sativa*) is a very sensitive plant to this gas, and was very severely injured at a concentration of 3,144 μg per m^3 (1.2 ppm) acting for one hour.

Extensive studies on the determination of a critical concentration of SO_2 under field conditions near the source of emission have been conducted for many crop and forest plants by R. Guderian & H. Stratmann (in Guderian, 1977). These were later used for the specification of air pollution standards in

Table II   *Lolium perenne* growth responses to different SO_2 concentrations.

SO_2 mean concentration (mg per m^3)	Crop reduction (%)	Authors
0.053	0	Bell & Clough (1973) Bell & Mudd (1976)
0.144−0.175	20−25	Bell & Mudd (1976)
0.310	25	Ashenden & Mansfield (1977)
0.027−0.057	16−57	Bleasdale (1973)
Daily mean peak 0.258		

West Germany. According to the data presented, it is obvious that the maximum permissible average concentration of this gas during the vegetative period (0.05 mg per m^3), as proposed by VDI Richtlinien (1978*a*, 1978*b*) and IUFRO (1978; Linzon, 1978), will be only slightly injurious. Therefore, it should be used for those air pollution standards where cultivation of the sensitive Norway Spruce, Scots Pine, Wheat (*Triticum sativum*), Rye (*Secale cereale*), and Barley (*Hordeum vulgare*), provide the basis of the economy.

Fluorides

Coniferous forest trees are particularly sensitive to injury by fluorides (Treshow *et al.*, 1967; Halbwachs, 1971; Treshow, 1971; Keller, 1975; Białobok & Karolewski, 1978) and action of HF at a concentration of 1.3 μg per m^3 for 408 hours has caused permanent necroses of needle tips in *Abies nordmaniana* and *Larix japonica*. A duration of 240 hours caused slight leaf chloroses in *Sorbus* sp.; 240 hours caused necroses of needle tips in *Picea abies* (Norway Spruce) and 168 *Pinus strobus* (Eastern White Pine). A 1,440-hours' exposure caused only slight necroses in *Fagus sylvatica* (European Beech) leaves (Guderian *et al.*, 1969), though it should be noted that fluorides can be responsible for a 50% reduction in wood production in the absence of visible leaf injury (Treshow *et al.*, 1967).

In studies on the selection of *Picea abies* (Norway Spruce) trees (Rohmeder & Schönborn, 1965), injuries to needles were first observed following a 10-days' exposure to 0.025 mg F per m^3. In this experiment, grafts of tolerant Spruce proved less sensitive to fluorides than did normal ones when exposed for 10 days to a concentration of 0.075–0.1 mg F per m^3.

The effects of fluoride on crop plant productivity and utility have been evaluated by Treshow & Harner (1968), who found that the fresh and dry weights of Pinto Beans (*Phaseolus vulgaris* cultivar) exposed for 3 weeks, and of Alfalfa exposed for 4 weeks, were positively correlated with an increase in fluoride content when the plants had been subjected to HF concentrations of 0.3–3.2 μg per m^3.

The groups of plants most sensitive to fluoride according to D. C. McCune (in Vostal, 1971) are species of *Gladiolus, Sorghum*, and conifers. He suggests that for these plants the adopted critical concentration be 0.5 μg per m^3 acting over short periods of time, and for *Citrus* plants 0.5 μg per m^3 over longer periods. According to data referring to the most sensitive plants, the maximum tolerable concentration of fluoride is below 0.5 μg per m^3. Thus the maximum concentration of this gas, namely 0.3 μg per m^3, proposed by IUFRO (1978) and VDI-Richtlinien (1978*b*), for Norway Spruce, Scots Pine, cereals, and fodder grasses, is presumably correct for plant protection. While most cereals and fodder plants *per se* are not sensitive, the accumulation of fluoride may be important as a poison when they are used for forage (e.g. Bolton, 1962).

Photochemical Oxidants

A great amount of damage in the environment is caused by ozone, peroxy-acetyl nitrate, and oxides of nitrogen, in countries where the climate is warm and automobile transport is much developed. Coniferous trees are especially sensitive to these substances. Selected sensitive seedlings of *Pinus strobus* (Eastern White Pine) have been injured by a concentration of 0.07 ppm acting over 4 hours (Hicks, 1978). However, broadleaved trees are much more resistant to the action of ozone (Treshow, 1970; Davis & Wood, 1972; Wood & Coppolino, 1972). Critical concentrations of ozone that cause injury are highly variable in the cases of different plants. Thus the doses for sensitive plants that would be critical would amount to about 0.05–0.10 ppm (98–196 μg per m^3) when fumigating for 2–4 hours (Hicks, 1978). The critical values of O_3 action on Sweet Corn (*Zea mays*) and Oats (*Avena sativa*) amount to 0.10–0.12 ppm (196–233 μg per m^3) when fumigating for 2 hours. Tobacco (*Nicotiana tabacum*), which is highly sensitive to this gas even at 0.05 ppm (98 μg per m^3) is seriously injured following a 4-hours' exposure (Hicks, 1978). Hill *et al.* (1961) considered that plants injured at ozone concentrations of less than 0.30 ppm were sensitive, and plants injured only by more than 0.40 ppm were tolerant, while plants of intermediate sensitivity were included between those two classes.

Peroxyacetyl Nitrate (PAN)

In general, lower doses of PAN than of ozone are required to injure plants, and the critical values for sensitive plants are of the order of 0.01–0.02 ppm (49–99 μg per m^3) acting over 4 hours (Hicks, 1978). When comparing the reactions of plants following simultaneous treatment with ozone and PAN, it appears that in some instances the effect of PAN on plants is less than that of ozone, while in others the opposite is true; (Davis, 1975; Friedlander, 1977).

Oxides of Nitrogen

These compounds demonstrate a strong synergism with SO_2 and O_3, and, as a result, participate in injury to plants caused by these gases. We do not have many data on the critical concentrations of NO_x that injure plants, but in general it is believed that a NO_2 concentration of 0.05 ppm (0.1 mg per m^3) for an annual average, or 0.13 ppm (0.244 mg per m^3) as a mean for 24 hours, is below the threshold for visible effects on vegetation (Crocker, 1977).

INTERACTION BETWEEN POLLUTANTS

Air quality standards of various countries give the maximum permissible concentrations for the most commonly occurring gases polluting the air. However, particularly in large industrial regions, it is not so much the individual gases that act on plants as their mixtures, or compounds forming newly in the atmosphere. Thus, it is essential to know something of the degree of harmfulness of mixtures of airborne pollutants on vegetation—which harmfulness is as yet little documented, in terms both of its nature and of the extent of injury to plant organs. The effect of gases may be simultaneous, sequential and/or intermittent (Reinert *et al.*, 1975; Hicks, 1978). Middleton *et al.* (1958) illustrated this by an example. If the ratio of SO_2 and O_3 mixture is 5 : 1 (e.g. 1.5 ppm SO_2 and 0.3 ppm O_3), the effect of ozone dominates, while if the mixture is 6 : 1, then symptoms of injury that are typical for both these pollutants appear; however, if the ratio in the mixture is 4 : 1, the symptom of ozone injury appear to interfere with those of the expected injury from sulphur dioxide.

Sulphur Dioxide and Ozone

Quite apart from the above indications, there is a wide range of possible effects from mixtures of SO_2 and O_3 (Dochinger & Seliskar, 1970), which can be greater-than-additive, additive, or less-than-additive (Tingey *et al.*, 1973; Reinert *et al.*, 1975). The injurious effect on *Medicago sativa* (Alfalfa), *Brassica oleracea* var. *botrytis* (Cauliflower, Broccoli), *B. oleracea* var. *capitata* (Cabbage), *Raphanus sativus* (Radish), *Lycopersicum esculentum* (Tomato), and *Nicotiana tabacum* (Tobacco), of exposure to $SO_2 : O_3$ combinations of 0.5 : 0.05, 0.1 : 0.1, 0.25 : 0.1, or 0.5 : 0.1, was greater-than-additive, additive, or less-than-additive, depending on the gas concentrations (Tingey *et al.*, 1973). Białobok & Karolewski (1978) reported that injury to needles of Scots Pine (*Pinus sylvestris*) caused by a mixture of these gases is greater-than-additive, while Houston (1974) reported a similar response in the case of *P. strobus* (Eastern White Pine).

Sulphur Dioxide and Nitrogen Dioxide

These two gases are emitted jointly in many industrial processes. The type of plant injury caused by a mixture of these gases is similar to that caused by ozone, particularly when the concentrations of both these gases are close to the critical values (Reinert *et al.*, 1975; Hicks, 1978). From observations on the effects of given concentrations of both these gases under controlled conditions, it appears that injury is greater-than-additive in Chard (*Beta vulgaris* var. *cicla*) and Oats (*Avena sativa*), Tobacco (*Nicotiana tabacum*), Pinto Bean (*Phaseolus vulgaris* cultivar), and Quaking Aspen (*Populus tremuloides*). Under field conditions, however, the effects are merely additive (Reinert *et al.*, 1975).

Sulphur Dioxide and Hydrogen Fluoride

This commonly occurring mixture from various technological processes is highly injurious to plants. The degree of harmfulness of these two gases combined is little known, but the interaction between SO_2 and HF is believed to have effects that are injurious to many plants (Treshow, 1971). Reinert *et al.* (1975) reported few data on this, though greater-than-additive effects on Maize (*Zea mays*) and Barley (*Hordeum vulgare*) were caused by a 27-days exposure to a mixture of SO_2 at 0.08 ppm and HF at 0.6 ppb. However, when SO_2 concentrations were higher, the effects were only additive.

ACID PRECIPITATION

As a result of chemical compounds formed in the atmosphere from a varied and increasing number of pollutants (in Europe, primarily SO_2 and NO_x) including dusts and metals, the pH of the air changes, and this is reflected in the precipitation (Ottar, 1980). As a result of these processes, secondary injury occurs, primarily among lichens, and a change in the vegetation cover is observable. The pH of water in reservoirs and rivers as well as in the soil may be altered, thus creating a great and difficult interdisciplinary problem of understanding the interactions and protecting the vegetation cover. Coniferous forests of the Boreal region in the northern hemisphere are particularly endangered by acid precipitation (Abrahamsen *et al.*, 1976).

Injury to conifer needles could lead to a reduction in height and girth-growth, and eventually to death of the trees (Cogbill, 1976; Legge *et al.*, 1976; Linzon, 1978). A reduction of the amounts of gases in the atmosphere by control of emissions at the source may lead to a reduction of the acidification of the atmosphere. This important topic is dealt with elsewhere in the present volume—particularly in Chapter 4.

MAXIMUM PERMISSIBLE LEVELS

Only general technical possibilities for reducing air pollution have been outlined, though a review has been attempted of the results of studies for determining the degree of sensitivity and tolerance of various groups of plants to air pollution. The results of these data range widely, but at least the complexity of the problems we are dealing with becomes clear. It can also be seen how great the need is to reduce toxic emissions in order to protect vegetation and maintain conditions for the productivity of crops. The need to search for new solutions is obvious if we are to reduce air pollution sufficiently to maintain recommended air quality standards.

The initiative of the *Verein Deutscher Ingenier* (Linzon, 1978; VDI-Richtlinien, 1978*a*, 1978*b*; Guderian & Küppers, 1979) to give instructions

Table III   Some dose-response relationships.*

	Criteria for ambient air pollution		
Sensitivity	Mean over 30 min. for single exposure (mg per m³)		Mean for growing-season (7 months− mg per m³)
Nitrogen dioxide (sensitive plants)	6.0		0.35
Sulphur dioxide	97.5 percentile for 30 min. mean mg per m³		
(Very sensitive)	0.25		0.05
(Sensitive)	0.40		0.08
(Tolerant)	0.60		0.12
Hydrogen fluoride	Mean over 24 h per μg per m³	Monthly mean μg per m³	
(Very sensitive)	2.	0.4	0.0003
(Sensitive)	3.	0.8	0.0005
(Tolerant)	4.	2.0	0.0014

* From VDI-Richtlinien (1978*a*, 1978*b*); Guderian & Küppers (1979).

on the maximum ambient air concentrations permissible to protect vegetation, including agricultural, horticultural, and forestry plants, from SO_2, HF and NO_x, will enable us to protect our wild and cultivated vegetation better than will the ambient air quality standards developed for broader regions where the need to protect vegetation is not the primary aim (Table III).

Establishment of the maximum concentrations of pollutants for the cultivation of herbaceous plants is generally much easier than for woody plants, but in the case of an ecosystem, very complicated indeed.

The extent of injury to spruce forests in central and western Europe caused by SO_2 and HF promoted the International Union of Forestry Research Organizations (IUFRO, 1978, Working party S2.09) to propose a resolution about the necessity to limit the emissions (ambient pollution standards, IUFRO, 1978). Restriction of the annual mean pollution level of SO_2 to 0.05 mg per m³, and of hydrogen fluoride to 0.3 μg per m³, will protect the normal spruce stand on most sites.

CHEMICAL TREATMENT OF PLANTS

The technical difficulties responsible for inadequately restricting air pollution at the source, have led to the development of studies to reduce the sensitivity of cultivated plants to injury by the use of chemical protectants. The results of some of these investigations have been sufficiently promising to put them into practice. However, using these methods of plant protection increases the amount of chemicals in the environment, as treatments can act unfavourably and introduce large and sometimes irreversible changes. We have too little experience as yet in this field, and thus one should approach the subject with due restraint.

The use of chemical compounds for protecting plants against air pollution will not replace efforts to reduce emissions at the source. The reduction of injury by toxic fumes with the help of chemical treatment of plants may prove possible only in exceptional situations, and practicable chiefly when no other means of protecting the plants exist. For example, protective chemicals might be used in greenhouses where, normally, the concentration of ozone and photochemical oxidants are filtered out, but a breakdown of the filtering devices requires that the plants be temporarily protected in a different manner.

Fertilization

The effect of mineral fertilization in reducing plant sensitivity to air pollution is dependent on many site, climatic, and genetic, factors. Not infrequently, results obtained from studies in this field are difficult to interpret and understand. Vigorously growing trees, agricultural crop plants, vegetables, and ornamental plants, are generally more sensitive to the action of SO_2 than are those which grow less satisfactorily (Hicks, 1978). But Heck *et al.* (1965) are of the opinion that plants may be more sensitive when grown under conditions of low nutrition than high. A similar opinion was also expressed by Setterstrom & Zimmermann (1939), and Rich (1975), on the basis of studies on Alfalfa (*Medicago sativa*) and Buckwheat (*Fagopyrum esculentum*).

Ilkun & Makhovskaya (1978) report, however, that *Tilia cordata* (Linden), *Aesculus hippocastanum* (Horse chestnut), and *Populus pyramidalis* (Lombardy Poplar), growing in containers, were least sensitive to SO_2 when fertilized with nitrogen, phosphorus, and potassium. Enderlein & Kästner (1967), when studying SO_2 resistance of Scots Pine (*Pinus sylvestris*) seedlings growing under deficient and normal levels of N, P, K, Ca, and Mg, found those on full nutrition to be slightly the more tolerant. Materna (1962) has shown that fertilization of Norway Spruce (*Picea abies*) stands in industrial regions increases their growth rate and may reduce losses due to air pollution—particularly by SO_2. However, it is difficult to claim that fertilization of these stands has reduced sensitivity to this gas. Guderian (1977) discusses the possibility of reducing injury to Norway Spruce, resulting from low concentrations of SO_2, by fertilizing with ammonium sulphate, while Will & Skelly (1974) have found it possible to increase the resistance of *Pinus strobus*

(Eastern White Pine) seedlings to the action of ozone by fertilizing them in autumn.

Nitrogen has a strong influence on the sensitivity of plants to air pollution, and thus has received much attention (Heck *et al.*, 1965; Friedlander, 1977). Guderian (1977) confirms from his studies that nitrogen has a distinct influence on the reduction of SO_2-sensitivity in Winter Wheat (*Triticum aestivale* cultivar), Summer Rape (*Brassica napus* var. *oleifera*), and Sunflower (*Helianthus annuus*), growing in various soil conditions.

In studying Tobacco and Tomatoes, Leone *et al.* (1966), Leone and Brennan (1972) have found that doses of nitrogen that are optimal for these plants increase their sensitivity to sulphur dioxide. Friedlander (1977), on the basis of existing studies, believes that the nitrogen-oxidant interaction has yet to be critically evaluated.

Fertilization of the soil with calcium in regions polluted primarily by SO_2 increases the soil pH and creates improved conditions for plant development. Calcium also reduces plant injury by fluorides (Pack, 1966; Treshow, 1971), while a calcium deficiency increased foliage injury by HF on tomatoes growing in water cultures besides reducing fruit size and development.

Guderian (1977) has shown that increasing doses of calcium (50–400 mg $CaCO_3$ per 100 g of soil) can reduce injury to *Hordeum vulgare* (Barley) caused by HF at 12.3 μg per m^3 over 192 hours. Under analogous experimental conditions, P_2O_5 and K_2O acted similarly.

Ozone reduced the dry weight of Radish (*Raphanus sativus*) when the nitrogen levels were either high or low at temperatures of 20°C and 30°C. Ozone was similarly influenced by high or low phosphorus levels at lower temperatures (Ormrod *et al.*, 1973). The scale of this problem is extensive, and there are few systematic investigations in this field that were conducted under varied experimental conditions. The effects of various elements on the reduction of plant injury by gases may be of either primary or secondary nature, and thus it is difficult to draw any clear conclusions.

Protectant Compounds

The use of chemical compounds in reducing plant injury caused by air pollution has been of greatest interest in the USA. Pollutant emissions are often very difficult to control at the source, and of necessity, endangered plants have to be protected at the site of cultivation. The mechanisms of action of chemical compounds protecting plants from air pollution, or reducing injury, are very diversified. Some break down ozone and other photochemical oxidants on leaf surfaces. This can be attained by using anti-oxidants and anti-ozonates. The protecting compounds may also reduce sensitivity of plants by changing their metabolism (Koiwai *et al.*, 1974).

Ozone destruction is achieved by spraying such compounds as kaolin, calcium and/or magnesium silicates, powdered ferric oxide, powdered charcoal, and others, onto leaf surfaces (Ilkun, 1978).

Other types of compounds react with oxides of sulphur and other gaseous substances to form weakly-dissociating salts (Ilkun, 1978). Obydiennyj (1977) proposes to protect pine and spruce forests in the USSR against oxides of nitrogen, sulphur dioxide, ammonium, and other gases, by spraying forest areas with solutions of compounds of mercury, silver, thorium, cadmium, or cobalt, or of potassium dichromate or vanadium pentoxide. The concentration of compounds for these treatments should be of the order of 0.1 to 0.3 g per litre of pure drinking water.

Great hopes of protecting plants against ozone have been placed by some authors on the use of compounds derived from methylenedioxyphenyl—such as piperonyl butoxide, a known basic protective compound (Koiwai *et al.*, 1974). These authors have tested the protective action of such compounds on *Nicotiana tabacum* var. Hicks-2. In the opinion of Koiwai *et al.* (1974), these compounds control the metabolism of lipids, which are components of the semipermeable cell membranes, and thus reduce the sensitivity of plants to the action of ozone.

The compounds that are most effective in reducing the injurious effects of ozone on plants, are derivatives of methylenedioxyphenyl and are shown in Table IV.

Some other cleavage substances of the methylenedioxy-ring are also able to protect leaves of Tobacco actively against ozone injury. The most intensively active substances in this group include: p-phenylenediamine, which protects leaves 100%, p-aminophenol (protecting leaves 93%), and m-aminophenol (protecting leaves 60%) (Koiwai *et al.*, 1974).

Currently, the derivatives of benzimidazole are also considered of importance in protecting plants against ozone and other photochemical oxidants. Much hope is attached to the fungicide Benomyl, which is increasingly investigated and used for the spraying of aerial plant parts. Benomyl is added to the soil as well as to water that is being used to water plants, and is

Table IV   Compounds effective in reducing injurious effects of ozone on plants.

Compound	Protective action (%)
Piperonyl butoxide	100
3,4-methylenedioxyphthaldehyde	100
Safroxane	100
Piperonylic acid	70
Piperonyl acetate	50

employed to reduce injury to many cultivated plants that may be caused by ozone or other photochemical oxidants.

G. S. Taylor (1970) has shown that Tobacco, sprayed with a Benomyl solution, had 80% fewer necrotic spots caused by ozone than had plants growing in control plots. Another fungicide, Zineb, was also tested in this experiment, and also reduced the number of spots on Tobacco leaves. Benomyl can also protect Grape Vines (*Vitus vinifera*) against air pollution. In experiments conducted by Kender *et al*. (1973), various doses were tested. These authors have shown that Benomyl, at a concentration of 3.537 g per l and applied six times, was most effective. The possibility of adding Benomyl to the soil was demonstrated by Pell (1976). A reduction in the injury caused by O_3 and PAN was observed in the Pinto Bean (*Phaseolus vulgaris* cultivar), but only on the older leaves.

G. S. Taylor & Rich (1974) have added Benomyl to the soil in Tobacco plantations. The plants proved more viable than the control ones, and the authors believe that the protective action of this compound operates through an effect on the plant's metabolism. However, this compound may also injure plants.

An increasing interest centres around the fungicide Cardoxin, which also has a protective influence on Tobacco against 'flecking' caused by ozone. It acts at concentrations of 5–10 μg per g of soil (G. S. Taylor & Rich, 1974), but also causes injury. In the studies of Moyer *et al*. (1974), Cardoxin reduced injury in *Rhododendron kurune* cultivar Snow. It was used as a spray and as a soil additive, but Benomyl acted more successfully. Also, Rich *et al*. (1974) have demonstrated the protective influence of Cardoxin in such crop plants as beans, cotton, tomato, and tobacco. This compound causes yellowing of tobacco leaves at some concentrations.

Ascorbic acid, used as an anti-oxidant, has reduced ozone injury to Kidney Bean (*Phaseolus vulgaris*) cultivar White by 75% (Dass & Weaver, 1968).

Injury caused by air pollution can also be prevented by using compounds that can temporarily cause a closure of stomata. Such compounds include phenylmercuric acetate and abscisic acid (Koiwai *et al*., 1974).

The protective value of various compounds against the injurious action of SO_2 has been studied by Karolewski (1978), who found that the action of $\alpha\alpha$-dipyridyl solutions at concentrations 0.01–1.0 mM per l substantially reduced the degree of injury to leaves of *Weigela florida* (Flowering Weigelia) from SO_2 at a concentration of 2.0 ppm. The protective action of this compound is associated with the changes effected in the leaves in the content of proline and hydroxyproline.

Friedlander (1977) states that little is known about the role of protectant substances in plant production. This concerns our understanding of the role which such protectants can play in reducing sensitivity to air pollution, the magnitude of doses of these substances that may need to be used, the selectivity of their action, the side effects, if any, and the duration of their persistence in the environment. With so many aspects of the subject still unknown, research efforts should obviously be intensified.

TOLERANT AND RESISTANT PLANTS

Tolerant and Resistant Plants

The third possibility of reducing air pollution injury to plants that will be discussed in this chapter, is the selection of tolerant or moderately resistant species and varieties of agricultural, horticultural, and forest, plants. In the first part of this chapter, concentrations of gases and their mixtures that are toxic to various crop plants were given. They define, in the most general limits, the possibility of selecting plants for various cultural conditions in polluted atmospheres.

Selection of plants for tolerance to air pollution, or attempts at breeding tolerant plants, is sometimes erroneously interpreted by industrial technicians responsible for the organization and planning of industry. The result of this is the opinion that a reduction in the concentration of air pollution is not a particularly urgent problem if possibilities exist for the selection of plants with a reduced sensitivity to a particular pollutant. But this is not the case: selecting tolerant species must be considered as only a partial, temporary remedy—an interim attempt at 'control'.

Evidence of interspecific variability of plants to air pollution was presented in the first part of this chapter. Extensive lists of cultivated agricultural, horticultural, and forest, plants demonstrating this variability will be given at its end (Table V).

Studies have also been conducted on the variability of response in sensitivity of varieties within species of some important crop plants. These studies include such forest tree species as Scots Pine, Norway Spruce, and Eastern White Pine. Within these species, the existence has been demonstrated of genetic variability in sensitivity to SO_2, O_3, and HF (Gerhold *et al.*, 1972; Tzschacksch & Weiss, 1972; Demeritt, 1977; Gerhold & Wilhour, 1977; Scholz *et al.*, 1978; Scholz, 1979).

Genetic Mechanisms of Plant Tolerance

Populations of forest trees are polymorphic, and thus adapted to changing environmental conditions. Populations have various adaptive strategies (Stern & Tigerstedt, 1974). Under the influence of changes taking place in the environment, such as the entry of toxic gases, forms tend to be favoured which fortuitously possess the genetic polymorphism that may make possible their survival. As an example, Stern & Tigerstedt (1974) cite the existence of

Table V   Relative sensitivity of plants (T = tolerant, S = sensitive, I = Intermediate, [2] = difference of opinion).

Scientific and common names of plants	SO$_2$	O$_3$	F	NO$_x$
CONIFERS				
Abies alba (Fir, Silver)	S	T	S	I
Abies balsamea (Fir, Balsam)	I	T	S	S[2]
Abies concolor (Fir, White)	T[2]	T	S[2]	
Chamaecyparis lawsoniana (Cypress, Lawson)	T			I
Ginkgo biloba (Ginkgo)	T			T
Juniperus communis (Juniper, Common)	T[2]		T	
Juniperus occidentalis (Juniper, Western)	T	T		
Larix decidua (Larch, European)		I	I	S
Larix leptolepis (Larch, Japanese)	T[2]	I	T	S
Picea glauca (Spruce, White)	I		S	T
Picea engelmannii (Spruce, Engelmann)	I		I	
Picea omorica (Spruce, Serbian)	I		S	
Picea pungens (Spruce, Colorado)	T		S	I
Pinus contorta (Pine, Lodgepole)	I	I	T[2]	
Pinus nigra (Pine, Austrian)	I[2]	S	S[2]	T[2]
Pinus ponderosa (Pine, Ponderosa)	I	I	S[2]	
Pseudotsuga menziesii (Douglas Fir)	I	T	T[2]	
Sequoia gigantea (Sequoia, Giant)	I			
Sequoia sempervirens (Redwood)	T			
Taxus baccata (Yew, English)	T			S
Taxus cuspidata (Yew, Japanese)			T	I
Thuja occidentalis (Arbor-vitae)	T	T	I	
Thuja plicata (Cedar, Western Red)	T			
Tsuga heterophylla (Hemlock, Western)	I		T	
Tsuga canadensis (Hemlock, Common)		T	T	
BROAD-LEAVED TREES AND SHRUBS				
Acer campestre (Maple, Hedge or Field)	T	I	I[2]	
Acer negundo (Box-elder)	I	I	I	
Acer platanoides (Maple, Norway)	T	T	I	I
Acer pseudoplatanus (Maple, Sycamore)	T	I	I	
Acer rubrum (Maple, Red)	I	T		
Acer saccharinum (Maple, Silver)	T		I	
Acer saccharum (Maple, Sugar)	T	T		
Aesculus hippocastanum (Horse-chestnut, Common)	I[2]	I		
Alnus glutinosa (Alder, Black)	I[2]		T[2]	
Betula pendula (Birch, European)	I[2]	I	T	S
Carpinus betulus (Hornbean, European)	T		I[2]	T
Cornus florida (Dogwood, White)		T		
Cornus stolonifera (Dogwood, Red-osier)	I			
Crataegus douglasii (Hawthorn, Black)	T			
Crataegus oxyacantha (Hawthorn)			T	
Fagus sylvatica (Beech, European)	I	T	I[2]	T
Forsythia intermedia (Forsythia)	I	I	T	
Fraxinus excelsior (Ash, European)	I[2]	T[2]	I	
Ligustrum vulgare (Privet, Common)	T		T	

Table V — continued

Scientific and common names of plants	SO$_2$	O$_3$	F	NO$_x$
Nyssa sylvatica (Gum, Black)	T		T	
Platanus acerifolia (Plane, London)	T		T	
Philadelphus coronarius (Mock-orange or Coronarius)	I			
Populus alba (Poplar, White)	T		T	
Populus deltoides (Cottonwood, Eastern)	T^2			
Populus × *canadensis* 'Robusta'	S^2		I	
Populus × *canadensis* 'Marilandica'	I		I	
Populus candicans (Balm of Gilead)	T		T	
Populus tremula (Aspen, European)	T^2	I	T	
Prunus avium (Mazzard or Sweet)	T		T^2	
Prunus mahaleb (Cherry, Mahaleb)	T		T	
Quercus petraea (Oak, Durmast)	T		I	
Quercus robur (Oak, English)	T^2	T^2	T	T
Quercus virginiana (Oak, Live)	T		T	
Quercus rubra (Oak, Northern Red)	T	T	T	
Quercus palustris (Oak, Pin)	T	I		
Robinia pseudoacacia (Locust, Black)	T	T	T	T
Sambucus nigra (Elder, European)	T		T	T
Salix caprea (Willow, Goat)			T	
Salix alba 'Tristis' (Willow, Shrubby)	T		T	
Sorbus aucuparia (Mountain-ash, European)	T	T	T	
Sophora japonica (Pagoda, Japanese)	T			
Spiraea vanhouttei (Spirea, Van Houttes)	I		I	
Symphoricarpos albus (Snowberry, Alba)	T	S	T	
Syringa vulgaris (Lilac, Common)	I^2	I	I^2	
Tilia americana (Linden, American)		T	T	
Tilia cordata (Linden, Little-leaf)	T	T	T	T
Ulmus americana (Elm, American White)	T			
Ulmus leavis (Elm, European White)			T	
Viburnum lantana (Wayfaring-tree)			T	

FIELD CROPS AND GRASSES

	SO$_2$	O$_3$	F	NO$_x$
Medicago sativa (Alfalfa)	S	S	T	S
cv. Atlantic		T		
cv. Vernal		T^2		
cv. Kanza		I		
cv. Team		I		
Poa annua (Annual Bluegrass)				I
Gossypium sp. (Cotton)	S	T	T	
Zea mays (Corn or Maize)				
Sweet		S	I^2	
Field		I	I	
Avena sativa (Oats)	S	S	T^2	T^2
Solanum tuberosum (Potato)	T	S	T	I
Secale cereale (Rye)				
Mature plants			T^2	

Table V — continued

Scientific and common names of plants	SO$_2$	O$_3$	F	NO$_x$
Young plants			I	
Glycine max (Soybean)		S	T^2	
Nicotiana tabacum (Tobacco)		S	T	S
cv. Bel. B			I	
cv. Bel. C			T	
Triticum sp. (Wheat)				
Young plants	S		I	
Mature plants	S		T	
Sorghum vulgare (Sorghum)		I	I^2	
GARDEN CROPS				
Allium cepa (Onion)	T	S^2		T
Apium graveolens (Celery)	T		T	I^2
Asparagus officinalis (Asparagus)			T	T
Beta vulgaris (Beet)	S	T		
Brassica oleracea cv. gongyloides (Kohlrabi)	S			T
Brassica oleracea cv. capitata (Cabbage)	T	I	T	T
Brassica rapa (Turnip)	S	I		
Cucumis sativus (Cucumber)	S	I	T	
Daucus carota (Carrot)	S	I	T	S
Lactuca sativa (Lettuce)	S	T		S
Lycopersicum esculentum (Tomato)	S	S	I	I
Petroselinum sativum (Parsley)		I^2		
Phaseolus vulgaris cv. Pinto (Pinto Bean)		I		S
Phaseolus vulgaris cv. Eagle		T		
Phaseolus vulgaris cv. Provider		T		
Pisum sativum (Pea, Garden or Field)	S	I	T	S
Spinacia oleracea (Spinach)	S	S	T	
Citrus paradisi (Grapefruit)			I	
Citrus sinensis (Orange)			I	I
Fragaria sp. (Strawberry)		T	I	
Malus sylvestris (Apple)	S		I	S
Prunus armeniaca (Apricot)		T	S	
Prunus avium (Cherry, Sweet)			I	
Prunus persica (Peach)				
Fruit			S	
Foliage			I	
Pyrus communis (Pear)			T	S
ORNAMENTAL FLOWERS				
Antirrhinum majus (Snapdragon)		T		S
Aster sp. (Aster)	S		I	
Begonia sp. (Begonia)	I	I		
Camellia sp. (Camelia)			T	
Chrysanthemum leucanthemum (Daisy)		T		T
Dianthus caryophyllus (Carnation)		I		
Dianthus barbatus (Sweet William)	S		I	

Table V — continued

Scientific and common names of plants	SO_2	O_3	F	NO_x
Dahlia variabilis (Dahlia)				I
Euphorbia pulcherrima (Poinsettia)		T^2		
Gardenia jasminoides (Cape Jasmine)				I
Gladiolus sp. (Gladiolus)	S	T	S	T
Petunia multiflora (Petunia)	S	I	T^2	I
Tagetes sp. (Marigold)		T		

populations of *Festuca ovina* (Sheep's Fescue) and *Argostis tenuis* (Common Bent grass) inhabiting copper mine spills containing the heavy metals lead, copper, and zinc, in Drws-y-coed in the United Kingdom.

To establish the genetic differences in populations of trees growing within the range of air pollution, electrophoretic studies of isozymes were undertaken (Mejnartowicz, 1978). This author compared the variability in the enzymes leucyloamino peptidase (LAP) and acid phosphotase (APH) in populations of Scots Pine from a region under the influence of air pollution with those from a region free of such influences, and showed that the individuals that were tolerant to sulphur dioxide had the allele APH-B5. Szmidt (1978) studied the polymorphism of catalase of three populations from industrial regions and five from unpolluted areas. Populations from pollution-free regions were characterized by a greater proportion of tree genotypes of the type C-1/C-1, compared with populations from regions that were exposed to industrial emissions.

Studies on the genetic regulation of tolerance of cultivated plants to gases or their mixtures are still less frequent than of wild ones. The tolerance of Onions (*Allium cepa*) to ozone is regulated by a single gene (Ryder, 1973), and in the case of Kidney Bean (*Phaseolus vulgaris*) the mechanism of tolerance to ozone is controlled by 2 or 3 genes having an additive effect (Butler *et al.*, 1979). Scholz *et al.* (1978) have found a fluoride-tolerant population of Norway Spruce (*Picea abies*) in which additive and dominance variation accounted for 60% of the phenotypic variation. Karnosky (1976) and Houston & Stairs (1973), have found the existence of genetic control of tolerance to SO_2 and O_3 in populations of Eastern White Pine (*Pinus strobus*).

Demeritt (1977) has found that the heritability of tolerance to ozone is low to medium. Karnosky (1976) has shown that a genetic control exists in *Populus tremuloides* (Quaking Aspen) and Białobok & Karolewski (1978) have demonstrated the existence of a genetic relationship between the tolerance of mother plants and their half-sib progenies in Scots Pine (*P. sylvestris*).

Notes on Breeding

The various means, discussed in this chapter, by which plant injury by air pollution can be reduced in plant communities and cultivars would not be complete without mentioning breeding studies that are aimed at the production of plants of increased resistance (Ryder, 1973).

Plant breeders continuously produce new varieties of cultivated agricultural, horticultural, and forestry, plants which are characterized by improved productivity, resistance to diseases and pests, and adaptation to mechanical cultivation and cropping techniques. Another factor needs to be added to the breeding programmes: tolerance to air pollution, so as to have plants that would be able to grow in conditions with a stable maximal dose of injurious pollutants or their mixtures (VDI-Richtlinien, 1978*a*, 1978*b*).

Already there are substantial achievements in the selection of cultivated plants that would be characterized by improved tolerance to air pollution within already-existing varieties. Currently, many of these are being cultivated (Ryder, 1973; Davis & Wilhour, 1976; Hicks, 1978; Friedlander, 1977; Jeffree, 1980). In breeding tolerant plant varieties, there have been only modest achievements in such plants as onions, tobacco and beans. In the breeding of trees, the achievements have been slight, though very significant for those most sensitive to air pollution, such as Scots Pine (Demeritt, 1977).

A method of breeding of plants tolerant to gases already exists (Ryder, 1973; Gerhold & Wilhour, 1977; Scholz, 1979; Scholz *et al.*, 1980). In studies on the variability in isozymes at various loci in populations of trees tolerant to air pollution, it would be necessary to undertake a search for genetic markers which would facilitate the selection of individuals and populations tolerant to injurious gases (Białobok, 1980). I am sure that this method of reducing losses caused by air pollution will also be employed to increase the production of biomass and enhance food production.

Relative Sensitivity of Plants

A list of the relative tolerance of plants to pollutants has been compiled from the literature (Table V). Since the experiments included were conducted in different climatic and soil conditions, as well as with the use of different methods and plant species, the list is only a rough approximation. Sensitivity of forest, agricultural, and garden, plants are presented. The following reports, etc., have been included in the present list: Hill *et al.* (1961), Vostal (1971), Mooi (1974), Davis & Wilhour (1976), Kluczyński (1976), Crocker (1977), Friedlander (1977), Biological Services Program (BSP) (1978), Hicks (1978), Jeffree (1980), and Białobok (1980).

SUMMARY

Various strategies exist for controlling atmospheric pollution. The most desirable is to eliminate, or at least reduce, the emission at the source. This is not always possible, and in any case is expensive. A second and sometimes necessary approach is to grow pollutant-tolerant plants. This can often be practicable in agricultural situations, but is chiefly an interim solution.

Cultural practices, including use of chemical protectant sprays, can also minimize crop losses. In natural systems it is more difficult to control effects of atmospheric pollution, although pollutants provide a strong selection-pressure against intolerant plant species, populations, and individuals. Presence of a few tolerant individuals will usually sustain a population, provided pollutant concentrations are not extreme. In Europe, pollutant tolerance has sometimes been incorporated as an important criterion in reforestation programmes.

REFERENCES

Abrahamsen, G., Horntved, R. & Tveite, B. (1976). Impacts of acid precipitation on coniferous forest ecosystems. Pp. 991–1009 in *Proc. First International Symposium on Acid Precipitation and the Forest Ecosystem* (Columbus, Ohio, 12–15 May, 1975.) USDA Forest Service Gen. Tech. Report NE-23, xiii + 1074 pp., illustr.

Ashenden, T. W. & Mansfield, T. A. (1977). Influence of windspeed on the sensitivity of ryegrass to SO_2 *J. Exp. Bot.*, **28**, pp. 727–35.

Bell, J. N. B. & Clough, W. S. (1973). Depression of yield in ryegrass exposed to sulphur dioxide. *Nature* (London), **24**, pp. 47–9.

Bell, J. N. B. & Mudd, C. H. (1976). Sulphur dioxide resistance in plants: A case-study of *Lolium perenne*. Pp. 87–103 in *Effects of Air Pollutants on Plants* (Ed. T. A. Mansfield). Cambridge University Press, Cambridge, England, UK: 209 pp., illustr.

Białobok, S. (1980). Identification of resistant or tolerant strains and artificial selection or production of such strains in order to protect vegetation from air pollution. Pp. 253–71 in *Papers Presented to the Symposium on the Effects of Airborne Pollution on Vegetation, Warsaw, Poland.* UN Econ. Comm. Europe: xx + 410 pp.

Białobok, S. & Karolewski, P. (1978). [Ocena stopnia odpornosci drzew mateczynch sosny zwyczajnej i ich potomstwa na dzialanie SO_2 i O_3 oraz mieszaniny tych gazów—in Polish.] *Arboretum Kornickie*, **23**, pp. 299–310, illustr.

Biological Services Program (1978). Impacts of coal-fired power-plants on fish, wildlife and their habitats. Pp. 187–91 in *Biological Services Program Appendix C. Sensitivity list for vegetation.* FWS/OB-78/29. US Department of the Interior.

Bleasdale, J. K. A. (1973). Effects of coal-smoke pollution gases on the growth of Rye-grass (*Lolium perenne* L.). *Environ. Pollut.*, **5**, pp. 272–85.

Bolton, J. L. (1962). *Alfalfa: Botany, Cultivation, and Utilization.* (World Crops Books, Ed. Nicholas Polunin). Leonard Hill Books, London and Interscience Publishers, New York: xiv + 474 pp. illustr.

Butler, L. K., Tibbitts, T. W. & Bliss, F. A. (1979). Inheritance of Resistance to Ozone in *Phaseolus vulgaris* L. *J. Amer. Hort. Sci.*, **164**, pp. 211–3, illustr.

Cogbill, C. V. (1976). The effect of acid precipitation on tree growth in Eastern North America. Pp. 1027–32 in *Proc. First International Symposium on Acid Precipitation and the Forest Ecosystem*. (Columbus, Ohio, 12–15 May 1975.) USDA Forest Service Gen. Tech. Report NE-23: xiii + 1074 pp., illustr.

Crocker, T. T. (1977). Effects of nitrogen oxides on vegetation. Pp. 197–214 in *Nitrogen Oxides* (Chairman of Subcommittee on Nitrogen Oxides, T. T. Crocker). National Academy of Sciences, Washington, DC: vii + 333 pp.

Dass, H. C. & Weaver, G. M. (1968). Modification of ozone damage to *Phaseolus vulgaris* by antioxidants, thiols, and sulfhydryl reagents. *Can. J. Plant. Sci.*, **48**, pp. 569–74.

Davis, D. D. (1975). Resistance of young Ponderosa Pine seedlings to acute doses of PAN. *Plant Dis. Reptr*, **59**, pp. 183–4.

Davis, D. D. & Coppolino, J. B. (1974). Relationship between age and ozone sensitivity of current needles of Ponderosa Pine. *Plant. Dis. Reptr*, **58**, pp. 660–3.

Davis, D. D. & Wilhour, R. G. (1976). Susceptibility of woody plants to sulfur dioxide and photochemical oxidants. *Corvallis Environ. Research Lab., U.S. EPA*-600/3-76-102, Oregon: viii + 71 pp.

Davis, D. D. & Wood, F. A. (1972). The relative susceptibility of eighteen coniferous species to ozone. *Phytopathology*, **63**, pp. 381–8.

Demeritt, M. E. (1977). *Genetic Evaluation of Two-year Height and Ozone Tolerance in Scotch Pine* (Pinus sylvestris *L.*). Ph.D. thesis, Pennsylvania State University, University Park, Pennsylvania, USA: 64 pp.

Dochinger, L. S. & Seliskar, C. E. (1970). Air pollution and the chlorotic dwarf disease of Eastern White Pine. *Forest Science*, **16**, pp. 46–55.

Enderlein, H. & Kästner, W. (1967). Welchen Einfluss hat der Mangel eines Nährstoffes an die SO_2-Resistenz 1 jahriger Kiefern. *Arch. Forswew.*, **16**, pp. 413–35.

Feder, W. A. (1973). Cumulative effects of chronic exposure of plants to low levels of air pollutants. Pp. 21–31 in *Air Pollution Damage to Vegetation* (Ed. J. A. Naegele). (Advances in Chemistry, Series 112.) American Chemical Society, Washington, DC, USA: xiii + 137 pp., illustr.

Friedlander, S. K. (1977). Ch. 11, pp. 459–585 in *Ozone and Other Photochemical Oxidants*. National Academy of Sciences, Washington, DC, USA; 719 pp.

Gerhold, H. D. & Wilhour, R. G. (1977). *Effect of Air Pollution on* Pinus strobus *L. and Genetic Resistance: A Life-nature Review*. Corvallis Environmental Research Laboratory Office of Research and Development, US Environmental Protection Agency, Corvallis, Oregon 97330, USA: 45 pp.

Gerhold, H. D., Palpant, E. H., Chang, W. U. & Demeritt, M. E., jr (1972). Tubing fumigation method for selection of pines resistant to air pollutants. *Mitt d. Forstl. Bundes-Versuchsanstalt* (Wien), **97**, pp. 511–20.

Grodziński, W. & Lesiński, J. A. (1978). [Podstawy i wsakazówki dla przebudowy kompleksu lésnego Puszczy Nie polomickiej–in Polish] *Krakow*, 1978, pp. 1–80.

Guderian, R. (1977). *Air Pollution*. Springer-Verlag, Berlin–Heidelberg–New York: 127 pp., illustr.

Guderian, R. & Küppers, K. (1979). Problems in determining dose–response-relationships as a basis for ambient pollutant standards. Paper presented to the *Symposium on the Effects of Airborne Pollution on Vegetation*. Warsaw, Poland, United Nations Economic Commission for Europe: 15 pp. (mimeogr.)

Guderian, R., Haut, H. van & Stratmann, H. (1969). *Experimentalle Untersuchungen über pflanzenschädigen Fluorwasserstoff-Konzentrationen*. Köln–Opladen. Westdeutscher Verlag Forsch. Ber. d. Landes Nordrhein-Westfalen, FRD No. 2017: 54 pp.

Halbwachs, G. (1971). Die Symptomatologie förstlicher Rauch-schäden bei Koniferen. *Mitt. Forstl. Bundes-Versuchsanstalt* (Wien), **92**, pp. 33–56.

Hartogensis, F. (1980). Criteria for establishing legislation, regulations, and planning guidelines concerning ambient concentrations of airborne pollutants which meet the need to protect vegetation. Pp. 361–7 in *Papers presented to the Symposium on the Effects of Airborne Pollution on Vegetation, Warsaw, Poland.* UN Econ. Comm. Europe: xx + 410 pp.

Heck, W. W. & Brandt, C. S. (1977). Effects on vegetation: Native, crops, forests. Pp. 157–230 in *Air Pollution*, Vol. II: *The Effects of Air Pollution* (Ed. A. C. Stern). Academic Press, New York, NY, USA: 684 pp.

Heck, W. W., Dunning, J. A. & Hindawi, I. J. (1965). Interactions of environmental factors on the sensitivity of plants to air pollution. *J. Air Pollut. Contr. Assoc.*, **15**, pp. 511–5.

Hicks, D. R. (1978). *Diagnosing Vegetation Injury Caused by Air Pollution.* Environmental Protection Agency, Washington, DC, USA: 182 pp.

Hill, A. C., Pack, M. R., Treshow, M., Downs, R. J. & Transtrum, L. G. (1961). Plant injury induced by ozone. *Phytopathology*, **51**, pp. 357–63.

Houston, D. B. (1974). Response of selected *Pinus strobus* L. clones to fumigations with sulphur dioxide and ozone. *Can. J. For. Res.*, **41**, pp. 65–8.

Houston, D. B. & Stairs, G. R. (1973). Genetic control of sulphur dioxide and ozone tolerance in Eastern White Pine. *Forest Science*, **19**, pp. 267–71.

Ilkun, G. M. (1978). *Zagrajazniteli atmosfery i rastenija.* Naukova Dumka, Kiev, USSR: 247 pp., illustr.

Ilkun, G. M. & Makhovskaya, M. A. (1978). Effect of mineral nutrition on the pollution resistance of urban tree plantings. *Fiziologya i Biokhimiya kulturnykh Rastenii*, **10**, pp. 199–203 (from *Hort. Abstr.*, **28**(9), p. 8471).

IUFRO (1978). Resolution ueber maximale Immissionswerte zum schutze de Wälder. *Organization der IUFRO-Fachtagung*, vol. **18**, pp. 1–2.

Jeffree, C. E. (1980). Plant damage caused by SO_2. Pp 328–54 in *Papers presented to the Symposium on the Effect of Airborne Pollution on Vegetation, Warsaw, Poland.* UN Econ. Comm. Europe: xx + 410 pp.

Juda, J. (1978). [Ochrona powietrza atmosferycznego.—in Polish.] Pp. 521–49 *Ochrona i ksztaltowanie środowiska przyrodniczego* (Ed. W. Michajtow & K. Zabierowski). PWN, Warszawa–Krakow, Poland: xv + 850 pp., illustr.

Juda, J. & Chróściel, S. (1974). [*Ochrona powietrza atmosferycznego.*—in Polish.] Wydaw. Naukowo-techniczne, Warszawa, Poland: 448 pp., illustr.

Karnosky, D. F. (1976). Threshold levels for foliar injury to *Populus tremuloides* by sulphur dioxide and ozone. *Can. J. Forest Res.*, **6**, pp. 166–9.

Karolewski, P. (1978). Effect of action of some chemical substances on the degree of injury caused by SO_2. Pp. 68–75 *Studies on the Effect of Sulphur Dioxide and Ozone on the Respiration and Assimilation of Trees and Shrubs in Order to Select Individuals Resistant to the Action of These Gases* (Ed. S. Białobok). Fourth Annual Report (Pl-Fs-74, Fg-Po-326): 100 pp., illustr. (mimeogr.)

Keller, J. (1975). Zur Phytotoxizität von Fluorimmissionen für Holzarten. *Mitt. Eidg. Aust. Forstl. Vers'wes.*, **51**, pp. 302–31.

Kender, W. J., Taschenberg, E. F. & Shaulis, N. J. (1973). Benomyl protection of grapevines from air pollution injury. *Hort. Science*, **81**, pp. 396–8.

Kickert, R. N. (1980). Ecosystem simulation modeling of mixed conifer forest under photochemical air pollution. Pp. 3–28 in *Photochemical Oxidant Air Pollution Effects on a Mixed Conifer Forest Ecosystem: Final Report* (Ed. O. C. Taylor), US EPA, Corvallis, Oregon 97330, USA: xvi + 196 pp.

Kluczynski, B. (1976). [Oddzialywanie fluoru i jego zwiazków na rośliny.—in Polish.] *Arboretum Kornickie*, **24**, pp. 401–18.

Knabe, W. (1970). Kiefernwaldverbreitung und Schwefeldioxid-Immissionen im Ruhregebiet. *Staub-Reinhaltung der Luft*, **30**, pp. 1–4.

Knabe, W. (1971). Methoden zur erkennung und beurtelung forstschädlicher luftverunreinigungen. *Mitt. Forstl. Bundes-Versuchs.* (Wien), **92**, pp. 128–50.

Knabe, W. (1972). Immissionsbelastung und immissionsgefährdung der Wälder im Ruhrgebiet. *Mitt. Forstl. Bundes-Versuchs.* (Wien), **97**, pp. 53–87.

Knabe, W. (1973). Zur Auswesung von Immissionssechutzwaldungen. *Forstarchiv.*, **44**, pp. 21–7.

Knabe, W. (1980). Capacity and efficiency of vegetation in reducing air-borne pollution in urban industrial areas. Pp. 278–291 in *Papers presented to the Symposium on the Effects of Airborne Pollution on Vegetation, Warsaw, Poland.* United Nations Economic Commission for Europe: xx + 410 pp.

Koiwai, A., Kitano, H., Fucuda, M. & Kisaki, T. (1974). Methylenedioxyphenyl and its related compounds as protectants against ozone injury to plants. *Agr. Biol. Chem.*, **38**, pp. 301–7.

Kozlowski, T. T. & Mudd, J. B. (1975). Introduction. Pp. 2–9 in *Responses of Plants to Air Pollution* (Ed. J. B. Mudd & T. T. Kozlowski). Academic Press, New York–San Francisco–London: xii + 383 pp., illustr.

Legge, A. H., Amundson, R. G., Jaques, D. R. & Walker, R. B. (1976). Field studies of pine, spruce, and aspen periodically subjected to sulphur gas emissions. Pp. 1033–61 in *Proc. First International Symposium on Acid Precipitation and the Forest Ecosystem.* (Columbus, Ohio, 12–15 May 1975.) USDA Forest Service Gen. Tech. Report NE-23: 1074 pp.

Leone, I. A. & Brennan, E. (1972) Modification of sulfur dioxide injury to tobacco and tomato by varying nitrogen and sulfur nutrition. *J. Air Pollut. Contr. Assoc.*, **22**, 544–550.

Linzon, S. N. (1972). Effect of sulphur oxides on vegetation. *Forestry Chronicle*, **48**, pp. 1–5.

Linzon, S. N. (1978). Effects of airborne sulphur pollutants on plants. Pp. 110–62 in *Sulphur in the Environment, Part II: Ecological Impacts* (Ed. J. C. Nriagu). John Wiley & Sons, Chichester, England, UK: xii + 482 pp., illustr.

Magill, P., Holden, F. R. & Ackley, C. (Ed.) (1956). *Air Pollution Handbook.* McGraw-Hill, New York–Toronto–London: x + 15 sections.

Materna, J. (1962). Auswertung von Düngungsversuchen in rauchgeschädigten Fichtenbeständen. *Wissenschaftliche Z. Techn. Univ. Dresden*, **11**, pp. 589–95.

Mejnartowicz, L. (1978). Genetic characteristic of some Scots Pine trees susceptible or somewhat resistant to the action of SO_2. Pp. 26–32. Studies on the Effect of Sulphur Dioxide and Ozone on the Respiration and Assimilation of Trees and Shrubs in Order to Select Individual Resistant to the Action of These Gases (Ed. S. Białobok). *Fourth Annual Report* (Pl-Fs-74, Fg-Po 326): 100 pp., illustr. (mimeogr.).

Middleton, J. T., Darley, E. F. & Brewer, R. F. (1958). Damage to vegetation from polluted atmospheres. *J. Air Pollut. Control. Assoc.*, **8**, pp. 9–15.

Miller, P. R. (1973). Oxidant-induced community change in a mixed conifer forest. Pp. 101–17 in *Air Pollution Damage to Vegetation* (Ed. J. A. Naegele). American Chemical Society, Washington, DC: xiii + 137 pp., illustr.

Miller, P. R. & McBride, J. R. (1975). Effects of air pollutants on forests. Pp. 196–230 in *Responses of Plants to Air Pollution* (Ed. J. B. Mudd & T. T. Kozlowski). Academic Press, New York–San Francisco–London: xii + 383 pp., illustr.

Mooi, J. (1974). Investigation of the susceptibility to and uptake capacity of HF for woody plants. Stichting Bosbouwproefstation, De Dorschkamp, Wageningen, Nederland, *Mededeling*, **140**, pp. 321–7.

Moyer, J. W., Cole, H., jr & Lacasse, N. L. (1974). Suppression of naturally occurring oxidant injury on azalea plants by drench or foliar spray treatment with benzimidazole or oxatin compounds. *Plant Disease Reptr.*, **58**, pp. 136–8.

Obydiennyj, P. T. (1977). Sochranienie lesa v uslovijach promyslennogo zagraznenija vozducha. *Lesnoe chozjajstvo*, **6**, pp. 35–8.

Ormrod, D. P., Adedipe, N. O. & Hofstra, G. (1973). Ozone effects on growth of Radish plants as influenced by nitrogen and phosphorus nutrition and by temperature. *Pl. Soil*, **39**, pp. 437–9.

Ottar, B. (1980). Air pollution, a survey of sources and dispersion modeling. Pp. 65–75 in *Papers Presented to the Symposium on the Effect of Airborne Pollution on Vegetation, Warsaw, Poland*. UN Econ. Comm. Europe: xx + 410 pp.

Pack, M. R. (1966). Response of tomato fruiting to hydrogen fluoride as influenced by calcium nutrition. *J. Air Pollut. Contr. Assoc.*, **16**, pp. 514–44.

Pell, E. J. (1976). Influence of Benomyl soil treatment on Pinto Bean plants exposed to peroxyacetyl nitrate and ozone. *Phytopathology*, **66**, pp. 731–3.

Reinert, R. A., Heagle, A. S. & Heck, W. W. (1975). Plant responses to pollutant combinations. Pp. 159–75 in *Responses of Plants to Air Pollution* (Ed. J. B. Mudd & T. T. Kozlowski). Academic Press, New York–San Francisco–London: xii + 383 pp. illustr.

Rich, S. (1975). Interactions of air pollution and agricultural practices. Pp. 335–53 in *Responses of Plants to Air Pollution* (Ed. J. B. Mudd & T. T. Kozlowski). Academic Press, New York–San Francisco–London: xii + 383 pp., illustr.

Rich, S., Ames, R. & Zukel, J. W. (1974). 1,4-oxatin derivatives protect plants against ozone. *Plant Disease Reptr*, **58**, pp. 163–4.

Rohmeder, E. & Schönborn, U. A. von (1965). Der einfluss von Umwelt und erbgut auf die Widerstandsfähigkeit der Waldbaume gegenüber Luftverunreinigung durch Industrieabgase. *Forstw. Centralblatt*, **8**, pp. 1–13.

Ryder, E. J. (1973). Selecting and breeding plants for increased resistance to air pollutants. Pp. 75–85 in *Air Pollution Damage to Vegetation* (Ed. J. A. Naegele). (Advances in Chemistry, Series 112.) American Chemical Society, Washington, DC, USA: xiii + 137 pp.

Scholz, F. (1979). Considerations about selection for air pollution resistance in polluted stands and consequences for correlated traits. Lecture at the *IUFRO-Joint Meeting S2.09.-/08.09.12/ on Physiological and Biochemical Effects of Air Pollution on Plants and Genetics of Resistance*, Zabrze, Poland, pp. 1–11, (mimeogr.).

Scholz, F., Timmann, T. & Krusche, D. (1978). Untersuchungen zur Variation der Resistenz gegen HF—begasung bei *Picea abies* familien. *X Internationale Arbeitstagung IUFRO Fachgruppe* S.2.09, pp. 1–10 (mimeogr.).

Scholz, F., Timmann, T. & Krusche, D. (1980). Genotypic and environmental variance in the response of Norway Spruce families to HF-fumigation. Pp. 277 in *Papers Presented to the Symposium on the Effects of Airborne Pollution on Vegetation, Warsaw, Poland*. UN Econ. Comm. Europe: xx + 410 pp.

Setterstrom, C. & Zimmerman, P. W. (1939). Factors influencing susceptibility of plants to sulphur dioxide injury. *Contrib. Boyce Thompson Inst.*, **10**, pp. 155–81.

Stern, K. & Tigerstedt, P. M. (1974). *Okologische Genetik*. G. Fisher, Stuttgart, West Germany: 211 pp.

Szmidt, A. (1978). Zmienność katalazy w populacjach sosny znajdujacych sie pod wplywem zanieczyszczeń przemyslowych. Pp. 13–4 in *Reakcje biologiczne drzew na emisje przemyslowe*. Summaries. Institute of Dendrology, Kórnik, Poland: 14 pp.

Taylor, G. S. (1970). Tobacco protected against fleck by Benomyl and other fungicides. *Phytopathology*, **60**, p. 578 (abstr.).

Taylor, G. S. & Rich, S. (1974). Ozone injury to Tobacco in the field influenced by soil treatments with Benomyl and Carboxin. *Phytopathology*, **64**, pp. 814–7.

Taylor, O. C. (1973). Acute responses of plants to aerial pollutants. Pp. 9–21 in *Air Pollution Damage to Vegetation* (Ed. J. A. Naegele). (Advances in Chemistry, Series 112.) American Chemical Society, Washington, DC, USA: xiii + 137 pp., illustr.

Tingey, D. T., Reinert, R. A., Dunning, J. A. & Heck, W. W. (1973). Foliar injury responses of eleven plant species to ozone-sulphur dioxide mixtures. *Atmos. Environ.*, **7**, pp. 210–8.

Treshow, M. (1970). Ozone damage to plants. *Environ. Pollut.*, **1**, pp. 155–6.

Treshow, M. (1971). Fluorides as air pollutants affecting plants. *Ann. Rev. Phytopathol.*, **9**, pp. 21–43.

Treshow, M. & Harner, F. M. (1968). Growth responses of Pinto Bean and Alfalfa to sublethal fluoride concentrations. *Can. J. Bot.*, **46**, pp. 1207–10.

Treshow, M., Anderson, F. K. & Harner, F. M. (1967). Responses of Douglas Fir to elevated atmospheric fluorides. *Forest Sci.*, **13**, pp. 114–20.

Tzschacksch, O. & Weiss, M. (1972). Die Variation der SO_2-Resistenz von Provenienzen der Baumart Fichte (*Picea abies* [L.] Karst.) *Beiträge f.d. Forstwirtschaft*, **3**, pp. 21–5.

VDI-Richtlinien (1978*a*). Maximale Immissions-Werte für Schwefeldioxid. VDI 2310, *Blatt* **2**, pp. 1–3.

VDI-Richtlinien (1978*b*). Maximale Immissions-Werte für Fluorwasserstoff. VDI 2310, *Blatt* **3**, pp. 1–3.

Vostal, J. J. (1971). Effects of fluoride on vegetation. Pp. 77–132 in *Fluorides* (Chairman, Panel on Fluorides, J. J. Vostal). National Academy of Sciences, Washington, DC, USA: xi + 295 pp., illustr.

Will, J. B. & Skelly, J. M. (1974). The use of fertilizer to alleviate air pollution damage to White Pine (*Pinus strobus*) Christmas trees. *Plant Dis. Reptr*, **58**, pp. 150–4.

Wood, F. A. & Coppolino, J. B. (1972). The influence of ozone on deciduous forest tree species. *Mitt. der Forstl. Bundes-Versuchs.* (Wien), **97**, pp. 232–53.

Subject Index

486